# Communications in Computer and Information Science

**2606**

Series Editors

Gang Li, *School of Information Technology, Deakin University, Burwood, VIC, Australia*

Joaquim Filipe, *Polytechnic Institute of Setúbal, Setúbal, Portugal*

Zhiwei Xu, *Chinese Academy of Sciences, Beijing, China*

## Rationale

The CCIS series is devoted to the publication of proceedings of computer science conferences. Its aim is to efficiently disseminate original research results in informatics in printed and electronic form. While the focus is on publication of peer-reviewed full papers presenting mature work, inclusion of reviewed short papers reporting on work in progress is welcome, too. Besides globally relevant meetings with internationally representative program committees guaranteeing a strict peer-reviewing and paper selection process, conferences run by societies or of high regional or national relevance are also considered for publication.

## Topics

The topical scope of CCIS spans the entire spectrum of informatics ranging from foundational topics in the theory of computing to information and communications science and technology and a broad variety of interdisciplinary application fields.

## Information for Volume Editors and Authors

Publication in CCIS is free of charge. No royalties are paid, however, we offer registered conference participants temporary free access to the online version of the conference proceedings on SpringerLink (http://link.springer.com) by means of an http referrer from the conference website and/or a number of complimentary printed copies, as specified in the official acceptance email of the event.

CCIS proceedings can be published in time for distribution at conferences or as post-proceedings, and delivered in the form of printed books and/or electronically as USBs and/or e-content licenses for accessing proceedings at SpringerLink. Furthermore, CCIS proceedings are included in the CCIS electronic book series hosted in the SpringerLink digital library at http://link.springer.com/bookseries/7899. Conferences publishing in CCIS are allowed to use our online conference service (Meteor) for managing the whole proceedings lifecycle (from submission and reviewing to preparing for publication) free of charge.

## Publication process

The language of publication is exclusively English. Authors publishing in CCIS have to sign the Springer CCIS copyright transfer form, however, they are free to use their material published in CCIS for substantially changed, more elaborate subsequent publications elsewhere. For the preparation of the camera-ready papers/files, authors have to strictly adhere to the Springer CCIS Authors' Instructions and are strongly encouraged to use the CCIS LaTeX style files or templates.

## Abstracting/Indexing

CCIS is abstracted/indexed in DBLP, Google Scholar, EI-Compendex, Mathematical Reviews, SCImago, Scopus. CCIS volumes are also submitted for the inclusion in ISI Proceedings.

## How to start

To start the evaluation of your proposal for inclusion in the CCIS series, please send an e-mail to ccis@springer.com

Noureddine Seddari · Mohammed Redjimi
Editors

# Modeling, Simulation and Computer Technology

First International Conference, ICMSCT 2024
Skikda, Algeria, November 5–6, 2024
Proceedings

 Springer

*Editors*
Noureddine Seddari
University 20 August 1955
Skikda, Algeria

Mohammed Redjimi
University 20 August 1955
Skikda, Algeria

ISSN 1865-0929     ISSN 1865-0937 (electronic)
Communications in Computer and Information Science
ISBN 978-3-032-01921-9     ISBN 978-3-032-01922-6 (eBook)
https://doi.org/10.1007/978-3-032-01922-6



This Springer imprint is published by the registered company Springer Nature Switzerland AG
The registered company address is: Gewerbestrasse 11, 6330 Cham, Switzerland

If disposing of this product, please recycle the paper.

# Preface

This book presents the proceedings of the first International Conference on Modeling, Simulation and Computer Technology (ICMSCT 2024), which took place at the University of Skikda, Algeria. This international conference was dedicated to exploring cutting-edge research in modeling, simulation, and other emerging computer technologies and their applications. This conference served as a global platform for experts, professors, scholars, and engineers from both national and international universities, research institutes, enterprises, and other research institutions. Participants had the opportunity to share professional experiences, expand industry networks, and display research results in computer application technologies. The ICMSCT 2024 Conference was a perfect platform to discuss new issues, tackle complex problems, and find advanced enabling solutions, with a view to promoting the development and application of theories and technologies in academia and industry. Additionally, participants could establish valuable business or research contacts and seek potential global partners for future collaborations in their research fields.

All submissions to the conference underwent a rigorous single-blind review process, ensuring that at least two independent experts evaluated each paper for its originality, scientific rigor, and relevance. Out of 133 submissions, only 45 were accepted for presentation and inclusion in the proceedings — resulting in an acceptance rate of approximately **33%**. This selective process highlights the quality and relevance of the research featured in this volume.

Over the course of the conference, attendees had the opportunity to engage in thought-provoking presentations, stimulating discussions, and collaborative workshops. We are proud to say that the diverse contributions showcased at the conference set a strong foundation for future collaborations and advanced our collective understanding of how modeling, simulation, and computer technology can transform industries and societies alike.

The papers presented in these proceedings reflect the breadth and depth of the topics discussed, encompassing both theoretical research and applied solutions to real-world challenges. They provide valuable insights into the current state of the art and offer glimpses into the innovative work being done at the forefront of these disciplines.

We would like to express our sincere gratitude to all the authors and participants, and particularly to the distinguished keynote speakers for their enlightening and inspiring talks:

- Bernard Phillip Zeigler, University of Arizona, USA
- Abdelghani Bouras, Al Faisal University, Saudi Arabia
- Mohamed Younis, University of Maryland, Baltimore County, USA
- Zahia Guessoum, University of Reims Champagne-Ardenne, France.

We would also like to thank the organizing committee, sponsors, and volunteers whose hard work and dedication made this event possible.

As we look ahead to future editions of the conference, we are confident that the knowledge shared here will continue to inspire new ideas, foster international collaboration, and drive the advancement of modeling, simulation, and computer technology for years to come.

We hope that this collection of papers will serve as a valuable resource for researchers, educators, and professionals in the field, and we look forward to seeing the continued impact of these works on the global community.

May 2025

Mohammed Redjimi
Noureddine Seddari

# Organization

## General Chairs

Mohammed Redjimi     20 Août 1955-Skikda University, Algeria
Noureddine Seddari     20 Août 1955-Skikda University, Algeria

## Program Committee Chairs

Kenza Redjimi     20 Août 1955-Skikda University, Algeria
Imen Boulnemour     20 Août 1955-Skikda University, Algeria

## Organizing Committee

Imen Boulenmour     University of Skikda, Algeria
Djamel Zeghida     University of Skikda, Algeria
Said Layadi     University of Skikda, Algeria
Toufik Laroum     University of Skikda, Algeria
Lazhar Benoudina     University of Skikda, Algeria
Sara Kerraoui     University of Skikda, Algeria
Samir Sellami     University of Skikda, Algeria
Sohaib Hamioud     University of Skikda, Algeria
Daouia Azzouz     University of Skikda, Algeria
Samia Bouarroudj     University of Skikda, Algeria
Houda Hamrouche     University of Skikda, Algeria
Abdelouahid Bouhouche     University of Skikda, Algeria
Sabrina Sahraoui     University of Skikda, Algeria
Kenza Redjimi     University of Skikda, Algeria
Abdelhakim Bouremel     University of Skikda, Algeria
Mohamed Cheribet     University of Skikda, Algeria
Walid Laouar     University of Skikda, Algeria
Amina Remichi     University of Skikda, Algeria
Fouzia Krim     University of Skikda, Algeria
Toufik Lachouri     University of Skikda, Algeria
Soumia Mameri     University of Skikda, Algeria
Yasmine Lala Bouali     University of Skikda, Algeria
Maroua Bouchaour     University of Skikda, Algeria
Sabah Lecheheb     University of Skikda, Algeria

| | |
|---|---|
| Amina Bouhadja | University of Skikda, Algeria |
| Manel Khadiche | University of Skikda, Algeria |
| Feriel Ghennai | University of Skikda, Algeria |
| Moufida Aibeche | University of Skikda, Algeria |
| Rahima Boukerma | University of Skikda, Algeria |
| Ihcene Zaidi | University of Skikda, Algeria |
| Aicha Boudjrida | University of Skikda, Algeria |

## Scientific Committee

| | |
|---|---|
| Mohamed Younis | University of Maryland Baltimore County, USA |
| Omer Rana | University of Cardiff, UK |
| Kuan Yew Wong | University of Technology, Malaysia |
| Reda Alhajj | University of İstanbul Medipol, Turkey |
| Asif ali laghari | Sindh Madressatul Islam University, Pakistan |
| Ahmed Tlili | Beijing Normal University, China |
| Ladjel Bellatreche | University of Poitiers, France |
| Esma Aimeur | University of Montreal, Canada |
| Khalil Drira | University of Toulouse, France |
| Mohamed Batouche | University of Princess Nourah bint Abdulrahman, Saudi Arabia |
| Eleonora Bottani | University of Parma, Italy |
| Bernard P. Zeigler | University of Arizona, USA |
| Marcello Fera | University of Campania, Italy |
| Abdelouahab Moussaoui | University of Sétif 1, Algeria |
| Salim Heddam | University of Skikda, Algeria |
| Salim Chikhi | University of Constantine 2, Algeria |
| Chawki Djeddi | University of Tébessa, Algeria |
| Allaoua Chaoui | University of Constantine 2, Algeria |
| Abdesslem Layeb | University of Constantine 2, Algeria |
| Khaldi Amine | University of Ouargla, Algeria |
| Hamri Maamar El Amine | Université d'Aix-Marseille, France |
| Abdelghani Bouras | University of Al Faisal, Saudi Arabia |
| Teresa Murino | University of Naples, Italy |
| Piera Centobelli | University of Naples, Italy |
| Agostino Bruzzone | University of Genoa, Italy |
| Zahia Guessoum | University of Reims Champagne-Ardenne, France |
| Abdelouahid Derhab | King Saud University, Saudi Arabia |
| Gabriel A. Wainer | Carleton University , Canada |
| Saber Darmoul | University of Twente, The Netherlands |
| Tarek Al, Hawari | University of Al Faisal, Saudi Arabia |
| Mohamed Belaoued | Université de Reims Champagne-Ardenne, France |

Amirhossein Panahi            University of California, San Diego, USA
Garine Ramakrishna           University of North Texas, USA
Paul-Antoine Bisgambiglia    University of Corsica, France
Raphaël Duboz                University of Montpellier, CIRAD, INRAE,
                                 France
Fabien Michel                LIIRM GRANCE, France
Aznam Yacoub                 University of Windsor, Canada
Mahdi Abbasi                 Umeå University, Sweden
Saikou Diallo                Old Dominion University, Darpa, USA
Faycal Hamdi                 CNAM Paris, France
Farrukh Khan                 King Saud University, Saudi Arabia
Waleed Halboob               King Saud University, Saudi Arabia
Khouloud Boukadi             University of Sfax, Tunisia
Jahan Hassan                 University of Central Queensland, Australia
Mohammed Almulla             University of Kuwait, Kuwait
Noha Mostafa                 British University in Egypt, Egypt
Virginia Yannibelli          ISISTAN Research Institute, UNCPBA, Argentina
Zaki Brahmi                  University of Taibah, Saudi Arabia
Fahed Alkhabbas              University of Malmö, Sweden
Samia All Chelloug           University of Princess Nourah bint Abdulrahman,
                                 Saudi Arabia
Tayeb Basta                  Al Ghurair University, UAE
Nessrine Trabelsi            University of Sfax, Tunisia
Irfan Javid                  University of Poonch Rawalakot, Pakistan
Nasredine Cheniki            Infeurope S.A., Luxembourg
Nassira Chekkai              University of Lyon 1, France
Mohamed Mazouzi              University of Sfax, Tunisia
Samiha Brahimi               Imam Abdulrahman bin Faisal University, Saudi
                                 Arabia
Mohamed Benmohammed          University of Constantine 2, Algeria
Mahmoud Boufaida             Constantine 2 University, Algeria
Mohamed-Khireddine Kholladi  University of El Oued, Algeria
Abdelkader Laouid            University of El Oued, Algeria
Bachir Boucheham             University of Skikda, Algeria
Hima Abdelkader              University of El Oued, Algeria
Zizette Boufaida             Constantine 2 University, Algeria
Salima Harrat                Higher School of Computer Science, Algiers,
                                 Algeria
Ridha Kelaiaia               University of Skikda, Algeria
Ammar Ladjailia              University of Souk Ahras, Algeria
Abdelouahab Attia            University of Bordj Bou Arreridj, Algeria
Smaine Mazouzi               University of Skikda, Algeria

| | |
|---|---|
| Nasro Zarour | Constantine 2 University, Algeria |
| Cherif Tolba | Annaba University, Algeria |
| Mohamed Belaoued | University of Skikda, Algeria |
| Adlen Kerboua | University of Skikda, Algeria |
| Mohammed Amin Tahraoui | University of Chlef, Algeria |
| Ahmed Alioua | University of Jijel, Algeria |
| Imene Djelloul | ENSTA, Algiers, Algeria |
| Yassmina Saadna | Batna 2 University, Algeria |
| Brahim Farou | University of Guelma, Algeria |
| Sofiane Maza | University of Bordj Bou Arreridj, Algeria |
| Mohammed El Amine Abderrahim | University of Ouargla, Algeria |
| Sofiane Zaidi | University of Oum El Bouaghi, Algeria |
| Mohamed Nadjib Kouahla | University of Guelma, Algeria |
| Aissam Belghiat | University of Jijel, Algeria |
| Mohamed Amine Boutiche | University of Science and Technology Houari Boumediene, Algiers, Algeria |
| Tahar Mekhaznia | University of Tébessa, Algeria |
| Said Labed | Constantine 2 University, Algeria |
| Bourouina Hicham | University of M'Sila, Algeria |
| Meftah Zouai | University of Biskra, Algeria |
| El Arkam Mechhoud | University of Skikda, Algeria |
| Bendib Issam | University of Tébessa, Algeria |
| Samy Mezhoud | Constantine 1 University, Algeria |
| Charafeddine Mechalikh | University of Ouargla, Algeria |
| Rafik Menassel | University of Tébessa, Algeria |
| Bendib Riad | University of Skikda, Algeria |
| Akram Zine Eddine Boukhamla | University of Ouargla, Algeria |
| Chikh Ramdane | University of Skikda, Algeria |
| Teta Ali | University of Djelfa, Algeria |
| Sadek Benhammada | Constantine 2 University, Algeria |
| Mawloud Mosbah | University of Skikda, Algeria |
| Sabri Lyazid | University of Bordj Bou Arreridj, Algeria |
| Yacine Kissoum | University of Skikda, Algeria |
| Nour El Houda Dehimi | University of Oum El Bouaghi, Algeria |
| Fouzia Benchikha | Constantine 2 University, Algeria |
| Hocine Guentri | University Center of El Bayadh Nour El Bachir, Algeria |
| Mohamed Assabaa | Constantine 1 University, Algeria |
| Zouaoui Samia | Batna 2 University, Algeria |
| Mourad Bouzenada | Constantine 2 University, Algeria |
| Sara Daas | University of Annaba, Algeria |

| | |
|---|---|
| Chahrazed Mediani | University of Sétif, Algeria |
| Guezouli Lyamine | University of Batna, Algeria |
| Rabéa Cheggou | University of Algiers, Algeria |
| Abdeldjalil Ledmi | University of Khenchela, Algeria |
| Mouaouia Cherif Bouzid | University of Algiers, Algeria |
| Sofiane Boukelkoul | Constantine 2 University, Algeria |
| Abdelhak Mansoul | University of Skikda, Algeria |
| Halima Salah | University of Guelma, Algeria |
| Yousra Ben Aissa | University of Biskra, Algeria |
| Hamza Djebli | Constantine 2 University, Algeria |
| Majda Maatallah | University of El Tarf, Algeria |
| Boucherit Ammar | University of El Oued, Algeria |
| Chafika Ramdane | University of Skikda, Algeria |
| Bounouni Mahdi | University of Sétif, Algeria |
| Said Brahimi | University of Guelma, Algeria |
| Salah Bougueroua | University of Skikda, Algeria |
| Chérifa Boudia | University of Mascara, Algeria |
| Ala Djeddai | University of El Tarf, Algeria |
| Fatima Guessoum | University of Guelma, Algeria |
| Belferdi Wassila | University of Batna, Algeria |
| Mohamed Sedik Chebout | University of Oum El Bouaghi, Algeria |
| Adil Chekati | Constantine 2 University, Algeria |
| Safa Attia | University of Bordj Bou Arreridj, Algeria |
| Assia Brighen | University of Jijel, Algeria |
| Chawki Djeddi | University of Tébessa, Algeria |
| Soundes Belkacem | Batna 2 University, Algeria |
| Imene Bentounsi | Constantine 2 University, Algeria |
| Ali-Abdelatif Betouil | University of El Tarf, Algeria |
| Ledmi Makhlouf | University of Khenchela, Algeria |
| Zertal Soumia | University of Oum El Bouaghi, Algeria |
| Khadidja Ameur | University of Ouargla, Algeria |
| Mohamed Ben Bezziane | University of Ouargla, Algeria |
| Riad Bouaita | ENSET Skikda, Algeria |
| Mihoub Mazouz | University of Ouargla, Algeria |
| Abdenacer Nafir | University of Skikda, Algeria |
| Amin Benabbou | University of Tiaret, Algeria |
| Sofiane Bourouz | Constantine 2 University, Algeria |
| Abdelhafid Zeroual | University of Skikda, Algeria |
| Mehdi Boulaiche | University of Skikda, Algeria |
| Chafika Ramdane | University of Skikda, Algeria |
| Mohamed Cheikh | University of Skikda, Algeria |

# Contents

# A Secure and Energy-Efficient IoT Protocol for Smart Cities Using Fuzzy Logic and Blockchain Technology


Abdesselem Beghriche[1,2]($\boxtimes$) and Billel Kenidra[3]

[1] Computer Science Department, Ferhat Abbas University (Sétif-1), Sétif, Algeria
abdesselem_beghriche@univ-setif.dz
[2] Misc Laboratory, Abdelhamid Mehri University (Constantine-2), Constantine, Algeria
[3] Ecole Nationale Supérieure d'Informatique (ESI), Laboratoire LMCS, Algiers, Algeria
b_kenidra@esi.dz



**Abstract.** The fast development of smart cities predominantly relies upon the consummate integration of Internet of Things (IoT) devices to manage the cities in a resourceful way and enhance the quality of life. However, massive deployments of IoT devices suffer from security and energy consumption drawbacks. This paper introduces a new IoT protocol enhanced with Fuzzy Logic and Blockchain technology to alleviate these issues. The presented protocol exploits the advantages of Fuzzy Logic for optimal dynamic management of energy consumption by considering real-time data in the decision-making process to improve the operational efficiency of IoT devices under various circumstances. Moreover, Blockchain technology is exploited to provide a secure and decentralized data transmission and storage platform, preventing unauthorized access or manipulation and reducing other risks in IoT applications such as cyber-attacks and privacy intrusion. By conducting comprehensive simulations and experimental work, it is shown that the proposed protocol performs better in improving the Quality-of-Service metrics, saving power consumption, prolonging network lifetime, reducing delivery response time, and ensuring high-level security compared to existing popular IoT routing protocols.

**Keywords:** Smart Cities · Internet of Things (IoT) · Fuzzy Logic · Blockchain Technology · Energy Efficiency · Security


## 1 Introduction

The Internet of Things (IoT) is a disruptive technology that connects things/devices and systems in the era of smart cities. Smart traffic management and connected home automation are all set to improve efficiency and make city life more convenient [1]. However, in an environment where IoT device proliferation is happening incredibly fast, two critical challenges are presented: providing robust security and managing energy efficiently [2].


© The Author(s), under exclusive license to Springer Nature Switzerland AG 2025
N. Seddari and M. Redjimi (Eds.): ICMSCT 2024, CCIS 2606, pp. 1–15, 2025.
https://doi.org/10.1007/978-3-032-01922-6_1

Among the primary concerns in smart cities is security, as IoT devices communicate and store vast troves of sensitive data. As the volume and complexity of such networks grow, more than traditional security solutions may be needed. Cyberattacks and data breaches are all too real dangers [3]. This issue may worsen within a smart city as its reliance on the infrastructure increases [4]. Therefore, designing a security model capable of dealing with the dynamic interconnections of the IoT world becomes extremely necessary.

On the other hand, another very critical issue is energy efficiency. Since many IoT devices are battery-powered, making them more energy-efficient can increase their operational lifespan and reduce maintenance costs. Smart cities' conditions are more dynamic and, as a result, unpredictable, which makes this situation even harder. New methods for managing energy in use by IoT devices must grapple with this challenge posed by these conditions, which remain dynamic and result in more energy wastage and, hence, a shorter battery life for an IoT device [5].

Given these challenges, this paper proposes an advanced IoT protocol integrating Fuzzy Logic [6] and Blockchain technology [7] to boost security and energy efficiency in the next-generation smart city. The idea is to follow the strategy below to address the outlined problems.

Firstly, Fuzzy Logic will be applied to optimize energy utilization, thus improving efficiency and intelligence in energy management. In addition, the proposed protocol can optimize energy use for IoT devices operated based on the existing environment by dynamically changing their operation parameters. Unlike existing approaches that cannot handle uncertainty [8–12], the one presented in this paper can optimize the use of energy for the Internet of Things devices operated based on the existing environment. Continuous monitoring and optimization of the operating device will enhance the lifetime of the battery and lower consumed energy, which is one of the most significant issues in managing IoT devices.

Secondly, Blockchain offers a basis for security regarding IoT networks. The decentralized and immutable characteristics of Blockchain make it a suitable foundation for ensuring security. It creates a permanent record of transmitted and stored data; thus, there are guaranteed data access restrictions to the wrong people or data tampering. Such decentralized architecture will not allow single points of failure, which means the system will become very resilient to cyber threats [11].

This paper aims to develop the Secure and Energy-efficient Smart City Protocol (SESCP), an IoT protocol designed specifically for smart cities, by integrating Fuzzy Logic and Blockchain technology.

The main technical contributions of our work can be summarized as follows:

a. Design a dynamic energy management system that can change with real-time conditions for optimization of energy use within the IoT by using Fuzzy Logic, which addresses uncertainty and approximate reasoning, thus fitting for the complex and variable conditions characterizing smart cities.
b. Establish a decentralized security framework with Blockchain technology to guarantee safe and secure data transmission and storage across IoT networks.
c. The simulation and realistic deployment in-depth examine the enhancement of the proposed security protocol and energy efficiency improvement.

The remainder of the paper is structured as follows: After this introduction, Sect. 2 focuses on a review of related IoT protocols and applications of Fuzzy Logic and Blockchain, identifying research gaps in terms of energy efficiency and security approaches. The architecture, components, and implementation of the new IoT protocol are provided in Sect. 3. The simulation and testbed setups used in evaluating the protocol are outlined in Sect. 4. In Sect. 5, there is a performance evaluation and findings on the improvement of energy efficiency and security. Finally, Sect. 6 concludes the paper by summarizing contributions, limitations, potential future research directions, and improvements in the protocol and its possible applications.

## 2   Related Work

In this section, existing literature on IoT protocols for smart cities will be criticized, pointing out security and energy efficiency limitations.

### 2.1   Existing IoT Protocols

Integrating the Internet of Things technologies in a smart city is well-researched, and different protocols have been developed to facilitate efficient and secure communication between devices. Unfortunately, many protocols face difficulties, especially in security and energy efficiency.

One such protocol is the Constrained Application Protocol (CoAP) [13], designed for constrained devices and networks. CoAP, in the research field, is considered lightweight and efficient for communication within IoT. Its significant features are simplicity and low overhead, which help save bandwidth and, most importantly, power (the two main reasons for small devices with batteries). This makes CoAP very well-suited for applications where minimal power consumption is a must. However, CoAP has its downsides. It needs some of the more robust security features that other protocols have; hence, it exposes networks to several categories of attacks. The protocol primarily relies on Datagram Transport Layer Security (DTLS) for encryption, which may be inadequate to defeat advanced threats. All such lightweight security measures make CoAP efficient but not fit for top applications where data security is of the essence.

Another standard IoT protocol is the Message Queuing Telemetry Transport (MQTT) [14]. It is designed to be minimal and code-efficient for low-bandwidth communication. It is mainly used for devices that have to send tiny amounts of data that are too large in frequency for polling to be cost-effective, such as in remote monitoring and control systems. MQTT is lightweight, so less data has to be transmitted; therefore, there is a cost-saving in data and power consumption. Its main weakness is its dependence on Transport Layer Security (TLS) and Secure Sockets Layer (SSL) for securing information transmission. Despite some bit of encryption provision by TLS/SSL, at times, it only caters to some of the security requirements, which can be of significant importance to areas with high-security requirements.

In addition to that, Low-power and Lossy Networks Routing Protocol (RPL) [15] is also very common in IoT networks as it is meant to save energy. It helps reduce energy consumption by some specified objective function and rank-based routing; thus, it is suitable for devices with a relatively short battery life. RPL is good in the network, so it extends the device's life because it selects the route that is energy efficient. This is crucial in applications of IoT, where it is impossible to change batteries constantly. Nevertheless, RPL comes with some quite severe security weaknesses. An adversary can quickly attack it using rank attacks, disrupting the network. Version number attacks can destabilize the network. These flaws make RPL less suited for use in security-requiring solid applications. These limitations indicate the need for improved protocols to handle security and energy efficiency in IoT networks effectively.

## 2.2  Fuzzy Logic in IoT

Recently, Fuzzy Logic has been increasingly applied in the IoT system to develop decisions about resource management. The most common studies have shown that it is effective in various IoT applications, especially in controlling energy consumption and improving system efficiencies.

In [16], the authors propose a smart home energy management system in which energy from renewable resources is maximally utilized through fuzzy logic. The management system of the proposed approach managed energy effectively in smart homes with integrated solar and wind energy while dependence on the grid was minimized. Under such energy generation and consumption uncertainties, the fuzzy logic approach would enable fair decision-making for energy distribution. The research shows that the system cuts energy costs and improves smart home sustainability. Relying on fuzzy logic in basing the features introduces complexity to the design. It is also susceptible to being precisely calibrated to ensure full function. Additionally, the effectiveness may sometimes lie in the intermittent character and variability of the resources. However, the approach is marked by its enormous potential and ability to bring greater energy efficiency into residential environments.

In [17], a fuzzy logic approach to optimal decision-making in agricultural environments is introduced. The developed system achieves precision in irrigation, fertilization, and crop monitoring based on sensor technology for obtaining real-time data, and it contributes to raising efficiency and yield. An essential feature of this method is that it manages uncertainty and imprecision in the data, which is usually the case with most data in agriculture, so it becomes applicable to many conditions. However, fuzzy systems are often designed and tuned within a complex and time-consuming process. Further, they have performed well in specific cases, but their scaling capabilities have only sometimes been satisfactory in large and diverse agricultural scenarios.

Although the research has been ongoing, the way Fuzzy Logic will be implemented in the IoT protocols for smart cities remains to be understood. Most current implementations are domain-specific and need a holistic way to manage smart cities' highly diversified and dynamic characteristic environments. This gap forms an opportunity to develop integrated solutions by exploiting Fuzzy Logic to carry out comprehensive energy management across various IoT applications in smart cities.

## 2.3  Blockchain Technology in IoT

In recent years, some analysis has been done of the integration of Blockchain technology into Internet of Things applications, a few examples of which are shown in the following paragraphs.

A new blockchain-based authentication framework to add security to IoT networks is presented in the work [18]. This framework is designed using blockchain technology to ensure decentralized, tamper-proof authentication. It deals with significant components in security matters, including unauthorized access and data integrity in IoT systems. The critical components of this framework are smart contracts used for automatization of the processes associated with authentication and cryptographic techniques that reinforce security protocols. Although this gives security through decentralization, it can scale to large IoT networks and reduce dependency on centralized authorities.

A Lightweight Blockchain Framework for Secure Transactions in Resource-Constrained IoT Devices is a solution proposed by the authors in [19] that boosts security in the face of the weaknesses typical in IoT devices with limited resources. It builds upon lightweight cryptographic features and optimized blockchain operations, aiming to reduce computational and storage requirements to a point where it can be implementable on low-power devices. The essential advantage of this approach is that it could move blockchain technology into IoT settings without severe performance deterioration, thereby enhancing the security of transactions within these re-source-constrained environments. The lightweight design may, however, compromise some security measures vis-à-vis more traditional, resource-intense blockchain systems, which may compromise its effectiveness in high-security setups.

The work mentioned in [20] proposes a novel design that integrates edge devices with blockchain technology for IoT systems to enhance security and effectiveness in data management. The approach uses edge computing to reduce latencies and lighten the loads on central servers; moreover, it includes blockchain so that transactions are tamper-resistant and decentralized, enhancing scalability and security within distributed environments. Nevertheless, challenges remain, such as the energy consumption of blockchain protocols and the limited computational power of edge devices, these require careful optimization to achieve security, performance, and resource efficiency within IoT networks.


## 2.4  Synthesis

The literature reviewed demonstrates gaps in IoT protocols under smart cities that could address security and energy efficiency. The architecture of smart city cloud computing should include horizontal implementations related to Fuzzy Logic and Blockchain, which allow for better energy management and solutions associated with security issues. Finally, the ways to end their integration must be made between the two technologies to develop a complete solution.

Our proposed protocol's contribution would fill this gap by introducing fuzzy logic's dynamic energy management methodologies into the scheme, combined with the robust security framework brought along by Blockchain technology.

## 3  Secure and Energy-Efficient Smart City Protocol

### 3.1  Overview of the Protocol "SESCP"

This section provides an overarching view of the protocol's architecture and its key components:

Protocol Architecture. The protocol design's modular architecture has transparent layers for data sensing, processing, communication, and storage. The functionalities of each layer integrate energy efficiency and security management concerns (see Fig. 1).

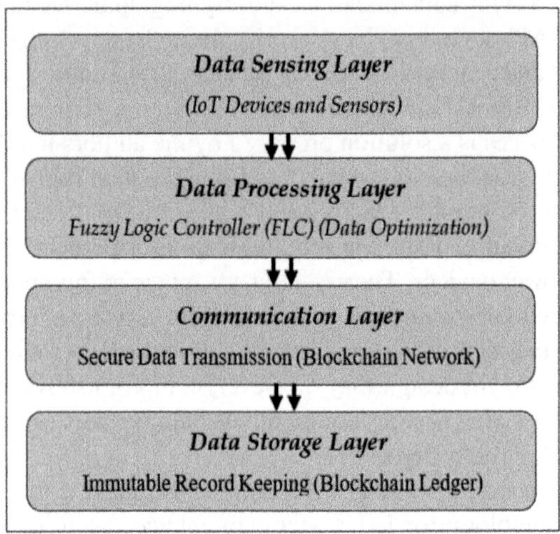


**Fig. 1.** Protocol Architecture Diagram.

- Data Sensing Layer: This layer uses IoT devices and sensors to supply real-time data from different city aspects, including, but not limited to, traffic, energy usage, environmental monitoring, etc.
- Data Processing Layer: This layer runs Fuzzy Logic algorithms to process the sensed data and make appropriate real-time decisions regarding energy consumption optimization.
- Communication Layer: Blockchain enables secure data transfer and reception between IoT devices and the central system.
- Data Storage Layer: The blockchain-based data is stored in this layer to provide secure storage by providing an immutable record of transactions and sensor readings.

Key Components.

*Fuzzy Logic Controller (FLC).* Embedded within the data processing layer, the FLC dynamically adjusts the operational parameters of IoT devices based on real-time data to optimize energy usage.

a. Design of Fuzzy Logic System: The pseudocode of the Fuzzy Logic Based Energy Management is given in Algorithm 1

```
Algorithm.1: Fuzzy Logic Based Energy Management

1. Initialize Fuzzy Logic Controller (FLC)
2. Define input variables:
   - Device Activity Level
   - Battery Status
   - Environmental Conditions
   - Data Transmission Frequency
3. Define output variable:
   - Energy Consumption Level
4. Create membership functions for input and output variables
5. Develop a set of fuzzy rules
6. While system is operational:
   a. Collect real-time data from IoT devices
   b. Convert crisp input values to fuzzy values (Fuzzification)
   c. Apply fuzzy rules to derive fuzzy output values (Inference)
   d. Convert fuzzy output values to crisp actions (Defuzzification)
   e. Adjust IoT device parameters based on crisp actions
   f. Monitor system performance and repeat
```

b. Implementation of Fuzzy Logic Algorithms

- Dynamic Energy Management: FLC continuously observes the operational environment and executes run-time device parameter adjustments, optimizing energy consumption without sacrificing performance.
- Algorithm Efficiency: Employ efficient fuzzy logic algorithms that can run in real-time, offering minimum computational overhead and fast responses.

*Blockchain Network.* It is integrated within the communication and data storage layers to provide decentralized security and data integrity.

a. Design of Blockchain Network

- Blockchain Architecture: Design a private or consortium blockchain tailored for IoT environments, focusing on scalability, low latency, and high throughput.

- Consensus Mechanism: Choose an appropriate consensus mechanism, such as Proof of Stake (PoS) [21] or Practical Byzantine Fault Tolerance (PBFT) [22], to ensure quick and secure validation of transactions without excessive energy consumption.

b. Implementation of Blockchain for Data Transmission

- Secure Data Transmission: Utilize Blockchain to create a secure communication channel between IoT devices, ensuring that all data transmissions are encrypted and authenticated.
- Decentralized Validation: Implement decentralized transaction validation to eliminate single points of failure and enhance the IoT network's robustness against cyber-attacks.

- Immutable Record Keeping: Store all sensor readings and transactions in the Blockchain to create an immutable ledger, ensuring data integrity and traceability.

The pseudocode of the Blockchain Based Security is given in Algorithm 2.

```
Algorithm.2: Blockchain Based Security

1. Initialize Blockchain Network
2. Define Blockchain architecture:
   - Private or Consortium Blockchain
3. Select consensus mechanism:
   - Proof of Stake (PoS)
   - Practical Byzantine Fault Tolerance (PBFT)
4. For each data transmission:
   a. Encrypt data before transmission
   b. Send data to the Blockchain network
   c. Validate transaction using the chosen consensus mechanism
   d. Record transaction in the Blockchain ledger
5. Ensure immutability and integrity of stored data
6. Optimize Blockchain operations for minimal latency and energy
   consumption
```

## 4  Implementation

### 4.1  Simulation Environment

In this section, we present the setup of the simulation environment, which is a controlled setting to test and evaluate the proposed protocol in IoT. (see Table 1).

- We used NS-3 [23] for network simulation and MATLAB [24] to implement and test the Fuzzy Logic-based energy management system.
- We implemented the Blockchain framework using Hyperledger Fabric [25], which features a modular architecture and is suitable for experiments.
- The simulation setup included smart meters, environmental sensors, and surveillance cameras.
- Network topology that allowed the active network of devices in multiple nodes, representing infrastructure components of a smart city, with gateways to manage data flow logically between IoT devices and the central system.
- We used the Data-Driven application tool to produce realistic traffic patterns for typical smart city applications: energy consumption monitoring, environmental data gathering, and video surveillance.
- Various scenarios were simulated to test the strength of the protocol in regular operation at times of peak usage and in the case of security breaches.

**Table 1.** Summary of Simulation Parameters.

| Parameter | Value |
| --- | --- |
| Simulation Tool | NS-3 |
| Energy Management Tool | MATLAB |
| Blockchain Framework | Hyperledger Fabric |
| Network Topology | Mesh topology with multiple gateways |
| Number of Nodes | 100 nodes |
| IoT Devices | Smart meters, sensors, cameras |
| Communication Range | 100 m |
| Data Transmission Rate | 1 Mbps |
| Energy Model | Basic energy model in NS-3 |
| Fuzzy Logic Rules | 10 rules based on energy states |
| Blockchain Consensus | Practical Byzantine Fault Tolerance (PBFT) |
| Security Scenarios | DoS attack, Man-in-the-Middle, data tampering |
| Simulation Duration | 24 h |
| Energy Consumption Metric | Average energy consumption per node |
| Data Integrity Check | Blockchain hash verification |
| Latency Measurement | End-to-end delay |
| Mobility Model | Static, Random Waypoint |
| Packet Size | 512 bytes |
| Initial Energy of Nodes | 1000 Joules |
| Propagation Model | Free-space, Two-ray ground |
| Power Source | Battery-powered, Solar-powered |

## 4.2  Real-World Deployment

In addition to simulations, we have also carried out tests in deployment in the real world to ensure the correct execution of the protocol.

- A testbed was set up in a smart city area with real IoT devices, including smart streetlights, weather stations, and public transport monitoring systems.
- The testbed was interfaced with a private Blockchain network to which secure data transactions are performed.
- All the IoT devices were equipped with sensors and actuators that could collaborate with the Fuzzy Logic and Blockchain systems.
- The energy management and security software modules were deployed on edge devices and gateways to ensure processing and decision-making at the edge.
- Data was collected over months, during which all the different operation conditions occurred, along with variations in external factors such as weather conditions and public events.

- The data gathered was also used to assess the protocol's real-time performance regarding energy consumption, security incidents, and system reliability.

## 5   Results and Discussion

The forthcoming section deals with the reliability of the results obtained after running and evaluating the proposed IoT protocol.

### 5.1   Comparative Analysis

The comparative analysis makes a judgment about the performance of the SESCP against some state-of-the-art protocols concerning energy efficiency, system uptime, security, and data integrity. It will include a Baseline Protocol with no advanced features "RPL, for example" [15], an Energy-Aware Routing Protocol (EARP) [26], an Adaptive Duty Cycling (ADC) Protocol [27], a Secure IoT Protocol (ECC-SIP) [28], and a Blockchain-Based IoT Protocol (BBIP) [29].

**Energy Efficiency.**
*Average Energy Consumption.* This metric measures the total energy consumption by each IoT device over a certain period, for instance, per day. This is a lower value, proving better in utilizing energy resources. Efficiency in energy use will help extend the lives of the devices by reducing total operation costs. At an average of 12.5 J/device/day, the energy consumption by SESCP is below that of the Baseline Protocol, which measured 18.0 J/device/day. Figure 2 shows that EARP and ADC protocols have an average consumption of 14.2 J/device/day and 13.1 J/device/day, respectively. The formula gives the Average Energy Consumption per device:

$$Average\ Energy\ Consumption = \frac{Total\ Energy\ Consumed}{Number\ of\ Devices * Time\ Period} \tag{1}$$

- Total Energy Consumed: The total amount of energy consumed by all devices in the system is typically measured in joules (J).
- Number of Devices: The total number of IoT devices in the system.
- Time Period: The duration over which the energy consumption is measured, typically in days.

*Energy Savings Percentage.*  The energy savings achieved by SESCP were calculated at 30.6%, compared to the Baseline. This is a notable improvement over EARP's 21.1% and ADC Protocol's 27.2%, highlighting the effectiveness of the Fuzzy Logic-based energy management system (see Fig. 3). The Energy Savings Percentage can be calculated by comparing the energy consumption of the SESCP with a Baseline protocol (or reference protocol). The formula is:

$$Energy\ Savings\ Percentage = \frac{Energy\ Consumption_{Baseline} - Energy\ Consumption_{Proposed}}{Energy\ Consumption_{Baseline}} \tag{2}$$

- Energy Consumption$_{Baseline}$: The average energy consumption per device for the Baseline protocol was measured in joules (J).
- Energy Consumption$_{Proposed}$: The average energy consumption per device for SESCP, measured in joules (J).

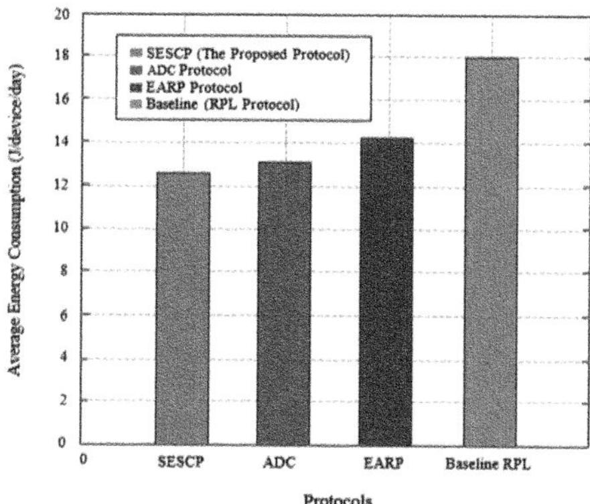


**Fig. 2.** Average Energy Consumption per Device per Day Across different IoT Protocols.

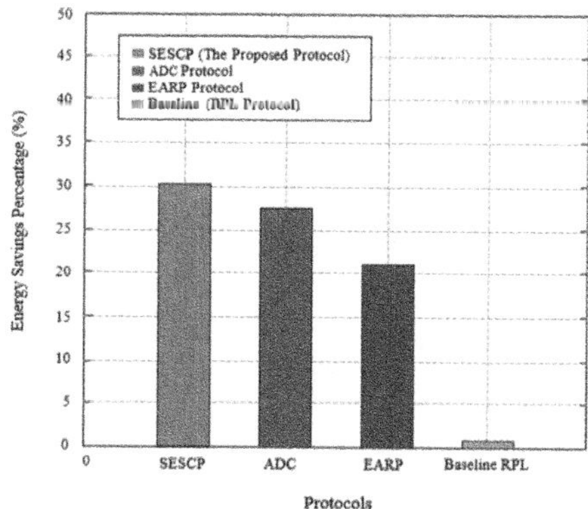


**Fig. 3.** Energy Savings Percentage Comparison Across different IoT Protocols.

System Uptime. System uptime refers to the percentage of time the IoT system remains operational without failures or interruptions. High uptime is essential for critical applications in smart cities, such as traffic management, public safety, and environmental monitoring, where continuous data collection and real-time responses are necessary.

As shown in Fig. 4, SESCP maintained a high system uptime of 99.9%, indicating excellent reliability and minimal interruptions. This performance slightly surpasses that of EARP (99.5%), ADC Protocol (99.7%), and Baseline RPL (99%), demonstrating the protocol's robustness in maintaining continuous operation.

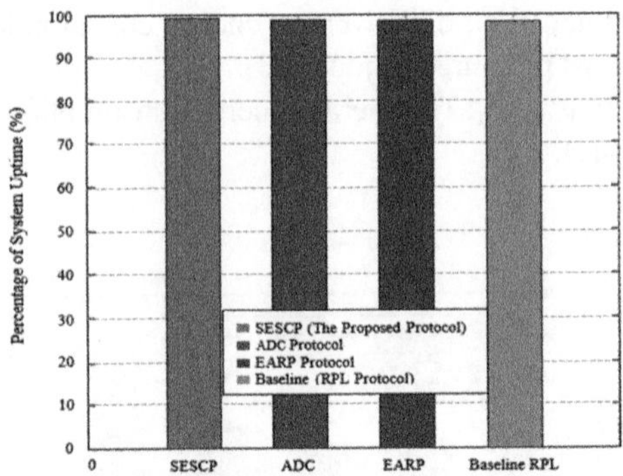

**Fig. 4.** Comparison of System Uptime Across different IoT Protocols.

### Security and Data Integrity.

*Data Integrity.* SESCP achieved 100% data integrity, ensuring all transmitted and stored data remained accurate and tamper-proof. In contrast, the ECC-SIP protocol occasionally exhibited vulnerabilities in key management, while the Baseline Protocol lacked specific security measures. The BBIP maintained 100% data integrity but with higher latency (see Fig. 5).

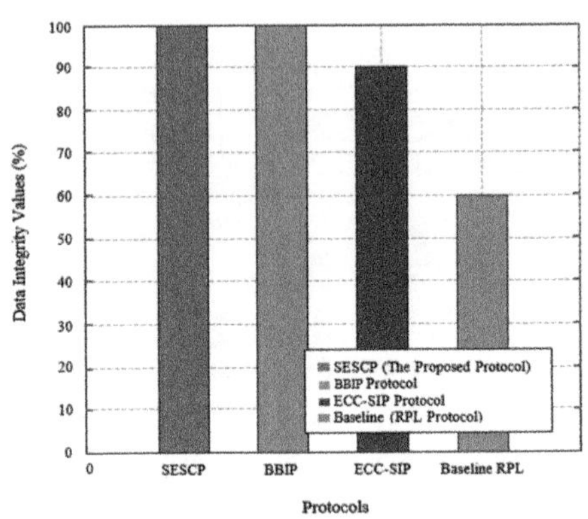

**Fig. 5.** Comparison of Data Integrity Across different IoT Protocols.

*Latency.* SESCP exhibited an average latency of 120 ms, primarily due to the Blockchain consensus process. While this latency is higher than protocols without security features (such as the Baseline Protocol at 50 ms), it is lower than the BBIP, which had a latency of 250 ms due to the proof-of-work mechanism. ECC-SIP had a latency of 90 ms, benefiting from centralized encryption mechanisms but with potential security trade-offs (see Fig. 6).

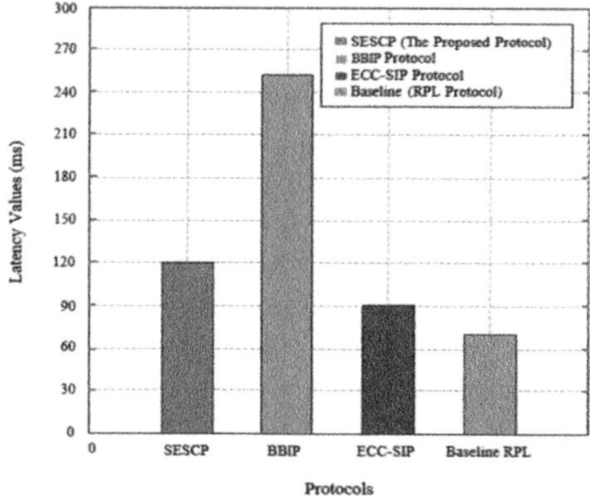


**Fig. 6.** Latency Comparison Across different IoT Protocols.

Resistance to Attacks. As shown in Table 2, SESCP strongly resisted various simulated attacks. Blockchain's decentralized nature minimized the risk of single points of
failure, making the system highly resilient to cyber threats. ECC-SIP was vulnerable to
centralized point failures, and the Baseline Protocol showed susceptibility to multiple
attack vectors. The table presents the number of successful attacks out of 100 attempts for
each protocol, highlighting their vulnerability levels to various cyber threats, including
DoS, data tampering, and man-in-the-middle attacks.

**Table 2.** Comparative Analysis of Successful Attack Attempts Across IoT Protocols.

| Protocol | DoS Attacks | Data Tampering | Man in the Middle |
| --- | --- | --- | --- |
| SESCP | 0 | 0 | 0 |
| ECC-SIP | 10 | 5 | 7 |
| RPL | 25 | 15 | 20 |
| EARP | 15 | 7 | 10 |
| ADC | 20 | 10 | 12 |
| BBIP | 2 | 1 | 1 |

## 6 Conclusions and Future Work

In this work, a new IoT protocol has been proposed for use in smart cities, utilizing Fuzzy
Logic for adaptive energy management and Blockchain technology for data security.
The protocol was quite energy efficient, with 30.6% energy savings and an average
consumption of 12.5 J/device/day. It also kept the system uptime high at 99.9% and data
integrity at 100%. Therefore, while there is a latency of 120 ms, the gains in security are
more than a makeup for it, and the protocol can be applied to real-time systems.

Future research may focus on Fuzzy Logic parameter optimization, more efficient mechanisms for Blockchain consensus, and advanced mechanisms for multi-factor authentication to enhance this security protocol. Real-world deployment tests the protocol's scalability and adaptability to fit the different urban settings.

Furthermore, this protocol application in other broader smart city applications, such as intelligent transportation systems and smart healthcare, will indicate how flexible the protocol is and how it can effectively contribute to sustainable urban development. Indeed, the proposed protocol has a solid and efficient solution to all open challenges in modern smart cities. Developing and fine-tuning would support a secure, reliable, and energy-efficient urban infrastructure.

## References


1. Yu, W., et al.: A survey on the edge computing for the internet of things. IEEE Access **6**, 6900–6919 (2018)
2. Alam, F., Mehmood, R., Katib, I., Albogami, N., Albeshri, A.: Data fusion and IoT for smart ubiquitous environments: a survey. IEEE Access **5**, 9533–9554 (2017)
3. Zhao, K., Ge, L.: A survey on the internet of things security. In: 2013 Ninth International Conference on Computational Intelligence and Security, pp. 663–667. IEEE, New York (2013)
4. Sicari, S., Rizzardi, A., Grieco, L.A., Coen-Porisini, A.: Security, privacy and trust in internet of things: the road ahead. Comput. Netw. **76**, 146–164 (2015)
5. Atzori, L., Iera, A., Morabito, G.: The internet of things: a survey. Comput. Netw. **54**(15), 2787–2805 (2010)
6. Ning, H., Liu, H.: Cyber-physical-social based security architecture for future internet of things. Adv. Internet Things **2**(1), 1–7 (2012)
7. Lu, R., Lin, X., Liang, X., Shen, X.: Secure provenance: the essential of bread and butter of data forensics in cloud computing. In: 5th ACM Symposium on Information. Computer and Communications Security, pp. 282–292. ACM, New York (2010)
8. Ferrag, M.A., Maglaras, L., Janicke, H., Jiang, J., Shu, L.: Authentication protocols for Internet of Things: A comprehensive survey. Secur. Commun. Netw. (2017)
9. Nakamoto, S.: Bitcoin: a peer-to-peer electronic cash system (2008). https://bitcoin.org/bitcoin.pdf. Last accessed 25 June 2024
10. Yuan, Y., Wang, F.-Y.: Blockchain and cryptocurrencies: model, techniques, and applications. IEEE Trans. Syst. Man Cyber. Syst. **48**(9), 1421–1428 (2018)
11. Sharma, P.K., Chen, M., Park, J.H.: A software defined fog node based distributed blockchain cloud architecture for IoT. IEEE Access **6**, 115–124 (2018)
12. Samaniego, M., Deters, R.: Blockchain as a service for IoT. In: 2016 IEEE International Conference on Internet of Things (iThings) and IEEE Green Computing and Communications (GreenCom) and IEEE Cyber, Physical and Social Computing (CPSCom) and IEEE Smart Data (SmartData), pp. 433–436 (2016)
13. Shelby, Z., Hartke, K., Bormann, C.: The constrained application protocol (CoAP). In: IETF RFC 7252 (2014). https://tools.ietf.org/html/rfc7252.
14. Banks, A., Gupta, R.: MQTT version 3.1.1. OASIS standard (2014). http://docs.oasis-open.org/mqtt/mqtt/v3.1.1/os/mqtt-v3.1.1-os.html.
15. Winter, T., et al.: RPL: IPv6 routing protocol for low-power and lossy networks. RFC 6550. Internet Engineering Task Force (IETF) (2012). https://datatracker.ietf.org/doc/html/rfc6550.
16. Zhang, R., Sathishkumar, V.E., Samuel, R.D.J.: Fuzzy efficient energy smart home management system for renewable energy resources. Sustainability **12**(8), 3115 (2020)

17. Abdullah, N., et al.: Towards smart agriculture monitoring using fuzzy systems. IEEE Access **9**, 4097–4111 (2021)
18. Al Hwaitat, A.K., et al.: A new blockchain-based authentication framework for secure IoT networks. Electronics **12**(17), 3618 (2023)
19. Raj, R., Ghosh, M.: A Lightweight blockchain framework for secure transactions in resource-constrained IoT devices. In: 7th International Conference on Recent Advances in Information Technology (RAIT), pp. 727–731. Dhanbad, India (2023)
20. Bisht, T., Dinesh, D., Usha, G., Gautam, K.: Edge devices and blockchain integration in IoT system: a novel design approach. In: 2023 International Conference on Intelligent Data Communication Technologies and Internet of Things (IDCIoT), pp. 218–223. IEEE (2023)
21. Saleh, F.: Blockchain without waste: proof-of-stake. Rev. Financ. Stud. **34**(3), 1156–1190 (2021)
22. Onireti, O., Zhang, L., Imran, M.A.: On the viable area of wireless practical byzantine fault tolerance (PBFT) blockchain networks. In: 2019 IEEE Global Communications Conference (GLOBECOM), pp. 1–6. IEEE, Waikoloa, HI, USA (2019)
23. NS-3 Network Simulator: (2016). http://www.nsnam.org. (Accessed 22 July 22 2024)
24. MATLAB: Version 9.14.0 (R2023a). The MathWorks Inc., Natick, Massachusetts (2023). Available at: https://www.mathworks.com/products/matlab.html
25. Androulaki, E., Barger, A., Bortnikov, V., Cachin, C., Christidis, K., De Caro, A., Enyeart, D., Ferris, C., Laventman, G., Manevich, Y., Muralidharan, S., Murthy, C., Nguyen, B., Sethi, M., Singh, G., Smith, K., Sorniotti, A., Stathakopoulou, C., Vukolić, M., Cocco, S., Yellick, J.: Hyperledger Fabric: A Distributed Operating System for Permissioned Blockchains. In: Proceedings of the Thirteenth EuroSys Conference, pp. 1–13. ACM, Porto, Portugal (2018)
26. Mann, R.P., Namuduri, K., Pendse, R.: Energy Aware Routing Protocol (EARP) for Ad-Hoc Wireless Sensor Networks. In: 2004 IEEE Global Telecommunications Conference Workshops (GLOBECOMW), pp. 1–6. IEEE, USA (2004)
27. Tong, F., Pan, J.: ADC: an adaptive data collection protocol with free addressing and dynamic duty-cycling for sensor networks. Mobile Netw. Appl. **22**(6), 983–994 (2017)
28. Dhillon, P.K., Kalra, S.: Secure and efficient ECC-based SIP authentication scheme for VoIP communications in internet of things. Multimedia Tools Appl. **78**, 22199–22222 (2019)
29. Alkalbani, M., Hamdaoui, B., Zorba, N., Rayes, A.: A blockchain-based IoT networks-on-demand protocol for responsive smart city applications. In: 2019 IEEE Global Communications Conference (GLOBECOM), pp. 1–6. IEEE, Piscataway (2019)

# Building a Secure Enterprise Network: A Blueprint for SDP and SDN Integration

Ilyas Merzouka[1], Abderrouaf Kirli[1], Rania Boudefla[1], Sarra Mamechaoui[1]([envelope]) [ORCID], and Ayoub Zahraoui[2]

[1] School of Telecommunications and Information and Communication Technologies (ENSTTIC), National High, Oran, DZ, Algeria
{ilyas.merzouka,abderraouf.kirli,boudefla.rania,
sarra.mamechaoui}@ensttic.dz
[2] Systems Engineer iServ Solution, Kouba, Algeria
ayoub.zahraoui@iservsolutions.com

**Abstract.** With the growing adoption of Software Defined Networking (SDN), its inherent security risks and vulnerabilities are becoming increasingly critical. These challenges encompass a range of threats, including controller hijacking, insecure communications, and application-layer vulnerabilities. This paper proposes an innovative approach to address these issues by integrating Software Defined Perimeter (SDP) technology into SDN infrastructures. By leveraging SDP, we aim to provide enhanced security measures that mitigate both internal and external threats, addressing limitations of existing solutions. This integration offers enterprises deploying SDN a more robust security framework, ensuring improved protection across all layers of the network architecture. Our proposed solution is designed to strengthen network defenses, safeguarding enterprise environments from sophisticated cyberattacks and reinforcing the overall security posture of SDN deployments. This study anticipates that this approach will substantially enhance the resilience of SDN environments, making them more secure and reliable for enterprise-level applications.

**Keywords:** SDN security · SDP integration · network vulnerabilities · enterprise cybersecurity · controller hijacking · secure communications · application security

## 1 Introduction

With the advent and increasing adoption of Software-Defined Networking (SDN), which offers enhanced flexibility and programmability to improve network performance while reducing human labor and error, many businesses and individuals are considering migrating to fully functional SDN networks to capitalize on this technology's potential benefits [1]. However, while adopting SDN, it is crucial to prioritize security and manageability in addition to its advantages. Enterprises must implement robust security measures to ensure the confidentiality, authenticity, and integrity of their data. This proactive approach is essential for safeguarding their reputation and maintaining service availability [2].


N. Seddari and M. Redjimi (Eds.): ICMSCT 2024, CCIS 2606, pp. 16–28, 2025.
https://doi.org/10.1007/978-3-032-01922-6_2

One concept that has emerged due to the efforts of specialists is Zero Trust Network Access (ZTNA) [3]. In this model, no entity whether inside or outside the network is trusted by default. Users and devices are granted only the minimum access necessary to perform their tasks, ensuring strict access control and reducing potential security risks.

The purpose of this paper is to implement Software Defined Perimeter (SDP), a real-world application of ZTNA, within an enterprise SDN. It aims to simulate the impact of SDP on various types of network attacks and examine how this affects the network's ability to withstand and mitigate these threats. Through these simulations, we aim to provide a comprehensive analysis of SDP's effectiveness in enhancing the security posture of enterprise SDN environments.

## 1.1  General Overview

A study conducted by Sandvine reveals that the global internet infrastructure facilitates the transmission of approximately 33 exabytes of data daily. Of this, 22 exabytes re transmitted through fixed networks and 11 exabytes through mobile networks. Major technology companies like Meta, Apple, Microsoft, Amazon, Alphabet, Netflix, Tik-Tok, and Disney+ are responsible for 65% of this traffic, positioning them as primary contributors to global internet usage [4]. For Internet Service Providers (ISPs), ensuring optimal performance of these applications, known as Application Quality of Experience (App QoE), is critical to maintaining customer satisfaction. Apps such as video conferencing, messaging, social media, and productivity tools are the primary means by which individuals interact with the internet, and service providers must ensure their seamless functionality. This requires robust infrastructure capable of handling the demands of modern applications.

Software Defined Networks (SDNs) have emerged as a powerful solution for network administration and automation, providing significant benefits to both end-user enterprises and ISPs. By reducing Capital Expenditures (CAPEX) and Operating Expenses (OPEX), while simultaneously increasing average revenue per user, SDN has gained widespread acceptance. However, with this increased adoption comes significant challenges that primarily arise from the reliance on APIs for orchestration, which introduces new vulnerabilities and attack vectors that open avenues for exploitation. Attackers can steal API credentials or exploit vulnerabilities within the API itself, potentially gaining unauthorized control over network infrastructure. Common methods include phishing, malware attacks, or exploiting vulnerabilities in applications that rely on the API [5].

From other perspectives, «authentication» remains a critical component of SDN security. Ensuring that only authorized users can access and control the network is essential. Centralized control and extensive API reliance introduce security risks that malicious actors can exploit. To mitigate these risks, access control mechanisms, such as Attribute-Based Access Control (ABAC) [6], can be employed to limit user actions after authentication based on predefined attributes like user roles or device characteristics.

«Data integrity» is another key concern, as SDN's flexibility allows for tailored network configurations, which can introduce vulnerabilities [7]. To preserve data integrity, methods like secure communication channels, data validation, secure boot procedures, and Intrusion Detection Systems (IDS) can be employed.

SDNs are also vulnerable to Denial of Service attacks due to their centralized control model. Compromising the controller can disrupt the entire network, while flow table limitations in SDN devices can be exploited by attackers to overload the network with fraudulent rules, resulting in congestion and interference with legitimate traffic [8].

Finally, while SDN's programmability offers flexibility, it also introduces potential confidentiality and integrity risks. Attackers gaining access to the controller can manipulate traffic invisibly, evading detection by traditional security measures like firewalls or IDS, which operate independently of the SDN controller's programming. Thus, while SDN enhances automation and flexibility, it introduces significant challenges in maintaining robust security [9].

### 1.2  Security Levels Overview

The segregation of the control plane on a controller significantly enhances network automation capabilities. However, it also introduces a range of security challenges, which can be categorized into two primary categories: outer level and inner level security.

**Outer Level (Infrastructure) Security.** Focuses on protecting physical network components, such as servers and switches, from threats like malware that can affect SDN applications. Key risks include metadata sniffing [10], where attackers analyze network traffic to extract critical information, and flooding attacks, such as Distributed Denial of Service (DDoS) [11], which can overwhelm devices and disrupt services [12]. Breaches at this level can lead to disrupted network operations, data theft [13], further attacks on other systems, and reputational damage.

**Inner Level (Controller) Security.** Centers on securing the SDN controller, a crucial single point of failure in the network. Threats include Man-in-the-Middle (MITM) attacks [14], zero-day exploits [15], and Denial of Service (DoS) attacks [16]. Compromising the controller can result in network chaos [17], data breaches [18], loss of control [19], business disruption, and reputational harm, similar to outer-level breaches.

## 2  The Proposed Solutions to Enhance SDN

SDN has garnered considerable attention within the technological landscape due to its flexibility, scalability, and centralized management capabilities. However, the security challenges associated with SDN have become a focal point of discussion in recent years. Addressing these concerns requires thorough planning, as well as the implementation of robust security solutions tailored to the specific vulnerabilities of SDN infrastructures. To this end, several industry-focused groups, including research teams from the NSA (Network Security Agency) and the ONF (Open Networking Foundation), have dedicated efforts to analyze these challenges and propose solutions aimed at mitigating risks. While current SDN security solutions provide partial mitigation against certain threats, they remain insufficient in addressing the full scope of vulnerabilities. For example, authentication mechanisms can prevent unauthorized access, and controller replication schemes enhance network resilience. However, a comprehensive security framework tailored specifically for SDN is necessary to achieve widespread adoption across diverse environments, beyond private data centers and isolated organizational deployments [20].

## 2.1 Categorization of SDN Security Challenges

SDN security challenges can be classified into two primary categories: the outer level and the inner level, as illustrated in Fig. 1. A range of existing security solutions corresponds to these two levels, with a summary provided in Table 1.

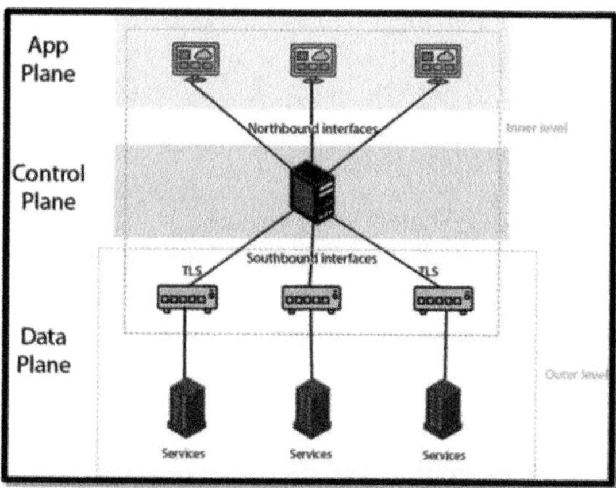


**Fig. 1.** Illustrates the rkevolutionary SDN architecture, highlighting the distinct positions of both the inner and outer security levels.

**Table 1.** Offers a brief description of each solution, categorized by the level type

| Scope | Targets | Challenges | Existing Solutions | Type | Drawbacks |
|---|---|---|---|---|---|
| Outer Level | -Services. <br> -Switches. | -Flow tables memory limitation. <br> -Performance degradation. <br> -Vulnerability of flooding attacks (i.e DoS). <br> -Physical security. | Security middle-boxes (firewall, IDP/IPS, …) | HW | -Cost. <br> -Difficult integration with the SDN. |
| | | | Micro-segmentation. | SW | -Complexity and reduced scalability. |
| | | | Rate limiting | SW | -False positives & granularity. |
| | | | SD-security platforms (vArmour, VMware vShield). | SW | -Expertise & cost. <br> -Resources consumption. <br> -Vendor lock-in. <br> -Poor performance against massive attacks. |
| | | | Machine learning classification techniques. | SW | -Cost. <br> -Data quality. <br> -Resources consumption. <br> -Security vulnerabilities. |
| | | | NTA (Network Traffic Analysis). | SW/HW | -Cost. <br> -Expertise & performance. <br> -False positives. |
| | | | Flow visor technology. | SW/HW | -Cost & vendor lock-in. <br> -Complexity & performance. |
| | | | Sandboxing. | SW/HW | -Cost. <br> -Management overhead. <br> -Performance. <br> -Resource consumption. |

(*continued*)

**Table 1.** (*continued*)

| Inner Level | -Controller.<br>-SBI/NBI.<br>-SDN Apps. | -Controller as a single point of failure.<br>-Network manipulation (controller hijacking).<br>-Lack of authorization & authentication.<br>-Lack of encryption.<br>-Performance degradation.<br>-Vulnerable to sniffing and spoofing attacks. | Encrypted channel. | SW/HW | -Complexity & performance.<br>-Not supported by all SDN switches & controllers.<br>-Not all transferred data is encrypted. |
| --- | --- | --- | --- | --- | --- |
| | | | Multi-Factor Authentication. | SW/HW | -Complexity, user experience & cost. |
| | | | Controller hardening. | SW | -Functionality limitations.<br>-Management overhead. |
| | | | ACLs. | SW | -Difficult to utilize and manage. |

## 2.2  Outer Level Security Solutions

These solutions can also be classified into three distinct categories based on their working principles:

**Traffic Inspection and Control.** [21] is crucial for SDN security, enabling precise management of network traffic to detect threats, optimize performance, and enforce security protocols through solutions like security middleboxes, rate limiting, and network traffic analysis (NTA). Unlike traditional methods that react to threats after they occur, SDN supports a proactive approach using technologies such as machine learning and sandboxing to prevent issues before they escalate.

**Micro-Segmentation and Access Control.** [22] further enhance security by providing granular control over network access and resources, with techniques such as Micro-Segmentation and FlowVisor Technology reducing the impact of breaches.

**Software Defined Security (SDS).** [23] utilizes SDN's programmability to automate and dynamically adjust security measures based on real-time data, offering centralized management through platforms like VMware vShield and vArmour. While SDS improves security flexibility and scalability, it also introduces challenges such as resource consumption, potential vendor lock-in, and varying implementation costs.

## 2.3  Inner Level Security Solutions

These solutions can be categorized into three distinct groups based on their operating principles:

**Encrypted Channels and Access Control Lists (ACLs).** Are vital for protecting communication between the SDN controller and network devices, preventing eavesdropping and data manipulation. Despite their importance, not all SDN controllers support encryption, and managing ACLs can be complex and prone to errors [24, 25]. **Controller Hardening** involves strengthening the SDN controller by disabling unnecessary services, implementing robust authentication mechanisms, and isolating it in a separate network segment to minimize vulnerabilities [26]. However, these practices may introduce performance overhead and operational challenges. Additionally, **Host-Based Security** solutions, such as anti-virus and anti-malware, are essential for protecting individual devices within the network [27]. While integrating these solutions with SDN controllers enhances security, it can also complicate management and be time-consuming across a range of devices.

In summary, securing SDN environments requires a multi-faceted approach that addresses both outer and inner level challenges. While current solutions provide significant improvements in areas such as traffic inspection, threat detection, and access control, a more comprehensive and scalable security framework is needed to ensure SDN's widespread adoption. Emerging technologies like machine learning, sandboxing, and SDS are promising, but they must be implemented with careful consideration of complexity, performance, and cost to achieve optimal security in SDN deployments.

## 2.4  Beyond Traditional SDN Security

Organizations must adopt more advanced solutions to ensure a secure environment. While conventional methods focus on traffic control and access policies, leveraging automation, intelligence, and self-defense is essential for achieving proactive and adaptive security. Automation helps streamline repetitive security tasks such as threat detection and incident response, allowing personnel to focus on strategic efforts. Intelligence, through tools like machine learning and data analytics, helps identify anomalies in network traffic, while self-defense systems can automatically respond to threats, reducing human intervention and mitigating damage swiftly. Perimeter approaches, like Zero Trust Network Access (ZTNA) [3], offer robust access control but are not a substitute for the capabilities of advanced security solutions. These solutions, such as SIEM (Security Information and Event Management) [28] and SOAR (Security Orchestration, Automation, and Response) [29], provide real-time threat intelligence and automated response but can suffer from challenges like alert fatigue and limited visibility. When integrated with perimeter defenses, these tools create a comprehensive security system, reducing the attack surface and improving response times.

SIEM acts as a security hub, collecting and analyzing data from multiple sources to detect threats, while SOAR automates responses based on SIEM's insights. This automation allows for faster containment of threats, freeing up security personnel for complex investigations. Further advancing SDN security, Intent-Based Networking (IBN) and self-defining networks represent a shift towards intelligent and adaptive solutions. IBN translates business goals into security policies, allowing administrators to define security objectives without technical knowledge. Self-defining networks, powered by AI and machine learning, continuously learn and adapt to threats, providing a proactive defense mechanism. Vendor-specific solutions, such as Cisco's ISE, Palo Alto's Prisma

Access, and MacAfee's MVISION Cloud, offer varying functionalities that can complement IBN or the perimeter approach. These tools provide features like identity management, zero-trust access, and extended threat detection, enhancing the overall security architecture.

Ultimately, the integration of advanced security solutions with perimeter approaches like SDP (Software Defined Perimeter) will revolutionize access control, creating a dynamic and scalable security environment that evolves with organizational needs [30].

## 3   Software Defined Perimeter

SDP emerged in 2007 through efforts by the Defense Info Systems Agency for the Global Info Grid initiative. SDP gained traction in 2014 when the Cloud Security Alliance (CSA) outlined its specifications, with updates following in 2018. SDP operates as a network-layer security framework, effectively hiding critical infrastructure from unauthorized users by establishing secure, one-to-one connections between verified users and applications, while preventing external visibility. Traditionally, organizations used firewalls to create network perimeters, but the rise of cloud computing and remote workforces made these static defenses inadequate. SDP addresses this challenge by implementing dynamic, software-based security that adheres to Zero Trust principles, ensuring users access only specific resources based on identity and context. The first open-source SDP modules were developed in 2016, signifying broader adoption of the technology [30].

### 3.1   Key Components of SDP

**SDP Client:**  Installed on user devices, it authenticates users and devices before allowing access to the network.

**SDP Controller:**  Acts as a policy engine, verifying user requests and enforcing security protocols.

**SDP Gateway:**  Establishes secure tunnels between authorized users and the applications they are allowed to access.

These components work together to create encrypted connections using mutual authentication methods like Mutual Transport Layer Security (mTLS) and Single Packet Authorization (SPA), significantly reducing the attack surface and making the network infrastructure invisible to attackers [30].

SPA operates by verifying user identity via cryptographic handshake, allowing access only through a single packet, thus adding an extra layer of security to the SDP model. This makes SDP particularly effective in high-security environments, reducing vulnerabilities associated with traditional methods such as port knocking [30].

While VPNs offer secure remote access, they do not provide granular control over what parts of the network are accessible to users. SDP, in contrast, allows for precise access control, limiting users to specific applications and providing enhanced security, flexibility, and visibility into network usage [20].

SDP closely aligns with Zero Trust Network Access (ZTNA), incorporating key zero-trust principles like least privilege and continuous verification. SDP creates secure micro-perimeters around critical resources, ensuring granular access control and reducing the risk of unauthorized access.

SDP continues to evolve, integrating with emerging technologies such as Secure Access Service Edge (SASE) and advancing in areas like machine learning, analytics, and automation. However, its adoption faces challenges, including migration complexity, vendor reliance, and knowledge gaps. By focusing on open standards, interoperability, and a phased approach, organizations can address these challenges and benefit from SDP's robust security framework. Leading providers in the SDP space include CloudBees, Appgate SDP, and Palo Alto Prisma SASE, offering both open-source and commercially licensed solutions tailored to various business needs [20].

## 4  Lab Simulation

This section presents a laboratory simulation aimed at evaluating the security advantages of integrating Software Defined Perimeter (SDP) with Software Defined Networking (SDN). SDN provides flexibility and programmability but is susceptible to security vulnerabilities, especially concerning controller weaknesses and client exposure. The simulation involves various attacks on the SDN architecture, followed by the incorporation of SDP to analyze its effect on enhancing security [20].

In an SDN architecture, attacks target different planes: the control plane (controller), the application plane, or the data plane (end-hosts and switches). Attackers often focus on the control plane by probing the controller's interfaces (northbound and southbound) to exploit vulnerabilities and gain access. Flooding the controller with packets can overwhelm it, causing a Denial of Service (DoS). Similarly, attackers target the data plane by overloading the flow tables of switches with excessive packets. End-hosts are vulnerable when an attacker spoofs legitimate users' identity by intercepting traffic, sometimes breaking encryption protocols if outdated. Traditional security measures become less effective, necessitating modern approaches.

### 4.1  Simulation Environment

The laboratory setup utilizes two network emulation platforms across two PCs.

*PC_1 (GNS3)* hosts the SDN architecture with the active controller (ODL), standby controller (HP VAN), and eight OpenVSwitches. The SDP architecture includes a controller, gateway, and client, all running on Ubuntu except the attacker VM, which operates on Kali Linux. The OpenVSwitches are Docker containers with flows forwarding traffic between nodes. The GNS3 software controls these components, excluding the attacker and legitimate user VMs, which are connected through a public internet network. Figure 2 outlines the component relationships. Both architectures are connected using an ad-hoc VPN.

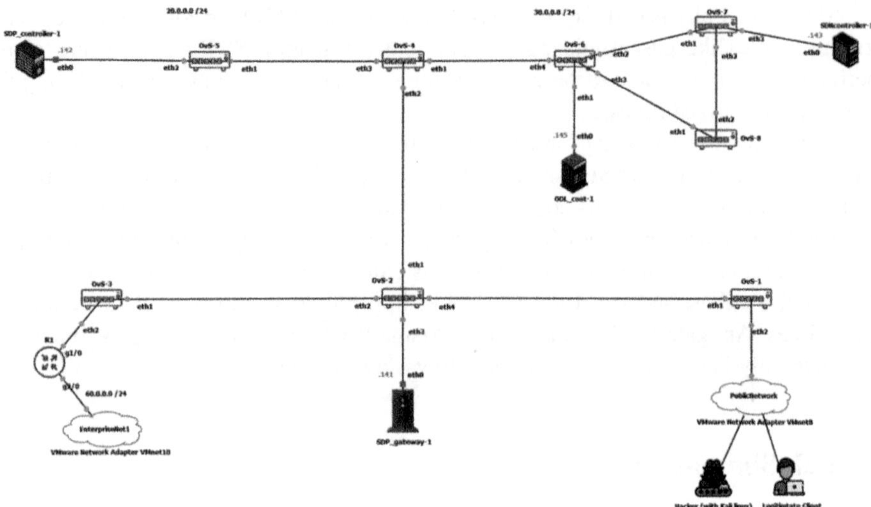

**Fig. 2.** SDP and SDN architecture

*PC_2 (PnetLAB)* represents a full enterprise network based on real Algerian enterprises, composed of three Autonomous Systems (AS): the data center (AS_100), the backbone (AS_200), and the enterprise office network (AS_300). The data center offers multiple network services, and the backbone uses BGP and MPLS for traffic routing. The AS_300 network connects employees to data center services. The VPN connection allows the routers in both architectures to communicate as if directly connected. Figure 3 presents the topology.

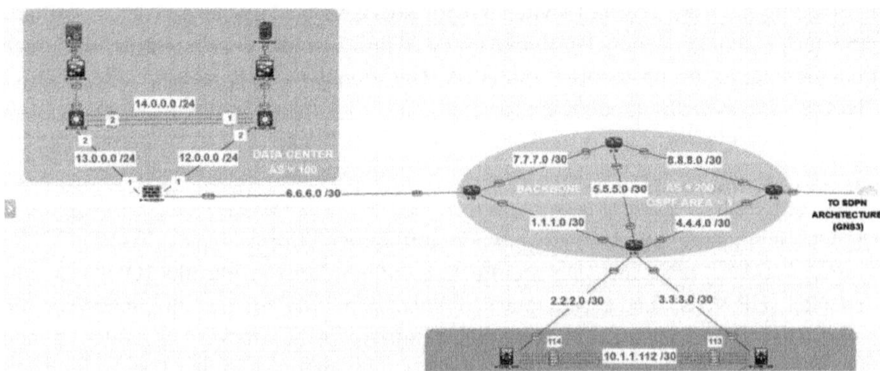

**Fig. 3.** PNET lab topology (enterprise topology).

## 4.2 Lab Simulation Process

The simulation evaluates attacks on both the control and data planes and introduces SDP as a security enhancement.

*Controller Attacks:*  The attacker begins by performing advanced enumeration of SDN controllers using tools like 'nmap' and 'Metasploit'. Vulnerability information is gathered, followed by attempted exploitation of the HP VAN controller. A DoS attack is launched against the ODL controller using 'hping3'. Figures 4 and 5 illustrate the sensor states of the ODL controller within the PRTG software, comparing its status before and after the initiation of the DoS attack.

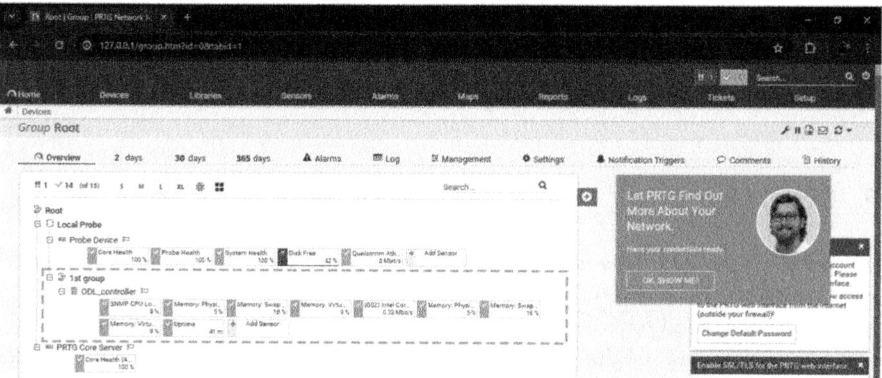

**Fig. 4.** ODL sensor states within the PRTG software prior to the initiation of the DoS attack.

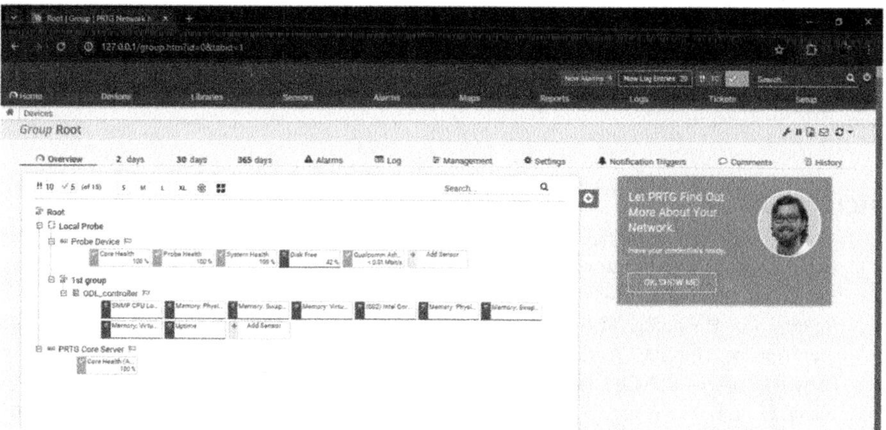

**Fig.5.** ODL sensor states within the PRTG software after the initiation of the DoS attack.

*User Attacks:*  Admin users can manage SDN controllers using the SNMP protocol, monitored via PRTG Network Monitor software. The attacker intercepts this traffic using 'ettercap' to obtain SNMP parameters. Additionally, a normal user attempting to access a file from a remote FTP server (Windows Server 2019) has their traffic sniffed by the attacker to obtain credentials.

*SDP Integration:* In this phase, the SDP components are activated, encrypting traffic and enforcing strict access control. The same attacks are repeated with SDP in place, resulting in significant improvements in security. SDP enhances SDN security by introducing a secure tunnel for client interactions with the SDN controller, encrypting traffic with distributed certificates and encryption keys. The centralized identity management system verifies user credentials stored in an SDP MySQL database. Simulation of the SDP components includes the use of FWknop and Node.js scripts to replicate expected behavior. When attacks were relaunched with SDP integrated, DoS attacks were blocked, and port scanning tools like 'nmap' were unable to detect OpenFlow ports, showcasing SDP's ability to obscure ports and mitigate security risks. Exploitation attempts also failed due to the SDP gateway intercepting malicious payloads.

## 5  Conclusion

As Software-Defined Networking (SDN) becomes more widely adopted, its inherent security risks and vulnerabilities are becoming increasingly significant. This paper examines the critical need for robust security within SDN frameworks, identifying existing security weaknesses and the necessity for a specialized security framework. The paper highlights Software-Defined Perimeter (SDP) as a powerful solution that employs Zero Trust Network Access (ZTNA) to enhance security and service availability. SDP's integration with SDN leverages SDN's flexibility and programmability while strengthening data and service security. We detail how SDP mitigates various attacks on SDNs, emphasize the security improvements from SDP implementation, and outline how strict access controls and filtering of OpenFlow ports reduce attack risks. We conclude by suggesting that integrating SDP with advanced security solutions, such as SIEM/SOAR and Identity-Based Networking (IBN), can further enhance SDN security, offering enterprises a robust security ecosystem.

## References


1. Believe Ayodele and Victor Buttigieg SDN as a defence mechanism: a comprehensive survey. Int. J. Inf. Secur. **23,**141–185 (2024). https://doi.org/10.1007/s10207-023-00764-1
2. Rezaei, G., Hashemi, M.R.: An SDN-based firewall for networks with varying security requirements. In: 2021 26th International Computer Conference, Computer Society of Iran (CSICC), pp. 1–7 (2021). https://doi.org/10.1109/CSICC52343.2021.9420571
3. Kang, H., Liu, G., Wang, Q., Meng, L., Liu, J.: Theory and application of zero trust security: a brief survey. Entropy **25**(12), 1595 (2023). https://doi.org/10.3390/e25121595
4. Sandvine. https://www.sandvine.com/phenomena Last accessed 03 Sep 2024
5. Ahmed Sallam, F., Ahmed Refey, S., Abdallah Shami, T.: on the security of SDN a complete secure and scalable framework using the SDP. In: IEEE (2019)
6. Bhatt, S., Sandhu, R.: Abac-cc: attribute-based access control and communication control for internet of things, of the 25th ACM Symposium on Access Control (2020). dl.acm.org.
7. Karimi, M., Krishnamurthy, P.: Software defined ambit of data integrity for the internet of things. 2021 IEEE/ACM 21st International (2021). ieeexplore.ieee.org.

8. Balarezo, J.F., Wang,, S., Chavez, K.G., Al-Hourani, A.: A survey on DoS/DDoS attacks mathematical modelling for traditional, SDN and virtual networks. Sci. Technol. Elsevier (2022)
9. Sayeed, M.A., Sayeed, M.A.: Intrusion detection system based on software defined network firewall. 2015 1st International. (2015). ieeexplore.ieee.org.
10. Chica, J.C.C., Imbachi, J.C., Vega, J.F.B.: Security in SDN: a comprehensive survey. J. Netw. Comput. Elsevier (2020)
11. Wang, J., Wen, R., Li, J., Yan, F., Zhao, B.: Detecting and mitigating target link-flooding attacks using SDN. IEEE Transactions on. (2018). ieeexplore.ieee.org.
12. Hirayama, T., Miyazawa, T., Furukawa, H.: Reconstruction of control plane of distributed SDN against large-scale disruption and restoration. 2017 IEEE Conference (2017). ieeexplore.ieee.org.
13. Zhao, Y., Yi, P., Zhang, Z., Hu, T.: Data theft attack detection method for SDN edge switch. Inf. Big Data (2022). ieeexplore.ieee.org.
14. Sebbar, A., Boulmalf, M.: Detection MITM attack in multi-SDN controller. 2018 IEEE 5$^{th}$. (2018). ieeexplore.ieee.org.
15. Al-Rushdan, H., Shurman, M.: Zero-day attack detection and prevention in software-defined networks arab conference on (2019). ieeexplore.ieee.org.
16. Dong, S., Abbas, K., Jain, R.: A survey on distributed denial of service (DDoS) attacks in SDN and cloud computing environments. IEEE Access (2019). ieeexplore.ieee.org.
17. Shi, Y., Zhang, H., Wang, J., Xiao, F.: Chaos: an SDN-based moving target defense system. Networks (2017). Wiley Online Library
18. Shaghaghi, A., Kaafar, M.A., Buyya, R., Jha, S.: Software-defined network (SDN) data plane security: issues, solutions, and future directions. Handbook of Computer, Springer (2020)
19. Oh, B.H., Vural, S., Wang, N.: Priority-based flow control for dynamic and reliable flow management in SDN. IEEE Transactions on. (2018). ieeexplore.ieee.org.
20. Ilyas Merzouka, F., Abderraouf Kirli, S., Mamechaoui, R.B.: Building a secure enterprise network: a blueprint for SDP and SDN integration, final study project. In: ENSTTIC, Oran DZ (2024)
21. Li, G., Dong, M., Ota, K., Wu, J., Li, J.: Deep packet inspection based application-aware traffic control for software defined networks. 2016 IEEE Global (2016). ieeexplore.ieee.org.
22. Al-Ofeishat, H.A., Alshorman, R.: Build a secure network using segmentation and micro-segmentation techniques. Int. J. (2023). journal.uob.edu.bh.
23. Coly, A., Mbaye, M.: S-SDS: a framework for security deployment as service in software defined networks and interdisciplinary solutions for underserved areas, Springer (2019)
24. Yigit, B., Gur, G., Tellenbach, B.: Secured communication channels in software-defined networks. IEEE Commun. (2019). ieeexplore.ieee.org.
25. Ramprasath, J., Seethalakshmi, V.: Mitigation of malicious flooding in software defined networks using dynamic access control list. Wireless Personal Communications (2021) Springer
26. Maziku, H., Shetty, S., Jin, D., Kamhoua, C.: Diversity modeling to evaluate security of multiple SDN controllers. 2018 International (2018). ieeexplore.ieee.org.
27. Vinnarasi, J., Sudha N.: Security solution for SDN using host-based IDSs over DDoS attack. Int. J. Emerg. (2019). papers.ssrn.com.
28. Radoglou-Grammatikis, P.: Securecyber: an SDN-enabled SIEM for enhanced cybersecurity in the industrial internet of things-MMTC. (2023). mmc.committees.comsoc.org.

29. Shea, S.: SOAR (security orchestration, automation and response) - techtarget. Com. (2019). https://www.techtarget.com.
30. Moubayed, A., Refaey, A., Shami, A.: Software-defined perimeter (sdp): State of the art secure solution for modern networks. IEEE Network (2019). ieeexplore.ieee.org

# A Comparative Study of Quantum Neural Networks and Compositional Models for Quantum Natural Language Processing

Yousra Bouakba[1]([⊠]) [iD] and Hacene Belhadef[2] [iD]

[1] LISIA Laboratory, University Abdelhamid MEHRI of Constantine 2, Constantine,
Algeria
`{yousra.bouakba,hacene.belhadef}@univ-constantine2.dz`
[2] NTIC Faculty, University Abdelhamid MEHRI of Constantine 2, Constantine,
Algeria

**Abstract.** Quantum Natural Language Processing (QNLP) has two primary research directions: compositional models like DisCoCat and quantum neural network (QNN)-based models. However, there is a gap in the literature regarding comparative studies between these approaches. This paper aims to synthesize the key differences in scalability, learning paradigms, and quantum implementation requirements of both models. We provide a comparative analysis of Quantum Recurrent Neural Network (QRNN) and DisCoCat, two of the most prominent models in the QNN and compositional categories, respectively. Our experiments, focusing on accuracy and training time, show that QRNN outperforms DisCoCat, achieving 30% higher accuracy and reducing training time by 1–2 hours. However, traditional RNN models surpass QRNN, offering 20% higher accuracy and a 200-second faster training time. These findings suggest that while DisCoCat is suitable for smaller, syntactically driven tasks, QRNN is more appropriate for larger datasets and tasks where data-driven learning is essential. This study provides a practical guide for choosing between these models based on task complexity, dataset size, and computational resources.

**Keywords:** Quantum Recurrent Neural Networks · Quantum Natural Language Processing · Quantum Machine Learning · DisCoCat

## 1  Introduction

Over the past century, information has emerged as one of society's most valuable resources, necessitating careful collection, storage, transmission, and encryption. Described by Norbert Wiener as distinct from matter or energy, information requires a carrier, whether material or energetic, to exist. Various computing paradigms have developed to process this information, with classical computing, based on transistors, dominating modern technology. However, due to the limitations of Transistor-based systems, alternative paradigms are proposed such as Optical, Neuromorphic, DNA, Molecular and Quantum computing.


N. Seddari and M. Redjimi (Eds.): ICMSCT 2024, CCIS 2606, pp. 29–40, 2025.
https://doi.org/10.1007/978-3-032-01922-6_3

Quantum computing, introduced by Richard Feynman, employs qubits and principles like superposition and entanglement to tackle complex problems beyond classical capabilities, offering substantial advancements in fields such as Artificial Intelligence (AI), Optimization, and Telecommunication. AI plays a pivotal role in driving innovations across various sectors, but as models increase in complexity, they encounter challenges such as high resource demands and slow training time. Traditional computing technologies struggle with problems requiring significant computational power, especially in optimization and high-dimensional search spaces. In contrast, quantum computing's unique abilities facilitate parallelism, enabling it to efficiently address such challenges, particularly within AI applications. This convergence of quantum computing and AI has given rise to Quantum Machine Learning (QML), a sub-field aimed at enhancing AI's efficiency and capabilities in advanced technological applications [16]. In this paper we investigate the application of quantum machine learning to natural language processing tasks. the main contributions are follows :

1. Explain the relationship between the two approaches in QNLP, Compositional and QNN models highlighting key similarities and points of difference in classical and quantum era.
2. Provide a practical case study demonstrating the application of quantum Recurrent Neural Network to a real-world NLP task, offering insights into the current feasibility and limitations of QRNN.
3. A comparison between RNN, DisCoCat and QRNN in term of accuracy and training time for MC and RP datasets .

This paper is organized as follows: Sect. 2 provides an in-depth overview of key distinctions and challenges in quantum computing Hardware. Section 3 identifies several under-explored areas in quantum machine learning. In Sect. 4, we examine how both compositional models and QNN-based models approach natural language understanding and processing, comparing their theoretical foundations and quantum implementations. Section 5 and 6 presents a practical use case, comparing QRNN, RNN, and DisCoCat in terms of accuracy and training time. Finally, we conclude with insights and future research directions.

## 2    Quantum Computing Hardware

Quantum hardware can be classified into several categories based on the level of development and the qubit topologies employed. A key distinction exists between fault-tolerant quantum computers, Noisy Intermediate-Scale Quantum (NISQ) devices, and simulators. **Fault-tolerant** quantum computers aim to maintain correct functionality despite component failures or errors, ultimately enabling unprecedented computational accuracy for complex problems in fields such as cryptography and materials science. However, these systems are still in the early development stages, as they require a substantial number of physical qubits to create logical qubits capable of stable long-term computations. In contrast,

**NISQ** devices, currently more prevalent, typically consist of 50 to a few hundred qubits and are more prone to noise, yet they play a vital role in near-term applications, such as quantum chemistry simulations and optimization problems [13]. **Simulators**, primarily classical, can emulate quantum circuits and algorithms, facilitating the testing and debugging of quantum algorithms prior to their implementation on actual quantum hardware, thereby addressing current hardware limitations, particularly during the NISQ era [9]. Various qubit topologies, including superconducting, trapped ion, topological, and photonic qubits, each present unique strengths and challenges in developing practical, scalable quantum systems.

# 3  Quantum Machine Learning

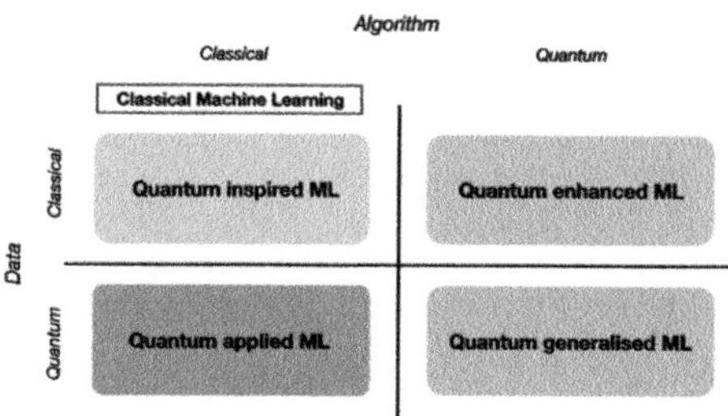


**Fig. 1.** Classification of machine learning (ML) approaches based on the nature of data and the algorithms (classical or quantum) used [6]

The emergence of Quantum Machine Learning (QML) has resulted in various approaches classified according to the nature of data (classical or quantum) and the algorithms employed (classical or quantum), as illustrated in Fig. 1. This classification comprises four distinct quadrants: the top-left quadrant represents **Traditional Machine Learning**, which utilizes classical algorithms, such as neural networks, on classical datasets, including images and text, while also encompassing **Quantum-Inspired Machine Learning**, where classical data is processed using algorithms influenced by quantum principles to enhance efficiency, exemplified by tensor networks in physics and chemistry. The top-right quadrant features **Quantum-Enhanced Machine Learning**, which employs quantum algorithms to process classical data, thereby improving performance beyond what classical methods can achieve, with notable examples including Quantum Support Vector Machines (QSVM) and Quantum Recurrent Neural Networks (QRNN) [17]. In the bottom-left quadrant, **Quantum-Applied Machine Learning** analyzes quantum data derived from systems like quantum sensors through classical algorithms to extract valuable insights, particularly in quantum chemistry applications. Lastly, the bottom-right quadrant showcases **Quantum-Generalized Machine Learning**, wherein both quantum data and algorithms are utilized, potentially offering significant speedups and enhanced accuracy for specific problems, such as simulating quantum systems. Notably, both Quantum-Enhanced and Quantum-Applied Machine Learning approaches necessitate an encoding phase to convert data into compatible formats for their respective algorithms, with quantum-to-classical encoding extracts meaningful classical information from quantum states, acknowledging the unique characteristics of quantum data, such as superposition and entanglement. While quantum data encoding embedding classical data into quantum systems, Fig. 2 represents the general pipeline for applying quantum computation to classical datasets. Firstly classical data is transformed into quantum states using quantum encoding method, then it processed using quantum algorithms, and finally measured to obtain classical results. [6].

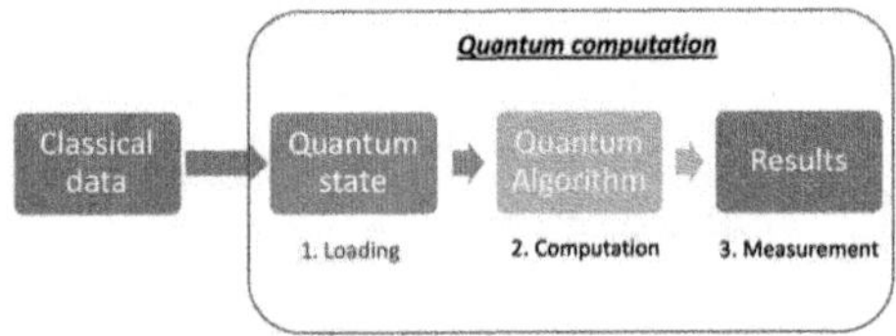


**Fig. 2.** Quantum-Enhanced Machine Learning general pipeline [17]

Figure 2 represents the general pipeline for applying quantum computation to classical datasets.

## 4   Quantum Natural Language Processing

Natural Language Processing (NLP) is a key field in AI that handles large-scale, complex problems like machine translation, sentiment analysis, and chatbots. It involves processing vast amounts of unstructured text data and understanding linguistic patterns, which becomes challenging as datasets grow. The high-dimensional nature of language and the need for real-time analysis make NLP computationally intensive, often overwhelming traditional computing systems. This make NLP one of the first AI applications that stands to benefit significantly, making it an important area for quantum exploration leading to emerge QNLP area of research.

In the literature, QNLP research follows two main directions: one based on compositional models like DisCoCat, and the other based on quantum neural networks (QNNs). Here, we present the fundamentals of both directions to enable a comprehensive comparison of their similarities and differences.

### 4.1   Compositional Models

In natural language processing (NLP), compositional models are approaches that derive the meanings of larger linguistic structures, such as sentences, by systematically combining the meanings of their individual components (words or smaller units). This methodology is grounded in the principle of compositionality, which posits that "the meaning of a complex expression is determined by the meanings of its constituent expressions and the rules used to combine them." [7]. Many compositional models leverage formal mathematical frameworks, such as category theory or vector spaces. A prominent example is the DisCoCat (Distributional Compositional Categorical Model), which integrates category theory, specifically pregroup grammars, with distributional semantics to construct a coherent model of meaning [4]. Empirical support for DisCoCat has been established, followed by its application to various NLP tasks [8]. In 2020, Bob Coecke outlined how DisCoCat could be extended to quantum computing, paving the way for quantum NLP. The framework has since been implemented on quantum hardware for practical NLP tasks [12]. To facilitate these advancements, Cambridge Quantum Computing developed Lambeq, the first library dedicated to building and utilizing compositional models like DisCoCat, Spiders and Cups [10]. Lambeq enables users to (see Fig. 3):

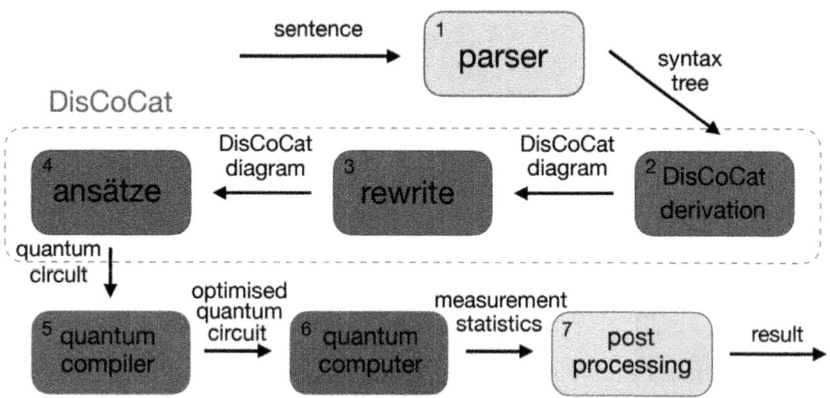


**Fig. 3.** The general quantum pipeline in Lambeq toolkit [11]

1. Parse sentences into syntactic structures using pregroup grammars, then select one of the five available compositional models to derive its diagram and apply rewriting rules to obtain a simplified version of the diagram.
2. Map these structures onto quantum circuits
3. Train models in a quantum-compatible way using simulated quantum hardware or actual quantum computers.

### 4.2   QNN Models

Quantum Neural Networks (QNNs) are an innovative approach that integrates principles of quantum computing into neural network architectures. QNNs represent and process information in a fundamentally different way and explore multiple solutions simultaneously due to the superposition of states. The architecture of QNNs is built using quantum gates that manipulate qubits. These gates can perform complex operations that are analogous to classical neural network layers but can potentially offer exponential speedups for certain computations. Many QNN approaches combine classical neural network techniques with quantum algorithms, creating hybrid models that take advantage of both classical and quantum computing strengths (see Fig. 4) [15]. Researchers have successfully applied QNNs to NLP diffrent tasks, such as sentiment analysis [1],named entity recognition [5] and POS tagging [14] tasks.

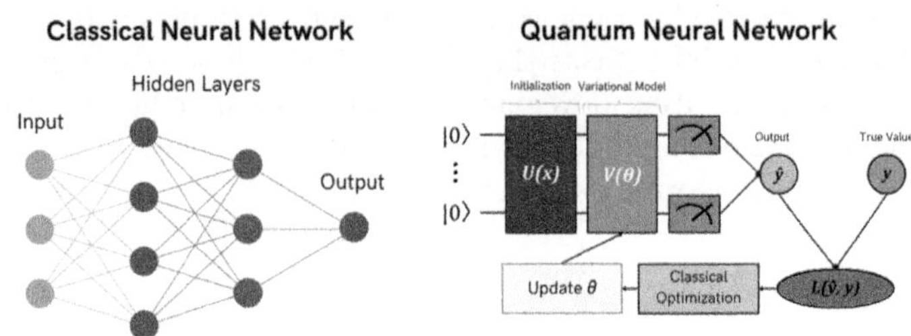


**Fig. 4.** NN vs QNN architecture for Quantum-Classical models [15]

### 4.3  Compositional Models vs Quantum Natural Networks for NLP Tasks

This comparison examines the similarities and differences between DisCoCat and Quantum Neural Networks (QNNs) within both classical and quantum contexts. DisCoCat is based on compositionality and grammar-driven methods, emphasizing how sentence meaning arises from the structure and relationships between words, while QNNs adopt a data-driven approach that learns representations directly from data without explicit syntactic constraints. In terms of syntax and semantics, DisCoCat explicitly incorporates grammatical rules for precise linguistic handling, whereas QNNs learn these aspects implicitly through exposure to large datasets. While DisCoCat is efficient for small, structured tasks, it may struggle with larger datasets, in contrast to QNNs, which scale effectively for modern NLP applications. Additionally, DisCoCat typically operates within a rule-based framework requiring minimal training, whereas QNNs rely on gradient-based optimization that necessitates substantial training data. In quantum implementation, DisCoCat maps grammatical structures to quantum circuits, leveraging their compositional nature, while QNNs employ more complex circuit designs optimized through quantum algorithms. Finally, DisCoCat is suited for structured tasks needing explicit syntactic handling, such as formal reasoning, while QNNs excel in flexible, large-scale NLP tasks like sentiment analysis and text generation, benefiting from their ability to recognize patterns in extensive datasets (Table 1).

**Table 1.** Comparison between DisCoCat and QNNs in the Classical and Quantum Era

| Aspect | Compositional models | Quantum Neural Networks (QNNs) |
| --- | --- | --- |
| Core Principle | Compositionality, grammar-based | Data-driven, learned representations |
| Syntax and Semantics | Explicitly handled through grammars | Learned implicitly from data |
| Scalability | Efficient for small, structured tasks | Scalable to large datasets |
| Learning Paradigm | Rule-based, minimal training | Data-driven, gradient-based learning |
| Quantum Implementation | Direct mapping of grammar to circuits | Complex circuit design for optimization |
| NLP Application Focus | Structured language tasks | Flexible, large-scale NLP tasks |

## 5   Methodology

### 5.1   Quantum Recurrent Neural Networks

Quantum Recurrent Neural Networks (QRNNs) integrate quantum computing principles with traditional Recurrent Neural Network (RNN) architectures,
enabling more efficient information representation and processing through quantum states. Utilizing quantum parallelism and entanglement via Parametrized
Quantum Circuits (PQC), such as Variational Quantum Circuits (VQC),
QRNNs replace the classical neural network component with a VQC. The formulation of a QRNN cell is defined by two equations: the hidden state $h_t$ is
calculated as $h_t = f(VQC(v_t))$, where the input vector $v_t$ is formed by concate-
nating the previous hidden state $h_{t-1}$ with the current input $x_t$. Additionally,
the output $y_t$ is produced by passing the hidden state $h_t$ through a classical neural network layer (denoted as NN), highlighting the hybrid nature of the QRNN
architecture. This innovative structure aims to leverage quantum capabilities for
enhanced performance in sequential data processing tasks [3].

### 5.2   Proposed Vanilla Quantum Recurrent Neural Networks

This study investigates the use of Quantum Recurrent Neural Networks
(QRNNs) for binary classification tasks in natural language processing (NLP)
and provides a comparative analysis against traditional Recurrent Neural Networks (RNNs) and the DisCoCat model. A Vanilla QRNN architecture is proposed (See Fig. 5), which employs the Tanh activation function and incorporates
a simple Variational Quantum Circuit (VQC) structure. This VQC consists of
three main components: a data encoding circuit utilizing H, Ry, and Rz gates,
a variational or parameterized circuit, and a measurement stage. The design is
adaptable, allowing adjustments in the number of qubits and measurements to
accommodate various input and output dimensions. Furthermore, the variational
layer can be iteratively enhanced to increase model size and parameter count,
contingent on the capabilities of the quantum machine or simulation software
employed in the experiments. Variational Quantum Circuit (VQC) structure, as
illustrated in Fig. 6 below.

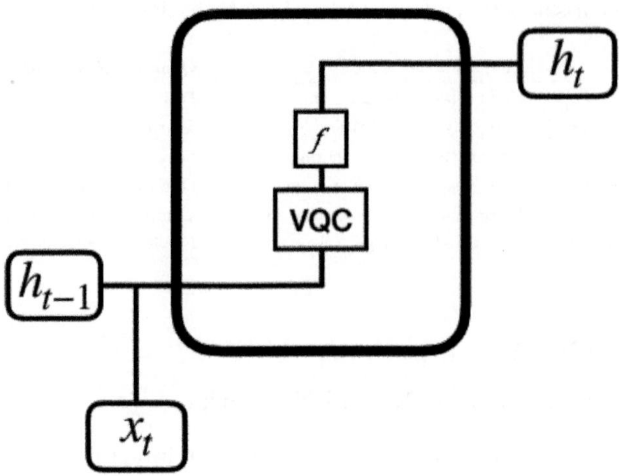

**Fig. 5.** The quantum recurrent neural networks (QRNN) architecture [3]

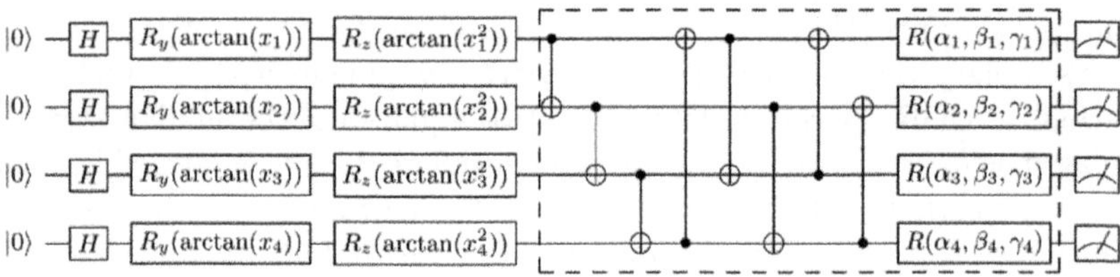

**Fig. 6.** Generic Variational Quantum Circuit structure

# 6    Results and Discussion

## 6.1    Datasets

This research utilizes the MC and RP datasets, which are specifically designed for QNLP tasks. The MC dataset consists of plain syntax generated from a 17-word vocabulary using a simple context-free grammar (CFG) and covers two topics: Food and IT. The RP dataset, derived from the original RelPron dataset, is used to predict whether a noun phrase contains a subject-based or object-based relative clause, and has a vocabulary of 115 words. The RP dataset was chosen due to its requirement for syntactic understanding. Additionally, its larger vocabulary and word sparsity present a more challenging benchmark compared to the MC dataset.

## 6.2    Experimental Setup

Using PennyLane and PyTorch, we developed basic architectures for both RNN and QRNN for binary classification of text data, evaluated on the MC and RP datasets. Preprocessing included removing special characters and padding sequences to four tokens. The models were trained with a learning rate of 0.001, a batch size of 10, and over 100 epochs, utilizing 10 qubits for the QRNN, with evaluation based on specified hyperparameters. As shown in Table 2.

**Table 2.** Hyperparameters for MC and RP Datasets

| Hyperparams/Dataset | MC | RP |
|---|---|---|
| vocab_size | 17 | 115 |
| embedding_size | 8 | 16 |
| hidden_size | 16 | 32 |

## 6.3   Results

Executing the QRNN and RNN models on the **'default.device'** simulator provided by PennyLane yielded the following results (See Figs. 7, 8, 9 and 10 ):

## 6.4   Discussion

Figures 7 and 8 compare the training and validation accuracy and loss of QRNN and RNN models for the MC dataset over 100 epochs. Both models show significant accuracy improvements in the early epochs, with RNN achieving faster progress initially. RNN reaches 1.0 accuracy more quickly, but both models eventually converge around 40 epochs with similar performance levels. Validation accuracy closely mirrors training accuracy for both models, stabilizing near perfect values.

Similarly, Figs. 9 and 10 present the same comparison for the RP dataset. Both models improve rapidly in the early stages, with RNN again progressing faster initially. Convergence occurs around 30 epochs, with both models achieving comparable final performance. Overall, while RNN has a slightly faster learning rate, QRNN and RNN exhibit similar effectiveness in binary classification tasks across both datasets.

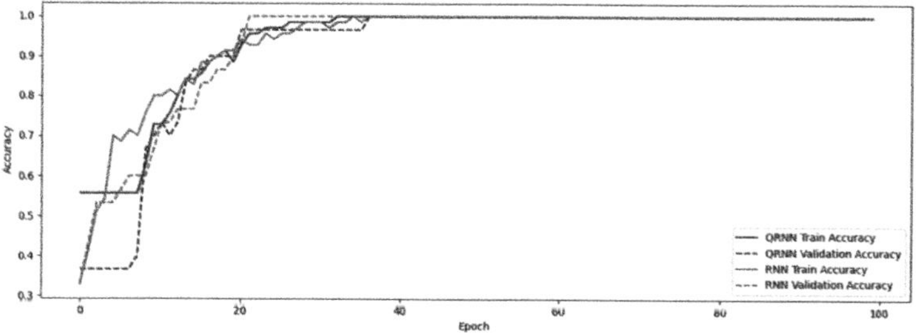


**Fig. 7.** QRNN and RNN Training and Validation Accuracy For MC data Over Epochs

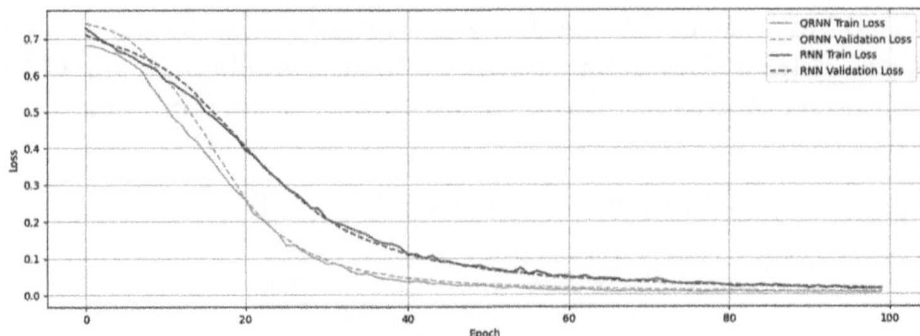

**Fig. 8.** QRNN and RNN Training and Validation Loss For MC data Over Epochs

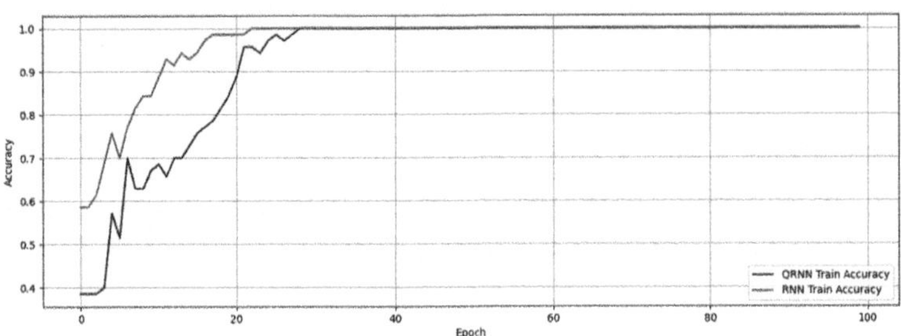

**Fig. 9.** QRNN and RNN Training Accuracy For RP data Over Epochs

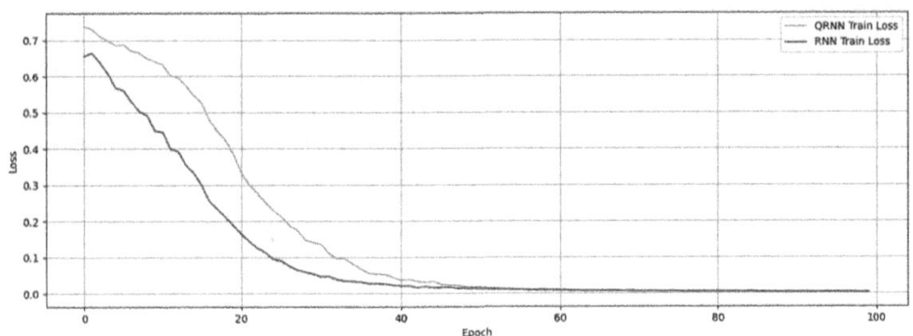

**Fig. 10.** QRNN vs RNN Training Loss For RP data Over Epochs

**Table 3.** Performance comparaison of RNN and QRNN against DisCoCat on MC and RP Datasets

| Dataset/Performance | Model | Test Accuracy | Training Time |
| --- | --- | --- | --- |
| MC | RNN | 100% | 4.31 s |
| | QRNN | 100% | 247.58 s |
| | DisCoCat [2] | 70% | 1 h |
| RP | RNN | 80% | 4.17 s |
| | QRNN | 76% | 197.07 s |
| | DisCocat [2] | 58% | 3 h |

The Table 3 presents the performance of RNN, QRNN, and DisCoCat models from our previous work in [2] on the MC and RP datasets in terms of test accuracy and training time. For the **MC dataset**, both RNN and QRNN achieve perfect accuracy (100%), but RNN is significantly faster (4.31 s) compared to QRNN (247.58 s). DisCoCat, on the other hand, achieves a lower accuracy of 70%, with a much longer training time of 1 h. This indicates that while DisCoCat is conceptually rich due to its compositional structure, it lags in efficiency and accuracy for this dataset compared to neural models. For the **RP dataset**, RNN again outperforms both QRNN and DisCoCat, with an 80% test accuracy and a very short training time (4.17 s). QRNN follows with 76% accuracy, but requires much more training time (197.07 s). DisCoCat performs the least well, achieving 58% accuracy and taking 3 h to train. This suggests that for more complex tasks requiring syntactic understanding, DisCoCat may struggle to generalize effectively, especially when compared to neural-based models like RNN and QRNN, which provide a better balance of accuracy and efficiency.

## 7 Conclusions

Quantum computing offers the potential to solve problems beyond the reach of classical computers, and Quantum Machine Learning (QML) has emerged as a promising field aimed at enhancing artificial intelligence (AI) by leveraging this computational power. QML approaches are categorized based on the nature of the data (classical or quantum) and the algorithms (classical or quantum) employed, with Quantum-Enhanced Machine Learning demonstrating significant potential by utilizing advanced datasets and machine learning techniques to further improve AI capabilities. Natural Language Processing (NLP), one of the most impactful AI applications, stands to benefit greatly from these advancements, leading to the rise of Quantum Natural Language Processing (QNLP). QNLP research typically follows two main directions: compositional models like DisCoCat, and quantum neural networks (QNNs). While both methods have been successfully applied to tasks such as sentiment analysis, named entity recognition, and part-of-speech tagging, there is a notable gap in understanding the relationship between them. This paper presents, for the first time, a detailed comparison between these two approaches through a practical binary text classification use case. The results show that QRNN outperforms DisCoCat, while classical RNN surpasses QRNN, likely due to the limitations of classical simulators used in training. This study offers a practical guide for selecting the appropriate model for specific NLP tasks: DisCoCat is better suited for structured language tasks requiring syntactic processing, while QNNs excel in large-scale tasks like text generation. Future work will focus on more complex Variational Quantum Circuits (VQCs) for QRNNs variants and exploring more challenging NLP tasks.

# References


1. Assiri, A., Gumaei, A., Mehmood, F., Ullah, S.: Social media user evaluation for quantum computing technology via sentiment analysis (2024)
2. Bouakba, Y., Belhadef, H.: Ensemble learning based quantum text classifiers. In: European Conference on Advances in Databases and Information Systems, pp. 407–414. Springer (2023)
3. Chen, S.Y.C., Fry, D., Deshmukh, A., Rastunkov, V., Stefanski, C.: Reservoir computing via quantum recurrent neural networks. arXiv preprint arXiv:2211.02612 (2022)
4. Coecke, B., Sadrzadeh, M., Clark, S.: Mathematical foundations for a compositional distributional model of meaning. arXiv preprint arXiv:1003.4394 (2010)
5. Day, W., Chen, H.S., Sun, M.T.: QNET: a quantum-native sequence encoder architecture. In: 2023 IEEE International Conference on Quantum Computing and Engineering (QCE), vol. 1, pp. 246–255. IEEE (2023)
6. Dunjko, V., Wittek, P.: A non-review of quantum machine learning: trends and explorations. Quantum Views **4**, 32 (2020)
7. Frege, G.: Über sinn und bedeutung. Philosophical Review **57**(a) (1948)
8. Grefenstette, E., Sadrzadeh, M.: Experimental support for a categorical compositional distributional model of meaning. arXiv preprint arXiv:1106.4058 (2011)
9. Jaradat, Y., Alia, M., Masoud, M., Mansrah, A., Jannoud, I., Alheyasat, O.: Roadmap for simulating quantum circuits utilising ibm's qiskit library: Programming approach. Eurasia Proc. Sci. Technol. Eng. Math. **26**, 624–632 (2023)
10. Kartsaklis, D., et al.: Lambeq: an efficient high-level python library for quantum NLP. arXiv preprint arXiv:2110.04236 (2021)
11. Lorenz, R., Pearson, A., Meichanetzidis, K., Kartsaklis, D., Coecke, B.: Qnlp in practice: running compositional models of meaning on a quantum computer. arXiv preprint arXiv:2102.12846 (2021)
12. Meichanetzidis, K., Gogioso, S., De Felice, G., Chiappori, N., Toumi, A., Coecke, B.: Quantum natural language processing on near-term quantum computers. arXiv preprint arXiv:2005.04147 (2020)
13. Nielsen, M.A., Chuang, I.L.: Quantum Computation and Quantum Information. Cambridge University Press, Cambridge (2010)
14. Pandey, S., Dadure, P., Nunsanga, M.V., Pakray, P.: Parts of speech tagging towards classical to quantum computing. In: 2022 IEEE Silchar Subsection Conference (SILCON), pp. 1–6. IEEE (2022)
15. Pandey, S., Pakray, P., Manna, R.: Quantum classifier for natural language processing applications. Computación y Sistemas **28**(2) (2024)
16. Sgarbas, K.N.: The road to quantum artificial intelligence. arXiv preprint arXiv:0705.3360 (2007)
17. West, M.T., et al.: Towards quantum enhanced adversarial robustness in machine learning. Nature Mach. Intell. **5**(6), 581–589 (2023)

# Vanilla Deep Learning Models in Session-Based Recommendation: A Comparative Study


Mouloud Amine Djenane[1]([envelope]) [iD], Boudjemaa Boudaa[2,3] [iD],
Abdelhafid Abouaissa[4] [iD], and Lhassane Idoumghar[4] [iD]

[1] Laboratoire de Génie Energétique et Génie Informatique (L2GEGI),
University of Tiaret, Tiaret, Algeria
`mouloudamine.djenane@univ-tiaret.dz`
[2] Computer Science Departement, University of Tiaret, Tiaret, Algeria
`boudjemaa.boudaa@univ-tiaret.dz`
[3] LabRI-SBA Lab., Ecole Superieure en Informatique, Sidi Bel Abbes, Algeria
[4] IRIMAS, University of Haute-Alsace, Mulhouse, France
`{abdelhafid.abouaissa,lhassane.idoumghar}@uha.fr`



**Abstract.** In this article we present a study comparing deep learning models. Multilayer Perceptron (MLP) Recurrent Neural Network (RNN) Bidirectional Long Short Term Memory (bi-LSTM) and Graph Neural Network (GNN). Within the realm of session-based recommender systems (SBRS). Our goal is to investigate how well these standard models perform in predicting user preferences based on interactions within sessions using real-time data. The assessment will focus on analyzing the strengths and weaknesses of each model in addressing the fleeting nature of user interactions commonly seen in SBRS. Our analysis reveals that bi-LSTM models generally exhibit the accuracy and reliability in capturing user preferences and session dynamics. Additionally, the simpler MLP model also demonstrates performance occasionally surpassing its complex counterparts. The findings shed light on the benefits and limitations of deep learning models for enhancing recommender systems, indicating promising avenues for future research and development in personalized content delivery systems.

**Keywords:** SBRS · MLP · RNN · Bi-LSTM · GNN


## 1 Introduction

In the evolving landscape of recommender systems [2], session-based recommender systems (SBRSs) have emerged as a pivotal strategy for capturing the transient interests of users within ongoing sessions. These systems rely on understanding and predicting user preferences in real-time based on a series of interactions [14], a challenge that deep learning techniques are particularly well-suited to address. While a variety of sophisticated deep learning models have been proposed for this purpose [12,21], there remains a significant value in revisiting the


© The Author(s), under exclusive license to Springer Nature Switzerland AG 2025
N. Seddari and M. Redjimi (Eds.): ICMSCT 2024, CCIS 2606, pp. 41–53, 2025.
https://doi.org/10.1007/978-3-032-01922-6_4

foundational, or 'vanilla', of this types of neural networks: Multilayer Perceptron [15], Recurrent Neural Network [3], and Graph Neural Network [1].

Several thoughts constitute the motivation behind the comparison of vanilla deep learning models. Fundamentally, because these models encapsulate simple deep learning mechanisms, they offer a pedantic experience for one to grasp the complex architectures layered above them. Many of the complex models discussed in SBRSs do not provide clear guidance on how to begin utilizing deep learning models in SBRSs, causing confusion for researchers on how to start. Secondly, while they are relatively simple, they often perform extremely competitively, if not occasionally outperform, the traditional approaches within SBRSs. Finally, there has been a comparative gap in analysing these models in the specified category context. Hence, it is insufficiently recognized how each model's strengths and weaknesses compare in this area.

This paper aims to fulfill the existing gap in session-based recommender systems by comparing MLP, RNN, bi-LSTM, and GNN models and conducting a complete comparative study. It considers two research questions – how to extract sequential interactions in a session to train and test them, and what advantages and disadvantages the chosen models have. Moreover, our code[1] for the comparison is available online, and it will help other researchers and developers in the session-based recommender system domain.

The structure of the remainder of this paper is as follows: Sect. 2 gives an outline of the experimental setup, which details the dataset used in it and its preprocessing, with descriptions of the deep learning models tested MLP, RNN, bi-LSTM and GNN. This part would also detail the evaluation metrics concerning which it would compare the performance of each model. Section 3 will illustrate the result and provide an analysis for comparing the effectiveness of the considered models in a SBRSs. Then Sect. 4 will end up by summarizing the findings with an indication of the implications for the development of SBRSs.

## 2   Methodology

In this section, we will discuss the dataset utilized, the preprocessing steps it underwent prior to being processed by the models, and provide brief descriptions of our models and their functionality within the realm of next-item recommendation, as well as the metrics employed to evaluate their performance.

### 2.1   Dataset and Preprocessing

SBRSs have become increasingly prevalent across a variety of industries. As a result, there is a multitude of datasets to select from, including those related to commerce, news, points of interest (POI), and more. For this study, we utilized the yoochoose dataset.

---

[1] https://github.com/Mouloud-Dj/Comparing-DL-models-in-SBRSs.

The YooChoose[2] dataset is a comprehensive e-commerce dataset that was utilized in the RecSys Challenge 2015. It contains about 33 million clicks, but for the purposes of this study, we will be focusing on a subset of 100,000 clicks with a total of 25292 sessions in order to optimize computational efficiency.

Each session is a list of items listed in chronological order based on the user's interactions with them. In SBRSs, our goal is to anticipate the user's next interaction. To achieve this, we approach the issue as a multi-class classification problem, with the number of classes corresponding to the unique items in our dataset.

A sub-session is constructed by taking all the sequences from a session up to the penultimate item, and the final item of the sequence is labeled as the sub-session label. The sub-session is then treated as a session and the last item is removed to serve as the label; the process is iteratively conducted until the sub-session consists of at least two items. This process is applied to all sessions within our dataset. A graphical representation of the above process has been provided in Fig. 1.

The statistics of the dataset before and after processing are summarized in Table 1.

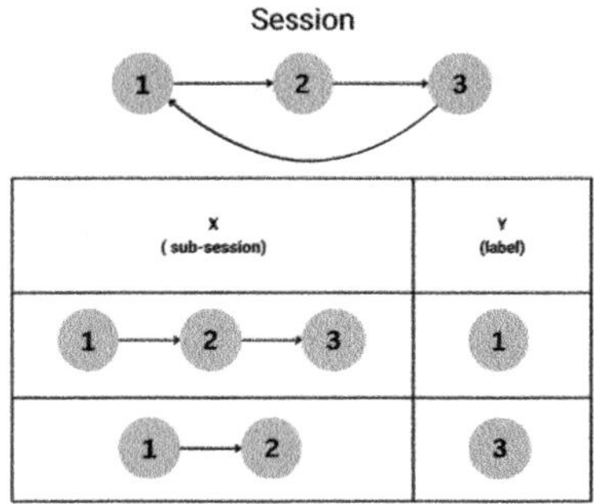


**Fig. 1.** Decomposition of Session into Sub-Sessions and Labels.

**Table 1.** Statistics of dataset after and before processing

| Statistics | Initial Dataset | After processing |
|---|---|---|
| # of sessions | 25292 | 51042 |
| Average length | 3.95 | 6.21 |

---

[2] https://www.kaggle.com/datasets/chadgostopp/recsys-challenge-2015/data.

## 2.2  Models' Descriptions

This section outlines each deep learning model utilized in our comparative study of session-based recommender systems. The selection of each model is justified based on its architectural features and relevance to handling session data in recommendation scenarios.

**Multilayer Perceptron** is a feedforward neural network architecture, consisting of an input layer, one or more hidden layers, and an output layer [4]. These network types are used for various tasks, including classification, regression, and function approximation. They also received recognition in the SBRS research area, MLP is particularly well-suited for handling unordered session data as it lacks the capability to effectively model sequence data [22].

In our study, we have organized the session data in a structured manner. To optimize compatibility with the Multi-Layer Perceptron model, we propose creating an input layer with a number of neurons equivalent to the unique items in our dataset. Each neuron will represent an individual item, with the activation of neurons corresponding to items present in the current session. These neurons will be assigned values based on their relative positions.

Subsequently, the processed data will be directed to a hidden layer with a relu activation function, before ultimately reaching the output layer. The output layer mirrors the input layer in terms of neuron count, with softmax activation utilized to generate probabilities indicating the likelihood of each item being selected by the user next.

In order to provide clarity, we have created Fig. 2 to illustrate the model and the process by which the session input will be utilized. It is evident from the figure that the next recommended item for the user is Item number 2, with a likelihood of 60% that it will be clicked on by the user.

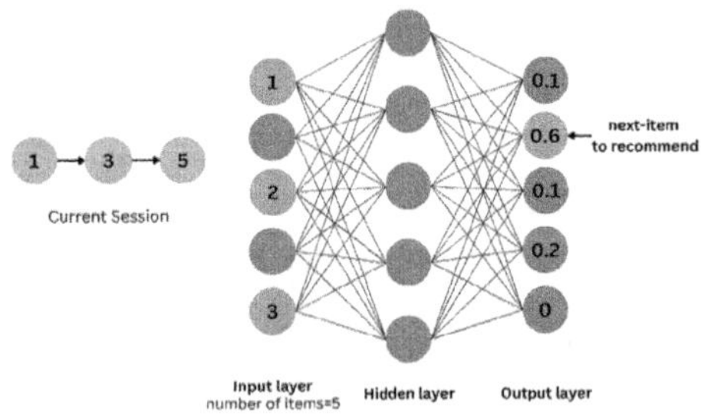


**Fig. 2.** Multi-Layer Perceptron Model for Next-Item Recommendation.

**Recurrent Neural Network** are a class of neural networks that have been specifically designed for the processing of sequences of data, such as text or time-series information [5]. Due to the previous inputs, RNNs know what it has been fed as a characteristic and can, therefore, encode dependencies and patterns from sequential data. RNN is, therefore, different from MLP the ability to use their inner state, hence processing sequences of inputs. This property makes these ideal for session-based recommendation systems when a context of a session matters most in predicting subsequent interaction [8].

In contrast to MLP, this approach will not involve an input layer that is equal to the number of items. Instead, we will implement recurrent cells that process sequences of items as show in Fig. 3. Rather than representing each item as a one-hot encoded vector of zeros and ones, we will utilize embedding due to the large number of items in our dataset, which can potentially reach into millions in real world datasets. The output will consist of a linear layer equal to the number of items, with a softmax activation function as demonstrated in the MLP example.

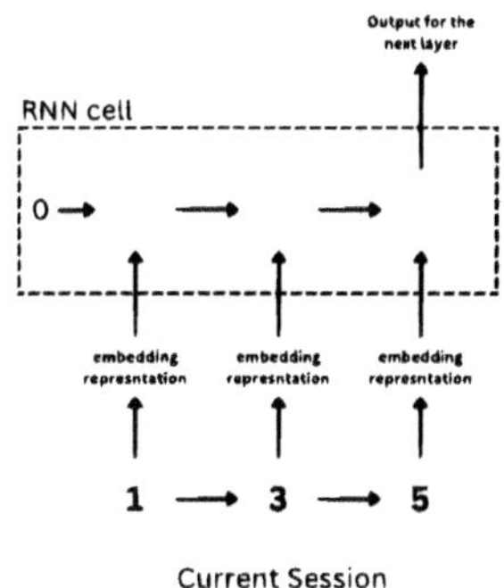


**Fig. 3.** RNN Cell with Embedding Representations for Item Sequence Processing.

**Bidirectional Long Short-Term Memory** are a variant of the standard LSTM architecture [10], which is a type of RNN so they both designed to process sequential data, in the same manner the deference is that LSTM are able to effectively model long-range dependencies in sequential data by using a unique cell structure that includes gates to control the flow of information [19].

Bi-LSTM networks can process sequential data in both forward and backward directions as shown in Fig. 4. This allows the model to capture contextual information from both past and future time steps, which can be particularly useful for tasks involving natural language processing or time series analysis [6]. they are noted for their efficacy in handling long-range dependencies and avoiding the vanishing gradient problem that can occur in standard RNNs.

In this study, the bi-LSTM was selected for its ability to effectively model the complete context of a session, including both past and future interactions, to enhance the recommendation quality. By leveraging the bidirectional nature of the architecture, the bi-LSTM can better capture the sequential dependencies and patterns within user sessions, which is crucial for accurate session-based recommendations.

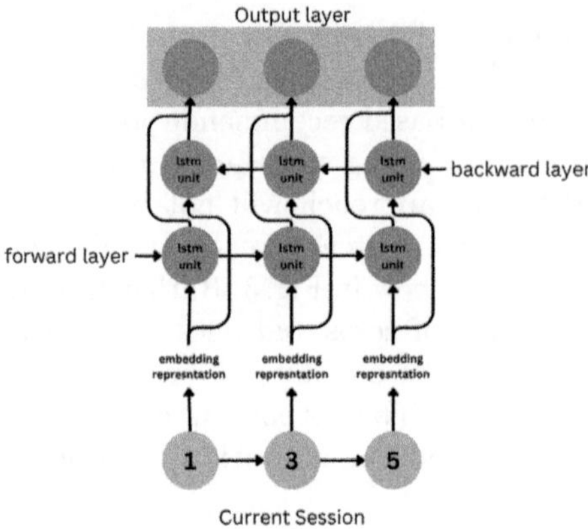


**Fig. 4.** Bi-LSTM Architecture for Modeling Session Context.

**Graph Neural Network** are a class of deep learning models designed to operate on graph-structured data [17]. GNNs can effectively capture the relational and structural information inherent in graph-based data, making them suitable for tasks involving social networks, knowledge graphs, and other interconnected data [9,18]. For session-based recommendations, where the relationship between items and user actions within a session can be graphically structured, GNNs offer a powerful tool for capturing complex patterns that are not easily modeled by sequence-based networks alone. They have demonstrated improved performance in the task of next item recommendation, as shown in the work of SR-GNN [23].

There are various variations of graph neural network (GNN) models such as Graph Convolutional Networks (GCN) which expand the convolution operation to graph-structured data [7], GraphSAGE which learns node embeddings in an inductive manner [9], and Graph Attention Networks (GATs) which introduce an attention mechanism to GNNs [20].

In this study, we will use only the vanilla GNN and the DeepWalk approach [13]. With the vanilla GNN approach, each session will be converted into a graph where items are represented as nodes and interactions between items are represented as edges. Prior to inputting items into the GNN, they will be encoded using an embedding as well. The GNN will then undergo message passing, aggregation, and update phases before reaching the final layer for item recommendation.

DeepWalk is another graph-based technique that can be valuable for learning node representations in a graph in an unsupervised manner. This method involves using a comprehensive graph of all sessions and item interactions, and then utilizing random walks to learn latent node representations that capture the graph's structural information. DeepWalk will determine if an item is likely to be recommended based on its similarity to other neighboring items.

## 2.3    Evaluation Metrics

In SBRSs, the commonly used metrics are HR@k and MRR@k for several reasons [11]. In this context, precision and recall essentially refer to the same concept. Initially referred to as precision@k and recall@k, the terminology was later revised to HR@k. This adjustment was made because, in the context of next-item recommendation, there is typically only one relevant item per session, serving as the ground truth for user interaction.

Therefore, the formulas for precision and recall are as follows:

$$\frac{\text{Number of sessions where the recommended next item is correct}}{\text{Total number of sessions}}$$

Due to these factors, using the F1 score is not applicable. Furthermore, accuracy cannot be relied upon either due to potential low model accuracy resulting from the large volume of items available for recommendation.

**Hit Rate at K (HR@K)** measures the proportion of sessions where the recommended next item was present within the top-k recommendations. It represents the likelihood of the correct next item being included in the system's top-k recommendations for each user session. HR@k is a useful metric for evaluating the relevance and coverage of the recommendations.

$$\text{HR@K} = \frac{n_{hit}}{|S|} \tag{1}$$

**Mean Reciprocal Rank (MRR@K)** It calculates the average of the reciprocal ranks of the correct next items across all sessions, considering only the top-k recommendations. The reciprocal rank is the inverse of the rank position of the correct next item, if it is present within the top-k recommendations. MRR@k provides a measure of the ranking quality of the recommendations, with higher values indicating that the correct next items are generally ranked higher in the list of recommendations.

$$MRR@K = \frac{1}{|S|} \sum_{i=1}^{|S|} \frac{1}{\text{rank}_i} \tag{2}$$

where,

- $|S|$ represents the total number of sessions.
- $n_{hit}$ indicate the frequency of the relevant item present within the top k recommendations.
- $\text{rank}_i$ indicates the position of the relevant item for the $i$-th session.

## 3    Experiments and Results

In this section, we outline the experimental framework, followed by an analysis and discussion of the results from the deep learning models in terms of comparison performance and sequence lengths efficiency.

## 3.1   Experimental Setup

All models underwent training and testing using the same dataset, as depicted in Table 2. They were all configured with identical hyperparameters, loss functions, and optimizers. Specifically, the Cross-entropy loss function was utilized, along side with the Adam optimizer set at a learning rate of 0.001. Each supervised model underwent 30 epochs of learning, processing the data in batches of size 32.

All experiments were conducted in the cloud utilizing a GPU to enhance the learning process, with a VRAM capacity of 15GB and a RAM capacity of 30GB.

**Table 2.** Statistics of training and testing data

| Statistics | Training | Testing |
| --- | --- | --- |
| # of sessions | 40834 | 10208 |
| # of items | 10067 | |

## 3.2   Performance Comparison

The performance of each deep learning model was assessed using HR and MRR metrics at various cutoff points (5, 10, and 20). In order to demonstrate that vanilla deep learning models can outperform traditional methods, we also included a baseline threw the test using item-KNN [16] as shown in Table 3.

**Table 3.** The performance of DL models and Item-KNN.

| Models | HR@5 | MRR@5 | HR@10 | MRR@10 | HR@20 | MRR@20 |
| --- | --- | --- | --- | --- | --- | --- |
| Item-KNN | 19.61 | 11.96 | 24.52 | 12.61 | 29.36 | 12.94 |
| MLP | 26.72 | 16.02 | 34.69 | 17.08 | 40.81 | 17.17 |
| RNN | 3.59 | 1.62 | 5.45 | 1.86 | 8.47 | 2.07 |
| Bi-LSTM | **33.03** | **21.55** | **40.37** | **22.54** | **48.01** | **23.53** |
| DeepWalk | 5.33 | 2.56 | 9.2 | 3.08 | 16.99 | 4.03 |
| GNN | 0.54 | 0.28 | 1.22 | 0.37 | 7.45 | 1.57 |

Based on the data presented in Table 3 and the bar chart depicted in Fig. 5, it is evident that the bi-LSTM had the best performance in all the measures compared to other models. For example, at the HR@5 cutoff, the bi-LSTM scored 33.03, which was significantly higher than that of all the other models, thereby signaling to it being better placed at making predictions with accuracy within the top five recommendations. This tendency is kept by HR@10 and HR@20, which shows that bi-LSTM well captures the sequential information required in session-based contexts.

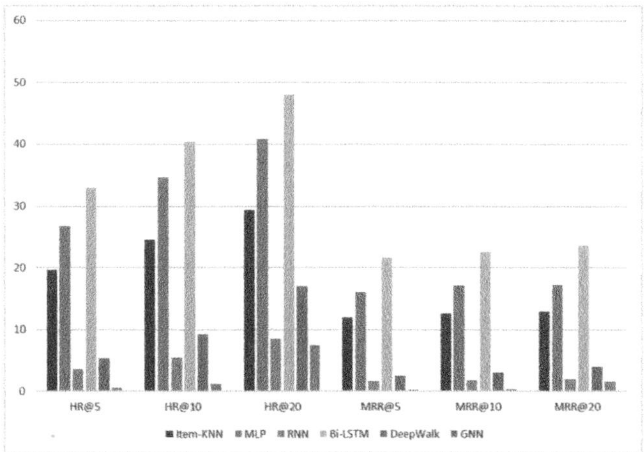


**Fig. 5.** The performance of DL models.

In contrast, GNN displayed lower performance, with its highest HR@20 reaching only 7.45. While theoretically, GNNs can model complex relationships between items in a session, as displayed in the SR-GNN [23]; from the results, it is insinuated that the simple GNN model cannot independently capture the sequential information within the graphs of sessions. However, when evaluating the performance of the unsupervised graph approach DeepWalk, it is evident that simple random walks within a general graph of all sessions successfully learned relationships between items and surpassed both GNN and RNN models in all metrics.

While the MLP, often considered a baseline in neural network applications, also gave strong performances, especially at higher cutoffs, where HR@20 at 40.81 was nearly equal to bi-LSTM. Pretty impressive, considering that the MLP model was so simplistic in every regard. This, therefore, means that while MLPs do not have the aspect of sequential data processing like in RNN-based models, it can still capture user preferences effectively provided that the recommendation set has a broad scope.

RNN model showed modest performance, with the highest HR@20 of 8.47. Lower scores of RNN compared to bi-LSTM and MLP do suggest that, though RNN can capture the sequence patterns, the capability may be having trouble meeting the complexity and variability of session data or needing more fine-tuning to reach performance equal to other models.

as anticipated, our analysis finds that very basic deep learning models, from MLP to even Bi-LSTM, beat item-KNN on key metrics such as HR and MRR at some cutoff points. This underlines the possibility that even rather simple architectures of neural networks can be capable of detecting complex patterns and relationships within the user's data session, avoiding traditional methods in some cases.

## 3.3   Session Sequence Lengths Analysis

This analysis evaluates how each of the four supervised models would respond to different user interaction patterns. From this point of view, the length of the session is a very important factor, as it largely varies from user to user and thus influences the predictive accuracy of the models in the recommendation systems based on sessions. Table 4 defines 'short' as sessions with less than 5 interactions, and 'long' as sessions with 5 or more interactions.

**Table 4.** Performance of DL models across different sequence lengths.

| Models | Short | | Long | |
|---|---|---|---|---|
| | HR@20 | MRR@20 | HR@20 | MRR@20 |
| MLP | 46.66 | 21.66 | 33.71 | 11.65 |
| RNN | 11.48 | 3.05 | 8.98 | 2.36 |
| Bi-LSTM | 47.69 | 22.99 | 45.78 | 22.23 |
| GNN | 9.7 | 2.01 | 4.75 | 0.96 |

The data indicates that both the MLP and Bidirectional Long Short-Term Memory bi-LSTM models exhibit robust performance across short and long sessions. The bi-LSTM, however, has a slight edge in longer sessions, with an HR@20 of 45.78 and an MRR@20 of 22.23, compared to the MLP's HR@20 of 33.71 and MRR@20 of 11.65 in long sessions. This suggests that the bi-LSTM's ability to process information bidirectionally allows it to better utilize the additional context provided in longer sessions, which could include a wider array of user actions and preferences.

Interestingly, the performance gap between the MLP and bi-LSTM is narrower in shorter sessions, with MLP achieving an HR@20 of 46.66 and bi-LSTM achieving an HR@20 of 47.69. This suggests that in shorter sessions, where less sequential information is available, the advantage of bi-LSTM's complex sequence modeling is less pronounced, and simpler models like the MLP can perform just as well. This observation is further illustrated in Fig. 6.

GNN and Recurrent Neural Networks RNN, however, showed decreased performance with increasing session length. For GNN, the performance in long sessions dropped significantly, with an HR@20 of 4.75, suggesting that the model may not effectively capture the additional complexity present in longer sessions. Similarly, the RNN's performance also declined in longer sessions, supporting the notion that standard RNNs may face challenges in capturing longer-term dependencies without advanced mechanisms like those present in bi-LSTM.

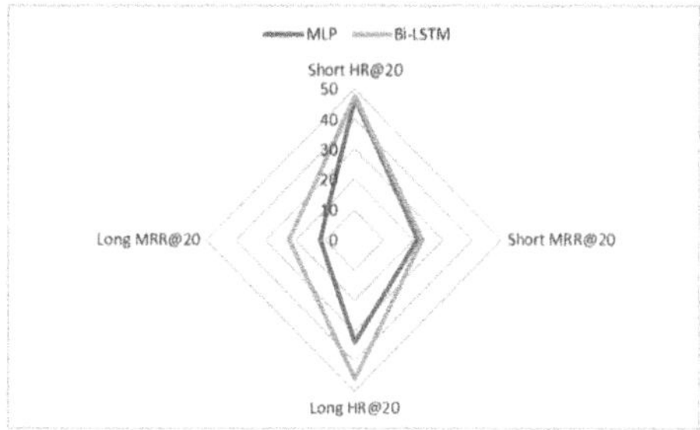


**Fig. 6.** MLP vs Bi-LSTM in long and short sessions.

## 4   Conclusion

Against this background, the present research carries out a detailed comparative performance evaluation of foundational deep learning models, including MLP, RNN, bi-LSTM, GNN, against traditional item-based k-Nearest Neighbors. Such interpretation was obvious in our finding that even simple deep learning architectures, such as MLP and bi-LSTM, not only compete but mostly outperform the KNN model with respect to many accuracy and precision metrics.

The bi-LSTM model had a better performance than all because it was very good at capturing serial dependencies and subtleness in user interactions in a session. This makes them perfect for systems where the precision of forecast in dynamic user environments is mandatory. Simple though not having any sequence modeling at all, MLP, on the other hand, was surprisingly strong, mostly on par with the more complex models. This would suggest that MLP could be successfully applied to reduced reliance on the order of events and would provide a stronger, easier alternative to the session-based recommendations.

The RNN and GNN models had their shortcomings too, where the RNN had problems with complex session data and, therefore, either further required tuning or used advanced sequence modeling approaches, like bi-LSTM. Although theoretically potent, GNN did not perform well to capture the sequential nature of session data, which probably may be due to different attention items: item interrelationships vs. sequence patterns.

## References

1. Abadal, S., Jain, A., Guirado, R., López-Alonso, J., Alarcón, E.: Computing graph neural networks: a survey from algorithms to accelerators. ACM Comput. Surv. (CSUR) **54**(9), 1–38 (2021)
2. Bobadilla, J., Ortega, F., Hernando, A., Gutiérrez, A.: Recommender syst. surv. Knowledge-based syst. **46**, 109–132 (2013)

3. Fang, W., Chen, Y., Xue, Q.: Survey on research of RNN-based SPATIO-temporal sequence prediction algorithms. J. Big Data **3**(3), 97 (2021)
4. Goodfellow, I., Bengio, Y., Courville, A.: Deep learning. MIT press (2016)
5. Graves, A., Graves, A.: Supervised sequence labelling. Springer (2012)
6. Graves, A., Jaitly, N., Mohamed, A.r.: Hybrid speech recognition with deep bidirectional LSTM. In: 2013 IEEE Workshop On Automatic Speech Recognition And Understanding, pp. 273–278. IEEE (2013)
7. Hamilton, W., Ying, Z., Leskovec, J.: Inductive representation learning on large graphs. Adv. Neural Inf. Proc. Syst. **30** (2017)
8. Hidasi, B., Karatzoglou, A., Baltrunas, L., Tikk, D.: Session-based recommendations with recurrent neural networks. arXiv preprint arXiv:1511.06939 (2015)
9. Kipf, T.N., Welling, M.: Semi-supervised classification with graph convolutional networks. arXiv preprint arXiv:1609.02907 (2016)
10. Lindemann, B., Müller, T., Vietz, H., Jazdi, N., Weyrich, M.: A survey on long short-term memory networks for time series prediction. Procedia Cirp **99**, 650–655 (2021)
11. Ludewig, M., Jannach, D.: Evaluation of session-based recommendation algorithms. User Model. User-Adap. Inter. , 331–390 (2018). https://doi.org/10.1007/s11257-018-9209-6
12. Ludewig, M., Mauro, N., Latifi, S., Jannach, D.: Empirical analysis of session-based recommendation algorithms: a comparison of neural and non-neural approaches. User Model. User-Adap. Inter. **31**(1), 149–181 (2021)
13. Perozzi, B., Al-Rfou, R., Skiena, S.: Deepwalk: Online learning of social representations. In: Proceedings of the 20th ACM SIGKDD international conference on Knowledge discovery and data mining, pp. 701–710 (2014)
14. Quadrana, M., Cremonesi, P., Jannach, D.: Sequence-aware recommender systems. ACM comput. surv. (CSUR) **51**(4), 1–36 (2018)
15. Rana, A., Rawat, A.S., Bijalwan, A., Bahuguna, H.: Application of multi layer (perceptron) artificial neural network in the diagnosis system: a systematic review. In: 2018 International Conference On Research In Intelligent And Computing In Engineering (RICE), pp. 1–6. IEEE (2018)
16. Sarwar, B., Karypis, G., Konstan, J., Riedl, J.: Item-based collaborative filtering recommendation algorithms. In: Proceedings of the 10th international conference on World Wide Web, pp. 285–295 (2001)
17. Scarselli, F., Gori, M., Tsoi, A.C., Hagenbuchner, M., Monfardini, G.: The graph neural network model. IEEE Trans. Neural Netw. **20**(1), 61–80 (2008)
18. Schlichtkrull, M., Kipf, T.N., Bloem, P., van den Berg, R., Titov, I., Welling, M.: Modeling Relational Data with Graph Convolutional Networks. In: Gangemi, A., Navigli, R., Vidal, M.-E., Hitzler, P., Troncy, R., Hollink, L., Tordai, A., Alam, M. (eds.) ESWC 2018. LNCS, vol. 10843, pp. 593–607. Springer, Cham (2018). https://doi.org/10.1007/978-3-319-93417-4_38
19. Van Houdt, G., Mosquera, C., Nápoles, G.: A review on the long short-term memory model. Artif. Intell. Rev. **53**(8), 5929–5955 (2020). https://doi.org/10.1007/s10462-020-09838-1
20. Velickovic, P., et al.: Graph attention netw. stat **1050**(20), 10–48550 (2017)
21. Wang, S., Cao, L., Wang, Y., Sheng, Q.Z., Orgun, M.A., Lian, D.: A survey on session-based recommender systems. ACM Comput. Surv. (CSUR) **54**(7), 1–38 (2021)

22. Wu, C., Yan, M.: Session-aware information embedding for e-commerce product recommendation. In: Proceedings of the 2017 ACM on conference on information and knowledge management, pp. 2379–2382 (2017)
23. Wu, S., Tang, Y., Zhu, Y., Wang, L., Xie, X., Tan, T.: Session-based recommendation with graph neural networks. In: Proceedings of the AAAI conference on artificial intelligence. vol. 33, pp. 346–353 (2019)

# A Novel Rule-Based Control Simulation Approach for Manufacturing Processes Optimization: Application to Industrial Production Systems


Naima Beladam[1]([✉]) and Abdelghani Ghomari[2]

[1] Department of Computer Science, Faculty of Science and Technology, University of Relizane, Relizane, Algeria
`naima.beladam@univ-relizane.dz`
[2] Department of Computer Science, Faculty of Exact and Applied Sciences, RIIR Laboratory, University Oran1 Ahmed Ben Bella, Oran, Algeria
`ghomari.abdelghani@univ-oran1.dz`



**Abstract.** Effective management of industrial production systems presents a major challenge in the modern industry due to the increasing complexity of manufacturing processes and the growing demands for higher product quantities and faster production rates. This paper proposes an innovative approach based on rule-based control simulation for the planning and control of manufacturing processes in industrial production systems. The presented method allows for the simulation of these processes and the monitoring of their performance while optimizing criteria such as downtime and resource utilization. The results are illustrated through a case study in an industrial production company, demonstrating the benefits of this approach in improving the overall performance of the organization.

**Keywords:** Simulation · industrial production · manufacturing processes optimization


## 1 Introduction

### 1.1 Research Context and Motivation

In the current industrial production environment, managing systems is becoming increasingly complex due to the multiple interactions and dynamic dependencies between various components, such as resources, processes, and products. This complexity presents significant challenges in representing manufacturing processes within these systems [1–5]. With rising market demands, particularly in terms of production volume and delivery deadlines [4], it is crucial for industrial enterprises to optimize their production systems.

The challenges posed by these systems can be summarized in three main points: (1) effective utilization of resources, whether material or human, which requires avoiding overloads while minimizing periods of non-productive inactivity; (2) minimizing delays



N. Seddari and M. Redjimi (Eds.): ICMSCT 2024, CCIS 2606, pp. 54–66, 2025.
https://doi.org/10.1007/978-3-032-01922-6_5

between different production stages, which not only accelerates production flow but also improves customer satisfaction by shortening delivery times; (3) effective management and coordination of production processes to avoid bottlenecks and ensure continuous flow.

To address these issues and overcome these challenges, computer simulation, particularly discrete event simulation, has proven to be a valuable method for managing and improving these systems [6]. The advantage of discrete event simulation is that it allows for experiments that cannot be conducted on real manufacturing systems, and designing a simulation model can provide insights that may lead to improvements in the actual system. Moreover, discrete event simulation is used in conjunction with other methods to solve various problems such as system conceptual design [7], production planning [8], scheduling issues [9], bottleneck detection [10], as well as analyzing efficiency [11, 17] and stability [12] of production systems.

## 1.2  Problem Definition

Simulation, as a dynamic modeling tool, allows for testing and evaluating various production system configurations before their actual implementation. However, to ensure that simulated systems adhere to the constraints imposed by technical specifications and performance objectives, it is essential to integrate control rules. These rules enable the monitoring of the production process during the simulation, adjusting certain parameters, and thereby ensuring the proper functioning of the systems.

The problem addressed by this research can be summarized by the following question: How can rule-based control simulation enhance the performance of industrial production systems by meeting increasing demands for quantity and production speed, while minimizing downtime and optimizing resource utilization?

## 1.3  Contribution

In this work, we propose a simulation approach based on mathematical control rules aimed at verifying compliance with specific constraints of industrial production systems. Our method provides the capability to effectively control production processes while offering precise estimates of expected outcomes.

We also develop a simulation tool guided by these rules, which assists decision-makers in analyzing and improving production system performance. This tool thereby facilitates more effective management of manufacturing processes, addressing the performance and efficiency needs of industrial production systems.

The remainder of the article is structured as follows: Sect. 2 reviews some related works in the field of performance optimization of manufacturing systems. In Sect. 3, we describe a comparative study of different approaches of the literature, while Sect. 4 describes the proposed approach, and Sect. 4 is dedicated to applying this approach to a case study, presenting the results of the initial simulation. Section 5 discusses the proposed solutions for improving performance. Finally, Sect. 6 concludes the article and suggests directions for future research.

## 2  Related Works

In recent years, simulation has taken a prominent role in optimizing manufacturing systems. Various approaches have been proposed to improve flexibility, resource allocation, and overall system efficiency. This section provides a summary of the most relevant related work in the field, focusing on simulation methods and optimization techniques used to enhance the performance of manufacturing systems.

The work of [12] proposes an advanced simulation approach, enabling the automatic experimentation of different input parameters to adjust the production system's flexibility in response to market changes. The Experiment Wizard and Analyze Experiment Results modules facilitate the testing of various scenarios without the need for real-world trials. This allows for optimized decision-making regarding scheduling, investment costs, and flexibility in meeting changing customer demands.

In [13], an approach combining combinatorial optimization, supervised learning, and discrete event simulation is presented to improve workshop scheduling. The authors use decision trees to dynamically generate allocation rules based on offline data exploration and tabu search. This combination allows for efficient optimization of operations while adapting task allocation.

The work of [14] addresses the problem of optimal scheduling in flexible manufacturing systems using metaheuristic techniques. The integration of supervisory control theory for discrete event systems allows for automatic correction of infeasible solutions and ensures robust control. This approach significantly enhances overall performance by combining the advantages of predictive scheduling with structural control.

In [15], the authors use agent-based simulation to model a flexible production system. They demonstrated that their model, implemented in Tecnomatix Plant Simulation, outperforms traditional models in terms of flexibility and adaptability. It proves the feasibility of integrating agent-based approaches in real industrial environments, offering scalable capabilities without imposing fixed production cycles.

The work of [16] proposes an approach that uses a SCADA system (Supervisory Control and Data Acquisition) based on discrete event simulation. This approach allows for validating simulation results through statistical methods and direct observations. The predictive model created has helped reduce production costs while maximizing resource utilization and anticipating potential system disruptions.

In [17], a computer simulation is used to model and analyze material flow in custom production. The authors demonstrate how simulation can identify bottlenecks and improve production efficiency by adjusting production parameters based on the specific needs of customers.

## 3  Comparative Table of Different Simulation Approaches for Manufacturing Processes

In Table 1, a comparative study of our approach with those of the literature is presented, where three different criteria are used.

**Table 1.** Comparative table of different simulation approaches for manufacturing processes optimization

| Authors | Used Approach | Optimized Performance | Constraint Control |
|---|---|---|---|
| [12] | Advanced simulation (Automatic experimentation) | Flexibility, resource optimization | No |
| [13] | Discrete simulation + Combinatorial optimization | Task allocation | No |
| [14] | Discrete simulation + Metaheuristic optimization | Scheduling, performance improvement | Yes, through supervisory control |
| [15] | Agent-based simulation | Flexibility, adaptation to changes | No |
| [16] | Simulation + SCADA | Resource maximization, cost reduction | Yes, through the SCADA model |
| [17] | simulation | Efficiency of custom production | No |
| Our Work | Process simulation + Control rules | Optimization of downtime and resources | Yes, through mathematical control rules |

The analyzed studies highlight the use of simulation to optimize various aspects of manufacturing systems, whether it is improving flexibility [12, 15], optimizing resource allocation [13, 14], or increasing overall efficiency [16, 17]. Unlike the work of [14, 16], most existing approaches focus on simulations without integrating explicit mechanisms to control the performance of production systems.

The approach proposed in this paper differs from those in the literature by integrating control rules into the simulation model, even though this is still performed in offline mode. Specifically, this means that when critical parameters exceed a predefined threshold (for example, maximum downtime or resource overload), our simulation tool automatically detects these exceedances and flags them as errors to be corrected. This enhances the robustness and accuracy of simulations by anticipating constraint violations without waiting for a post-simulation analysis.

By integrating these control rules into the simulation flow, our approach not only allows for the evaluation of the performance of the simulated systems but also ensures that processes adhere to the established limits regarding resource utilization and waiting times. This way, we can identify and flag bottlenecks or inefficiencies before they impact the final outcomes, facilitating faster and more informed decision-making.

## 4  Description of the Adopted Approach

The adopted approach follows a three-step process, illustrated in Fig. 1. The first step involves modeling the production system as a generic simulation model, capturing the key characteristics of the system. The second step involves integrating the proposed control rules into this generic model to ensure compliance with specific constraints. Finally, the third step entails simulating the model to monitor and optimize production processes, with the goal of enhancing the overall performance of the industrial system.

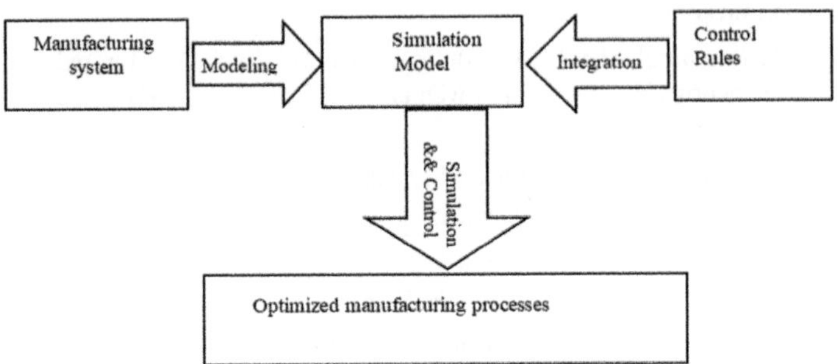


**Fig. 1.** Flow Chart of the Adopted Approach.

### 4.1  Production System Modeling

The proposed model is based on three generic elements: the product, the resource, and the processing entity, which interact with each other. Each element is defined by various attributes related to its identification and characteristics that allow for performance evaluation; specific functions also facilitate the management of the entity. The elements: product, resource, and processing exchange knowledge about product processing on resources: each product $P_i$ undergoes one or more operations $T_j$; each resource $R_m$ handles one or more operations $T_j$, and each operation $T_j$ is handled by a single resource.

### 4.2  Integration of Control Rules

Control rules are used to monitor the manufacturing processes of the production system and ensure that it adheres to imposed constraints. This work focuses on controlling resource utilization and inactivity times, which allows for optimal and efficient resource management.

**Resource Utilization Control.** Resource utilization control relies on formulas (1), (2), (3), (4) and (5). These formulas are used to calculate and monitor the occupancy rate of a resource $R_w$ at the end of each processing of a product $P_i$. At time V1, product $P_i$ begins processing $T_k$ on resource $R_w$ with an initial processing time $R_w.time_m$. Then, at time V2, product $P_i$ completes this processing, and the accumulated processing time for resource $R_w$ becomes $R_w.time_{m+1}$.

$$\left(R_w.time_{m+1} = R_w.time_m + T_k(P_j, R_w).time\_max\right) \tag{1}$$

Formula (1) determines the new accumulated processing time for resource $R_w$ ($R_w.time_{m+1}$). It is obtained by adding the previous processing time of the resource ($R_w.time_m$) to the maximum processing time of product Pi on resource $R_w$.

$$\left(V2 = V1 + T_k(P_j, R_w).time_{max}\right) \tag{2}$$

Formula (2) calculates the end time of processing for product $P_j$(V2). It is obtained by adding the start time of processing for product $P_j$ on resource $R_w$(V1) to the maximum processing time of product $P_i$ on resource $R_w$.

$$(V1 = R_w.time\_availab + R_w.time\_prep\_max) \tag{3}$$

Formula (3) determines the start time of processing for product $P_i$ on resource $R_w$ (V1). It is calculated by adding the availability time of resource $R_w$ ($R_w.time\_availab$) to the maximum preparation time for this resource.

$$\left(R_w.rate = \frac{R_w.time_{m+1}}{V2}\right) \tag{4}$$

Formula (4) calculates the occupancy rate of resource $R_w$. This rate is obtained by dividing the accumulated processing time of resource $R_w$($R_w.time_{m+1}$) by the total simulation time (V2), where V2 corresponds to the end time of processing for product $P_j$ on resource $R_w$.

$$(R_w.rate \leq Rate\_max) \tag{5}$$

Formula (5) sets the limit for the accumulated occupancy rate of resource $R_w$. This rate should not exceed the maximum allowed occupancy rate. In case of a violation of this formula, a specific error message is generated to indicate that the occupancy rate of resource $R_w$ is not being met.

**Resource Inactivity Control.** Resource inactivity control is based on formulas (6), (7), (8) and (9). These formulas are used to calculate and monitor the inactivity time of a resource Rw at a time V after its availability. At time V1, product Pi completes its processing T on resource Rw. Then, at time V2, the next product Pj, begins processing Tk on the same resource Rw, and the accumulated inactivity time of Rw (Rw.time_inac) is reset to zero.

$$(R_w.time\_inac = V - V1) \tag{6}$$

Formula (6) determines the inactivity time of resource $R_w$. It is obtained by subtracting the end time of processing of product Pi on resource $R_w$ (V1) from a specific time V.

$$(V2 = V3 + R_w.time\_prep\_max) \tag{7}$$

Formula (7) calculates the start time of processing for the next product $P_j$ on resource $R_w$ (V2). This time is obtained by adding the maximum preparation time of resource $R_w$ to the resource's availability time (V3), which is the moment when the resource becomes available and product $P_j$ is waiting. If product $P_j$ arrives after the availability of resource $R_w$, V3 then corresponds to the arrival time of product $P_j$.

$$(V1 < V \leq V2) \tag{8}$$

Formula (8) indicates that V represents a specific time during the availability period of resource $R_w$.

$$(R_w.time\_inac \leq time\_inac\_max) \tag{9}$$

Formula (9) sets the limit for the accumulated inactivity time of resource $R_w$ (V5). This time should not exceed the maximum allowed inactivity time. In case of a violation of this formula, a specific error message is generated to indicate that the inactivity time of resource $R_w$ is not being met.

### 4.3  Simulation and Process Control

Once the model is established and the control rules defined, the simulation is executed to evaluate the system's performance as follows: Based on the adopted organization, the evaluation strategy is based on the following principle: at time t, product $P_i$ arrives and checks the availability of the resource $R_w$ involved in the first operation T1. If resource $R_w$ is busy, product $P_i$ waits in the queue for that resource; otherwise, it proceeds with its processing and then moves to operation T2, and so on, until all its operations are completed.

At each stage, an analysis is performed using the integrated control rules to identify bottlenecks, resource over-utilization's, or deviations from specifications. This analysis allows for adjustments to the system's parameters to optimize performance and improve efficiency.

## 5  Application to an Industrial Enterprise

### 5.1  System Overview

To illustrate the effectiveness of our approach, we apply our method to a production unit of a real industrial company specializing in the manufacture of boilers of various capacities and uses. To achieve its production plans, the unit has technical and industrial production means divided between two main workshops: a boiler body workshop equipped with technical facilities for constructing essential boiler components, and a machining workshop with machines and tools for machining some of the boiler accessories.

The production process of a boiler involves two major phases: the production of semi-finished products necessary for its construction, and the assembly, finishing, and final performance testing phase. This work focuses on the seven semi-finished products required for the manufacture of a single steam boiler, which are: the "casing", the "firebox", the "two front and rear tubular plates", the "combustion chamber", and the "two combustion chamber plates".

- The casing and the firebox go through the following operations: oxy-cutting of the sheet metal, molding of the sheet metal ends to remove burrs, bending, and two welding operations.
- The two front and rear tubular plates undergo the following operations: oxy-cutting of two sheets of similar dimensions and drilling of the two plates on the drilling machine.
- The combustion chamber goes through the following operations: oxy-cutting of the sheet metal according to the dimensions specified in the manufacturing plan, molding at the ends, and longitudinal welding outside the cylinder manually.
- The two combustion chamber plates undergo the operation of oxy-cutting of two sheets of similar dimensions.

## 5.2 System Modeling

In the context of the generic model, the product entity encompasses the 7 semi-finished products identified earlier, labeled P1…P7. The resource entity comprises the 5 resources R1…R5, which correspond to the involved machines: oxy-fuel cutting machine, bending machine, 2 welding machines, and the drilling machine. The processing entity includes the 7 operations T1…T7 necessary for the production of the semi-finished products. The information about the resources (Fig. 2) includes the identifier (Ri), the name of the resource, the type (machines), the resource state (available), and both minimal and maximal preparation times, which are set to 1 unit of time (UT) for all resources. Resource constraints include a processing rate that must not exceed 95%, and a maximum allowed inactivity time of 100 units of time.

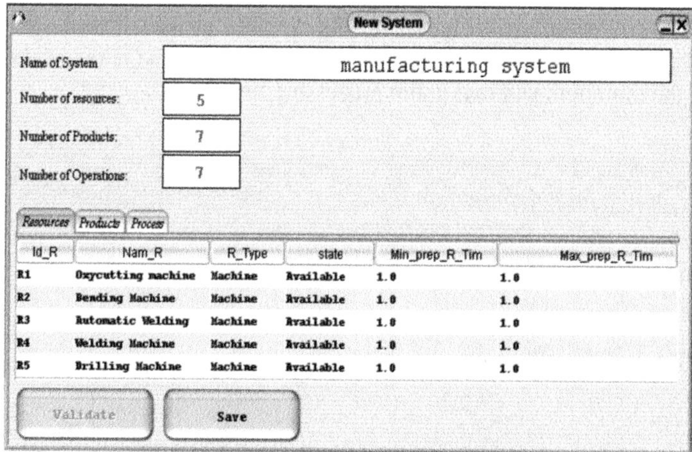


**Fig. 2.** Knowledge of Resource Entities.

Following the principle of communication between entities, the 7 products (Fig. 3) exchange processing knowledge such as the product's arrival time, the number of operations performed for processing the product, and the corresponding processing sequence. For example, the semi-finished product P1 (combustion chamber) undergoes treatments T1, T2, and T7 according to the relevant operations.

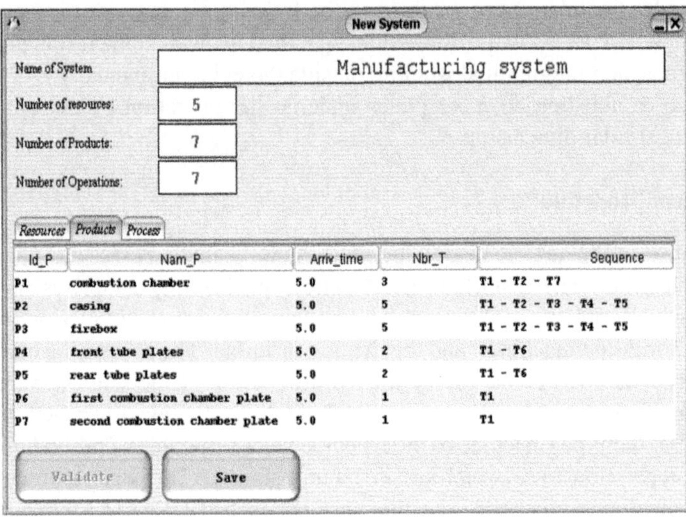

**Fig. 3.** Knowledge of Product Entity.

Similarly, the 7 treatments (Fig. 4) exchange knowledge with the resources, including its identification, the identification of the relevant resource, the name of the treatment, as well as the minimum and maximum processing times.

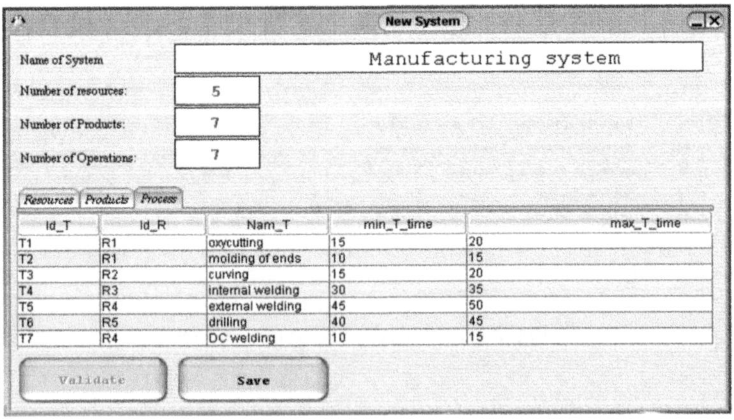

**Fig. 4.** Knowledge of Treatment Entity.

## 5.3   Results and Discussion

In light of the defined performance indicators and within the context of assessing the company's performance through the efficiency of the semi-finished product manufacturing process, we adopt the following three performance criteria:

– Work time
– Resource utilization rate
– Resource processing time

**Initial Performance Related to Resources and Work Time.**  By applying the simulation principle, the initial evaluation of performance criteria for the 5 resources (Fig. 5) yields the following results concerning processing time and utilization rates:

| Id Resource | Nam Resource | Process Time | occupancy Rate(%) |
|---|---|---|---|
| R1 | Oxycutting machine | 185.0 | 54.58 |
| R2 | Bending Machine | 40.0 | 11.8 |
| R3 | Automatic Welding | 70.0 | 20.65 |
| R4 | Welding Machine | 115.0 | 33.93 |
| R5 | Drilling Machine | 90.0 | 26.55 |

**Fig. 5.** Initial Performance Related to Resources.

The simulation was halted at time 339.0. The resource occupancy rate ranges between 11.8% and 54.58%, with the highest rate being for resource R1 (oxy-cutting machine). This high rate is attributed to the number of treatments performed by this machine and the large number of products processed on this machine, as all products undergo the oxy-cutting operation.

**Reported Errors.**  The resource utilization control generated an initial error message indicating that the inactivity time for resources R2, R3, and R4 exceeded the maximum allowable inactivity threshold at time 101 (see Fig. 6).

| Resource | process time | Occupancy Rate (%) |
|---|---|---|
| R1 | 92.0 | 91.09 |
| R2 | 0.0 | 0.0 |
| R3 | 0.0 | 0.0 |
| R4 | 0.0 | 0.0 |
| R5 | 13.0 | 12.88 |

**Fig. 6.** Reported Errors.

**Interpretation.** The inactivity duration of machines R2, R3, and R4 has been exceeded because their operations occur after the completion of tasks T1 and T2, which are assigned to machine R1. Furthermore, since all products arrived simultaneously, machine R1 must first complete all T1 tasks for each product before proceeding with T2 for product P1. Only after this task is completed can machine R4 begin its first operation T7 for product P1, followed by the other machines R2 and R3.

## 6 Proposed Solution

Following the initial error detected in the resource inactivity control within the studied system, we have decided to distribute the tasks for products P2 and P3 into multiple phases. Operations T1-T2 and T3-T4 will be completed the day before, while operation T5 will be carried out on the day of manufacturing the other products related to boiler production. On the same day, operations T1-T2 and T3-T4 for products P2 and P3 will also be executed to prepare for the production of the next boiler (see Fig. 7).

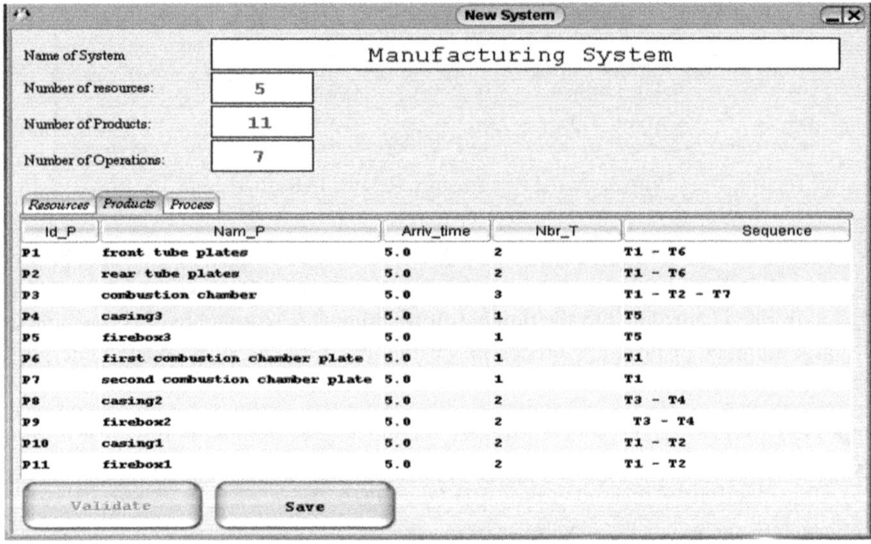



| Id_P | Nam_P | Arriv_time | Nbr_T | Sequence |
| --- | --- | --- | --- | --- |
| P1 | front tube plates | 5.0 | 2 | T1 – T6 |
| P2 | rear tube plates | 5.0 | 2 | T1 – T6 |
| P3 | combustion chamber | 5.0 | 3 | T1 – T2 – T7 |
| P4 | casing3 | 5.0 | 1 | T5 |
| P5 | firebox3 | 5.0 | 1 | T5 |
| P6 | first combustion chamber plate | 5.0 | 1 | T1 |
| P7 | second combustion chamber plate | 5.0 | 1 | T1 |
| P8 | casing2 | 5.0 | 2 | T3 – T4 |
| P9 | firebox2 | 5.0 | 2 | T3 – T4 |
| P10 | casing1 | 5.0 | 2 | T1 – T2 |
| P11 | firebox1 | 5.0 | 2 | T1 – T2 |



**Fig. 7.** Proposed Task Distribution.

### 6.1 Performance Relative to Working Time

The simulation was stopped at 199 UT, representing a reduction of 140 UT compared to the previous situation.

## 6.2  Performance Relative to Resources

With the implementation of our approach, the occupancy rate of the 5 resources has improved, as illustrated in Fig. 8 below. This figure presents the occupancy rates of the 5 resources before and after applying the control rules detailed in Subsects. 4.2.

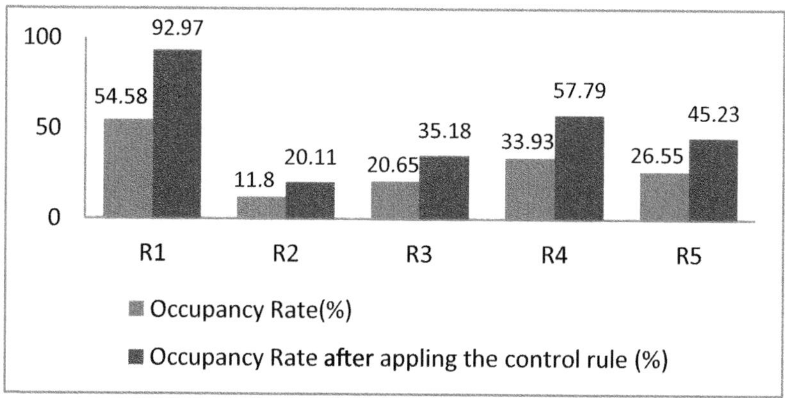


**Fig. 8.** Improved ooccupancy rates.

The resource occupancy rates in the studied system have been significantly improved. Specifically, the rates increased by 38.39% for machine R1, 8.31% for machine R2, 14.53% for machine R3, 23.86% for machine R4, and 18.68% for machine R5.

## 7  Conclusion

This research presents an innovative approach to optimizing industrial manufacturing systems through simulation guided by control rules. Our approach enables the monitoring of the manufacturing process, identification of resource inefficiencies, and enhancement of resource management to ensure system efficiency. We applied our approach to an industrial manufacturing company in Algeria, demonstrating its potential to deliver significant improvements in production system performance. This approach contributes to more effective and reliable management of manufacturing processes within industrial systems.

For future research, it would be valuable to: (1) extend this approach to other industrial sectors and types of complex systems; (2) use this approach with artificial intelligence algorithms to improve process optimization and real-time decision-making; (3) propose new rules to control other constraints. These perspectives would provide new insights and further improvements in terms of performance and reliability.

## References

1. Deshmukh, A.V., Talavage, J.J., Barash, M.M.: Complexity in manufacturing systems, Part 1: analysis of static complexity. IIE Trans. **30**(7), 645–655 (1998)

2. Perona, M., Miragliotta, G.: Complexity management and supply chain performance assessment. a field study and a conceptual framework. Int. J. Prod. Econ. **90**(1), 103–115 (2004)
3. Jacobs, M.A.: Product complexity: a definition and impacts on operations. Decis. Line (2007)
4. Efthymiou, K., Mourtzis, D., Pagoropoulos, A., Papakostas, N., Chryssolouris, G.: Manufacturing systems complexity analysis methods review. Int. J. Comput. Integr. Manuf. **29**(9), 1025–1044 (2016)
5. Herrera Vidal, G., Coronado Hernández, J.R.: Complexity in manufacturing systems: a literature review. Prod. Eng. Res. Devel. **15**, 321–333 (2021)
6. Kampa, A., Gołda, G., Paprocka, I.: Discrete event simulation method as a tool for improvement of manufacturing systems. Computers **6**(1), 10 (2017)
7. Xie, C., Allen, T.T.: Simulation and experimental design methods for job shop scheduling with material handling: a survey. Int. J. Adv. Manuf. Technol. **80**, 233–243 (2015)
8. Krenczyk, D.: Automatic generation method of simulation model for production planning and simulation systems integration. Adv. Mater. Res. **1036**, 825–829 (2014)
9. Kampa, A.: Planning and scheduling of work in robotic manufacturing systems with flexible production. J. Mach. Eng. **12**, 34–44 (2012)
10. Krenczyk, D., Olender, M.: Production planning and control using advanced simulation systems. Int. J. Mod. Manuf. Technol. **6**, 38–43 (2014)
11. Ingemansson, A., Bolmsjö, G.S.: Improved efficiency with production disturbance reduction in manufacturing systems based on discrete-event simulation. J. Manuf. Technol. Manag. **15**(3), 267–279 (2004)
12. Burduk, A.,Burduk, A.: Stability analysis of the production system using simulation models. Process Simulation and Optimization in Sustainable Logistics and Manufacturing, pp. 69–83 (2014)
13. Shahzad, A., Mebarki, N.: Learning dispatching rules for scheduling: a synergistic view comprising decision trees Tabu search and simulation. Computers **5**(1), 3 (2016)
14. Pena, P.N., Costa, T.A., Silva, R.S., Takahashi, R.H.: Control of flexible manufacturing systems under model uncertainty using supervisory control theory and evolutionary computation schedule synthesis. Inf. Sci. **329**, 491–502 (2016)
15. Büth, L., Broderius, N., Herrmann, C., Thiede, S.: Introducing agent-based simulation of manufacturing systems to industrial discrete-event simulation tools. In: 2017 IEEE 15th International Conference on Industrial Informatics (INDIN), pp. 1141–1146 (2017)
16. Mousavi, A., Siervo, H.R.A.: Automatic translation of plant data into management performance metrics: a case for real-time and predictive production control. Int. J. Prod. Res. **55**(17), 4862–4877 (2017)
17. Rosova, A., Behun, M., Khouri, S., Cehlar, M., Ferencz, V., Sofranko, M.: Case study: the simulation modeling to improve the efficiency and performance of production process. Wireless Netw. **28**(2), 863–872 (2022)

# Advancing Early Detection of Type 2 Diabetes: A Machine Learning Approach with Combined Clinical Datasets

Amani Mankouri[1]([⊠]) [ID], Abdeslem Dennai[2] [ID], and Farid Kacimi[1] [ID]

[1] Laboratoire LITAN, École Supérieure en Sciences et Technologie de l'Informatique et, du Numerique RN 75 Amizour 06300, Béjaia, Algeria
mankouri@estin.dz

[2] SGRE Laboratory, Department of Computer Science, University of Tahri Mohammed, Bechar, Algeria

**Abstract.** Early detection of type 2 diabetes mellitus (T2DM) is crucial for preventing severe complications but remains challenging due to the disease's complex nature. This study aims to improve T2DM prediction by combining the Pima Indians Diabetes Dataset (PIDD) with a larger Microsoft Azure dataset, totaling 15,768 instances, and applying advanced machine learning (ML) techniques. Comprehensive range of ML algorithms was employed, including traditional methods like decision trees, random forests, and support vector machines, as well as advanced models such as XGBoost, CatBoost, LightGBM, and TabNet. Multiple imputation methods (KNN, Hot Deck, MICE) were explored for missing data handling. Performance was assessed using accuracy, precision, F1-score, and AUC-ROC, with SHAP analysis for interpretability. Ensemble methods and advanced models demonstrated superior performance on the combined dataset, with CatBoost and LightGBM achieving up to 94.39% accuracy. The integration of the larger Azure dataset consistently improved model performance across all algorithms. Hot Deck imputation generally yielded the best results on the combined dataset. This comprehensive approach shows significant potential for improving early T2DM detection, offering promising tools for clinical decision support and enabling more timely interventions.

**Keywords:** Type 2 diabetes mellitus · machine learning · early detection · Deep learning · Missing Data Imputations · SHapley Additive exPlanations

## 1 Introduction

T2DM is the most common form of diabetes, accounting for around 90–95% of all diabetes cases worldwide. It is rapidly increasing globally, driven by factors like obesity, sedentary lifestyles, and an aging population. Uncontrolled T2DM can lead to serious complications including cardiovascular disease, nerve damage, kidney disease, eye damage, skin conditions, hearing impairment and Alzheimer's disease. It significantly increases the risk of premature death. T2DM requires lifelong management through a combination of lifestyle changes, medications, regular monitoring of blood sugar levels, and screening for complications [1]. The imperative for early detection and timely diagnosis of type 2 diabetes cannot be overstated, as it holds profound implications for mitigating the risk of severe complications and preserving quality of life. Alarmingly, women with T2DM exhibit a substantially higher relative risk of cardiovascular disease (CVD) and mortality compared to their male counterparts [2].

Accurately diagnosing diabetes at an early stage is critical for initiating timely interventions and mitigating the risk of severe complications. However, this task is intrinsically complex due to the multifaceted nature of the disease, involving complicated interactions between various risk factors and clinical measurements. Machine learning techniques offer a promising avenue to unravel the underlying patterns and develop robust predictive models for early diabetes detection. By using the power of machine learning algorithms, diverse datasets encompassing a wide array of risk factors can be integrated and analyzed, clinical biomarkers, and demographic information. This data-driven approach enables the identification of subtle patterns and non-linear relationships that may be overlooked by traditional statistical methods or human analysis alone.

Furthermore, combining multiple datasets can potentially enhance the general applicability and robustness of the developed models. By combining data from various sources and populations, a broader spectrum of disease manifestations and risk profiles can be captured, thereby improving the models' ability to generalize across diverse demographic and clinical contexts.

## 1.1  Contribution

This study makes several key contributions to the field of early T2DM detection:

1. Dataset Integration: The widely used Pima Indians Diabetes Dataset (PIDD) was combined with a larger Microsoft Azure dataset, creating a more comprehensive and diverse dataset for analysis.
2. Comprehensive ML Approach: 12 different traditional machine learning algorithms and 3 advanced models (CatBoost, LightGBM, TabNet) were employed and compared, identifying the most effective models for T2DM prediction.
3. Missing Data Handling: Multiple imputation techniques (KNN, Hot Deck, MICE) were systematically explored to address the challenge of missing data in clinical datasets, finding Hot Deck generally most effective for the combined dataset.
4. Performance Improvement: Significant improvements in predictive performance across all tested algorithms were demonstrate when using the combined dataset, with advanced models achieving up to 94.39% accuracy
5. Interpretability Analysis: SHAP (SHapley Additive exPlanations) was used to provide insights into feature importance and model predictions, enhancing the interpretability of our results.
6. Comparative Analysis: A detailed comparison of traditional and advanced ML models was provided, highlighting the strengths and potential clinical applications of each approach in T2DM prediction.

## 1.2  Paper Structure

The subsequent sections of this paper are structured to provide a comprehensive overview of our research. A thorough literature review of machine learning approaches in T2DM detection is presented in Sect. 2, highlighting recent methodologies and identifying current research gaps. Our study's methodology, including dataset integration, preprocessing procedures, and the implementation of various machine learning models, is delineated in Sect. 3. The experimental results, offering a comparative analysis of model performance and examining the impact of dataset fusion and imputation methods, are elucidated in Sect. 4. A critical discussion of our findings, interpreting the results within the context of early T2DM detection and their potential clinical implications, is provided in Sect. 5. Finally, Sect. 6 concludes the study by synthesizing key findings, acknowledging limitations, and proposing future research directions in this field. Figure 1 presents the Paper organization.

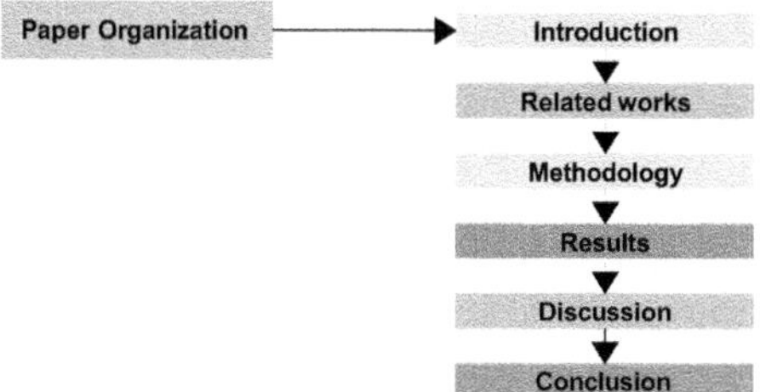


**Fig. 1.** Paper Organization

## 2  Related Work

This section provides an overview of recent studies in ML-based T2DM detection, highlighting their approaches, achievements, and limitations. It sets the context for our work and demonstrates the need for our approach using combined datasets.

Chang et al. [3] used Naive Bayes, random forest, and J48 decision tree models on the PIDD, achieving accuracy rates close to 80%. Building on this work, Reza et al. [4] employed more advanced stacking ensemble approaches, including a DNN method, on multiple datasets. Their DNN stacking ensemble achieved a higher accuracy of 95.50% on simulated data.

In a similar vein, Noviyanti et al. [5] applied Random forests to the PIDD, handling missing values with mode imputation, and achieved 87% accuracy. Expanding on this, Hama et al. [6] conducted a comparative study of GBC, DTC, ETC, and ABC on both PIDD and BRFSS datasets. Their results showed that ETC performed best with AUC of 0.96% for PIDD and 0.99% for BRFSS.

Taking a different approach, Reza et al. [7] proposed an improved SVM kernel, achieving 85.5% accuracy on the PIDD after median imputation of missing values. Meanwhile, Kuriakose et al. [8] explored the combination of multiple ML algorithms on PIDD and Health Facts Database, with Random Forest achieving 80% accuracy on the latter.

Further advancing the field, Alnowaiser et al. [9] introduced a novel Tri-Ensemble model (Extra Tree, XGBoost, Random Forest) on PIDD, achieving an impressive 97.49% accuracy after using KNN imputation for missing values.

These studies predominantly utilized the PIDD dataset, with some incorporating additional data sources to enhance model performance. They explored a range of ML techniques, from traditional algorithms to sophisticated ensemble methods, with accuracies varying from 80% to 97.49%.

Despite the valuable contributions of previous studies, several issues persist that limit the development of predictive models for early T2DM detection. A notable constraint is the reliance on the relatively small PIDD, which may compromise the models' ability to capture diverse population characteristics and disease manifestations. Additionally, many studies have explored a limited scope of machine learning algorithms, overlooking the potential that can arise from combining diverse techniques. Furthermore, the handling of missing data, a pervasive challenge in clinical datasets, has often been addressed through simplistic approaches, such as value removal, which can lead to information loss and biased estimates.

## 3  Methodology

This section outlines our approach to data preprocessing, feature engineering, and the implementation of various machine learning models for T2DM prediction.

### 3.1  Datasets

Two primary datasets were used in this study.

**Pima Indian Diabetes Dataset (PIDD)**
The PIDD is a popular benchmark dataset used for diabetes classification research. The dataset was originally donated by Johns Hopkins University to the UCI Machine Learning Repository in 1990. It contains health data from a population of Pima Indian women aged 21 and older living in Phoenix, Arizona. The Pima Indians are a Native American group with a high incidence rate of diabetes mellitus, so this dataset was deemed significant for global health research related to this disease [10].

The dataset comprises 768 instances and 9 features, including: Pregnancies, which records the number of times pregnant; Glucose, measuring plasma glucose concentration after a 2-h oral glucose tolerance test; BloodPressure, indicating diastolic blood pressure in mm Hg; SkinThickness, measuring triceps skin fold thickness in mm; Insulin, recording 2-h serum insulin in mu U/ml; BMI, calculated as weight in kg divided by height in meters squared; DiabetesPedigreeFunction, a score assessing the likelihood of diabetes based on family history; Age in years; and Diabetic, a class variable where 0 indicates non-diabetic and 1 indicates diabetic.

The goal is to use the 8 feature variables to predict whether a person has diabetes (class 1) or not (class 0) based on this medical diagnostic data. Out of 768 instances, there are 268 cases labeled as positive for diabetes and 500 cases labeled as negative. This represents a class imbalance, with around 65% non-diabetic cases and 35% diabetic cases.

**Microsoft Learning Azure Dataset**
A new dataset is obtained from Microsoft Learning Azure which contains the same features as the PIDD. The dataset contains 15,000 instances, significantly larger than the original PIDD. The larger dataset size may provide better generalization capabilities and more robust model training, given the increased diversity of instances [11]. The dataset consists of 5000 cases labeled as positive for diabetes (class 1) and 10000 cases labeled as negative (class 0). This represents a class imbalance, with around 66.66% non-diabetic cases and 33.33% diabetic cases.

**Dataset Fusion**
In this study, PIDD and Microsoft Azure dataset were combined to create a larger and more comprehensive dataset for analysis. The datasets were merged using the row concatenation technique, resulting in a combined dataset with a total of 15,768 instances. After combining the datasets, duplicate instances were checked and found no duplicates present. To ensure consistency and compatibility between the datasets, certain features were renamed in the Azure and PIDD datasets.

## 3.2  Data Preprocessing and Feature Engineering

**Handling Missing Values**
During the initial data exploration, certain features, such as 'Glucose', 'BloodPressure', 'SkinThickness', 'Insulin', 'BMI', 'DiabetesPedigree', and 'Age', were identified as having missing values represented as zeros. To address this issue, the zero values in these features were replaced with NaN (Not a Number) values, allowing for proper handling of missing data. Multiple imputation techniques were explored to handle these missing values including K-Nearest Neighbors (KNN) imputation, Hot Deck (HD) imputation, and Multiple Imputation by Chained Equations (MICE) [12–14]. Each imputation method resulted in a separate imputed dataset: D_MICE, D_KNN, and D_HD, respectively.

**Data Exploration and Visualization**
Exploratory data analysis was performed on the imputed datasets, including calculating descriptive statistics and visualizing feature correlations using heatmaps. Additionally, the distribution of individual features was inspected to identify potential outliers, such as instances with unusually high values.

**Feature Scaling Techniques**
To mitigate any patterns or biases present in the original data order, randomness was introduced into the imputed datasets by shuffling the rows using the sample (frac = 1) method. Feature scaling was then applied to the imputed and randomized datasets

to ensure that all features were on a consistent scale. Two scaling techniques were employed, Min-Max scaling, and Standardization (Z-score normalization).

These preprocessing steps were crucial in preparing our data for machine learning model training, ensuring that the input features were properly formatted, scaled, and free from missing values or obvious biases. The resulting preprocessed datasets formed the foundation for our subsequent analysis and modeling efforts (Fig. 2).

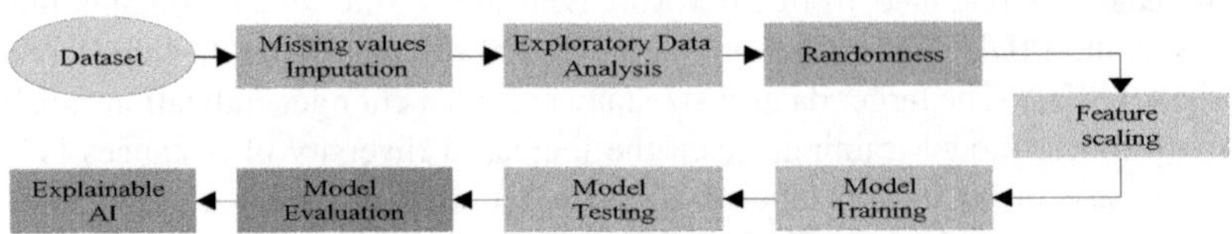


**Fig. 2.** Data Processing and Machine Learning Pipeline

### 3.3  Machine Learning Methods

To address the complex task of early T2DM detection, a comprehensive array of machine learning techniques was used. Our approach incorporated both well-established algorithms and noval models designed for tabular data. This diverse selection aimed to capture various aspects of the T2DM prediction problem, allowing us to compare and contrast different methodologies.

**Traditional ML Algorithms**

A set of diverse machine learning (ML) algorithms was employed to capture various aspects of the T2DM prediction problem. Our study included Decision Trees, Random Forests, AdaBoost, XGBoost, Logistic Regression, Support Vector Machines (SVM), Naïve Bayes, and Feedforward Neural Network (FFNN).

Decision Trees can capture non-linear relationships and interactions between features, which may be present in diabetes risk factors. Random Forests are robust to noise and outliers, which are common in healthcare datasets like the PIDD. Boosting methods such as AdaBoost and XGBoost can often achieve better predictive accuracy compared to individual models, making them valuable for the diabetes prediction task [15].

Logistic Regression models the probability of the target class (diabetic or non-diabetic) directly, which is useful for understanding the risk factors associated with diabetes. SVMs can effectively handle high-dimensional data, which is often the case in healthcare datasets with multiple features. Naïve Bayes can provide insights into the conditional probabilities of features given the class labels, which can be useful for understanding the risk factors associated with diabetes [15]. FFNN can potentially capture intricate interactions between the various risk factors associated with diabetes [15].

**Advanced ML Algorithms**

Three advanced models were demonstrated to evaluate whether more recent developments in machine learning could provide improved performance in T2DM prediction compared to traditional algorithms. TabNet, which uses sequential attention for complex feature interaction learning; CatBoost, efficient in handling mixed numerical and categorical data without extensive preprocessing and LightGBM, designed for speed and efficiency with large datasets. These models complement our traditional approaches by offering improved handling of complex data interactions, categorical variables, and large-scale data processing. [16–18].

**Model Training Overview**
In total, 90 different ML models configurations were trained during this study. Figure 3 represents the overall methodology of this study, illustrating our approach to model selection, training, and evaluation.

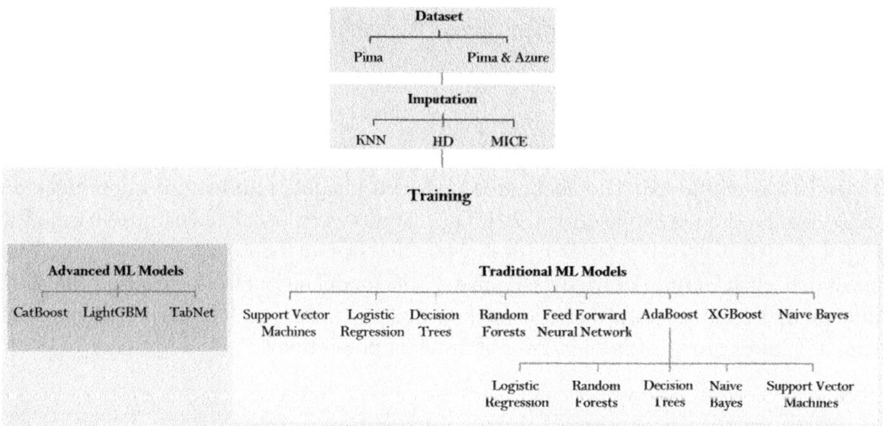


**Fig. 3.** Methodology flowchart for T2DM prediction using combined datasets and multiple ML algorithms

## 3.4  Model Training and Evaluation

This section details our approach to training and evaluating the machine learning models, including the metrics used to assess their performance.

**Training Process**
Most of the used models were trained on a dataset split with 30% of the data allocated for testing and a fixed random state of 42 for reproducibility. For the Random Forest model, a random search was performed to identify the optimal number of estimators within the ensemble. Feature importance was computed and visualized using a bar plot to highlight the most influential predictors for the diabetes classification task.

Prior to XGBoost training, a grid search was conducted with stratified k-fold cross-validation to optimize the maximum depth of the decision trees. The optimal hyperparameters were then used to train the XGBoost model, with early stopping based on the validation set to mitigate overfitting.

For neural networks (FFNN), a sequential architecture with an input layer was created, two hidden layers with ReLU activation and dropout regularization, and an output layer with sigmoid activation for binary classification. The model was compiled with the Adam optimizer and binary cross-entropy loss. Early stopping was employed during training to prevent overfitting.

For TabNet, CatBoost, and LightGBM, similar training was followed and evaluation procedures as with the other models. However, the default hyperparameters provided by their respective libraries was used for initial training.

**Evaluation Metrics**

The models' performance was evaluated on both training and test sets, computing various metrics, including accuracy, precision, and F1-score. The ROC curve was plotted with AUC value [19]. These metrics were chosen to provide a balanced view of model performance, particularly important given the potential class imbalance in diabetes datasets.

**Interpretability Analysis**

To enhance model interpretability, particularly crucial in healthcare applications, SHAP analysis was employed. This technique provided insights into feature importance and individual prediction explanations [20]. The insights were visualized using two key plots, Fig. 4. SHAP waterfall plot for both diabetic and non-diabetic cases: This figure illustrates how each feature contributes to pushing the model output from the base value to the final prediction for specific diabetic and non-diabetic instances. It helps in understanding which features are most influential in individual predictions.

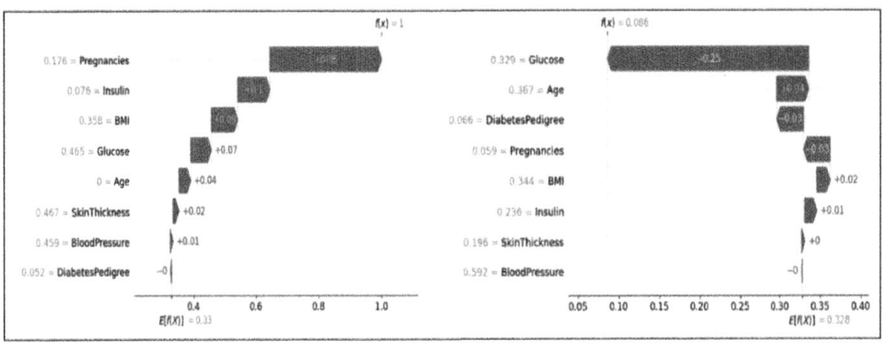

**Fig. 4.** SHAP waterfall plot for both diabetic and non-diabetic cases

Figure 5 plot demonstrates how the model arrives at its decisions across multiple instances, showing the cumulative impact of features on model output. It provides a holistic view of feature interactions and their effects on predictions across the dataset.

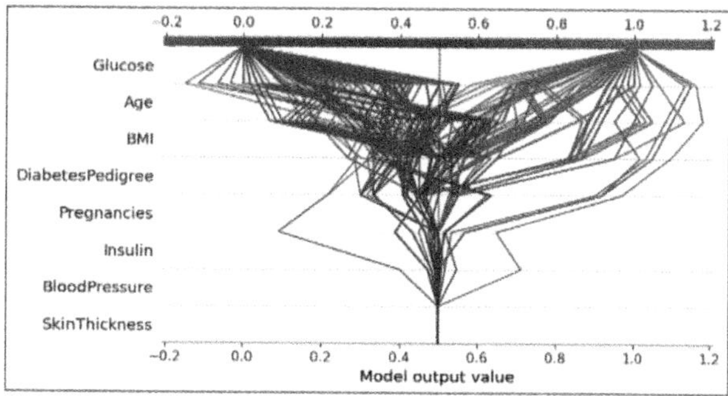


**Fig. 5.** SHAP decision plot for Decision tree on PIDD

These visualizations offer valuable insights into the model's decision-making process, enhancing transparency and trust in the predictions, which is particularly important in healthcare applications.

To provide a clear visual representation of our model's decision-making process, a decision tree based on the combined dataset was constructed (Fig. 6). This tree illustrates the hierarchical nature of feature-based decisions in classifying diabetes cases.

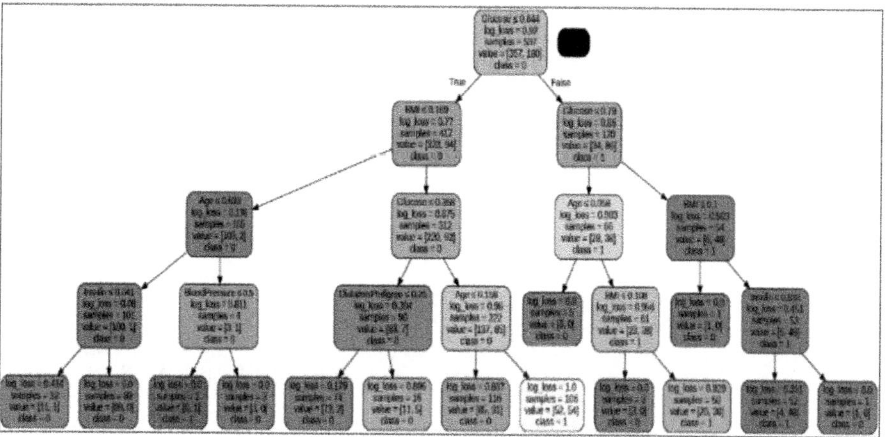

**Fig. 6.** Decision tree on the combined dataset

To quantify and rank the relative importance of different features in our predictive model, a Random Forest algorithm was used (Fig. 7). Notably, glucose levels emerge as the most significant predictor, followed by BMI and age.

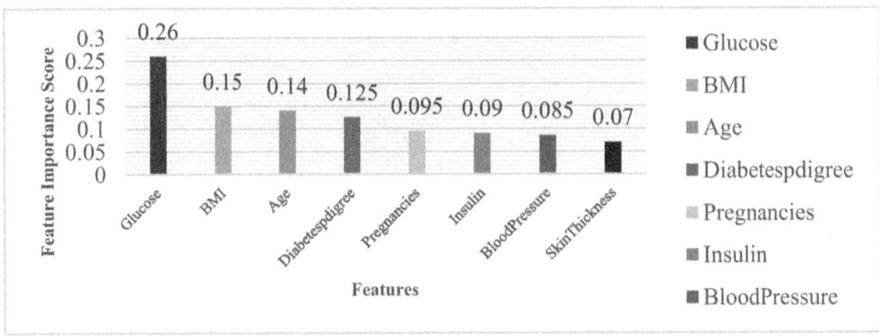

**Fig. 7.** Feature Importance ranking by Random forest model

# 4 Results

The following figures provide a comprehensive overview of our model performance across different metrics and datasets. Figures 8, 9, 10 and 11 illustrate the accuracy, precision, F1-score, and Area Under the Curve (AUC) respectively for various machine learning models, comparing their performance on both the PIDD and the Combined Dataset.

Figure 12 focuses on the error rate progression of one of our top-performing models, AdaBoost with Random Forest. These visualizations form the basis of our subsequent analysis, demonstrating the impact of dataset combination, the relative performance of different models, and the effects of various imputation techniques.

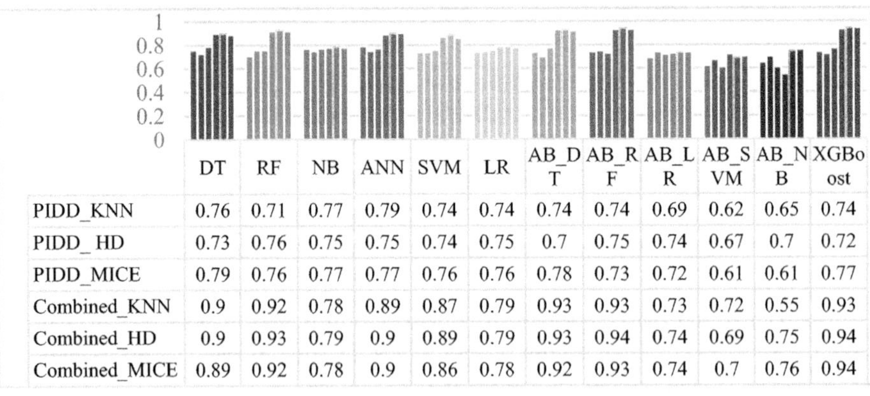

| | DT | RF | NB | ANN | SVM | LR | AB_DT | AB_RF | AB_LR | AB_SVM | AB_NB | XGBoost |
|---|---|---|---|---|---|---|---|---|---|---|---|---|
| PIDD_KNN | 0.76 | 0.71 | 0.77 | 0.79 | 0.74 | 0.74 | 0.74 | 0.74 | 0.69 | 0.62 | 0.65 | 0.74 |
| PIDD_HD | 0.73 | 0.76 | 0.75 | 0.75 | 0.74 | 0.75 | 0.7 | 0.75 | 0.74 | 0.67 | 0.7 | 0.72 |
| PIDD_MICE | 0.79 | 0.76 | 0.77 | 0.77 | 0.76 | 0.76 | 0.78 | 0.73 | 0.72 | 0.61 | 0.61 | 0.77 |
| Combined_KNN | 0.9 | 0.92 | 0.78 | 0.89 | 0.87 | 0.79 | 0.93 | 0.93 | 0.73 | 0.72 | 0.55 | 0.93 |
| Combined_HD | 0.9 | 0.93 | 0.79 | 0.9 | 0.89 | 0.79 | 0.93 | 0.94 | 0.74 | 0.69 | 0.75 | 0.94 |
| Combined_MICE | 0.89 | 0.92 | 0.78 | 0.9 | 0.86 | 0.78 | 0.92 | 0.93 | 0.74 | 0.7 | 0.76 | 0.94 |

**Fig. 8.** Accuracy among different Machine learning models

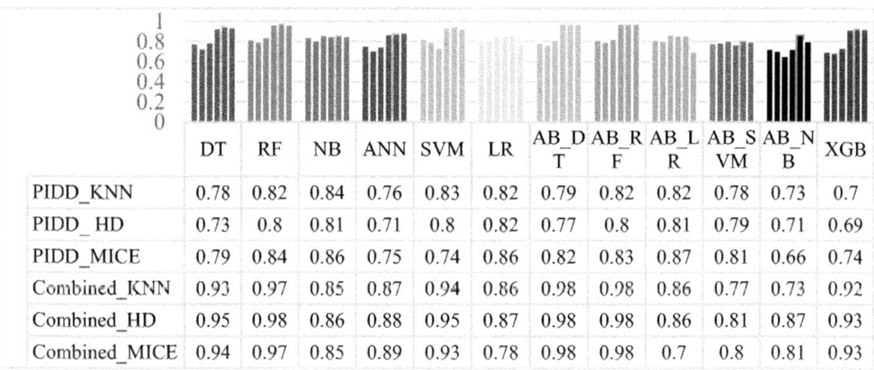

| | DT | RF | NB | ANN | SVM | LR | AB_D T | AB_R F | AB_L R | AB_S VM | AB_N B | XGBo ost |
|---|---|---|---|---|---|---|---|---|---|---|---|---|
| PIDD_KNN | 0.77 | 0.7 | 0.76 | 0.79 | 0.74 | 0.74 | 0.73 | 0.74 | 0.71 | 0.77 | 0.69 | 0.74 |
| PIDD_ HD | 0.72 | 0.76 | 0.75 | 0.63 | 0.73 | 0.74 | 0.7 | 0.75 | 0.74 | 0.78 | 0.68 | 0.73 |
| PIDD_MICE | 0.79 | 0.76 | 0.77 | 0.76 | 0.77 | 0.77 | 0.78 | 0.73 | 0.78 | 0.76 | 0.77 | 0.79 |
| Combined_KNN | 0.9 | 0.92 | 0.78 | 0.86 | 0.87 | 0.78 | 0.93 | 0.93 | 0.77 | 0.72 | 0.67 | 0.93 |
| Combined_HD | 0.91 | 0.93 | 0.79 | 0.89 | 0.88 | 0.78 | 0.93 | 0.94 | 0.76 | 0.71 | 0.78 | 0.94 |
| Combined_MICE | 0.89 | 0.92 | 0.78 | 0.86 | 0.86 | 0.78 | 0.92 | 0.93 | 0.76 | 0.72 | 0.76 | 0.94 |

**Fig. 9.** Precision among different Machine learning models

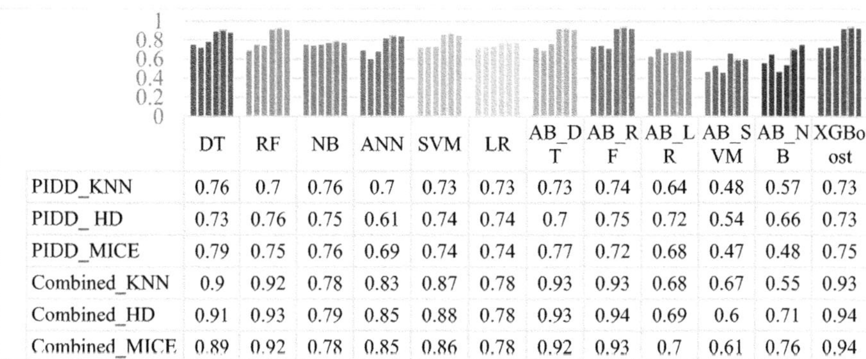

| | DT | RF | NB | ANN | SVM | LR | AB_D T | AB_R F | AB_L R | AB_S VM | AB_N B | XGB |
|---|---|---|---|---|---|---|---|---|---|---|---|---|
| PIDD_KNN | 0.78 | 0.82 | 0.84 | 0.76 | 0.83 | 0.82 | 0.79 | 0.82 | 0.82 | 0.78 | 0.73 | 0.7 |
| PIDD_ HD | 0.73 | 0.8 | 0.81 | 0.71 | 0.8 | 0.82 | 0.77 | 0.8 | 0.81 | 0.79 | 0.71 | 0.69 |
| PIDD_MICE | 0.79 | 0.84 | 0.86 | 0.75 | 0.74 | 0.86 | 0.82 | 0.83 | 0.87 | 0.81 | 0.66 | 0.74 |
| Combined_KNN | 0.93 | 0.97 | 0.85 | 0.87 | 0.94 | 0.86 | 0.98 | 0.98 | 0.86 | 0.77 | 0.73 | 0.92 |
| Combined_HD | 0.95 | 0.98 | 0.86 | 0.88 | 0.95 | 0.87 | 0.98 | 0.98 | 0.86 | 0.81 | 0.87 | 0.93 |
| Combined_MICE | 0.94 | 0.97 | 0.85 | 0.89 | 0.93 | 0.78 | 0.98 | 0.98 | 0.7 | 0.8 | 0.81 | 0.93 |

**Fig. 10.** AUC among different Machine learning models

| | DT | RF | NB | ANN | SVM | LR | AB_D T | AB_R F | AB_L R | AB_S VM | AB_N B | XGBo ost |
|---|---|---|---|---|---|---|---|---|---|---|---|---|
| PIDD_KNN | 0.76 | 0.7 | 0.76 | 0.7 | 0.73 | 0.73 | 0.73 | 0.74 | 0.64 | 0.48 | 0.57 | 0.73 |
| PIDD_ HD | 0.73 | 0.76 | 0.75 | 0.61 | 0.74 | 0.74 | 0.7 | 0.75 | 0.72 | 0.54 | 0.66 | 0.73 |
| PIDD_MICE | 0.79 | 0.75 | 0.76 | 0.69 | 0.74 | 0.74 | 0.77 | 0.72 | 0.68 | 0.47 | 0.48 | 0.75 |
| Combined_KNN | 0.9 | 0.92 | 0.78 | 0.83 | 0.87 | 0.78 | 0.93 | 0.93 | 0.68 | 0.67 | 0.55 | 0.93 |
| Combined_HD | 0.91 | 0.93 | 0.79 | 0.85 | 0.88 | 0.78 | 0.93 | 0.94 | 0.69 | 0.6 | 0.71 | 0.94 |
| Combined_MICE | 0.89 | 0.92 | 0.78 | 0.85 | 0.86 | 0.78 | 0.92 | 0.93 | 0.7 | 0.61 | 0.76 | 0.94 |

**Fig. 11.** F1-score among different Machine learning models

In addition to the traditional machine learning algorithms, the performance of three advanced models: TabNet, CatBoost, and LightGBM was evaluated. These models demonstrated competitive performance across various metrics. Table 1 presents the accuracy scores of these advanced models on both the PIDD and the Combined Datasets, using different imputation methods (KNN, Hot Deck, and MICE).

**Table 1.** Accuracy scores of advanced models on PIDD and Combined Datasets

|  |  | KNN | Hot Deck | MICE |
| --- | --- | --- | --- | --- |
| Pima Dataset | **TabNet** | 76.62% | 68.83% | 72.08% |
|  | **CatBoost** | 79.22% | 70.13% | 72.08% |
|  | **LightGBM** | 75% | 73% | 71% |
| Combined Datasets | **TabNet** | 90.42% | 89.92% | 90.49% |
|  | **CatBoost** | 93.53% | 94.39% | 93. 66% |
|  | **LightGBM** | 94% | 94% | 94% |

### 4.1  Impact of Combined Dataset on Models Performance

As evident from Figs. 8, 9, 10 and 11, the Combined Dataset improved model performance across all metrics. Tree-based models showed significant gains: Decision Trees hotincreased from 73–79% to 89–90% accuracy, while Random Forests jumped from 71–76% to 92–93%. Among traditional algorithms, Naive Bayes and Logistic Regression saw modest improvements, while ANN and SVM showed more substantial progress. Ensemble methods, particularly AdaBoost variants and XGBoost, demonstrated impressive gains, with XGBoost improving from 72–77% to 93–94% accuracy.

### 4.2  Comparison of Traditional and Advanced ML Models

For traditional ML Models, AdaBoost with random forest was a top performer, achieving 94% in accuracy, precision, and F1-score, with 0.98 AUC and 0.06 final error on the combined dataset (Fig. 12).

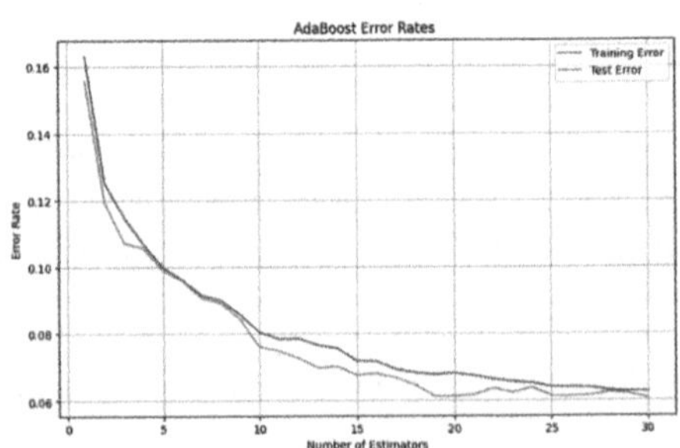


**Fig. 12.**  AdaBoost with Random Forest setting error rates

XGBoost, Random Forests, and AdaBoost with Decision Trees followed closely. FFNN and Decision Trees showed strong performance (0.89–0.90), while SVMs improved to 0.86–0.89. Simpler models like Naive Bayes and Logistic Regression underperformed, rarely exceeding 0.79. Advanced models excelled on the combined dataset, with CatBoost and LightGBM reaching 94.39% and 94% accuracy respectively, while TabNet showed competitive performance with 0.96 AUC-ROC (Table 1).

### 4.3  Effect of Imputation Techniques on Model Performance

K-Nearest Neighbors (KNN), Hot Deck (HD), and Multiple Imputation by Chained Equations (MICE) were explored for data imputation. For the PIDD, MICE often performed best (Decision Trees: 79% accuracy with MICE vs. 77% KNN, 72% HD). On the combined dataset, Hot Deck frequently outperformed others (AdaBoost with Random Forest: 94% HD vs. 93% KNN/MICE) (Fig. 13).

| Model | PIDD KNN | PIDD HD | PIDD MICE | Combined KNN | Combined HD | Combined MICE |
|---|---|---|---|---|---|---|
| Decision Trees | 77 | 72 | 79 | 90 | 91 | 89 |
| Random Forests | 71 | 76 | 76 | 92 | 93 | 92 |
| AdaBoost RF | 74 | 75 | 73 | 93 | 94 | 93 |
| XGBoost | 74 | 73 | 79 | 93 | 94 | 94 |
| CatBoost | 79 | 70 | 72 | 94 | 94 | 94 |
| LightGBM | 75 | 73 | 71 | 94 | 94 | 94 |

**Fig. 13.** Imputation Techniques Performance Heatmap

Advanced models showed similar trends, with CatBoost reaching 94.39% accuracy using Hot Deck on the combined dataset. LightGBM maintained 94% accuracy across all imputation methods, demonstrating robustness. These results suggest that while MICE may be more effective for smaller datasets like PIDD, Hot Deck imputation appears particularly well-suited for larger, combined datasets in T2DM prediction.

## 5  Discussion

Our study reveals interesting patterns in the performance of various machine learning models for T2DM prediction. Notably, the advanced models CatBoost and LightGBM demonstrated performance on par with, or slightly better than, our previously best-performing models, AdaBoost with Random Forest and XGBoost. This suggests that these advanced models are indeed capable of capturing the complex patterns inherent in the diabetes prediction task.

As illustrated in Fig. 14 below, CatBoost and LightGBM achieved the highest AUC-ROC scores of 0.99, marginally outperforming the traditional ensemble methods. Interestingly, while TabNet showed lower accuracy, it still demonstrated competitive performance. This underscores the potential of these advanced models in enhancing T2DM detection capabilities. In terms of specific performance metrics, CatBoost and LightGBM achieved the highest accuracy at 94.39% and 94% respectively on the combined dataset, outperforming most traditional models. This marginal improvement over AdaBoost RF and XGBoost (both at 94%) suggests that while advanced models offerbenefits, the performance gap may be narrowing at the higher end of the accuracy spectrum.

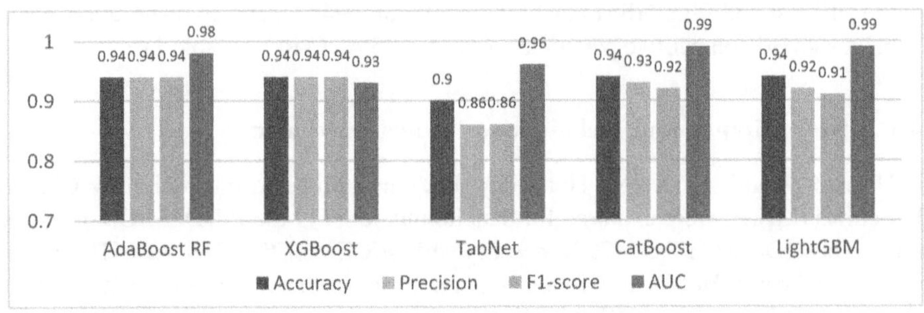

**Fig. 14.** Performance comparison of top traditional models and advanced models.

The combined PIDD-Azure dataset consistently led to performance improvements across all machine learning models tested. However, the magnitude of improvement varied among different algorithms. Ensemble methods and tree-based models generally showed the most substantial gains, highlighting the potential of the combined dataset to enhance the predictive power of complex model architectures. This suggests that these advanced models are indeed capable of capturing the complex patterns inherent to the diabetes prediction task.

Regarding data preprocessing, our findings with imputation methods were consistent across both traditional and advanced models. The Hot Deck (HD) imputation method generally yielded the best results on the combined dataset. This consistency across different model types suggests that HD imputation may be particularly well-suited for larger, more diverse datasets in the context of T2DM prediction.

## 6   Conclusion

The early detection of T2DM remains critical for mitigating severe complications and preserving quality of life. This study demonstrated the significant potential of advanced machine learning techniques combined with larger, more diverse datasets for improving T2DM early detection.

The integration of the Pima Indians Diabetes Dataset with a larger Microsoft Azure dataset consistently improved model performance across all tested algorithms. Ensemble and advanced models consistently outperformed traditional algorithm. This superior performance likely stems from their ability to capture complex, non-linear relationships between T2DM risk factors. Furthermore, Hot Deck imputation generally yielded the best results on the combined dataset.

These results underscore the value of comprehensive datasets and sophisticated modeling techniques in capturing complex T2DM onset patterns, suggesting potential for more accurate clinical risk assessment tools. However, it's crucial to acknowledge the limitations of this study. The combined dataset may not fully represent all populations, and its retrospective nature doesn't account for evolving data collection methods. Despite high accuracy, balancing model performance with clinical interpretability remains a challenge, requiring explainable AI approaches. Looking ahead, future work should address these limitations and advance the field. External validation with diverse datasets is crucial for model generalizability. Prospective studies will evaluate real-world impact on patient outcomes and costs. Additionally, investigating deep learning for temporal T2DM progression and federated learning techniques could yield valuable insights while maintaining data privacy.

In conclusion, this study represents a significant step forward in machine learning for T2DM prediction. However, realizing its full potential requires careful consideration of clinical integration, model interpretability, and ethical implications. As these techniques are being refined, improved T2DM detection and management could transform diabetes care and prevention.

## References


1. Ogurtsova, K., et al.: IDF diabetes Atlas: global estimates of undiagnosed diabetes in adults for 2021. Diabetes Res. Clin. Pract. **183**, 109118 (2022)
2. Kautzky-Willer, A., Leutner, M., Harreiter, J.: Sex differences in type 2 diabetes. Diabetologia **66**(6), 986–1002 (2023)
3. Chang, V., Bailey, J., Xu, Q.A., Sun, Z.: Pima Indians diabetes mellitus classification based on machine learning (ML) algorithms. Neural Comput. Appl. **35**(22), 16157–16173 (2023)
4. Reza, M.S., Amin, R., Yasmin, R., Kulsum, W., Ruhi, S.: Improving diabetes disease patients classification using stacking ensemble method with PIMA and local healthcare data. Heliyon (2024)
5. Noviyanti, C.N., Alamsyah, A.: Early detection of diabetes using random forest algorithm. J. Inf. Syst. Explor. Res. (2024)
6. Hama Saeed, M.A.: Diabetes type 2 classification using machine learning algorithms with up-sampling technique. J. Electr. Syst. Inf. Technol. **10**(1), 8 (2023)
7. Reza, M.S., Hafsha, U., Amin, R., Yasmin, R., Ruhi, S.: Improving SVM performance for type II diabetes prediction with an improved non-linear kernel: Insights from the PIMA dataset. Comput. Methods Programs Biomed. Update **4**, 100118 (2023)
8. Kuriakose, S.M., Pati, P.B., Singh, T.: Prediction of diabetes using machine learning: analysis of 70,000 clinical database patient record. In: 2022 13th International Conference on Computing Communication and Networking Technologies (ICCCNT), pp. 1–5, IEEE (2022)
9. K. Alnowaiser, "Improving Healthcare Prediction of Diabetic Patients Using KNN Imputed Features and Tri-Ensemble Model," IEEE Access, 2024
10. Chang, V., Bailey, J., Xu, Q.A., Sun, Z.: Pima Indians diabetes mellitus classification based on machine learning (ML) algorithms. Neural Comput. Appl. **35**(22), 16157–16173 (2023). https://doi.org/10.1007/s00521-022-07049-z
11. Microsoft Learning GitHub. Accessed: 16 May 2024. https://github.com/MicrosoftLearning.
12. Samad, M.D., Abrar, S., Diawara, N.: Missing value estimation using clustering and deep learning within multiple imputation framework. Knowl.-Based Syst. **249**, 108968 (2022)
13. Pujianto, U., Wibawa, A.P., Akbar, M.I.: K-nearest neighbor (k-NN) based missing data imputation. In: 2019 5th International Conference on Science in Information Technology (ICSITech), pp. 83–88, IEEE (2019)
14. Andridge, R.R., Little, R.J.: A review of hot deck imputation for survey non-response. Int. Stat. Rev. **78**(1), 40–64 (2010)

15. Sarker, I.H.: Machine learning: algorithms, real-world applications and research directions. SN Comput. Sci. **2**(3), 160 (2021)
16. Arik, S.Ö., Pfister, T.: Tabnet: attentive interpretable tabular learning. Proc. AAAI Conf. Artif. Int. (2021)
17. Liudmila, P., Gleb, G., Aleksandr, V., et al.: CatBoost: unbiased boosting with categorical features. Adv. Neural Inf. Process. Syst. (2018)
18. Junliang, F., Xin, M., Lifeng, W., et al.: Light gradient boosting machine: an efficient soft computing model for estimating daily reference evapotranspiration with local and external meteorological data. Agric. Water Manage. **225**, 105758 (2019)
19. Rainio, O., Teuho, J., Klén, R.: Evaluation metrics and statistical tests for machine learning. Sci. Rep. **14**(1), 6086 (2024)
20. Prendin, F., Pavan, J., Cappon, G., Del Favero, S., Sparacino, G., Facchinetti, A.: The importance of interpreting machine learning models for blood glucose prediction in diabetes: an analysis using SHAP. Sci. Rep. **13**(1), 16865 (2023)

# An Efficient Age-Based Caching Policy in NDN Cluster


Ibrahim Khalil Douis[(✉)] [ID], Hafida Bouziane [ID], and Abdallah Chouarfia [ID]

Department Computer Science, University of Science and Technology of Oran (USTO),
Bir El Djir, Algeria
{ibrahimkhalil.douis,hafida.bouziane,
abdallah.chouarfia}@univ-usto.dz



**Abstract.** In Named Data Networking (NDN), in-network caching is crucial in optimizing network performance and enhancing data delivery efficiency. This paper introduces a novel age-based caching policy specifically designed for NDN clusters. The proposed policy effectively manages content placement and replacement by considering content age, popularity, and size. By minimizing content redundancy, increasing the cache hit ratio, and reducing the frequency of caching operations, this policy offers a significant improvement over traditional caching strategy.

**Keywords:** Named Data Networking (NDN) · In-network caching · Clustering


## 1 Introduction

In the last 20 years, the use of the Internet has greatly increased due to advances in communication and information technologies, especially mobile devices. However, the current Internet structure, which relies on host-to-host communication, is no longer enough for content distribution. Therefore, there is a need for a new Internet architecture based on content distribution.

Named Data Networking (NDN) aims to develop a new Internet architecture that prioritizes data by naming it rather than focusing on its location. NDN makes data secure by separating trust in data from trust in hosts, and it enables efficient communication methods [1].

Named Data Networking (NDN) employs caching at multiple levels using various strategies. However, many NDN caching policies, such as Leave Copy Everywhere (LCE) the default NDN caching policy [2], suffer from limitations including inefficient utilization of router cache storage, high cache redundancy, and significant replacement overhead. These issues can lead to reduced efficiency in NDN systems.

This paper introduces a novel age-based caching policy designed to mitigate content redundancy and latency, thereby enhancing overall network performance.



N. Seddari and M. Redjimi (Eds.): ICMSCT 2024, CCIS 2606, pp. 83–90, 2025.
https://doi.org/10.1007/978-3-032-01922-6_7

## 2  NDN Overview

NDN project has its roots in a previous project called Content-Centric Networking (CCN), which was first presented by Van Jacobson publicly in 2006, aligning with the principles of Information-Centric Networking (ICN) [3].

In NDN architecture there 2 types of packets: Interest packet and Data packet, the interest packet consumer side places the name of the resulting data chunk in the INTEREST packet, which is then sent to the network. Network routers utilize specific naming information to forward the packet further in the network towards the Provider. When an INTEREST packet arrives at the producer node and contains the requested data, a DATA packet is sent back. The DATA packet includes the name and content of the data, as well as the producer's signature. Similar to INTEREST packets, DATA packets follow the same reverse path [4] (Fig. 1).

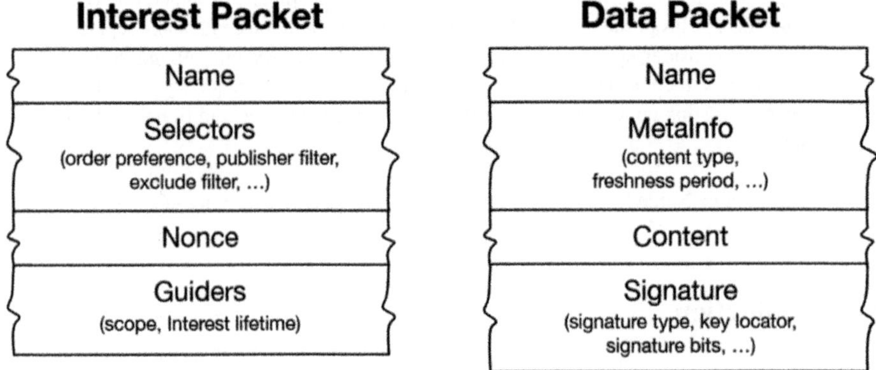


**Fig. 1.**  Packets in Named Data Networking [3]

In NDN each router has 3 data structures:

- *Forwarding Information Base (FIB)*: routes Interest packets toward the potential source(s) of matching Data. It closely resembles an IP FIB but allows for a list of outgoing faces instead of a single one [5].
- *Content Store (CS):* is an in-network data object cache. It can fulfill interest packets on behalf of a producer if it has previously cached the necessary data objects. Put, CS represents a transient cache for data objects received by the router [5].
- *Packet Interest Table (PIT):* serves the purpose of recording the path traversed by the INTEREST packet. Each entry in the PIT table consists of the INTEREST name, a collection of incoming faces representing the previous hops, and a set of outgoing faces indicating the subsequent hops taken by INTERESTs with that specific name. Additionally, the PIT table can be utilized for diverse functions, including aggregating INTEREST packets and congestion management [5] (Fig. 2).

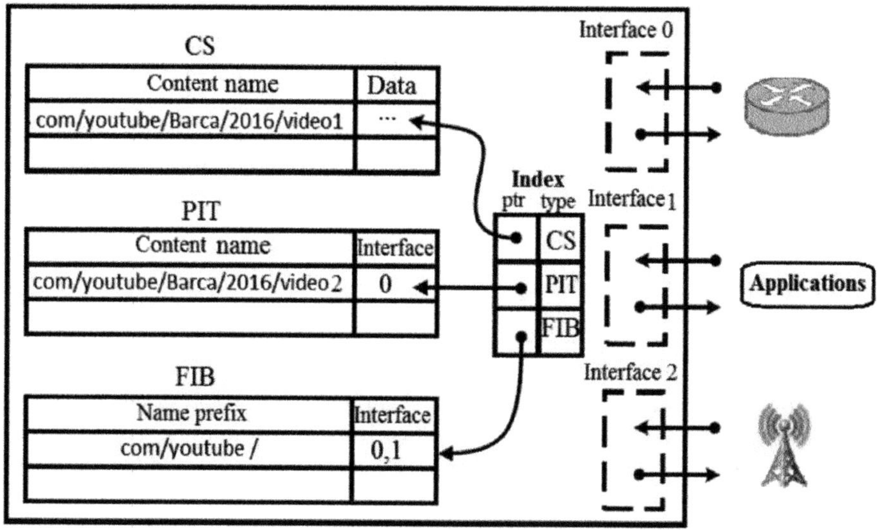


**Fig. 2.** NDN router components [6]

## 3    Related Work

Leave Copy Everywhere (LCE) [2] is the default NDN caching policy. This policy entails caching a copy on every router Content Store (CS) along the path. While this increases the hit ratio, it also leads to redundancy of cached contents and caching operations.

Least Recently Used (LRU) [7] is a replacement caching policy that increases cache hit probability by storing the most recent data for longer and deleting the least recently used content upon replacement.

Probabilistic in-network caching [8] estimates the caching capability of a path and caches content probabilistically. This method allows for a balanced distribution of cached content.

The Efficacious Content Caching and Eviction Priorities (CCEP) [9] is a strategy that selects the best router to cache content along the path based on its popularity. The content is cached on the selected router, and for replacement, eviction priorities are calculated based on the content popularity and other significant factors. Content with high eviction priority is replaced with new content in cases where there is no space on the selected router.

An Efficient Distributed Content Store-Based Caching Policy [10] coordinates the CS of a node with its neighboring nodes in a distributed manner, aiming to cache more popular content objects near the node. Each node contributes to the goal of its neighborhood by maximizing the number of unique popular content objects cached in its storage and not cached in the storage of its neighbors.

NECS-based Cache Management [11] is a caching strategy in NDN clusters. When content arrives at the main router, it is cached directly on it. In cases where there is no free space in the main router, the LFU [2] (Less Frequently Used) replacement method is applied to add new content, and less frequently used contents are discarded first. This strategy utilizes Multi-Attribute Decision-Making methods (MADM) to select the main router and two others as backups in the event of the main router's failure.

New cache replacement policy in named data network based on FIB table [12] is a new cache replacement policy called Discard of Fast Retrievable Content (DFRC). In DFRC, the retrieval time of the content is evaluated using the FIB table information, and the content with less retrieval time receives more discard priority new cache replacement policy in named data network based on FIB table information.

## 4   Proposed Caching Policy

### 4.1   Age-Based Caching Policy in NDN Cluster

Our caching policy focuses on efficient caching through a novel data structure called the Content Store Station (CSS) implemented on the cluster head. The system is organized as follows: each cluster consists of routers, a cluster head, replacement1, and replacement2, collectively forming the store station.

Each item in the store station has a defined cache size (in GB). When content arrives at the station, it is stored on the cluster head, replacement1, or replacement2 based on our caching policy (Algorithm 1).

The CSS data structure on the cluster includes attributes such as content name, age, size, and placement (Table 1).

**Table 1.** List of Acronyms

| Acronym | Meaning |
| --- | --- |
| CS | Content Store |
| PIT | Pending Interest Table |
| CSS | Content Store Station |
| Req $(c_i)$ | Number of requests of content ci |
| Tot.Req | Total number of requests of cached contents on the station |
| Size $(c_i)$ | Size of content $c_i$ in GB |
| Tot.Size | Total size of cached contents on the station in GB |
| Size.Pen $(c_i)$ | Size Penalty of content $c_i$ |

Upon receiving new content, the cluster head calculates the content's age (using Eq. 3) based on its popularity and size. The cluster head first attempts to cache the content if there is available space in the store station. If the cache is full, the cluster head attempts to free up space by deleting content with the smallest ages compared to the new content. If, after this deletion, sufficient space is available, the new content is added to the station store. If there is still no available space after deletion, the content is forwarded to its destination without being cached. After each content addition, the CSS is updated with the new content data, and an update operation is performed to refresh content ages (using Eq. 3). This update ensures that content with the greatest age is stored in the cluster head, followed by replacement1 and replacement2, in that order, to facilitate prompt content delivery.

$$\text{Weight}(c_i) = \begin{cases} \frac{Req(c_i)}{Tot.Req}, & \text{if } c_i \text{ already cached} \\ \frac{Req(c_i)}{Req(c_i)+Tot.Req}, & else \end{cases} \tag{1}$$

$$Size.Pen(c_i) = \frac{Size(c_i)}{Tot.Sizes} \times \alpha \tag{2}$$

$$Age(c_i) = Weight(c_i) - Size.Pen(c_i), \tag{3}$$

$$Where\ Age(c_i) < 0, Then\ Age(c_i) = 0$$

The penalty coefficient ($\alpha$) is represented as follows:

- If the storage used is less than 25% of the total station size, $\alpha$ equals 0.
- If the storage used is between 25% and 50% of the total station size, $\alpha$ is 0.75.
- If the storage used exceeds 50% of the total station size, $\alpha$ is equal to 1.

---

**Algorithm 1** Add New Content ($c_i$)

---

**Require:** $c_i$ (new content)
1: Calculate the age of $c_i$ using Equation 3
2: **if** free space is available on $CS$, $CS_1$, or $CS_2$ **then**
3:     Cache $c_i$ on the first available storage
4: **else**
5:     Initialize $sizes\_to\_delete = 0$, $to\_delete\_contents = \{\}$
6:     Initialize $i \leftarrow 1$
7:     **while** $i \leq |CSS|$ **and** $sizes\_to\_delete < size(c_i)$ **and** $CSS[i].age < age(c_i)$ **do**
8:         Let $item \leftarrow CSS[i]$
9:         $sizes\_to\_delete \leftarrow sizes\_to\_delete + item.size$
10:        $to\_delete\_contents \leftarrow to\_delete\_contents \cup \{item\}$
11:        $i \leftarrow i + 1$
12:    **end while**
13:    **if** $free\_space + sizes\_to\_delete > size(c_i)$ **then**
14:        Delete $to\_delete\_contents$,
15:        Cache $c_i$ on the first available storage
16:    **else**
17:        Forward $c_i$ to destination without caching
18:    **end if**
19: **end if**
20: Add $c_i$ to $CSS$
21: Call **Update CSS** (Algorithm 2)

---

## 4.2  Case Study

In this section, we will outline a simple scenario of our caching policy. Let's assume we have a cluster, $Cl$, which contains $n$ routers: $R_h$ (cluster head) with a size of 1 GB, $R_1$

(replacement 1) with a size of 0.6 GB, and $R_2$ (replacement 2) with a size of 0.4 GB. Additionally, we have contents $C_1$, $C_2$, and $C_3$ which will be presented in Table 2.

---

**Algorithm 2** Update CSS

---

1: **for all** content in CSS **do**
2:     Calculate and update the age of the content
3: **end for**
4: $head\_size\_tmp \leftarrow head\_size$
5: $replacement1\_size\_tmp \leftarrow replacement1\_size$
6: $replacement2\_size\_tmp \leftarrow replacement2\_size$
7: $i \leftarrow 1$
8: **for all** content in CSS **do**
9:     Select content with the $i^{th}$ highest age
10:     **if** content.size $\leq$ head_size_tmp **then**
11:         Move content to head CS
12:         $head\_size\_tmp \leftarrow head\_size\_tmp - content.size$
13:     **else if** content.size $\leq$ replacement1_size_tmp **and** content.placement $\neq$ Replacement1 CS **then**
14:         Move content to Replacement1 CS
15:         $replacement1\_size\_tmp \leftarrow replacement1\_size\_tmp - content.size$
16:     **else if** content.size $\leq$ replacement2_size_tmp **and** content.placement $\neq$ Replacement2 CS **then**
17:         Move content to Replacement2 CS
18:         $replacement2\_size\_tmp \leftarrow replacement2\_size\_tmp - content.size$
19:     **else**
20:         **Skip content** (no action if it doesn't fit)
21:     **end if**
22:     Update the placement of the content in the CSS
23:     $i \leftarrow i + 1$
24: **end for**

---

Table 2. Case study contents

| Content name | Nbr of requests | Size (GB) |
| --- | --- | --- |
| C1 | 12 | 0.2 |
| C2 | 23 | 0.4 |
| C3 | 45 | 0.5 |

When $C_1$ arrives at the cluster head, it is assumed that the station is empty (no content is cached). The age of $C_1$ is calculated using Eq. 3, and the results after caching are recorded in Table 3.

**Table 3.** CSS after caching C1

| Content name | Age | Size (GB) | Placement |
|---|---|---|---|
| C1 | 1 | 0.2 | Head |

After applying our caching policy, $C_1$ was cached on the cluster head. The penalty coefficient $\alpha$ remains at 0 because less than 25% of the station's total storage (2 GB) has been utilized (Table 4).

**Table 4.** CSS after caching C2

| Content name | Age | Size (GB) | Placement |
|---|---|---|---|
| C1 | 0.268 | 0.2 | Head |
| C2 | 0.507 | 0.4 | Head |

Following our caching policy, $C_2$ is also cached on the cluster head. The penalty coefficient $\alpha$ increases to 0.75, as the used storage now ranges between 25% and 50% of the station's total storage (2 GB) (Table 5).

**Table 5.** CSS after caching C3

| Content name | Age | Size (GB) | Placement |
|---|---|---|---|
| C1 | 0.050 | 0.2 | Replacement1 |
| C2 | 0.088 | 0.4 | Head |
| C3 | 0.313 | 0.5 | Head |

After applying our caching policy, $C_3$ is cached on the cluster head. Due to the new content, $C_1$ is evicted from the cluster head and cached on Replacement1, as its age is smaller compared to $C_2$ and $C_3$. The penalty coefficient $\alpha$ increases to 1 because the used storage exceeds 50% of the station's total storage (2 GB).

### 4.3  Results and Discussion

Based on the scenario described above, our caching policy involves caching the contents with the big ages on the cluster head to deliver popular content within the cluster quickly. This approach aims to increase the hit ratio, reduce content redundancy (i.e., having content stored only on the store station), and prioritize popular content. If there is no free space on the store station, content with the shortest ages will be deleted from the station to make room for more popular content within the cluster.

## 5  Conclusion

This paper presents an efficient age-based caching policy in the NDN cluster aimed at improving network performance. Our caching policy utilizes a storage station to store cluster contents based on their age, which is calculated using the content's popularity on the cluster and its size. This station is managed by the cluster head using a novel data structure called CSS to ensure that the most popular contents are stored on the cluster head for quick delivery. Our caching policy aims to reduce content redundancy, latency, and network operations while also increasing the cache hit ratio.

## References


1. NDN Project Overview. https://named-data.net/project/. Last accessed 10 Aug 2024
2. Saxena, D., Raychoudhury, V., Suri, N., Becker, C., Cao, J.: Named data networking: a survey. Comput. Sci. Rev. **19**, 15–55 (2016). https://doi.org/10.1016/j.cosrev.2016.01.001
3. Zhang, L., et al.: Named data networking. ACM SIGCOMM Comput. Commun. Rev. **44** (2014)
4. Tariq, A., Rehman, R.A., Kim, B.-S.: Forwarding strategies in NDN-based wireless networks: a survey. IEEE Commun. Surv. Tutorials. **22**, 68–95 (2020). https://doi.org/10.1109/COMST.2019.2935795
5. Jacobson, V., Smetters, D.K., Thornton, J.D., Plass, M.F., Briggs, N.H., Braynard, R.L.: Networking Named Content. ACM CoNEXT'09 (2009)
6. Majed, A., Wang, X., Yi, B.: Name Lookup in Named Data Networking: A Review (2019)
7. Li, Y., Yu, M., Li, R.: A cache replacement strategy based on hierarchical popularity in NDN. In: 2018 Tenth International Conference on Ubiquitous and Future Networks (ICUFN), pp. 159–161. IEEE, Prague, Czech Republic (2018). https://doi.org/10.1109/ICUFN.2018.8436597
8. Psaras, I., Chai, W.K., Pavlou, G.: Probabilistic in-network caching for information-centric networks. In: Proceedings of the second edition of the ICN workshop on Information-centric networking, pp. 55–60. ACM, Helsinki Finland (2012). https://doi.org/10.1145/2342488.2342501
9. Alkhazaleh, M., Aljunid, S.A., Sabri, N.: An efficacious content caching and eviction priorities (CCEP) for in-network caching high performance in information-centric networking. IAENG Int. J. Appl. Math. **53**, 169–182 (2023)
10. Dinh, N.-T., Kim, Y.: An efficient distributed content store-based caching policy for information-centric networking. Sensors **22**, 1577 (2022). https://doi.org/10.3390/s22041577
11. Fethellah, N.E.H., Bouziane, H., Chouarfia, A.: NECS-based cache management in the information centric networking. Int. J. Interact. Mob. Technol. **15**, 172 (2021). https://doi.org/10.3991/ijim.v15i21.20011
12. Hosseinzadeh, M., Moghim, N., Taheri, S., Gholami, N.: A new cache replacement policy in named data network based on FIB table information. Telecommun. Syst. **86**, 585–596 (2024). https://doi.org/10.1007/s11235-024-01140-7

# An Optimized LFM in Collaborative Localization

Can Zhang[1(✉)], Qun Li[1], YongLin Lei[1], and WeiCheng Lun[2]

[1] College of Systems Engineering, National University of Defense Technology,
Changsha 410072, China
`zhangcan536@nudt.edu.cn`
[2] Basic Education College, National University of Defense Technology,
Changsha 410072, China

**Abstract.** Community detection is an important task for identifying the structure and function of flying sensor networks. In this paper, we present a novel overlapping community detection method, O-LFM, to reuse overlapping beacon nodes in UAV flying sensor networks. In order to fully exploit the importance of nodes located at the edge of the interference area, we designed a collaborative localization scheme based on LFM community detection and centralized MDS collaborative localization algorithm. This solution solves the problem of reasonable allocation of beacon node resources at the edge of the interference area, improves the overlap rate of beacon nodes between communities, and thus enhances the overall positioning accuracy of the UAV swarm. It also improves the informationization and intelligence level of navigation and positioning of the swarm as a whole. The efficacy of the proposed method has been established through experiments on various UAV flying sensor network scenarios.

**Keywords:** Overlapping Community Detection · Local Fitness Method · Multi Dimensional Scaling · Collaborative Localization

## 1 Introduction

In recent years, UAV (unmanned aerial vehicle) swarm systems have been increasingly applied in both civilian and military fields, such as reconnaissance patrols, precision strikes, disaster relief, and agricultural irrigation [1]. Collaborative localization is achieved by exchanging information and data fusion between nodes in flying sensor network, resulting in better positioning performance than individual nodes. With the increasing complexity and variability of application scenarios and task requirements, efficient and accurate positioning services are the foundation for achieving drone swarm control. The emergence of collaborative positioning technology provides a new solution for highly reliable and accurate positioning of UAV swarm in complex flight environments, and has become a research hotspot in the field of UAV swarm collaboration [2].


N. Seddari and M. Redjimi (Eds.): ICMSCT 2024, CCIS 2606, pp. 91–102, 2025.
https://doi.org/10.1007/978-3-032-01922-6_8

In large-scale GNSS denial/restriction scenarios, GNSS signals are often unable to be received due to various factors such as multipath, interference, deception attacks, or other catastrophic events causing GNSS failures [3]. In this case, the MDS (Multi Dimensional Scaling) algorithm can be used for centralized collaborative localization with clustered nodes [4]. As a classic algorithm for node position perception in wireless sensor network, MDS algorithm can obtain the relative positions between nodes in the entire network with a small number of beacon nodes, and restore the shape between positioning nodes through rigid transformation (including rotation, translation, reflection and their combinations) (rigid transformation does not change the Euclidean distance matrix among clustered nodes), thus obtaining the absolute position of the node to be located.

When handling collaborative localization problem with large-scale nodes, due to the centralized processing of MDS, the algorithm complexity is $O(N^3)$. When the number of nodes $N$ is large, the computational efficiency will be greatly reduced. However, the flying sensing network which composed of large-scale nodes, belongs to the category of complex networks. Therefore ,community detection method can be used to detect and recognize the network structures, which assist better cooperative localization. Furthermore, community structure is of great significance for the resilience, robustness, and stability of complex networks, as well as for knowledge discovery and data mining based on big data. Obviously, it is necessary to perform community detection on flying sensor networks by dividing the network into smaller scale units, which can effectively limit the number of dimention in the MDS algorithm, reduce the time consumption of the algorithm, and improve the real-time performance of collaborative calculations.

The community detection method is an important research direction in network science, aiming to discover the community structure in complex networks, that is, the grouping of nodes in the network, so that nodes within the same group have stronger connections than nodes between different groups [5]. In general, finding the exact solution to such problems is an NP hard problem, when the swarm size is large, there is no effective exact solution. Currently , there are various community detection methods have already been published, from Modular optimization method [6–8] to Clique based filtering method [9–12] , and to Label propagation based method [13,14], and to Edge clustering based methods [15], and to Random walk based method [16,17].

Modular optimization method, which detects community structure by optimizing the modularity Q. Modularity is an important indicator for measuring the quality of network partitioning, which considers the difference between the density of edges within a community and the density of edges of the same size in a random network. For example, BGLL [18] is a modular optimization method that uses heuristic methods such as simulated annealing or spectral optimization to find the optimal partition. However, such methods may generate one or more larger communities, requiring further refinement of the community structure through multiple iterations and optimizations.

In those clique based filtering method, clique refers to a fully connected subset in the network, which is a complete graph. And the community structure is discovered by searching cliques in the network. This type of method is simple and effective, but may not be able to handle nodes that are not in cliques. Furthermore, there are improved CPMw [19] can handle overlapping communities in both weighted and unweighted networks.

The basic idea of Label propagation based method (LPA) is to iteratively update the labels of nodes, use the similarity between nodes to propagate the labels, and ultimately form a "community" structure of the same label. The advantage of algorithms lies in their simple thinking and low time complexity, making them suitable for handling large and complex networks.

Edge clustering based methods discover community structures by clustering edges in the network. For example, the extended link clustering ELC method calculates the similarity of extended links and uses the extended modularity as the basis for dividing the hierarchical tree, in order to find the best community division result. This method can handle the similarity between non adjacent edges and is suitable for discovering overlapping communities.

Random walk based method utilizes random walks on the network to detect community structure. Due to the sparse connections between online communities, random walks tend to wander within communities. Therefore, it can be achieved through the probability distribution of random walks. This type of method relies on the entire network structure, but can effectively reveal the community structure of the network.

However, an important prerequisite for the above community detection methods is the need to know the topology of the entire complex network, but this limitation is quite difficult for UAVs to grasp the constantly changing topology of network in real-time.

Scholars have proposed various community detection method based on local information to address this issue. The more commonly used method is the Local Fitness Method (LFM) [20], which is an overlapping community detection method based on local expansion. It uses the fitness function and local optimization method to obtain the network partition, which can discover both the overlapping community structure and the hierarchical structure of the network. It controls the size of communities by adjusting the parameter $\alpha$, thereby achieving the discovery of community hierarchy.

In this article, we mainly focused on UAV flying sensor network in GNSS denial/restricted scenarios. From the perspective of complex network with large-scale, we proposed a novel community detection method, O-LFM, which assisted the MDS collaborative positioning algorithm to balance the problems of node size and algorithm real-time performance, as well as improved collaborative positioning accuracy.

## 2   Overall Framework of Collaborative Localization

Be geared to the needs of GNSS denial/restriction scenarios, we consider the following optimization directions:

(1) The size of sub community is evenly. Beacon nodes are allocated reasonably. And the scale of sub community can be adjusted. The sub community includes partially beacon nodes and partially non-beacon nodes.
(2) The community classification results support that beacon nodes have a high overlap rate, while non-beacon nodes have a low overlap rate.
(3) Within the same community, there is a high density of connections (clustering coefficient).

Based on the optimizations and objectives mentioned above, we combined the MDS algorithm and LFM to design a collaborative localization scheme. The basic process of the system is shown in the Fig. 1.

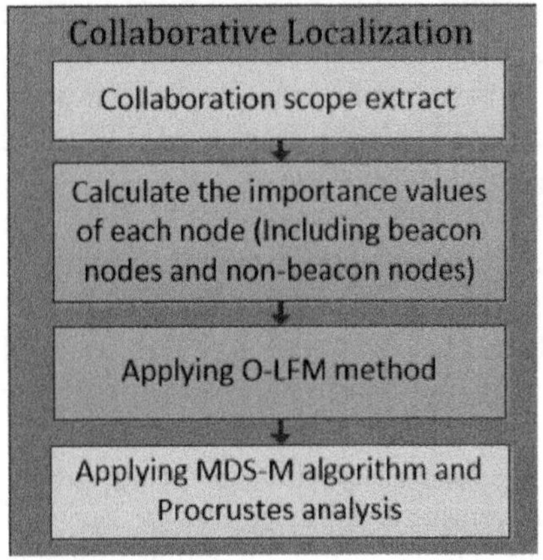


Fig. 1. Overall Framework of Collaborative localization.

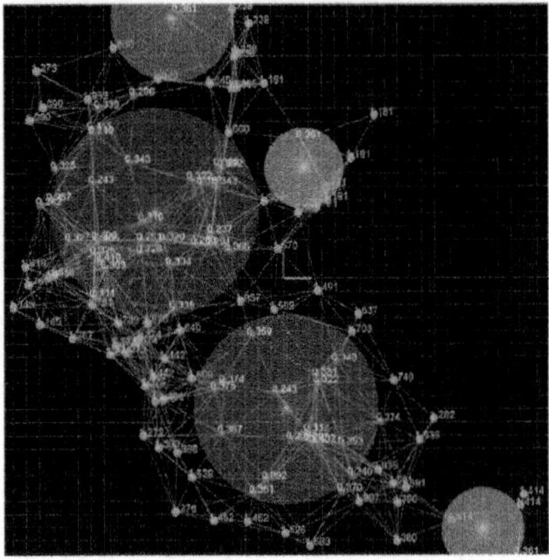

Fig. 2. Scope Extract and Node-Importance result.

## 2.1 Collaboration Scope Extract

In real-world outdoor swarm flight scenarios, GNSS signal shielding areas often appears in chunks, such as continuous canyon terrain and urban buildings. In this paper, we use circular interference areas instead of GNSS signal shielding areas to run our experiments (Fig. 2).

In order to facilitate subsequent processing and ensure that the topology input in the subsequent node-importance modeling algorithm is a connected graph, We have preprocessed the scope of nodes, that is, extract nodes within and around the interference area.

The pseudocode of the algorithm is shown as Algorithm 1.

---

**Algorithm 1.** Collaborative Scope Extract Algorithm

---

**Input:** Non-beacon nodes Set $N_1$,Beacon nodes Set $N_2$, Adjacency matrix $M$
**Output:** Sub communities and the number of sub communities n
$S = \Phi$
**for all** $i \in N_1$ **do**
   $S = S + \{i\}$
   $S = S + Neighbor(i)$
**end for**
**for all** $i \in S$ **do**
  **if** $j \in N_2$ $and$ $Neighbor(j) \subseteq N_2$ **then**
    $S = S - \{j\}$
  **end if**
**end for**
**if** $G(S)$ $is$ $not$ $a$ $connected$ $graph$ **then**
  $G_1, G_2, ...G_n \longleftarrow G(S)$
**end if**
**return** $n, G_1, G_2, ...G_n$

---

## 2.2 Node-Importance Modeling

Aims at numerical quantification of the importance, as an attribute of one node, which lays stress on nodes who located at the edge of the interference area. We conduct the following consistency distributed control protocol with two phase to calculate the node-importance of each node.

Firstly, initialize the node-importance of each node to 1.0, and we treat normal non interfering nodes as beacon nodes, treat interfered nodes as non-beacon nodes.

A system containing $n + m$ UAV nodes, where the first $n$ are non-beacon nodes, $N_1$ is the set of non-beacon nodes, the last $m$ are beacon nodes, $N_2$ is the set of beacon nodes, $N_{1i}$ is the set of non-beacon nodes in the neighboring nodes of node $i$, $N_{2i}$ is the set of beacon nodes in the neighboring nodes of node $i$, $P_i(t)$ is the node-importance of node $i$ at time $t$, and $| * |$ is the number of nodes in one set.

Next, we strengthen the role of beacon nodes in the allocation phase.

$$\begin{cases} P_i(t) = P_i(t) + \sum_{j=1}^{N_{1i}}[\frac{1}{2}(\frac{1}{|N_{2j}|} + \frac{1}{|N_{1j}|+|N_{2j}|})P_j(t)] & i \in N_2 \\ P_i(t) = P_i(t) - \frac{1}{2}[\frac{1}{|N_{2i}|} + \frac{1}{|N_{1i}|+|N_{2i}|}]P_i(t) & i \in N_1 \end{cases} \tag{1}$$

And then, we reassign the node-importance of each node in the correction phase.

$$\begin{cases} P_i(t+1) = P_i(t) - \frac{(P_i(t)-1.0)^2}{4} & i \in N_2 \\ P_i(t+1) = P_i(t) + \frac{1}{|N_1|}\sum_{j=1}^{N_2} \frac{(P_j(t)-1.0)^2}{4} & i \in N_1 \end{cases} \tag{2}$$

According to the Lyapunov stability theory [21], when the sensor network is a connected graph, the node-importance of each node can converge to a specific value. The protocol ensure that the node-importance of each node can converge to a specific value. Further more, the node-importance of all beacon nodes converge to a specific value greater than 1.0, and the node-importance of all non-beacon nodes converge to a specific value less than 1.0.

## 2.3   Optimized Local Fitness Method

In reality, however, for a single UAV nodes, it is more common to know only part of the flying network, either because of the network size or because the topology dynamic. The basic advantage behind O-LFM is that communities are essentially local structures, involving the nodes belonging to the modules themselves plus at most an extended neighborhood of them. This is certainly plausible for large-scale flying sensor networks, where each node does not depend on most of its peers. We set the form of the tness and obtained the best results with the simple expression

$$f_g = \frac{k_{in}^g}{(k_{in}^g + k_{out}^g)^\alpha} \tag{3}$$

where $k_{in}^g$ and $k_{out}^g$ are the total internal and external node-importance of the nodes in subgraph, and $\alpha$ is a positive real-valued parameter, controlling the size of the sub communities. The expression is similar to the classic LFM algorithm. By setting the $\alpha$ resolution adjustment parameter,it can effectively meets the requirements of adjustable size in large-scale scenarios. Large values of $\alpha$ yield very small communities, small values instead deliver large modules. If $\alpha$ is small enough, all nodes end up in the network itself. We have found that, in most cases, for $\alpha < 0.5$ there is only one community, for $\alpha > 2$ one recovers the smallest communities. A community consists of a set of internal nodes that can maximize the fitness function, meaning that adding any neighboring nodes to the community will decrease fitness.

$$g = argmax(f_g) \tag{4}$$

And the fitness contribution of a node to a subgraph is defined as the difference between the fitness of the community that included or not. We call such subgraph the natural community of node A.

$$f_g^A = f_{g+A} - f_{g-A} \tag{5}$$

The iterative steps of a node $A$ to find its natural community are the same as those of classical LFM. The iteration stops when the nodes examined all have negative tness. This procedure corresponds to a sort of greedy optimization of the tness function, as at each move one looks for the highest possible increase. We dene a cover of the graph as a set of communities such that each non-beacon node is assigned to at least one community. Different from classical LFM ,the process of find a cover can be summarized as follows:

(1) Pick a beacon node A at random, and detect the natural community of node A;

(2) Pick at random a beacon node B not yet assigned to any group. Detect the natural community of B, exploring all nodes regardless of their possible membership to other groups;
(3) Repeat 2 until all neighbors beacon node has been choosen;
(4) Pick at random a non-beacon node C not yet assigned to any group. Detect the natural community of C;
(5) Repeat 4 until all neighbors node has been choosen;

The algorithm stops when all non-beacon nodes have been assigned to at least one group.

## 2.4   Improved MDS-M Algorithm with Euclidean Distance Matrix

MDS is a nonlinear dimensionality reduction technique that preserves pairwise geodesic distances between the nodes. It aims to nd the intrinsic geometry of the deployed nodes accurately, using the geodesic distances between them. For nodes within their communication range, Euclidean distance obtained through P2P-ranging represents the geodesic distance. For non-neighboring nodes, geodesic distance is represented by the shortest path between them.

In this paper, we approximate the observation distance matrix using P2P (peer to peer) ranging distance and estimated distance between nodes. That is, between node i and node j, if there is a P2P-ranging, then fill the distance matrix with measured distance. Otherwise, the estimated distance is used to fill the distance matrix.

$$d_{ij}^k = \{ \begin{array}{l} Rang_{ij}^k \\ d_{ij}^{k-1} + \bar{v}_{ij}^k \cdot dt \end{array} \tag{6}$$

Among them, $d_{ij}^k$ denotes the approximate Euclidean distance between nodes $i$ and $j$ at time $k$, $Rang_{ij}^k$ denotes the $P2P$ distance between nodes $i$ and $j$ at time $k$, $\bar{v}_{ij}^k$ denotes the relative velocity between nodes $i$ and $j$ at time $k$, and $dt$ denotes the time interval.

The goal of classical MDS is to preserve the similarity of data points in the embedding space as it was in the input space.

Assuming there are $n$ nodes $X_i(i = 1,...,n)$ within the sub community, the approximate Euclidean distance $d_{ij}$ between adjacent nodes $i$ and $j$.

$$d_{ij}^2 = (x_i - x_j)^T (x_i - x_j) \tag{7}$$

Let matrix $A = a_{ij} = -\frac{1}{2}d_{ij}^2$, and then, the inner product matrix $B$ can be expressed as $B = HAH$ ,where $H = (E - n^{-1} \cdot 1)$ is the centralization matrix, $E$ is the n-th order identity matrix, 1 is the vector of ones. The matrix form of B is $B = X \cdot X^T$, and the rank of the $B$ represents the coordinate dimension $p$, that is, $r(B) = r(X \cdot X^T) = r(X) = p$. If we decompose B using eigenvalue decomposition, and sort the eigenvectors from leading (largest eigenvalue) to trailing (smallest eigenvalue). We take the top p maximum eigenvalues $\lambda_1, \lambda_2, ..., \lambda_p$ to form a diagonal matrix $\Lambda$, and combine the corresponding eigenvectors $e_1, e_2, ..., e_p$ into an $N \times P$ dimensional matrix $V$. Namely, $B = V\Lambda V^T$, $\Lambda = diag(\lambda_1, \lambda_2, ..., \lambda_p)$, $V = [e_1, e_2, ..., e_p]$. And then, the coordinates of all nodes in the embedding space is $X = V\Lambda^{1/2}$.

Since MDS only calculates the relative positions of each node, in order to obtain the absolute coordinate information of the nodes [22], it is necessary to use the known absolute coordinates and existing relative coordinates of the beacon nodes to obtain the linear transformation relationship, and apply it among non-beacon nodes to obtain the absolute coordinates of non-beacon node. Normally, a p-dimensional space requires $p+1$ absolute coordinates of beacon nodes. In this article, we uses Procrustes analysis [2] to convert coordinates between relative and absolute coordinate systems.

## 3   Simulation Analysis

To demonstrate and verify the proposed methods, simulation and experiments in different scenarios are performed in this research.

### 3.1   Simulation Experiment Design

In this paper ,we evaluate the performance of the proposed method mainly from the positioning accuracy. The root mean square error is used to measure the positioning accuracy, and the formula is as follows.

$$RMSE = \sqrt{\frac{\sum_{i=1}^{m} \left[ (x_i - x_b)^2 + (y_i - y_b)^2 + (z_i - z_b)^2 \right]}{m}} \tag{8}$$

where, $(x_i, y_i, z_i)$, $(x_b, y_b, x_b)$ are the real coordinate $(NED)$ and evaluated coordinate $(NED)$ of UAV nodes respectively. And $m$ represents the number of non-beacon nodes.

We use the GNSS system $(30GPS + 15BD + 21Glonass + 24Galileo)$ to conduct the experiments, and we set the satellite node elevation angle to be more than $10°C$ to be visible. We take the precision orbit data during the time period from 13:00 to 15:00, June 17, 2020.

The UAV swarm consist of 196 fixed-wing MAV. And UAVs is designed to be deployed in the range of 5000 m × 5000 m area which centered on the coordinate origin of the local navigation system. The initial start point of each UAV are randomly and evenly distributed in the range of 300–500 m height. The flight path of each UAV includes straight flight and curve flight path, which are evenly distributed throughout the field. GNSS/INS integrated navigation is adopted inside each beacon node. GNSS signal frequency is set to 1HZ. The peer to peer wireless range between UAVs is set to 500 m.

We chose two typical scenarios with five chuncks of interference area to run our experiments respectively, and the interference area settings is shown in the Table 1. The snapshot of the flight experiment is shown in Fig. 3 and Fig. 4.

**Table 1.** Interference area settings.

| Experiment 1 | Settings (R,x,y) | Experiment 2 | Settings (R,x,y) |
| --- | --- | --- | --- |
| Area 01 | (800,500,−1000) | Area 01 | (1500,0,0) |
| Area 02 | (476,1981,−862) | Area 02 | (600,1600,1600) |
| Area 03 | (724,−999,21) | Area 02 | (600,1600,−1600) |
| Area 04 | (322,−1923,1574) | Area 04 | (600,−1600,1600) |
| Area 05 | (309,840,158) | Area 05 | (600,−1600,−1600) |

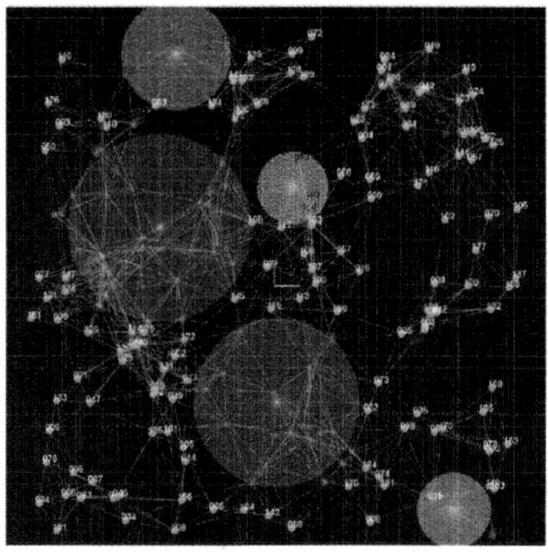

Fig. 3. Snapshot of Experiment 1.        Fig. 4. Snapshot of Experiment 2.

## 3.2 Numerical Results

We designed and ran five comparative method experiments, as shown in the Table 2. And the last column in the Table 2 means the ability of finding overlapping communities.

We extensively tested our method in two experiment. Simulation results are shown in the figures below. Figure 5 and Fig. 6 reveals the RMSE in each experiment respectively. Figure 7 and Fig. 8 illustrate the comparative performance of each method, in which Cumulative Distribution Function (CDF) can be more intuitionistic to the differences between each method. Figure 9 reveals the average RMES values of non-beacon nodes in each method. The results indicate that O-LFM exhibited a great superiority over other four method on the level of positioning accuracy. Figure 10 records the average members per community in each method. Compared with the classic LFM, particularly in experiment 1, the O-LFM showcased its capability in overlapping rate of beacon node, and achieved a high average of 64.3% which surpassed the baseline method LFM (51.2%). The superior performance of O-LFM is due to its design of node-importance and the preferential selection strategy, it can effectively improves the accuracy of collaborative localization.

**Table 2.** Five comparative method of community detection.

| Name | Abbreviation | Brief description | Ability |
|---|---|---|---|
| Classical LFM | LFM | Use degrees of the nodes to calculate fitness ($\alpha = 0.8$) [20] | yes |
| Optimized LFM | O-LFM | Optimized LFM in this paper ($\alpha = 0.8$) | yes |
| None Method | ONE | Treat the collaborative scope as a whole | no |
| Label propagation method | LPA | Label propagation method in [14] | no |
| Modular optimization | louvain method | Fast modular optimization method [8] | no |

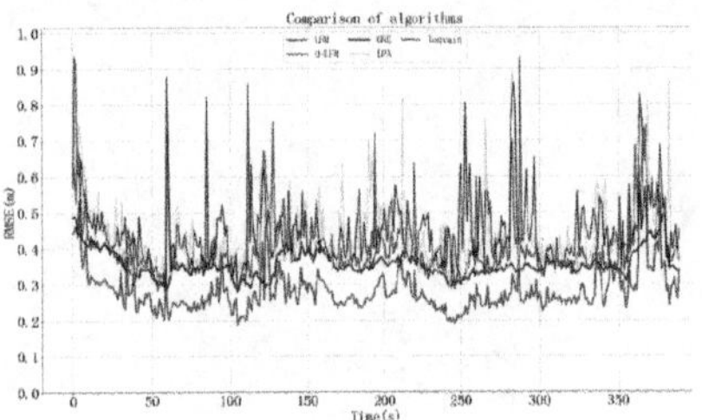

**Fig. 5.** Comparison of algorithm results in Experiment 1.

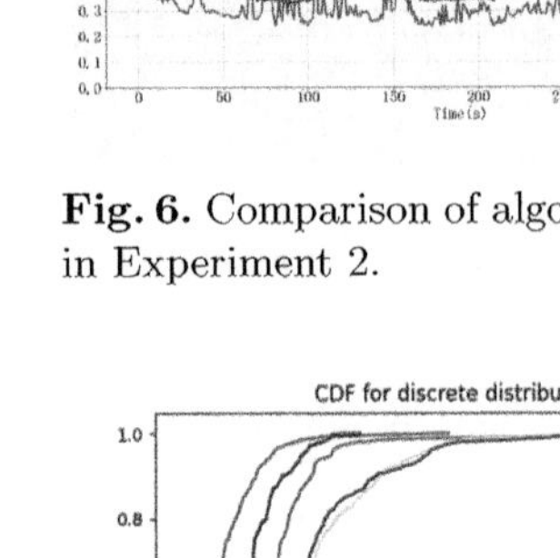

**Fig. 6.** Comparison of algorithm results in Experiment 2.

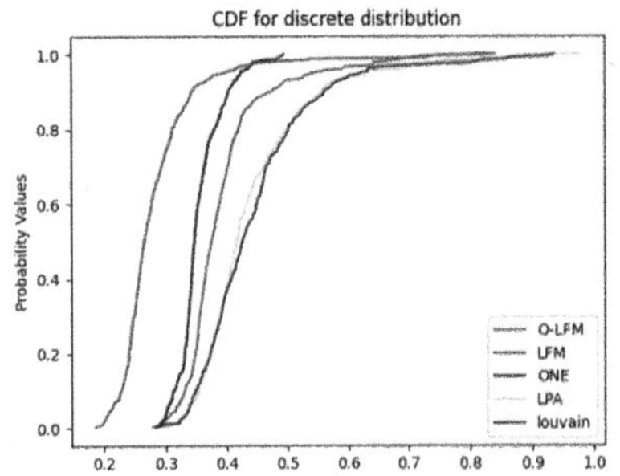

**Fig. 7.** CDF of each method in Experiment 1.

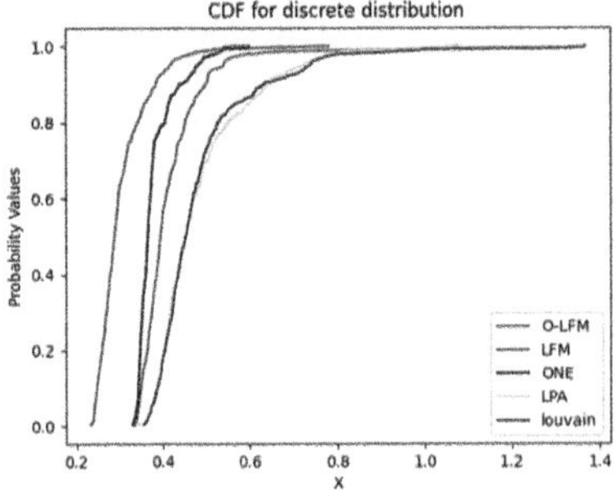

**Fig. 8.** CDF of each method in Experiment 2.

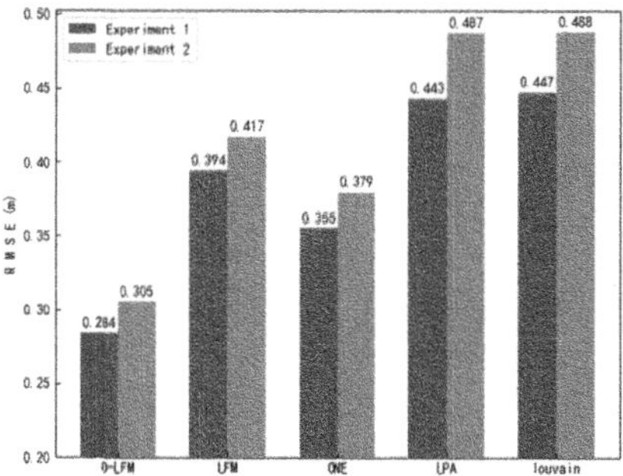

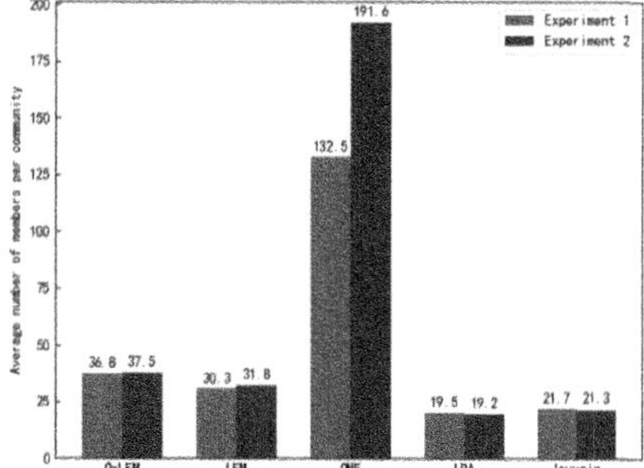

**Fig. 9.** Average RMES values of non-beacon nodes.

**Fig. 10.** Average number of members per community in each method.

# 4   Conclusion

In this research, a novel community detection method of collaborative localization algorithm is proposed. We choose dense UAV swarm with five chuncks of interference area as our target scenario. We studied the node-importance strategy and optimized LFM, in an effort to improve overlap rate of beacon nodes and reduce MDS algorithm dimensions. The simulation results shows that the collaborative localization scheme applying the O-LFM can greatly improves the positioning accuracy, and is suitable for the UAV swarm application with large-scale.

In the next work, we will consider using intelligent AI algorithm to self-adaption the number of community nodes, and will conduct in-depth research and discussion on the time consumption of each method.

**Acknowledgments.** This study was partly supported by National Natural Science Foundation of China (62003359). The authors would like to thank the timely help given by Guo Dong in programming. In addition, the authors wants to thank, in particular, the care, patience, support and help from Hongtao Hou over the passed years.

# References


1. Wu, C.F., Cheng, J., Guo, X.Y., Xu, C., Hu, E.W., Q H.: Development and prospect of aircraft clusters cooperative positioning and navigation countermeasures technology. J. Astronaut. **43**(2), 131–142 (2022)
2. Rui, C., Bin, Y., Wei, Z.: Distributed and collaborative localization for swarming UAVs. IEEE Internet Things J. **8**(6), 5062–5074 (2021)
3. Ruan, L., Li, G.X., Dai, W.H., et al.: Cooperative relative localization for UAV swarm in GNSS-denied environment: a coalition formation game approach. IEEE Internet Things J. (2021)
4. Xu, H., Li, Q.Y., Wang, J.P., Chen, K.Q., Sun, W.: An improved MDS localisation algorithm for a WSN in a sub-surface mine (Article). Int. J. Sens. Networks **29**(1), 58–73 (2019)

5. Baltsou, G., Christopoulos, K., Tsichlas, K.: Local community detection: a survey. IEEE Access **1**, 99 (2022)
6. Maoyu, Z., Jingfei, Z., Wenlin, D.: Fast community detection in dynamic and heterogeneous networks. J. Comput. Graphical Stat. **33**(2), 487–500 (2024)
7. Furkan, O., Ali, K.: KO: modularity optimization in community detection. Neural Comput. Appl. **35**(15), 1–15 (2023)
8. Meo, P.D., Ferrara, E., Fiumara, G., et al.: Fast unfolding of communities in large networks (2008). https://doi.org/10.1088/1742-5468.
9. Hao, F., Min, G.Y., Pei, Z., Park, D.S., Yang, L.T.: k-Clique Community Detection in social networks based on formal concept analysis. IEEE Syst. J. **11**(1), 250–259 (2017)
10. Gregori, E., Lenzini, L., Mainardi, S.: Parallel k-clique community detection on large-scale networks. IEEE Trans. Parallel Distrib. Syst. **24**(8), 1651–1660 (2013)
11. Xiaoyanga, L., Yudiea, W.,et al.: Heterogeneous graph community detection method based on K-nearest neighbor graph neural network. Intell. Data Anal. 1–22 (2024,)
12. Pavla, D., Petr, P., Jan, P., Václav, S.: A hierarchical overlapping community detection method based on closed trail distance and maximal cliques. Inf. Sci. **662**, 120271 (2024)
13. Saif, M., Samie, M.E., Hamzeh, A.: Detecting overlapping communities in complex networks: an evolutionary label propagation approach. Int. J. Inf. Technol. Decis. Making **23**(1), 327–360 (2024)
14. Yue, Y., Wang, G., Jun, H., et al.: An improved label propagation algorithm based on community core node and label importance for community detection in sparse network. Appl. Intell. **53**(14), 17935–17951 (2023)
15. Chitra, K., Tamilarasi, A.: Community detection using Jaacard similarity with SIM-edge detection techniques. Comput. Syst. Sci. Eng. **44**(1), 327–337 (2022)
16. MarjanaCAa, M., Mahdib, K.: Community detection by influential nodes based on random walk distance. Concurrency Comput. Pract. Experience **36**(17), e8135 (2024)
17. Christian, T.C., Denis, H., Bernhard, C.: Synwalk: community detection via random walk modelling. Data Min. Knowl. Disc. **36**(2), 739–780 (2022)
18. Chaturvedi, P., Dhara, M., Arora, D.: Community detection in complex network via BGLL algorithm. Int. J. Comput. Appl. **48**(1), 32–42 (2012)
19. Zhang, F., Cai, W., Song, Y., Lee, M.Z., Shan, S., Dagan, D.: Overlapping node discovery for improving classification of lung nodules. **2013**, 5461–5464 (2013)
20. Lancichinetti, A., Santo,et al.: Detecting the overlapping and hierarchical community structure in complex networks. New J. Phys. (2009). https://doi.org/10.1088/1367-2630/11/3/033015.
21. Hatvani, L.: Aleksandr Lyapunov, the man who created the modern theory of stability. Electron. J. Qual. Theory Differ. Equ. **2019**(26), 1–9 (2019)
22. Liu, Y.P., Wang, Y.L., Wang, J., et al.: Distributed 3D relative localization of UAVs. IEEE Trans. Veh. Technol. **69**(10), 11756–11770 (2020)

# Methodological Guide for Evaluating a Knowledge Management System

Hamza Ladeg[✉] and Rachid Chalal

Laboratory of Systems Design Methods, Higher National School of Computer Science (ESI),
BP 68M, 16309 Algiers, Algeria
`{h_ladeg,r_chalal}@esi.dz`

**Abstract.** Knowledge Management Systems (KMS) are crucial for leveraging the intellectual capital of organizations, but their success depends on their alignment with the specific context of each organization.

This research aims to address the issue of selecting the most appropriate evaluation model for a given KMS, given the diversity of existing models and varied organizational needs.

We propose a four-step methodological approach, based on a detailed comparative table of six influential evaluation models. This table, serving as a decision support tool, analyzes the theoretical foundations, methodologies, and evaluation dimensions of each model.

The validation of our approach is illustrated by a case study evaluating the Enterprise Social Network (ESN) Workplace in a university context.

**Keywords:** knowledge management systems · evaluation models of knowledge management systems

## 1 Introduction

Knowledge Management Systems (KMS) are essential tools that facilitate the effective and efficient management of knowledge within companies [1, 2]. The success of a KMS primarily relies on the organization's ability to exploit information technologies to leverage the knowledge held by its employees.

However, despite considerable investments in KMS, many organizations struggle to fully benefit from them. Studies have shown that the failure rate of knowledge management initiatives can reach 50 to 70% [3]. This observation underlines the need for rigorous evaluation of KMS to identify success factors and obstacles to their adoption.

The literature review on KMS evaluation models allows us to understand the different dimensions such as (technical quality, user satisfaction, or impact on organizational performance,…) to be considered to optimize the performance and adoption of KMS within organizations.


N. Seddari and M. Redjimi (Eds.): ICMSCT 2024, CCIS 2606, pp. 103–113, 2025.
https://doi.org/10.1007/978-3-032-01922-6_9

However, choosing an appropriate model, adapted to the specificities of the organization and its KMS, is not trivial to obtain a relevant and reliable evaluation.

The main objective of our research is to propose a methodological approach to select the most appropriate evaluation model as a tool for evaluating a KMS according to the specific context of each organization.

Our research work thus aims to provide researchers and practitioners with a methodological approach to choose and adapt a KMS evaluation model, taking into account the specificities of the context of use.

We will illustrate our methodological approach through a practical case. The practical case focuses on enterprise social networks (ESN[1]). It involves evaluating the success of the Workplace ESN within the university community.

The article is structured as follows: after the introduction, Sect. 2 presents the main KMS evaluation models. In Sect. 3, we present a comparative study of KMS evaluation models. Section 4 presents our methodology for selecting the KMS evaluation model best suited to the organization. Section 5 presents an illustration of the proposed approach on a case study, and we will conclude our article with a conclusion.

## 2   KMS Evaluation Models

The success of a KMS largely depends on its ability to allow an organization to capture, store, and effectively reuse the knowledge held by its employees. As mentioned, "the success of KMS depends on an organization's ability to use information technologies to effectively exploit the knowledge that resides in the minds of its employees". However, the human factor plays a central role in this process. Indeed, the success of KMS depends on users' commitment to share and retrieve this knowledge via the system, which requires not only adapted technology but also organizational support in terms of culture, governance, and employee motivation [4, 5]. According to DeLone and McLean [6], the success of KMS is often measured by the impact it has on organizational performance, user satisfaction, and the quality of shared information.

The evaluation of KMS has become a crucial issue for organizations seeking to optimize their use of knowledge as a strategic resource. Over the years, several models have been developed to evaluate the success of KMS. Each proposed evaluation model brings its own perspective on the key factors of KMS success. We have chosen to present the six influential models of the moment:

- Jennex and Olfman Model (2006);
- Wu and Wang Model (2006);
- Kulkarni et al. Model (2007);
- Halawi et al. Model (2007);
- Hwang et al. Model (2008);
- Nantapanuwat et al. Model (2010).

---

[1] The Enterprise Social Network (ESN) provides support for professional communities in shaing best practices, strategic monitoring and reflection, the emergence of initiatives and innovations, and knowledge transfer.

1. Jennex and Olfman Model (2006):

   Jennex, Olfman, Pituma, and YongTae [7] initially proposed the model that would later be known as Jennex and Olfman [8], adopting the generic framework of DeLone and McLean's information systems success model [9]. They customized the dimensions to reflect the necessary constructs of system quality and use for the organizational memory information system. The model was modified by Jennex and Olfman [10] following a five-year longitudinal study on knowledge management (KM) in an engineering organization, based on DeLone and McLean's revised IS success model [6]. This final model was developed to incorporate the experience of using the model to design KMS and to incorporate other research on KMS success factors from the literature.

2. Wu and Wang Model (2006):

   Wu and Wang conducted a study on Taiwanese companies using knowledge management systems [11]. They modified DeLone and McLean's IS success model [9] to replace the concept of system use with the perceived benefits of the knowledge management system. Individual impact and organizational impact were interpreted as the use of the knowledge management system. Thus, the concept of system use was retained in the model, but it was positioned differently from DeLone and McLean's original IS success model. Based on 204 responses, they found that the effects of knowledge quality on user satisfaction and perceived benefits were more important than the effects of system quality. This suggests that users rely more on the quality of content provided by the knowledge management system than on the system's performance and functions.

3. Kulkarni et al. Model (2007):

   Kulkarni et al. conducted a study with mid-level executives enrolled in a part-time MBA program at a US university to learn about the use of information technology in knowledge management practice within their organization [12]. They extended DeLone and McLean's IS success model [9] to include organizational support: leadership, incentives, colleague, and supervisor. Based on 111 responses, they found that organizational factors (leadership, incentive, colleague, and supervisor) indirectly affected knowledge use.

4. Halawi et al. Model (2007):

   Halawi et al. conducted a study on US organizations that have implemented knowledge management systems [13]. They used DeLone and McLean's IS success model [6] and interpreted net benefits as KMS success. Based on 117 responses, they found that the multiple regression of KMS success on knowledge quality, system quality, service quality, intention to use, and user satisfaction resulted in 66% of the variance of knowledge management system success accounted for.

5. Hwang et al. Model (2008):

   Hwang et al. conducted a study on KMS users in Taiwan hospitals [14]. They modified DeLone and McLean's IS success model [6] by replacing system use with perceived usefulness. Net benefits were interpreted as reduced knowledge transfer costs and improved knowledge worker efficiency. Based on 163 responses, they discovered that system quality and information quality indirectly affected net benefits.

6. Nantapanuwat et al. Model (2010):

   Nantapanuwat et al. developed a knowledge management system success model [15]. The model consists of four main constructs that lead to KMS success, namely technical, social, use, and user satisfaction. The technical construct includes system quality, knowledge quality, and service quality. The social construct includes trust.

Their research added trust to the social factor [15]. Alavi and Leidner commented on the fact that trust can be developed to improve knowledge use by individuals in a KMS [4]. Trust has been proven to affect KMS success as a social factor [15].

## 3  Comparative Study of KMS Evaluation Models

This comparative study is seen as a support tool for our methodological approach to select the KMS evaluation model best suited to the context of use of the KMS to be evaluated. Thus, the tool will provide a summary of each retained model.

The comparative study will focus on the six retained KMS evaluation models.

The comparison elements cover several dimensions. They are considered the most important in the evaluation of information systems and KMS:

The chosen comparison aspects cover the main dimensions, namely:

- Theoretical foundation: This aspect is crucial as it shows the conceptual basis on which each model is built.
- Methodology: This indicates how the model was developed and validated, which is important for evaluating its scientific rigor.
- System quality: This is a fundamental aspect of any information system, including KMS.
- knowledge quality: Essential for KMS, as the quality of knowledge is at the heart of their function.
- System use: This measures the adoption and effective use of the system.
- User satisfaction: A key indicator of system success from the end-user's perspective.
- Net benefits: This evaluates the overall impact of the system on the organization or individuals.
- Trust: An important factor influencing the success of the KMS. The high degree of confidence encouraged by the use of the knowledge management system, thus contributing to user satisfaction
- Domain of use: This allows understanding of the specific context for which each model was designed.

They are considered important in the evaluation of information systems and KMS.

The comparative table (Table 1) presented offers an overview of influential KMS evaluation models in the KM field.

The analysis of the comparative table of KMS evaluation models reveals both fundamental similarities and significant differences between approaches:

In terms of similarities, we note that all models share a common theoretical base, often relying on existing theories or adapting previous models, notably that of DeLone and McLean [6]. This common base testifies to a certain continuity in the field of KMS evaluation. Moreover, all presented models have been empirically validated, whether through case studies or large-scale surveys, which reinforces their credibility and practical applicability. The models also converge on several key evaluation dimensions. They all incorporate measures of system quality, knowledge quality, system use, and user satisfaction. These dimensions seem to constitute the core of KMS evaluation, reflecting the importance given to both technical and human aspects of these systems. Similarly, all models seek to measure net benefits or system impact, whether at the individual or organizational level, underlining the importance of demonstrating the added value of KMS.

However, the models present notable differences that reflect the evolution of the field and the diversity of application contexts. A major difference lies in the specificity of the context. While some models, such as that of Jennex and Olfman [8] or Wu and Wang [11], adopt a more general approach, others are specifically adapted to particular sectors. For example, Hwang et al.'s model [14] is designed for the medical context, while that of Nantapanuwat et al. [15] is adapted to the banking sector.

The models also differ in their approach to different evaluation dimensions. For example, regarding system quality, some emphasize ease of use, others security, or accuracy in specific tasks. Similarly, the approach to system use varies, with some models distinguishing between intention to use and effective use, while others focus on use for specific tasks.

More recent models tend to introduce new concepts or emphasize particular aspects. For example, Kulkarni et al. introduce the concept of KM organizational capability [12], while Nantapanuwat et al. include trust as an evaluation factor [15]. These additions reflect a more nuanced and in-depth understanding of the factors that influence KMS success. Finally, the models differ in their approach to net benefits. Some focus on overall organizational effectiveness, others on more specific aspects such as improved decision-making or diagnostic accuracy in the medical context. This diversity underlines the variety of expectations and objectives associated with KMS in different organizational contexts.

In conclusion, this analysis reveals that, despite a common foundation, KMS evaluation models have evolved to reflect the growing complexity of the domain and the diversity of application contexts. This evolution testifies to the maturity of the research field on KMS and the growing recognition of the importance of adapting evaluation models to the specific needs of different organizations and sectors.

**Table 1.** Comparative table of KMS evaluation models

| Aspects | Jennexet Olfman (2006) | Wu et Wang (2006) | Kulkarni et al. (2007) | Halawi et al. (2007) | Hwang et al. (2008) | Nantapanuwat et al. (2010) |
|---|---|---|---|---|---|---|
| Theoretical foundation | Based on the theory of organizational learning and organizational memory | Adaptation of DeLone and McLean's IS success model | Integrates socio-cognitive theory and social exchange theory | Builds on the DeLone and McLean model and resource-based theory | Adapts the DeLone and McLean model to the medical context | Combines elements from several previous models |
| Methodology | Developed from a literature review and validated by case studies | Developed and empirically tested through a survey of 50 companies | Developed theoretically and validated by a survey of 150 participants | Empirically tested with a survey of 150 organizations | Case study in a Taiwanese hospital | Survey of 550 bank employees |
| System quality | Includes measures of response time, ease of use, and system reliability | Emphasizes ease of learning and flexibility of the system | Evaluates availability, reliability and ease of use | Measures functionality, usability and performance | Focuses on accuracy and speed of classification | Evaluates accessibility, stability and security |
| knowledge quality | Evaluates accuracy, completeness, consistency and relevance | Measures the accuracy, completeness and format of information | Focuses on the relevance and usefulness of KM content | Evaluates accuracy, timeliness and completeness | Examines the quality of medical knowledge | Measures accuracy, timeliness and relevance to the banking industry |
| System use | Measures frequency of use and purpose of use | Distinguishes intention to use from actual use | Evaluates usage for different knowledge-related tasks | Measures the frequency and nature of use | Examines the use for classification of diseases | Evaluates usage for various banking tasks |
| User satisfaction | Evaluates satisfaction with features and effectiveness | Measures overall and specific satisfaction | Focuses on satisfaction with content and knowledge sharing | Evaluates overall satisfaction and by function | Measures satisfaction in the context of disease classification | Evaluates bank employee satisfaction |
| Net benefits | Includes measures of individual productivity and organizational effectiveness | Assesses the impact on individual and organizational performance | Measures improvement in organizational performance | Examines the impact on decision making and productivity | Evaluates the improvement in diagnostic accuracy | Measures impact on employee and bank performance |
| Trust | Not supported | Not supported | Not supported | Not supported | Not supported | Pris en charge |
| Domain of use | KMS | KMS | KMS | KMS | KMS | KMS |

# 4  Methodological Approach for Selecting a KMS Evaluation Model

The general approach to choosing a KMS evaluation model includes four steps:

1. **Step 1: Context Definition**

   This initial step is essential to understand the KMS to be evaluated and its specific objectives, thus laying the foundations for the evaluation. To do this:

   - Present the KMS to be evaluated
   - Identify the objectives of the KMS to be evaluated

2. **Step 2: Identification of Key Dimensions**

   This analytical phase aims to determine the crucial dimensions of the KMS to be evaluated, based on the previously defined objectives. It breaks down as follows:

   - Analyze the objectives of the KMS to be evaluated to determine the important dimensions
   - Associate evaluation sub-dimensions with each dimension

3. **Step 3: Choice of Evaluation Model**

   This step is decisive in the evaluation process as it involves selecting the most appropriate model to evaluate the KMS. To facilitate the execution of this 3rd step, we deemed it useful to accompany our approach with a comparative table of the six best evaluation models from the literature. This comparative table constitutes a support tool for our approach. Also, for this step, here are the tasks to accomplish:

   - Consult the comparative table of existing evaluation models
   - Analyze the relevance of the models in relation to the context based on the dimensions and their sub-dimensions identified in step 2
   - Select and justify the choice of the most appropriate model

4. **Step 4: Implementation of KMS Evaluation**

   This final step consists of concretely applying the chosen model, adapting it if necessary, and carrying out the actual evaluation of the KMS. It unfolds as follows:

   - Adapt the chosen model to the specific context if necessary
   - Plan the empirical evaluation
   - Conduct the evaluation
   - Interpret the results

# 5  Case Study

## 5.1  Step 1: Context Definition

- Presentation of the KMS to be evaluated:

   Our research work focuses on improving the quality of knowledge shared through ESNs in an organization.

   With the introduction of active pedagogy at the Higher National School of Computer Science (ESI), ESI encourages teamwork with quality supervision and access to various pedagogical tools adapted to active pedagogy, such as the Workplace ESN.

Workplace is an ESN developed by Meta. It is among the best on the market. Like Facebook, Workplace allows a company's employees to stay in touch easily and be informed of what's happening within the company.

ESNs can play a role in facilitating collaboration and communication between students, as well as between students and teachers.

To do this, we proposed to revisit the architecture of the Workplace ESN for a significant improvement in the quality of knowledge shared through the new architecture of the Workplace ESN constitutes the KMS to be evaluated [16].

Fourth-year computer science students in the "Information Systems and Technology" specialty work in teams on a project. Within each team, students submit knowledge for certification before sharing it among themselves. The pedagogical team plays the role of "knowledge manager" who proceeds to certify or not certify the knowledge. Certified knowledge can be shared and reused, while non-certified knowledge cannot be but is not destroyed to serve as lessons later.

- Identify the objectives of the KMS to be evaluated:

  - The revisited Workplace ESN will allow ESI to:
  - Improve the quality of shared knowledge;
  - Promote trust in knowledge sharing and reuse;
  - Promote user satisfaction with the revisited Workplace ESN.

### 5.2  Step 2: Identification of Key Dimensions

The analysis of the objectives of the KMS to be evaluated will allow us to highlight the most important dimensions that the evaluation model of the KMS to be evaluated must integrate.

Here is the detailed analysis of each of the objectives of the KMS to be evaluated:

- Objective of the KMS to be evaluated: Improve the quality of shared knowledge. To judge the quality of shared knowledge, we will therefore retain the sub-dimensions accuracy and relevance.
- Objective of the KMS to be evaluated: Promote trust in knowledge sharing and reuse. To judge trust in knowledge sharing and reuse, we will therefore retain the trust dimension.
- Objective of the KMS to be evaluated: Promote user satisfaction with the revisited Workplace ESN. To judge user satisfaction with the revisited Workplace ESN, we will therefore retain the user satisfaction dimension.

This step of analyzing the objectives of the KMS to be evaluated has highlighted three dimensions (the quality of shared knowledge, trust, and user satisfaction).

### 5.3  Step 3: Choice of Evaluation Model for the Revisited Workplace ESN

For our case, we must identify an evaluation model for the revisited Workplace ESN that integrates the three dimensions (the quality of shared knowledge, trust, and user satisfaction).

To facilitate the execution of this 3rd step, we will use our tool to identify the evaluation model for the revisited Workplace ESN. The result of exploiting our tool is given in (Table 2). You will notice that only the lines corresponding to the retained dimensions appear there.

It emerges from our analysis that to evaluate our KMS, the revisited Workplace ESN, the model of Nantapanuwat et al. [15] stands out among the six evaluation models studied.

**Table 2.** Table for identifying the evaluation model

| Aspects | Jennex et Olfman (2006) | Wu et Wang (2006) | Kulkarni et al. (2007) | Halawi et al. (2007) | Hwang et al. (2008) | Nantapanuwat et al. (2010) |
|---|---|---|---|---|---|---|
| knowledge quality | Evaluates accuracy, completeness, consistency and relevance | Measures the accuracy, completeness and format of information | Focuses on the relevance and usefulness of KM content | Evaluates accuracy, timeliness and completeness | Examines the quality of medical knowledge | Measures accuracy, timeliness and relevance to the banking industry |
| User satisfaction | Evaluates satisfaction with features and effectiveness | Measures overall and specific satisfaction | Focuses on satisfaction with content and knowledge sharing | Evaluates overall satisfaction and by function | Measures satisfaction in the context of disease classification | Evaluates bank employee satisfaction |
| Trust | Not supported | Not supported | Not supported | Not supported | Not supported | Supported |

## 5.4  Step 4: Implementation of the Evaluation of the Revisited Workplace ESN

- Adapting the retained model to the context of the revised Workplace ESN:

During this step, the evaluator must decide whether to retain all the dimensions of the selected model or eliminate some of them. In our case, we have kept all the dimensions. This decision was made because all dimensions contribute to a complete understanding of the effectiveness and value of the revised Workplace ESN for the organization.

- Planning the empirical evaluation:

Our evaluation model for the revised Workplace ESN is based on the model of Nantapanuwat et al. [15]. Our evaluation model for the revised Workplace ESN includes three dimensions (the quality of shared knowledge, trust, and user satisfaction). We still need to:

– Design a questionnaire based on the selected model;
– Plan data collection from student users of the revised Workplace ESN;
– Choose quantitative and qualitative methods for analyzing the results;
– Analyze the results.

This final step allowed us to complete our evaluation of the revised Workplace ESN.

The evaluation of the revised Workplace ESN using the model of Nantapanuwat et al. [15] allowed us to observe the success of the revised Workplace ESN in the field. This led us to make the following recommendations:

- Integrate the knowledge certification module into the architecture of ESNs.
- Designate a knowledge manager to ensure the certification work.

## 6  Conclusion and Perspectives

Effective evaluation of knowledge management systems (KMS) remains a major challenge for organizations seeking to optimize their intellectual capital. Our research focused on developing a methodological approach to select the most appropriate evaluation model for a given KMS, taking into account the specificities of each organizational context.

Our methodological approach is structured in four steps: defining the context, identifying key dimensions, choosing the evaluation model, and implementing the evaluation.

To facilitate the implementation of our approach, we developed a decision support tool for the evaluator. This is a comparative table of six of the most influential evaluation models.

The validation of our approach was demonstrated through a case study on the evaluation of an ESN in a university context. This involved the revised Workplace ESN.

Our research contributes to the literature on KMS evaluation by providing a structured and adaptable approach for selecting an evaluation model. It emphasizes the importance of considering the unique context of each KMS in the evaluation process.

In terms of perspectives, our methodology paves the way for future improvements and extensions. It can be adapted to integrate new evaluation models, continuously enriching our decision support tool and ensuring its relevance in the face of developments in the field. Moreover, the development of a software tool based on our methodology could facilitate the selection of the most appropriate model based on the specific criteria of each organization, thus making the process more efficient and accessible.

## References


1. Tsai, C.-H., Chen, H.: Assessing knowledge management system success: an empirical study in taiwan's high-tech industry. J. Am. Acad. Bus. Cambridge **10**(2), 257–262 (2007)
2. Lai, J.-Y.: How reward, computer self-efficacy, and perceived power security affect knowledge management systems success: an empirical investigation in high-tech companies. J. Am. Soc. Inform. Sci. Technol. **60**(2), 332–347 (2009)
3. Akhavan, P., Jafari, M., Fathian, M.: Exploring the failure factors of implementing knowledge management system in the organizations. J. Knowl. Manage. Pract. **6** (2005)
4. Alavi, M., Leidner, D.: "'Knowledge management and knowledge management systems'": conceptual foundations and research issues. MIS Q. **25**(1), 107–136 (2001)
5. Davenport, T.H., Prusak, L.: Workingknowledge: "How Organizations Manage Whatthey Know?" Harvard Business SchoolPress, USA (1998)

6. Delone, W.H., McLean, E.R.: The Delone and McLean model of information systems success: a ten-year update. J. Manage. Inform. Syst. **19**(4), 9–30 (2003)
7. Jennex, M.E., Olfman, L., Pituma, P., YongTae, P.: An organizational memory information systems success model: an extension of DeLone and McLean's I/S success model. In: Proceedings of the 31st Annual Hawaii International Conference on System Sciences (1998)
8. Jennex et Olfman 2006Jennex, M.E. and Olfman, L., "A Model of Knowledge Management Success", International Journal of Knowledge Management, 2(3), pp. 51–68 (2006)
9. Delone, W.H., McLean, E.R.: Information systems success: the quest for the dependent variable. Inform. Syst. Res. **3**(1), 60–95 (1992)
10. Jennex, M.E., Olfman, L.: Organizational memory/knowledge effects on productivity: a longitudinal study. In: Proceedings of the 35th Annual Hawaii International Conference on System Sciences (2002)
11. Wu, J.H., Wang, Y.M.: Measuring KMS success: a respecification of the DeLone and McLean's model. Inform. Manage. **43**(6), 728–739 (2006)
12. Kulkarni, U.R., Ravindran, S., Freeze, R.: A knowledge management success model: theoretical development and empirical validation. J. Manag. Inf. Syst. **23**(3), 309–347 (2007)
13. Halawi, L.A., Mccarthy, R.V., Aronson, J.E.: An empirical investigation of knowledge management systems success. J. Comput. Inform. Syst. **48**(2), 121–135 (2007)
14. Hwang, H.-G., Chang, I.C., Chen, F.-J., Wu, S.-Y.: Investigation of the application of KMS for diseases classifications: a study in a Taiwanese hospital. Expert Syst. Appl. **34**(1), 725–733 (2008)
15. Nantapanuwat, N., Peter, R., Laddawan, K.: An investigation of the determinants of knowledge management systems success in banking industry. World Acad. Sci. Eng. Technol. **71**(2010), 588–595 (2010)
16. Ladeg, H., Chalal, R.: New architecture of enterprise social networks for knowledge sharing within engineering school. J. Inform. Knowl. Manage. **15**(1), 16–34 (2021)

# Improving the Performance of Collaborative Filtering Systems Using Artificial Neural Networks

Lalia Benathmane[1(✉)] and Sahraoui Kharroubi[2]

[1] Laboratoire de Génie Energétique et Génie Informatique (L2GEGI), University of Tiaret, 14000 Tiaret, Algeria
l_benathmane@esi.dz
[2] Laboratoire de Recherche en Intelligence Artificielle et Systèmes, University of Tiaret, 14000 Tiaret, Algeria

**Abstract.** Collaborative filtering systems are widely used to provide personalised recommendations in various domains such as e-commerce, streaming services and social media platforms. Despite their effectiveness, these systems face significant challenges, including data sparsity, cold start problems, and popularity bias, which can limit their recommendation quality. This article proposes a novel approach to improve the performance of collaborative filtering systems by using Artificial Neural Networks (ANNs). We propose the formation of a network of experts, similar to an artificial neural network, whose members provide ratings for different items, thereby increasing the available data. We will analyse how this network can improve the quality of recommendations and how it can be integrated with neural networks to improve accuracy and efficiency.

**Keywords:** Collaborative filtering · artificial neural networks · user ratings · performance enhancement · expert network

## 1 Introduction

Recommender Systems (RSs) are software tools and techniques providing suggestions for items to be of use to a user [7] in various domains, including e-commerce, streaming services and social media. These systems work by analyzing user interactions and preferences to suggest items that similar users have enjoyed, thereby increasing user satisfaction and engagement.

However, collaborative filtering systems face significant challenges, the most prominent being data sparsity, where a lack of sufficient user ratings can lead to inaccurate or irrelevant recommendations. This problem is particularly acute in novel user or item scenarios, where limited data hampers the system's ability to make informed predictions.

To address these challenges, the integration of Artificial Neural Networks (ANNs) has emerged as a promising solution [1]. ANNs can effectively model complex relationships in data and improve the predictive capabilities of recommendation systems by learning from diverse inputs. By harnessing the power of neural networks, this paper


N. Seddari and M. Redjimi (Eds.): ICMSCT 2024, CCIS 2606, pp. 114–127, 2025.
https://doi.org/10.1007/978-3-032-01922-6_10

aims to improve collaborative filtering systems and mitigate the impact of data sparsity through innovative approaches, including the formation of a community network of experts to provide additional ratings.

This article is structured as follows: In the second section, we will present the impact of the lack of evaluations on the quality of recommendations in different systems and the solutions adopted to mitigate its negative influence. Then, in the third section, we present our approach in a step-by-step and detailed manner. The fourth section deals with the evaluation and discussion, and the fifth section presents the conclusions and future perspectives.

## 2  State of the Art

The state of the art in collaborative filtering systems highlights recent advances in the field, particularly the integration of neural networks to overcome traditional challenges [6, 10]. Collaborative filtering systems are generally divided into two categories: user-based filtering and item-based filtering. However, the limitation of user ratings remains a major obstacle, resulting in low accuracy of recommendations.

Recent research has focused on using deep learning techniques to improve the quality of recommendations. For example, Shivangi et al. [5] demonstrated how deep neural networks can model the complex interactions between users and items, thereby improving the predictive ability of recommendation systems. In addition, the hybrid approach, which combines collaborative filtering with other methods, has been highlighted as an effective solution to the problem of data sparsity [11].

Other solutions to the problem of sparsity have also been explored. Here are some notable approaches since 2017:

- *Content-based models*: These models use item characteristics to generate recommendations even in the absence of data. Zhang et al. [8] investigated approaches that integrate contextual information to enrich recommendations.
- *Graph-based recommendation systems:* These systems use graphs to represent relationships between users and items [9]. The use of graphs can significantly improve the quality of recommendations by exploiting implicit links.
- *Reinforcement learning based filtering:* This approach uses reinforcement learning techniques to learn from real-time interactions [6].
- *Federated learning:* This technique allows models to be trained without centralizing user data, thus preserving privacy [4].

The integration of social networks to enrich recommendation systems is another emerging trend. Liu et al. [2] proposed a model that incorporates social relationships between users to improve recommendation accuracy. This approach highlights the importance of social context in user decision making and opens new avenues for collaborative filtering.

### Evaluation of Solutions

While these solutions offer promising ways to mitigate the problem of data sparsity, it is important to recognise their strengths and weaknesses:

- ***Content-based models:*** These models are effective when rich metadata is available, but can struggle to capture the nuances of user preferences that are not reflected in item attributes. In addition, they often lead to a "filter bubble" effect, limiting users to items similar to those they have already interacted with.
- ***Graph-based recommendation systems:*** While these systems can improve recommendations by exploiting relationships, they can also become computationally intensive, especially for large datasets. The complexity of the graph structure can also pose challenges for real-time recommendations.
- ***Reinforcement learning based filtering:*** This approach allows for dynamic adaptation to user preferences over time, but requires significant computational resources and careful tuning of hyper-parameters. In addition, training on sparse data can lead to suboptimal policies.
- ***Federated learning:*** This method improves user privacy by avoiding centralised data storage. However, it can face challenges related to communication efficiency and model convergence over heterogeneous data distributions.
- ***Social network integration:*** While incorporating social context can improve recommendation accuracy, it may raise privacy concerns and require effective data management strategies. In addition, the success of this approach depends on the availability of relevant social data.

There are a number of other limitations to user involvement in the recommendation process:

- Users can sometimes give a rating or take an action that is misinterpreted by the system.
- The ratings given depend on the level and culture of the user, which raises doubts about their credibility.
- Ratings often do not provide enough context to properly understand the situation.
- User ratings can change over time based on their experiences

We have proposed an approach based on deep learning that aims to increase user interaction and participation to increase both the quantity and quality of reviews. This approach thus improves the recommendation system's ability to provide more accurate and reliable recommendations. The objectives of our proposal are

- ***Improving the quality of recommendations:*** By leveraging expert knowledge and reviews, we aim to provide more relevant recommendations tailored to users' needs.
- ***Address sparsity and cold start issues:*** By increasing the amount of available reviews through the active participation of experts, we aim to mitigate the problem of data sparsity and improve the performance of recommendation systems when they are still in the start-up phase or have limited initial data.

## 3  Proposed Approach

The main challenge for collaborative filtering systems is the lack of ratings. To address this problem, we proposed the construction of a network of experts to provide some of the necessary information for the prediction task. This network works on the same principle as an artificial neural network, which is a computational system consisting of multiple layers of interconnected neurons. Each neuron receives input signals, processes them and passes the results on to neurons in subsequent layers, allowing the network to tackle complex problems such as image processing.

Our approach is divided into two steps, as shown in Fig. 1.

- The first step is to provide the necessary information on all available items (ratings, psychological state before using the item). To achieve this, we started to build an expert network consisting of
- Several layers of communities, where each community in layer "X" is connected to all communities in layer "X + 1". The names and characteristics of the communities are defined on the basis of domain ontology.
- Domain experts who are gradually integrated into the communities.

  The expert network works by receiving items, transforming them through several layers of non-linear processing, and producing recommendation information.
- The second step is to use this information to generate recommendations for the users of our system.

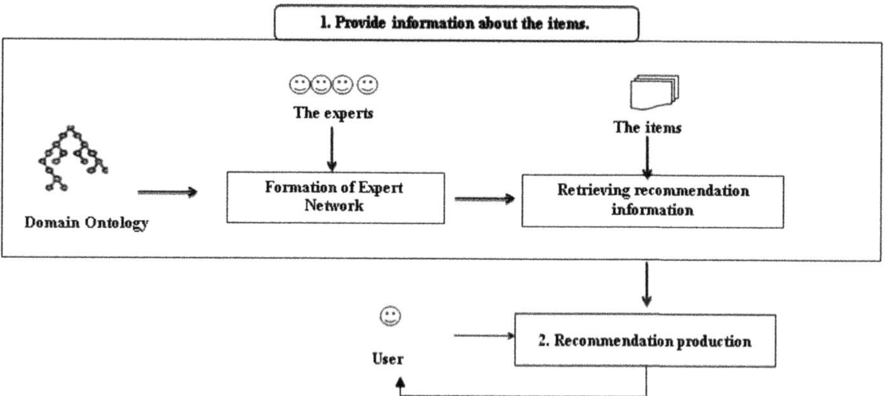


**Fig. 1.** General architecture

## 3.1 Formation of the Expert Network (EN)

The expert network is similar in structure and operation to an artificial neural network. Initially, the network consists of only two layers: an input layer and another containing community. The communities contain experts who are responsible for training the other layers; they have a name, properties and a weight. Weights are determined when inputs are received. The weight plays an important role in determining the community's responsibility for producing the output (the recommendation information) and the transmission output. The transmission output is information that is sent to the next layer and is considered an input for all of these communities, playing the same role as the inputs of the first layer.

### 3.1.1  Construction of the Expert Network (EN)

To construct the EN, we must elaborate on the steps illustrated below (Fig. 2):

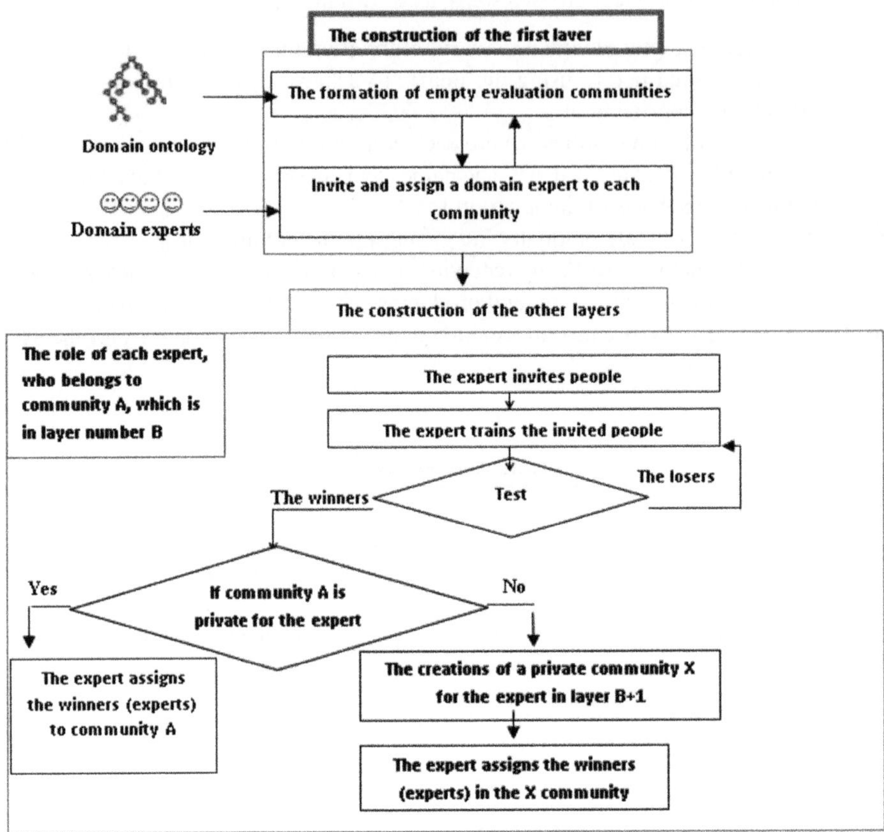


**Fig. 2.** The steps for building an expert network

### A.  The Formation of Evaluation Communities

Ontology's are often used to specify domains of interest. The categorical classes in ontology are groupings of concepts or entities that share similar characteristics or a common set of properties. These categorical classes help to organize and structure knowledge within a given domain. In our approach, we use the name and properties of each category as the name and properties of a second-layer community in our network.

### B.  Invitation and Assignment of Domain Experts in Communities

There are industry experts in every field. An expert is often knowledgeable and passionate about his or her field. They take on the role of a consultant, pointing out "best practices", as well as that of an evaluator and auditor, creating quality and performance indicators in the name of objectivity. The aim of this step is to invite and appoint a domain expert, called an "expert father", in each "CcatI" community, who will be called "Cexpert1".

## C. Reforming Evaluation Communities

The primary role of the experts in each community is to define the characteristics (or acceptance criteria) that individuals must have in order to evaluate items within that community. They must also determine whether it is necessary to divide the main community into sub-communities. The criteria for division (CD) may include factors such as age, gender, language and region. Sub-communities are always in the second layer and replace the main community.

The properties of the sub-communities = properties of the main communities + the division criteria.

Over time, the expert's decision may change. It is possible to decide to cancel the division, which only affects the divided communities, or to subdivide communities that have not yet been divided.

## D. Reassigning Domain Experts in the Communities

For each divided community, the expert decides whether they can act as an expert for one or more sub-communities. The expert will be assigned to the sub communities they have chosen, while other experts will be assigned to the remaining sub-communities.

## E. Constructing the Other Layers

At this stage, the communities are filled with experts and additional layers are added. Initially, it is the role of the parent expert 'Expert I' of each second-layer community '$C_{expert\,I}$' to invite individuals to participate in the evaluation process according to the acceptance criteria they have set. For each person who agrees to participate, the Parent Expert ensures that they meet the criteria. They then train these individuals to become domain experts. At the end of the training, individuals must pass a test to be accepted as an expert. The parent expert assigns the newly selected experts to their community.

The selected individuals are considered to be the child experts of Expert I. The children can also become parents, provided they invite, train and assign other individuals. Each expert belonging to the community "$C_{expert\,I}$" who becomes a parent has his name added to the table PEREC $_{expertI}[]$ of each community. They get their own community with their name, which belongs to the next layer (see Fig. 3). The number of communities in each layer is not necessarily the same.

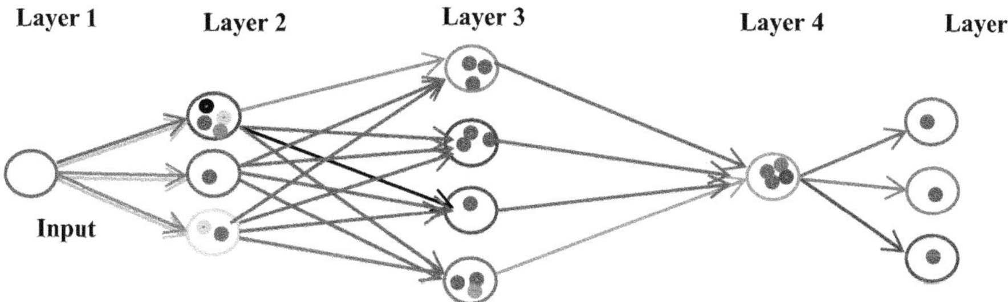


**Fig. 3.** A network of experts

### 3.1.2   Define the Input and Output of Each Community

Each community has input data, a weight, as well as a transmission output and outputs (see Fig. 4).

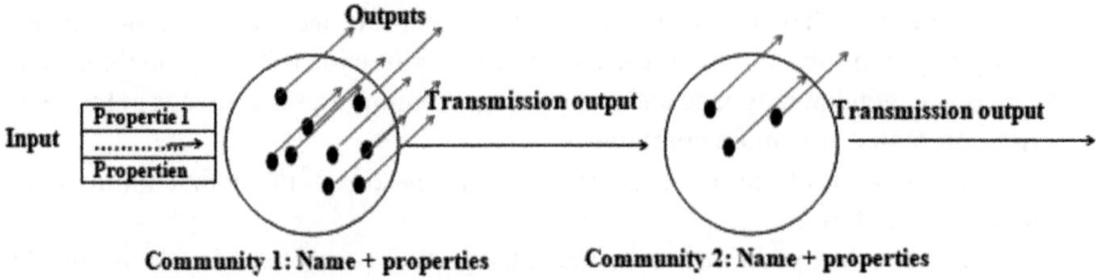


**Fig. 4.**  The Input and Outputs of the Community

- **The input:** There are two types of input:

  - For the first layer, it is an array TABI that containing the properties of item "I" where:$TABI[] = [P_1, P_2, ..., P_n]$ where

    $P_i$: the properties of item I
    n: the number of properties

  - For the other layers, the input consists of the transmission outputs sent by the communities of the previous layer.

- **Weight:** It is a numerical variable equal to 1 or 0. It will be calculated each time an input is received (either a transmission output or the item's properties). Its value determines whether the experts in the active community are responsible for evaluating the item (weight = 1) or not (weight = 0).
- **Transmission Output:** This is information determined on the basis of the value of the weight. It is sent to all communities in the next layer. It is presented in an array containing the names of the parent experts, those who belong to the community that has their own community. It is used to calculate the weight of the receiving community. To define the weight and the transmission output, there is a difference between the communities of the second layer and the other layers.

   For the second layer, the algorithm output-transmission (TAB[n], $TABC_{expertI}$ [m], $PEREC_{expertI}$[T]) is used, and for the others, the algorithm output-transmission2 (Name, Output-transmission[t]) **is** used.

*C $_{expert\ I}$:* the active community.

*Name:* this is "expert I," which represents the name of the expert who formed the community (C $_{expertI}$).

*TABC$_{expertI}$[m]:* this is an array of size 'm' that contains the properties of the community (C $_{expert\ I}$).

*PEREC $_{expert\ I}$ [t]:* this is an array of size 't' that contains the names of the parent experts in the community (C $_{expert\ I}$) who have their own communities bearing their names.

---

***Second Layer:***
**Algorithm output-transmission (TABI[n], TABCexpertI[m], PERECexpertI[t]);**
**Var**
Int S = 0;
// determining the weight value
Int weight = 0;
**Begin**
**// Checking if all the properties of the community exist in the input array**
**for (Int i ;i<=n; i++)**
**for (Int j ;j<=m; j++)**
**If (TABC[j] == TABX[i]) S = S + 1;**
**If (S == m) {**
**// In this case, all experts of this community are responsible for evaluating the item; thus, the value of the weight variable is changed to 1, and the output-transmission array is filled with the names of the parent experts from this community .**
**weight = 1;**
**for (Int i ;i<=t; i++) {**
**output-transmission1[i] = PEREC$_{expertI}$[i];}}**
**End.**

---

***The Other layers:***
**Algorithm output-transmission2 (Name, output-transmission1[t]);**
**// Determining the weight value**
**Var  Int weight = 0, i = 1;**
**Begin**
**// Checking if the name of the active community parent (Name) is present in the received output-transmission1array.**
**While (i <= t) {**
                **if (Name == output-transmission1 [i]) {**
                                **weight = 1;**
                                **i = t + 1 ;}**
                **i++;}**
**If (weight == 1)**
**for ( Int i ; i<=t ; i++) { output-transmission1[i] = PEREC $_{expertI}$[i];}**
**End.**

---

- **Output:** Represents the information needed to recommend items. There are 2 pieces of information determined by the experts (both parents and children):

- **Psychological state** before seeing the item, denoted by "EP," selected from the state table.
- **Numerical evaluation**, denoted by "EN," given after using the item. This value varies between 0 and 5. If the expert likes the item very much, the gives it a 5, otherwise the value decreases until it reaches 0, which means that he dislikes it.

### 3.2  Retrieve Recommendation Information

This is the most important step. The purpose of this step is to retrieve the four recommendation information (output). This process is triggered for two reasons:
Receipt of a new item and assignment of a new expert.

A. *Receiving a new item*: as soon as a new item is received, before sending it, we form TABI[] which represents the input and contains the properties of the received item.Each item is characterized by properties:

ItemProperties = {$Propertie_{I1}$, $Propertie_{I2}$, …, $Propertie_{In}$}
TABI[n] = [$P_1$, $P_2$, …, $P_n$] where $P_1$ = $Propertie_{I1}$, $P_2$ = $Propertie_{I2}$, …, $P_n$ = $Propertie_{In}$.

- For the second layer: TABI[] is sent to all the communities in this layer. The algorithm is triggered for each community:

  Output-transmission (TABI[n], $TABC_{expertI}$[m], $PEREC_{expertI}$[t]);
  To determine:

1. The weight value. If the value of this variable equals 1, then the experts (parents and children) of this community are invited to provide the output (i.e., the recommendation information) for this item. Otherwise, nothing needs to be done.
2. The output of the transmission that is sent to all the communities in the third layer, which represents the input for them.
- From the third layer: For each input received by a community, the algorithm output-transmission2 (Name, output-transmission1 [t]) will be triggered, in the same way as for the second layer, to determine the weight value and the output of the transmission. The latter will be sent to the communities in the next layer.
B. *Assigning a new expert to a community*: After registering a new expert "A" in a community "B," the parent expert "PB" of B sends all the items INB to "A" to retrieve the recommendation information. The order of transmission is defined as follows:
C. Initially, the items are classified into three groups:

- Group 1: Contains items that the majority of experts in "B" liked. That is:
- The number of experts who disliked it < the number of experts who liked it.
- Group 2: Contains items where the number of experts in "B" who liked it is equal to the number of experts in "B" who disliked it.
  The number of experts who disliked it = the number of experts who liked it.

- Group 3: Contains items that are disliked by the majority. That is: The number of experts who did not like it > the number of experts who liked it.
- If the expert likes the item, they give a positive numerical rating (above a given threshold), for example in the range [3–5]. Otherwise they give a negative numerical rating [0–2].

After the parent expert "PB" has started to send the items of group 2 in order to reclassify them into either group 1 or group 3, he then sends the items of group 3, with the possibility of transferring them to the second group. Finally, he sends the items from group 1, so that they either remain in their current group or move to the second group. Finally, after each use of the expert network, each item is categorised and it will have one or more values of EP and EN. It will be assigned to group 1, 2 or 3 based on the average of the numerical ratings. The formula for calculating the average is as follows

$$\text{Average} = \frac{\sum_{i=1}^{n} xi}{n} \qquad (1)$$

where xi: represents each individual value, and n: is the total number of values.

## 3.3  Production of Recommendations

At this stage, there are several scenarios based on the user's needs:

- **Scenario 1:** When the user searches for a category, the system sends him all the items from the first group that belongs to this category, ranked in descending order according to the average numerical ratings of the items.
- **Scenario 2:** If the user includes a psychological state in their search to express their need, the system compares this state with the Psychological State (PS) values of each item using a Boolean function. It will send all items that have an EP value equal to the given state, in order. The system starts with items from Group 1, ranked in descending order according to the ratings, followed by Group 2 and finally Group 3.
- **Scenario 3:** If the user uses numerical ratings in their search, the system will send all items that have the same numerical rating.
- **Scenario 4:** If the user is looking for a specific item, there are two options:

  a. If the item has already been sent to the recommendation engine (RE), it belongs to a category. In this case, the system will recommend all similar items from the same category, ranked in descending order according to the average numerical ratings of the items. The cosine similarity function is used to calculate the similarity

$$.sim(i;j) = \cos(\vec{i},\vec{j}) = \frac{\vec{i}.\vec{j}}{\|\vec{l_i}\|, 2 * \|\vec{j}\|2} \qquad (2)$$

  where $\vec{i},\vec{j}$ are the property vectors of item i and item j, respectively

  b. Otherwise, after determining the category of the item, the system recommends all similar items in the same category, ranked in descending order according to the average numerical ratings of the items, and sends the item to the recommendation engine (RE) for evaluation.

- **Scenario 5:** When the user combines multiple criteria, the system sends items that meet the requested criteria, grouped and in descending order.

## 4  Evaluation and Discussion

This phase aims to analyse various aspects of the idea, including the effectiveness of the model and the quality of the reviews provided. We want to understand how this structure can help to improve the evaluation processes and the interaction between members, while also providing a solution that helps to reduce the problem of insufficient evaluations.

### 4.1  Evaluation of the Model and Techniques Used

- **Network structure:** The artificial neural network model provides an effective framework for organizing the film evaluation process, facilitating the smooth transfer of information between levels for proper processing. Compared to other approaches, this network offers greater flexibility and the ability to handle complex data, thus improving recommendations.
- **Impact of Training on the Quality of Evaluations:**

  - *Increased understanding:* As the level of training increases, members become better equipped to analyse items in depth and to understand the different elements that influence the quality of evaluations. This aspect highlights the superiority of the model over some traditional techniques that lack focus on member training.
  - *Standardisation of criteria:* Good training helps to standardise scoring criteria across members, increasing the accuracy and reliability of scores. Many existing systems rely solely on individual assessments, which can lead to inconsistencies.
  - *Develop critical skills:* Training members to provide constructive feedback can lead to more objective and professional reviews, contributing to improved review quality.
  - *Increase specialisation:* If the training is tailored to each specialisation, it will enable members to focus on all aspects related to the items, leading to more accurate and in-depth assessments.
  - *Increased understanding:* As the level of training increases, members become better equipped to analyse items in depth and to understand the various elements that influence the quality of evaluations. This aspect highlights the superiority of the model over some traditional techniques that lack focus on member training.
  - *Standardisation of criteria:* Good training helps to standardise scoring criteria across members, increasing the accuracy and reliability of scores. Many existing systems rely solely on individual assessments, which can lead to inconsistencies.
  - *Develop critical skills:* Training members to provide constructive feedback can lead to more objective and professional reviews, contributing to improved review quality.
  - *Increase specialisation:* If the training is tailored to each specialisation, it will enable members to focus on all aspects related to the items, leading to more accurate and in-depth assessments.

## 4.2  Evaluation of Quality of the Evaluations Provided

The creation of a network of film rating experts is a strategic step to improve the accuracy and reliability of the recommendations provided to users. The success of this network depends on several key factors:

- *Improving the quality of the ratings:* With a variety of experts involved, the system is able to provide more accurate and comprehensive ratings. Each expert brings a unique perspective that increases the reliability of the recommendations.
- *Increase in true positives:* The use of a network of experts is expected to increase the number of true positives. When films are rated by a wide range of specialists, it is more likely that the recommendations will highlight films that better match the interests of the audience.
- *Reducing false positives:* There is likely to be a reduction in the number of false positives. By providing multiple expert reviews, the quality of recommendations can be improved and inappropriate suggestions can be avoided.
- *Diversity of perspectives:* The presence of experts from different backgrounds and specialities allows for ratings that reflect a diversity of opinions and interests. This diversity can help improve the overall accuracy of recommendations.
- *Address the lack of ratings:* We will use the sparsity rate metric to assess the performance of the system: This metric is commonly used to assess the sparsity of a data matrix, calculated using the formula:

$$\text{Sparsity Rate} = 1 - \frac{\text{Number of non} - \text{empty entries de M}}{n \times m} \times 100 \qquad (3)$$

where M: represents the user-item matrix of dimensions (n × m)

To build our expert network, we used the u.genre file from the MovieLens database (http://www.movielens.org/) to identify the list of available communities. This file outlines the different film genres in the database, covering 18 genres: Action, Adventure, Animation, Comedy, Crime, Documentary, Drama, Children's, Fantasy, Film Noir, War, Horror, Unknown, Musical, Mystery, Western, Thriller and Romance. These 18 genres make up the 18 communities in the second layer of our network; details of how the experts were added and how the other layers were defined can be found in Sect. 3. When new items are introduced, we apply the algorithms described in Sect. 3 to retrieve their recommendation information.

Assuming we have 1,682 items and 918 experts, distributed as follows:

- Action: 682 items and 100 experts
- Adventure: 500 items and 260 experts
- Animation: 500 items and 558 experts

For example, let's calculate the Sparsity Rate for the Action Community by taking a set of scenarios (Table 1):

**Table 1.** Sparsity Rate for the action community.

| Scenario | Sparsity% |
| --- | --- |
| In each community, all experts evaluate 10% of the items | 90% |
| In each community, all experts evaluate 30% of the items. | 70% |
| In each community, all experts evaluate 50% of the items. | 50% |
| In each community, all experts evaluate 70% of the items. | 30% |
| In each community, all experts evaluate 100% of the items. | 0% |

The table shows the variation of the sparsity rate based on the percentage of items evaluated by experts in the action community. The higher the percentage, the denser the matrix and the less sparse it becomes, which has a direct impact on the quality of the recommendations generated by the recommendation system.

*When is this idea effective?*
The idea of increasing the number of expert ratings is most effective when the percentage of rated items exceeds 30%. At this level, the amount of information available is sufficient to overcome sparsity problems and significantly improve the accuracy of recommendations. In addition, the diversity and specialisation of the raters can improve the quality of the ratings, making the system more efficient and reliable.

## 5 Conclusion

In this paper, we presented an overview of collaborative filtering and its drawbacks, in particular the problem of insufficient training data. In recent years, deep learning has attracted considerable interest in various research areas. We proposed to use it to optimise the performance of collaborative filtering systems by constructing a network of experts, which operates similarly to an artificial neural network and plays a role in providing evaluations to generate relevant recommendations. The creation of a network of trained experts to rate films is an innovative step towards improving the quality of ratings and increasing the available data pool. By addressing the problem of poor ratings, this idea can improve the accuracy of the recommendations provided to users, thus contributing to better viewing decisions. In the future, intelligent algorithms can be developed to provide more accurate recommendations. And new tools and methods can be innovated to enhance the film rating process, such as using virtual reality or artificial intelligence for a richer interactive experience.

## References


1. Chae, D.-K., Shin, J.A., Kim, S.-W.: Collaborative adversarial autoencoders: an effective collaborative filtering model under the GAN framework. IEEE Access **7**, 37650–37664 (2019)
2. Chen, R., et al.: A survey on recommendation methods based on social relationships. Electronics **12**, 4564 (2023)

3. Chen, X., Yao, L., McAuley, J., Zhou, G., Wang, X.: Deep reinforcement learning in recommender systems: a survey and new perspectives. Knowl.-Based Syst. **264** (2023)
4. Chronis, C., et al.: A survey on the use of federated learning in privacy-preserving recommender systems. IEEE Open J. Comput. Soc. **5** (2024)
5. Gheewala, S., Xu, S., Yeom, S.: In-depth survey: deep learning in recommender systems— exploring prediction and ranking models, datasets, feature analysis, and emerging trends. Neural Comput. Appl. (2025)
6. Li, C., et al.: Deep learning-based recommendation system: Systematic review and classification. IEEE Access **11** (2023)
7. Ricci, F., Rokach, L., Shapira, B.: Recommender Systems Handbook. Springer (2015)
8. Wu, L., He, X., Wang, X., Zhang, K., Wang, M.: A survey on accuracy-oriented neural recommendation: from collaborative filtering to information-rich recommendation. IEEE Trans. Knowl. Data Eng. (2023)
9. Yang, K., Toni, L.: Graph-based recommendation. In: IEEE Global Conference on Signal and Information Processing (GlobalSIP) (2018). https://doi.org/10.1109/GlobalSIP.2018.864 6359
10. Zhang, S., Yao, L., Sun, A., Tay, Y.: Deep learning based recommender system: a survey and new perspectives. ACM Comput. Surv. **1**(1), Article 1 (2018). https://doi.org/10.1145/000 0001.0000001
11. Zhang, S., et al.: Hybrid recommendation system combining collaborative filtering and content-based recommendation with keyword extraction. Appl. Comput. Eng. **2**(1), 927–939 (2023). https://doi.org/10.54254/2755-2721/2/20220579

# Intrusion Detection in IoMT: DRL Approach for Transmission-Level Security

Benhamza Karima[1]([✉]) [iD], Benselim Rania[2], and Seridi Hamid[1]

[1] LabSTIC, University of 08 Mai 1945, Guelma, Algeria
{Benhamza.karima,Seridi.Hamid}@univ-guelma.dz
[2] Telecommunications Institute, University of 08 Mai 1945, Guelma, Algeria

**Abstract.** As the Internet of Medical Things (IoMT) expands, securing healthcare data has become increasingly vital due to the inherent vulnerabilities associated with wireless communication. Traditional security measures often fall short in effectively mitigating sophisticated cyber threats that target IoMT networks. This study proposes a Deep Reinforcement Learning (DRL)-based Intrusion Detection System (IDS) specifically designed to enhance transmission-level security for IoMT applications. The proposed system utilizes an advanced Deep Q-learning algorithm to develop optimal strategies for dynamic decision-making, allowing for real-time adaptation to emerging threats. Through comprehensive experiments and comparative analyses across IoMT datasets, the proposed DRL-based IDS demonstrates significant improvements in detection accuracy, precision, recall, and response time when compared to conventional IDS methodologies. The findings highlight the potential of employing DRL techniques to substantially strengthen the security posture of IoMT ecosystems, thereby ensuring the integrity and safety of sensitive healthcare data.

**Keywords:** Medical Device Security · IoMT Security · Intrusion Detection Systems (IDS) · Deep Reinforcement Learning (DRL)

## 1 Introduction

The Internet of Medical Things (IoMT) is transforming healthcare by enabling remote, real-time patient monitoring through devices such as wearable sensors and implantable medical devices (Fig. 1). However, IoMT introduces substantial security challenges due to its reliance on inherently vulnerable wireless communication. Major concerns include protecting patient privacy, ensuring data accuracy, and maintaining continuous access to records. To address these challenges, IoMT systems must meet strict security requirements, including freshness to prevent replay attacks, scalability for network growth, non-repudiation to ensure information transmission, and forward secrecy to safeguard data both retrospectively and prospectively [1]. Securing IoMT networks is challenging due to the resource constraints, computational limits, and energy demands of IoMT devices, which increase their vulnerability to malicious activities that can compromise patient data integrity [2, 3].


N. Seddari and M. Redjimi (Eds.): ICMSCT 2024, CCIS 2606, pp. 128–138, 2025.
https://doi.org/10.1007/978-3-032-01922-6_11

Intrusion Detection Systems (IDS) are vital for monitoring network traffic and addressing security threats. Transmission Level Security is essential to protect the confidentiality, integrity, and availability of patient data, critical for maintaining care quality and patient trust [3, 4].

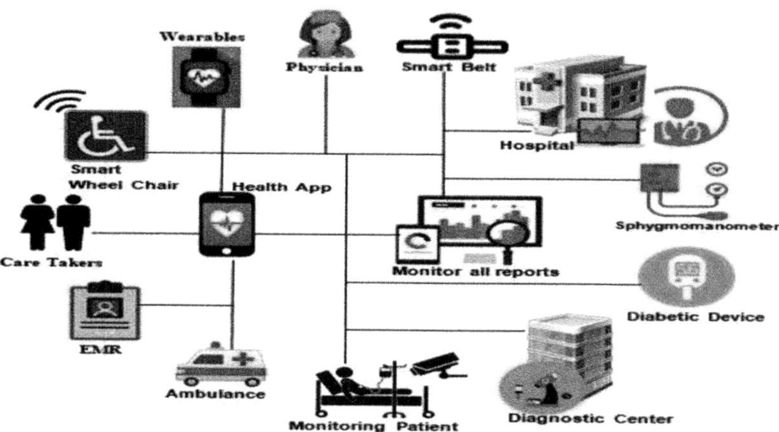


**Fig. 1.** Internet of Medical Things (IoMT) [2].

Traditional security measures, such as signature-based IDS and firewalls, are inadequate for the dynamic and diverse nature of IoMT networks, often failing to detect sophisticated threats like malware and unauthorized access [5, 6]. Furthermore, transmission-level security is crucial for real-time monitoring, particularly in remote settings [7]. The weaknesses in wireless technology and the diverse range of devices increase the attack surface, necessitating robust security mechanisms to counter advanced threats such as eavesdropping, man-in-the-middle attacks, and denial-of-service (DoS) attacks [8, 9].

To address these challenges, several IDS approaches have been proposed to focus on mitigating security risks at the transmission level in IoMT networks [1, 4]. Gao and Thamilarasu developed a Decision Tree-based IDS for anomaly detection in medical devices, achieving high accuracy and faster training but lacking mechanisms to stop ongoing attacks [10]. Al-Shaher et al. designed an Intelligent Healthcare Security System integrating a firewall, IDS, and web filter, achieving a 93% accuracy rate but failing to detect internal attacks [11]. Schneble and Thamilarasu proposed a federated learning-based IDS that enhances privacy but leaves anomaly detection notification unspecified [12]. Other approaches include a stacked Autoencoder for anomaly detection [13], a multi-agent framework using non-linear SVM for remote healthcare systems [14], and deep learning-based IDS [15–18]. While these methods offer strengths like high accuracy and low false positive rates, they also face limitations such as high computational demands and limited privacy protection.

The challenges in IoMT security highlight the need for advanced Intrusion Detection Systems (IDS). We propose a Deep Reinforcement Learning (DRL) approach to enhance IDS adaptability and robustness, specifically for transmission-level security.

The paper covers DRL principles in Sect. 2, the IDS development methodology in Sect. 3, performance evaluation in Sect. 4, and concludes with contributions and future research directions in Sect. 5.

## 2  Deep Reinforcement Learning (DRL)

Reinforcement Learning (RL) [19] is a powerful technique for decision-making in uncertain environments, allowing an agent to learn by interacting directly with its surroundings. Unlike traditional machine learning, RL emphasizes learning from experiences. Key concepts in RL include State (S), Action (A), and Reward (R). Q-learning, a well-known RL method, focuses on maximizing cumulative rewards over time using the Bellman equation [19].

$$Q(S_t, A_t) = E[R_{t+1} + \gamma R_{t+2} + \gamma^2 R_{t+3} + \ldots | S_t, A_t|_i] \tag{1}$$

The discount factor $\gamma \in [0,1]$ manages the importance levels of future rewards. The action value function (Q function) is then defined by:

$$Q^\pi(S, A) = E_\pi \{R_t | S_t = S, A_t = A\} \tag{2}$$

where, $E\pi$ represents the expectation under stochastic policy $\pi$.

The Q function, denoted as $Q^\pi(S,A)$, embodies the anticipated sum of discounted rewards when the agent performs action A while in state S and then adheres to the specified policy $\pi$. In scenarios where states (S) and actions (A) are finite and discrete, a straightforward approach to represent the Q function involves utilizing a table of values for every state-action pair. Initially, this table is arbitrarily initialized, and it evolves through updates based on the agent's experiences as follows:

$$Q(S, A) \leftarrow Q(S, A) + \alpha(R + \gamma^{max}Q * (S', A') - Q(S, A)) \tag{3}$$

where, $Q^*$ is the Q function under the optimal policy $\pi^*$ and $0 < \alpha \leq 1$ is the learning rate.

Tabular Q-learning faces significant convergence challenges in large or continuous state and action spaces. Deep Q-learning (DQN) addresses this by using a deep neural network to approximate Q-values, enabling effective decision-making in high-dimensional environments. DQN represents the Q-function as $Q(s,a,\theta)$, where $s$ is the state, $a$ is the action, and $\theta$ are the network parameters. The network is trained to minimize a loss function based on the Bellman equation [19].

$$L(\theta) = E(s, a, r, s') \sim D[(r + \gamma \max_{a'} Q(S', a'; \theta^-) - Q(s, a; \theta))^2] \tag{4}$$

here $\theta^-$ are the parameters of a target network, and $D$ represents the replay buffer.

In Deep Q-learning (DQN), the replay buffer is essential for storing past experiences $(s,a,r,s')$, which are randomly sampled during training to disrupt correlations between consecutive experiences and enhance learning stability. Through iterative adjustments of network parameters using stochastic gradient descent, DQN effectively learns optimal policies in complex environments [20]. This adaptability makes DQN suitable for various applications, including game playing, robotic control, and network security intrusion detection [21–23].

# 3  Methodology

This research develops and evaluates a Deep Reinforcement Learning-based Intrusion Detection System (DRL-IDS) for the Internet of Medical Things (IoMT). The methodology consists of two key steps: first, a data preprocessing module ensures data quality; second, the processed data is input into the DRL-based anomaly detection system to perform the intrusion detection task (Fig. 2).

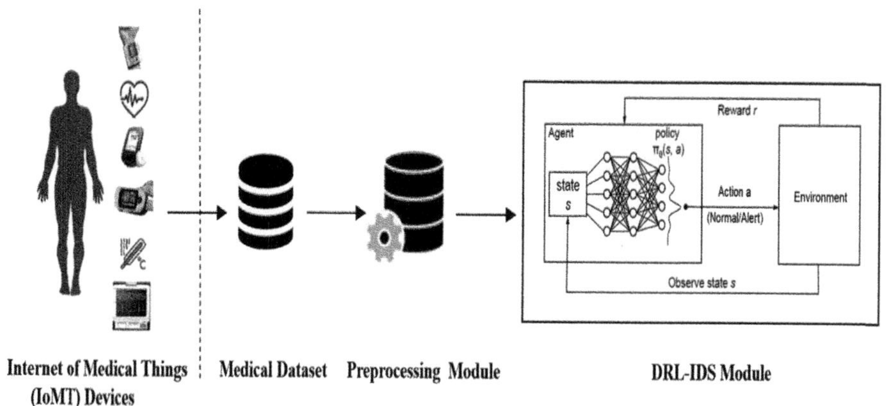


**Fig. 2.** Proposed System Architecture

1. ***Preprocessing Module***: In order to enhance intrusion detection system accuracy and minimize noise and redundancy within medical data, the subsequent steps are undertaken:
   a. *Handling Missing Values:* Deleting data with missing values exceeding a certain threshold (70% in this case) is a valid approach, maintaining data integrity and model performance.
   b. *Correlation-Based Feature Elimination:* Using the Pearson correlation coefficient [24] to remove one feature from strongly correlated pairs helps reduce multicollinearity, improving model performance.
   c. *Feature Importance Ranking:* LightGBM algorithm [25] is used for feature selection due to its ability to handle large datasets efficiently. Incrementally adding features based on importance scores until performance plateaus is a sound method [26].
   d. *Normalization:* Min-max normalization is a standard practice to ensure all features are on the same scale, which is crucial for many machine learning algorithms [27].
      Once the data preprocessing step has been meticulously completed to diminish noise, redundancy, and maintain the integrity and consistency of medical data, the proposed DRL-IDS system is deployed.
2. ***DRL-IDS Module***: The proposed Deep Reinforcement Learning-based Intrusion Detection System (DRL-IDS) is designed to detect intrusions in IoMT networks using a Deep Q-learning Network framework. This system operates in two primary

phases: the *Learning Phase* and the *Detection Phase,* which allow the DRL-IDS to dynamically adapt to real-time threats and changing network conditions.

The architecture processes both network and biometric data, providing comprehensive analysis and accurate intrusion detection capabilities. The DRL framework phases are described below:

- **Learning Phase**

  - The DRL agent interacts with historical IoMT data, which includes features such as network traffic metrics and biometric data.
  - The agent's goal is to learn an optimal policy for classifying data as Normal or Alert (intrusion detected).
  - A reward system guides the learning process:

    - Positive rewards for correct classifications.
    - Negative rewards for incorrect classifications.

  - The policy is updated using the Bellman equation to adjust Q-values for state-action pairs.
  - Experience replay is used to stabilize training by sampling batches of stored experiences (state-action-reward tuples) in a replay buffer. The agent samples batches of experiences randomly during training, breaking the correlation between consecutive experiences and improving generalization.

- **Detection Phase**

  - The trained agent monitors live IoMT data in real-time, evaluating state variables to detect anomalies.
  - The system dynamically adapts to emerging threats by periodically updating its policy based on new data.
  - Adaptive thresholding helps mitigate false positives by adjusting detection sensitivity according to recent alert patterns.

The system architecture is built on deep Q-learning with a Multi-Layer Perceptron (MLP) neural network, featuring an input layer, three hidden layers with ReLU activation functions [28], and an output layer for attack identification. This configuration ensures the neural network's efficiency and effectiveness in detecting intrusions:

- *Input Layer*

  - Receives a vector of features from network and biometric data.
  - Input data is preprocessed using normalization techniques to standardize feature ranges.

- *Hidden Layers*

  - The architecture includes three hidden layers with ReLU activation functions:

- These layers enable the network to learn complex patterns indicative of intrusions.

- *Output Layer*

  - Consists of two neurons corresponding to possible actions: Normal and Alert while retaining flexibility in handling emerging threats (Data Breach and Denial of Service). The system can continue to learn and improve its ability to detect various types of intrusions as new data and threats emerge in the IoMT environment.

## 4   Results and Discussion

The performance of the proposed Deep Reinforcement Learning-based Intrusion Detection System (DRL-IDS) is evaluated in this section, which includes a detailed description of the datasets utilized, comparative analyses with state-of-the-art methods, and insights gained from extensive experimentation. The experiments were conducted on a Windows 10 system equipped with an Intel Core i5-7200U CPU operating at 2.50 GHz and 16 GB of RAM.

### 4.1   Dataset Description

For this study, we utilized two publicly available IoMT datasets, each characterized by distinct features and attack scenarios:

a. *WUSTL-EHMS-2020:* Created by Hady et al. [3], this dataset was derived from a specialized testbed designed to simulate an IoMT environment. The testbed incorporates a multi-sensor board to measure various biometric parameters of the patient's body. The WUSTL-EHMS-2020 dataset comprises 14,272 normal instances and 2,046 attack instances, with a total of 36 features, including both network-related and biometric data. The dataset encompasses two categories of attacks: spoofing attacks and data alteration, where attackers intercept, modify, or spoof packets and redirect the altered data to the server.

b. *ECU-IoHT:* Developed by Ahmed et al. [29], this dataset was generated using the healthcare monitoring kit MySignals, which comprises multiple components and sensors that collect various biometric measurements, such as temperature, blood pressure, and heart rate, and transmit this data to a cloud server for analysis. The ECU-IoHT dataset contains 87,754 normal instances and 23,453 attack instances, with seven network-related features, including source IP, destination IP, protocol type, and attack type. It covers four types of attacks: ARP spoofing, Denial of Service (DoS), Nmap port scanning, and Smurf attacks, which were simulated by targeting the IoMT devices within the test environment.

The datasets were divided into training and validation subsets, with 70% of the data used for training the model and 30% for validating its performance. The model architecture was iteratively refined based on prediction errors computed through the loss function.

## 4.2   Performance Evaluation Metrics

To ensure a comprehensive assessment of the proposed DRL-based Intrusion Detection System (IDS), we employed six key evaluation metrics, which are widely used in the field of intrusion detection:

- *Accuracy:* Measures the proportion of correctly classified instances (normal or attack) out of the total instances. It provides an overall indication of the model's effectiveness.
- *Precision:* Indicates the ratio of correctly predicted attack instances to the total predicted attack instances, thus measuring the system's ability to minimize false alarms.
- *Recall:* Reflects the model's capability to detect all actual attack instances. High recall indicates a low rate of false negatives, which is crucial in healthcare contexts where undetected threats could compromise patient safety.
- *F1 Score:* The harmonic mean of precision and recall, providing a balance between the two metrics, especially useful in cases of class imbalance.
- *False Positive Rate (FPR):* Represents the proportion of normal instances that were incorrectly classified as attacks, with a lower FPR indicating better model reliability.
- *Area Under the Receiver Operating Characteristic Curve (AUC-ROC):* Evaluates the model's ability to distinguish between attack and normal instances, with higher values reflecting superior discrimination capabilities.

The proposed DRL-based IDS was tested on two datasets, WUSTL-EHMS-2020 and ECU-IoHT, which are representative of real-world IoMT environments, as detailed in previous studies by Hady and Ahmed. These datasets include a variety of attack types and features, providing a challenging environment for testing the system's robustness. Table 1 presents the performance metrics for the proposed DRL-IDS across the two datasets.

**Table 1.** Performance Evaluation

| Dataset | Accuracy | Precision | Recall | F1 Score | False Positive Rate | AUC-ROC |
|---|---|---|---|---|---|---|
| WUSTL-EHMS | 0.980 | 0.981 | 0.976 | 0.978 | 0.102 | 0.964 |
| ECU-IoHT | 0.978 | 0.976 | 0.974 | 0.975 | 0.104 | 0.961 |

The results demonstrate consistent performance across both datasets, validating the robustness and generalization capabilities of the proposed DRL-IDS in handling diverse attack types and network conditions. The high accuracy and F1 scores indicate that the system effectively balances the trade-off between detecting intrusions and minimizing false alarms. The AUC-ROC values above 0.96 further underscore the model's strong ability to differentiate between normal and malicious traffic.

## 4.3   Sensitivity Analysis of Hyperparameters

We conducted a sensitivity analysis to evaluate the impact of key hyperparameters on the performance of the proposed DRL-IDS. The hyperparameters assessed include the learning rate, discount factor, and the number of hidden layers. The aim was to identify the optimal settings that maximize model performance across various metrics, such as accuracy, precision, and AUC.

- *Learning Rate ($\alpha$):* We tested learning rates ranging from 0.001 to 0.01. The analysis showed that a learning rate of 0.005 provided the best balance between convergence speed and stability, leading to an improvement in accuracy by 2.4% over the initial settings.
- *Discount Factor ($\gamma$):* The discount factor was varied from 0.8 to 0.99 to observe its effect on reward accumulation. A discount factor of 0.95 yielded the most favorable results, providing a 1.8% increase in AUC by achieving a suitable trade-off between long-term and immediate rewards.
- *Number of Hidden Layers:* Configurations with one to three hidden layers were examined to assess the network's ability to learn complex patterns in the data. The best results were achieved with three hidden layers, resulting in a 3.2% improvement in precision, as this configuration allowed the model to capture intricate relationships without overfitting.

The results of the sensitivity analysis are summarized in Table 2, showing the range of values tested, the optimal settings, and the observed improvements in performance metrics.

**Table 2.** Sensitivity Analysis Results

| Hyperparameter | Range Tested | Optimal Value | Improvement Observed |
|---|---|---|---|
| Learning Rate ($\alpha$) | 0.001–0.01 | 0.005 | +2.4% in accuracy |
| Discount Factor ($\gamma$) | 0.8–0.99 | 0.95 | +1.8% in AUC |
| Number of Hidden Layers | 1–3 | 3 | +3.2% in precision |

## 4.4  Comparative Analysis

The comparative analysis presented in Table 3 evaluates the performance of the proposed Deep Reinforcement Learning Intrusion Detection System (DRL-IDS) against existing state-of-the-art methods, including Random Forest, Support Vector Machine (SVM), Artificial Neural Network (ANN), and XGBoost, as applied to the WUSTL-EHMS-2020 datasetThese methods were selected due to their frequent use in intrusion detection research and their relevance as benchmarks.

**Table 3.** Comparative Analysis

| Model | Accuracy | Predictive time |
|---|---|---|
| Random forest | 0.9389 | 2,35 |
| SVM | 0.9274 | 1,89 |

*(continued)*

Table 3. (continued)

| Model | Accuracy | Predictive time |
|---|---|---|
| ANN | 0.9218 | 1,65 |
| XGBoost | 0.9381 | 1,62 |
| **Proposed model** | **0.9801** | **1.23** |

The findings clearly demonstrate that the proposed DRL-IDS outperforms existing methods in both accuracy and predictive time. The accuracy improvement ranges from 4.20% to 5.83% across the benchmarks, with a reduction in predictive time from 0.39 to 1.12 s. This indicates a more efficient detection process suitable for real-time IoMT environments. The significant enhancement in accuracy validates the system's ability to detect and classify intrusions accurately while maintaining quick response times, which is critical for ensuring the security of healthcare systems.

### 4.5  Discussion of Results

The consistent performance across multiple datasets and the comparative advantage over traditional machine learning techniques suggest that the proposed DRL-IDS is robust and adaptable. The WUSTL-EHMS-2020 dataset, which combines network and biometric data, poses unique challenges due to the diversity of features and attack types. The proposed model's high recall and low false positive rate indicate effective detection of complex attacks. Similarly, the ECU-IoHT dataset, which features common IoMT cyber threats demonstrates that the DRL-based IDS can generalize well to different attack scenarios.

The results indicate that the proposed DRL-IDS provides a significant advantage over conventional approaches, allowing for the dynamic adaptation of the model to evolving threats. This adaptability is especially beneficial in IoMT settings, where cyber threats continuously change.

## 5  Conclusion

The Internet of Medical Things (IoMT) presents significant security challenges, including real-time attack detection, safeguarding patient privacy, and identifying novel threats. These complexities highlight the need for ongoing research and innovation in Intrusion Detection Systems (IDS). To address these challenges, this study introduces a Deep Reinforcement Learning (DRL)-based approach for enhancing transmission-level security in IoMT networks. The proposed DRL-IDS is designed to improve the adaptability and robustness of IDS solutions, effectively countering modern intrusion threats. Compared to existing state-of-the-art methods, the proposed model demonstrated notable improvements in performance, particularly in detection accuracy and response time. This research contributes to advancing security solutions specifically tailored to the unique demands of IoMT environments, thereby enhancing the overall safety and reliability of healthcare systems.

Deploying the DRL-IDS in real-world healthcare environments involves several challenges, such as computational limitations of resource-constrained IoMT devices, network latency, and device heterogeneity, which can affect real-time performance. The proposed system addresses these issues by employing lightweight architectures, edge computing strategies, and adaptive thresholding to dynamically adjust detection sensitivity and reduce false positives. Additionally, techniques like experience replay and model optimization ensure the system meets the demands of real-time operation while maintaining high detection accuracy. These strategies make the DRL-IDS a flexible solution capable of integrating seamlessly across diverse IoMT networks.

Future research could focus on further enhancing the system's capabilities by combining the DRL-IDS with other security mechanisms, such as encryption and authentication, to create a multi-layered defense strategy. Additionally, deploying the system in real-world healthcare settings and conducting extensive testing under various operational conditions will be crucial for evaluating its practical effectiveness and refining its performance. These efforts will pave the way for a more comprehensive and scalable security solution for the rapidly evolving IoMT ecosystem.

# References


1. Si-Ahmed, A., Al-Garadi, M.A., Boustia, N.: Survey of Machine Learning based intrusion detection methods for Internet of Medical Things. Appl. Soft Comput. **140**, 110227 (2023)
2. Dilawar, N., Rizwan, M., Ahmad, F., Akram, S.: Blockchain: securing internet of medical things (IoMT). Int. J. Adv. Comput. Sci. Appl. **10**(1), 82–89 (2019)
3. Hady, A.A., Ghubaish, A., Salman, T., Unal, D., Jain, R.: Intrusion detection system for healthcare systems using medical and network data: a comparison study. IEEE Access **8**, 106576–106584 (2020)
4. Hernandez-Jaimes, M.L., Martinez-Cruz, A., Ramírez-Gutiérrez, K.A., Feregrino-Uribe, C.: Artificial intelligence for IoMT security: a review of intrusion detection systems, attacks, datasets and Cloud-Fog-Edge architectures. Internet Things 100887 (2023)
5. Binbusayyis, A., Alaskar, H., Vaiyapuri, T., Dinesh, M.J.T.J.O.S.: An investigation and comparison of machine learning approaches for intrusion detection in IoMT network. J. Supercomput. **78**(15), 17403–17422 (2022)
6. Zachos, G., Essop, I., Mantas, G., Porfyrakis, K., Ribeiro, J.C., Rodriguez, J.: An anomaly-based intrusion detection system for internet of medical things networks. Electronics **10**(21), 2562 (2021)
7. Koutras, D., Stergiopoulos, G., Dasaklis, T., Kotzanikolaou, P., Glynos, D., Douligeris, C.: Security in IoMT communications: a survey. Sensors **20**(17), 4828 (2020)
8. Ghubaish, A., Salman, T., Zolanvari, M., Unal, D., Al-Ali, A., Jain, R.: Recent advances in the internet-of-medical-things (IoMT) systems security. IEEE Internet Things J. **8**(11), 8707–8718 (2020)
9. Perwej, Y., Akhtar, N., Kulshrestha, N., Mishra, P.: A Methodical Analysis of Medical Internet of Things (MIoT) security and privacy in current and future trends. J. Emerg. Technol. Innov. Res. **9**(1), d346–d371 (2022)
10. Gao, S., Thamilarasu, G.: Machine-learning classifiers for security in connected medical devices. In: 2017 26th International Conference on Computer Communication and Networks (ICCCN), pp. 1–5. IEEE (2017, July)
11. Al-Shaher, M.A., Hameed, R.T., Țăpuș, N.: Protect healthcare system based on intelligent techniques. In: 2017 4th International Conference on Control, Decision and Information Technologies (CoDIT), pp. 0421–0426. IEEE (201)

12. Schneble, W., Thamilarasu, G.: Attack detection using federated learning in medical cyber-physical systems. In: Proceedings of the 28th International Conference on Computing Communication Network (ICCCN), vol. 29, pp. 1–8 (2019)

13. He, D., et al.: Intrusion detection based on stacked autoencoder for connected healthcare systems. IEEE Netw. **33**(6), 64–69 (2019)

14. Begli, M., Derakhshan, F., Karimipour, H.: A layered intrusion detection system for critical infrastructure using machine learning. In: 2019 IEEE 7th International Conference on Smart Energy Grid Engineering (SEGE), pp. 120–124. IEEE (2019)

15. Rm, S.P., et al.: An effective feature engineering for DNN using hybrid PCA-GWO for intrusion detection in IoMT architecture. Comput. Commun. **160**, 139–149 (2020)

16. Gupta, L., Salman, T., Ghubaish, A., Unal, D., Al-Ali, A.K., Jain, R.: Cybersecurity of multi-cloud healthcare systems: a hierarchical deep learning approach. Appl. Soft Comput. **118**, 108439 (2022)

17. Singh, P., Gaba, G.S., Kaur, A., Hedabou, M., Gurtov, A.: Dew-cloud-based hierarchical federated learning for intrusion detection in IoMT. IEEE J. Biomed. Health Inform. **27**(2), 722–731 (2022)

18. Hameed, S.S., Selamat, A., Latiff, L.A., Razak, S.A., Krejcar, O.: WHTE: Weighted Hoeffding tree ensemble for network attack detection at Fog-IoMT. In: International Conference on Industrial, Engineering and Other Applications of Applied Intelligent Systems, pp. 485–496. Springer International Publishing, Cham (2022)

19. Szepesvári, C.: Algorithms for reinforcement learning. Springer nature (2022)

20. Hester, T., et al.: Deep q-learning from demonstrations. In: Proceedings of the AAAI Conference on Artificial Intelligence, vol. 32, no. 1 (2018)

21. Hausknecht, M., Stone, P.: Deep recurrent q-learning for partially observable mdps. In: 2015 AAAI Fall Symposium Series (2015)

22. Tai, L., Liu, M.: A robot exploration strategy based on q-learning network. In: 2016 IEEE International Conference on Real-time Computing and Robotics (RCAR), pp. 57–62. IEEE (2016)

23. Alavizadeh, H., Alavizadeh, H., Jang-Jaccard, J.: Deep Q-learning based reinforcement learning approach for network intrusion detection. Computers **11**(3), 41 (2022)

24. Sedgwick, P.: Pearson's correlation coefficient. Bmj 345 (2012)

25. Ke, G., et al.: Lightgbm: A highly efficient gradient boosting decision tree. In: Advances in Neural Information Processing Systems, vol. 30 (2017)

26. Li, Z.S., Yao, X., Liu, Z.G., Zhang, J.C.: Feature selection algorithm based on LightGBM. J. Northeastern Univ. (Nat. Sci,) **42**(12), 1688 (2021)

27. Patro, S.: Normalization: A preprocessing stage (2015). arXiv preprint arXiv:1503.06462

28. Agarap, A.F.: Deep Learning Using Rectified Linear Units (ReLU) (2018). arXiv preprint arXiv:1803.08375

29. Ahmed, M., Byreddy, S., Nutakki, A., Sikos, L.F., Haskell-Dowland, P.: ECU-IoHT: a dataset for analyzing cyberattacks in Internet of Health Things. Ad Hoc Netw. **122**, 102621 (2021)

# ECG-Based Person Identification Using the Shape Exchange Algorithm

Imen Boulnemour[1,2(✉)], Abdelmadjid Lahreche[3], Kenza Redjimi[1], Seloua Hadiby[1], Noureddine Seddari[1,4], and Chaima Ghis[1]

[1] LICUS Lab, Department of Computer Science, Université 20 Août 1955, Skikda, Algeria
{i.boulnemour,k.redjimi,n.seddari}@univ-skikda.dz
[2] LRES Lab, Department of Computer Science, Université 20 Août 1955, Skikda, Algeria
[3] Department of Computer Science, University of Science and Technology Houari Boumédiène, Algies, Algeria
[4] Future Technology Lab, University of Parma, Parco Area Delle Scienze 181/A, Parma, Italy

**Abstract.** Electrocardiogram (ECG) signal represents a promising and emerging technology for person identification. In this work, several similarity measures with the K-nearest neighbours (KNN) classifier in biometrics will be applied and tested. The tested measures include Shape Exchange Algorithm (SEA), Complexity Invariant Distance, Dynamic Time Warping (DTW) and its variations CIDDTW (Complexity Invariant Distance Dynamic Time Warping) and QPDTW (Quasi-Periodic Time Warping). The biometric techniques primarily rely on using these similarity measures to compare ECG signals without the need to segment them into cycles or extract their features. The proposed biometric algorithms using an international database of ECG signals called MIT/BIH are validated. The obtained results show that the SEA method provides the best results, followed by the hybrid methods QP-DTW and the CID-DTW, which yield better results compared to the DTW and CID methods.

**Keywords:** Biometrics · Similarity measure · ECG · SEA · QPDTW · CIDDTW · DTW · CID · KNN

## 1 Introduction

The biometrics community is booming in its search for new features that are difficult to imitate. One of these features is the electrocardiogram (ECG) signal. ECG is an internal bio-signal, which uniquely characterizes each person and which is difficult to modify manually. The use of the ECG as a biometric measurement is relatively new. The first approach was proposed in 1999 by Biel et al. [1]. This method is based on analyzing the landmarks of the ECG and has enabled the implementation of the SIMCA (Soft Independent Modeling of Class Analogy) method for classification. Later, in 2006, Wang et al. [2] introduced a novel biometric identification method. It relies on fiducial detection using the Discrete Cosine Transform (DCT) of the autocorrelation function. This method incorporates analytical and appearance features extracted from ECG signals for

© The Author(s), under exclusive license to Springer Nature Switzerland AG 2025
N. Seddari and M. Redjimi (Eds.): ICMSCT 2024, CCIS 2606, pp. 139–149, 2025.
https://doi.org/10.1007/978-3-032-01922-6_12

human identification [2]. Recently, researchers have been focused on improving classification methods. Several studies have relied on machine learning algorithms. Samaneh Kouchaki et al. [3] utilized a nearest neighbour classifier (1NN) employing empirical mode decomposition (EMD) and Hilbert transforms for ECG-based identification. Sidek et al. [4] also employed the KNN classifier using the Normalize Convolute-Normalize (NCN) method to identify individuals with abnormal cardiac conditions using electrocardiogram (ECG) signals in network environments. Rui Lia et al. [5] utilized the k-nearest neighbour's algorithm in conjunction with two techniques: generalized non-negative matrix factorization (GNMF) and sparse representation. GNMF enables the reduction of signal or data dimensionality by eliminating redundant or insignificant elements.

In conjunction with the K-nearest neighbours (KNN) classifier for person identification by ECG, in this study, various similarity measures are examined and evaluated. The similarity measures under investigation include the Shape Exchange Algorithm (SEA) [6], Complexity Invariant Distance (CID) [7], Dynamic Time Warping (DTW) [8, 9], along with its variations CID-DTW (Complexity Invariant Distance Dynamic Time Warping) [10] and QP-DTW (Quasi-Periodic Time Warping) [11] These measures represent advancements over DTW and the Euclidean distance. To assess the performance of the proposed biometric algorithms, we will validate them using the MIT/BIH international database, which contains ECG signals.

The rest of this paper is organized as follows: Sect. 2 exposes some backgrounds,and Sect. 3 is devoted to presenting the related work. The results and discussion are presented in Sect. 4 and Sect. 5 concludes this work.

## 2  Background

### 2.1  The SEA Method

Only two levels of headings should be numbered. Lower level headings remain unnumbered; they are formatted as run-in headingsThe Shape Exchange Algorithm (SEA) method, introduced by Boucheham [6], is renowned for its efficacy in aligning quasi-periodic time series (QPTS). This technique utilizes signature exchange between the series, where signatures are generated by sorting the values of the time series based on their amplitudes. In the subsequent discussion, we will outline the general functioning of the SEA method. Consider two time series, X and Y. The underlying principle of the SEA method is as follows:

- Rearrange the elements of X and Y based on their magnitude indexes, considering that each magnitude entry corresponds to a temporal entry.
- Perform an exchange of sorted magnitude indexes between X and Y while preserving their original temporal indexes, which are currently in a disordered state. In this process, linear interpolation is used to adjust the magnitude indexes and align them with the desired target temporal indexes, especially when the lengths of X and y are unequal.
- Create reconstructed versions of X and Y, denoted as XREC and YREC, respectively, by sorting them based on their temporal indexes.
- Compare X and XREC, as well as Y and YREC, by utilizing a distance measure such as the Euclidean Distance, which is employed in our approach.

## 2.2   The CID Method

The CID method (Complexity-Invariant Distance) [7, 10] calculates the dissimilarity between two-time series of equal length by computing a correction to the Euclidean distance based on the estimation of the complexity of the series. It involves the following steps:

- First, the Euclidean distance is calculated between the two numerical series. The Euclidean distance measures the similarity between the series but does not take into account their respective complexity.
- Next, series complexity is estimated using methods like Kolmogorov complexity, entropy measures, or fractal dimension, allowing for relative complexity comparison.
- The Euclidean distance is adjusted to accommodate complexity differences between series, ensuring its invariance concerning series complexity.
- The CID distance is obtained by multiplying the Euclidean distance by a complexity correction factor (Eq. 1), and it measures dissimilarity by considering numerical and complexity differences.

$$CID(Q, C) = ED(Q,\ C) \times CF(Q, C) \tag{1}$$

where CF is a complexity correction factor defined as:

$$CF(Q, C) = \frac{\max(CE(Q), CE(C))}{\min(CE(Q),\ CE(C))} \tag{2}$$

And ED is the Euclidean distance defined as:

$$ED(Q, C) = \sqrt{\sum_{i=1}^{n} (Q_i - C_i)^2} \tag{3}$$

CE (T) is an estimation of the complexity of a time series T.

$$ED(x, y) = \sqrt{\sum_{i=1}^{n} (x_i - y_i)^2} \tag{4}$$

where: x and y are vectors.

## 2.3   The C. DTW Method

The DTW (Dynamic Time Warping) [8, 9] is considered the best similarity measure for many applications in the field of time series [12]. It minimizes the effects of time shift and temporal distortion by allowing an "elastic" transformation of time series. This transformation enables the detection of similar patterns with different phases. Instead of comparing each point of one series to the corresponding point in the other series at the same time instant, as in Euclidean distance, DTW allows the comparison of each point in one series with one or more points in the other series.

## 2.4   The QP-DTW Method

The QP-DTW method proposed by Boulenemour and Boucheham [11] is a fusion of the DTW and SEA methods. It combines the capacity of the SEA method in aligning quasiperiodic time series and the alignment power of DTW for time series shifted along the Xaxis [11]. This led us to test the QP-DTW in the field of biometrics. QP-DTW method is described in Fig. 1.

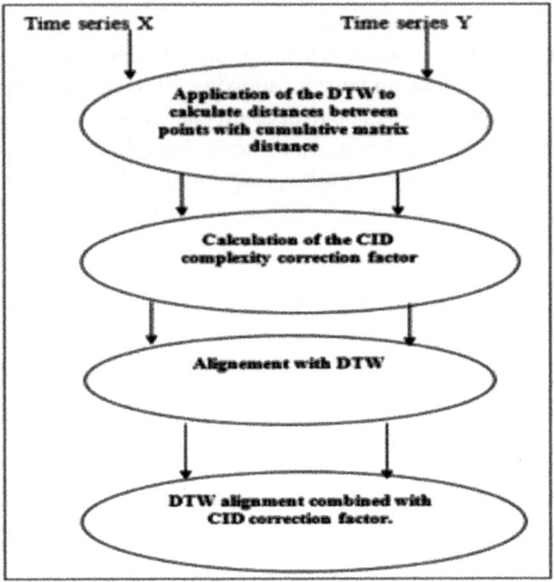


**Fig. 1.** Illustrative diagram of QP-DTW method.

## 2.5   The CID-DTW

The CID-DTW (Complexity-Invariant Dynamic Time Warping) is a method that combines the DTW distance measure with the complexity correction factor of the CID method [7, 10]. It enables a fair comparison of complex time series data, like ECG, by aligning and adjusting the series in a non-linear way [10]. The CID-DTW method is described in Fig. 2.

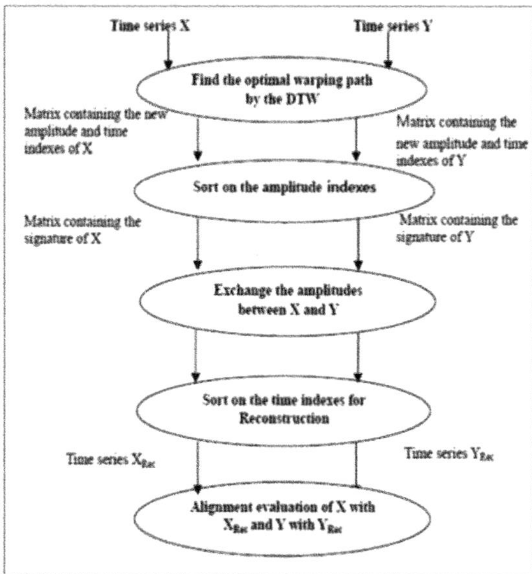



**Fig. 2.** Illustrative diagram of CID-DTW method.

# 3   Related Work

## 3.1   The DTW-FLDA

The FLDA (Fisher's Linear Discriminant Analysis) [13] aims to identify a set of projection vectors W that optimally distinguish between different classes. This is accomplished by maximizing the Fisher criterion, which involves the ratio of the determinant of the betweenclass scatter matrix $S_w$ (4) and (5).

$$W = \arg \max \frac{\left|w^t s_b w\right|}{\left|w^t s_w w\right|} \tag{5}$$

Sw and Sb for L classes are defined as follows (5) and (6):

$$S_w = \sum_{i=1}^{L} \sum_{k \in C_i} (x_k - m_i)^{\tau} \tag{6}$$

$$S_b = \sum_{i=1}^{L} n_i (m_i - m)(m_i - m)^{\tau} \tag{7}$$

In this context, x is the training vector, m is the overall dataset mean, and $m_i$ represents the mean ECG feature set for the class $c_i$, which contains $n_i$ samples. The projection matrix W can be determined from the eigenvectors of $S_w^{-1} S_b$. [13]. In this study, the researchers extracted nine feature parameters from ECG signals in the spatial domain for classification purposes. For person identification, they employed Dynamic Fisher's Linear Discriminant Analysis (FLDA) combined with a K-Nearest Neighbor Classifier (K-NNC) in a single-stage classification, achieving a recognition accuracy of 96%.

To further enhance the system's performance, a two-stage classification technique was adopted. In the first stage, FLDA was used with K-NNC, followed by a Dynamic Time Warping (DTW) classifier in the second stage, resulting in a 100% recognition accuracy. For person authentication, a QRS complex-based threshold technique was utilized. The overall system performance was verified on the MIT-BIH normal database, consisting of 375 recordings from 15 individuals, achieving 96% accuracy in both legitimate and intruder scenarios [14].

## 3.2 The DTW- PLR

The Piecewise Linear Representation (PLR) [15] method effectively extracts the key features of time series data and reduces its dimensionality. The core principle of PLR is to ignore minor fluctuations while preserving significant minima and maxima. The parameter Z regulates the compression rate: increasing Z results in fewer selected points, and decreasing Z leads to more points being chosen. By linking these important points with line segments, we obtain a piecewise linear representation of the time series.

In this study, the researchers propose a PLR-DTW method for ECG biometrics. The Piecewise Linear Representation (PLR) is employed to retain essential information of an ECG signal segment while reducing the data dimension when necessary. Dynamic Time Warping (DTW) is then used to measure the similarity between two signal segments. For the MIT-BIH Normal Sinus Rhythm database, the researchers selected subjects without significant arrhythmias, with the database sampled at 128 Hz. Fourteen subjects were chosen based on the length of the records. Since each subject has only one record, the first 20 s of each record were used to create the enrollment template, and the last 20 s were reserved for testing. Performance evaluation was conducted on three ECG databases, comparing the proposed method with an existing method that uses wavelet coefficients and has proven accuracy. The analysis results demonstrate that the PLR-DTW method achieves an identification accuracy rate of 100% [16].

## 3.3 The Multilayer Perceptron and Radial Basis Function Neural Networks

The authors of this study have developed a new method for identifying individuals usingtheir electrocardiograms (ECGs), specifically utilizing the QRS complex as a biometric feature. They extracted 324 QRS complexes from 18 subjects in the Normal Sinus Rhythm Database (NSRDB) and employed Multilayer Perceptron (MLP) and Radial Basis Function (RBF) neural networks to classify these QRS complexes. The method demonstrated classification accuracy rates of over 98% with MLP and 97% with RBF, provided that the training data was carefully selected to cover a wide range of input values.

This ECG-based identification method offers advantages over existing methods. However, there are limitations and drawbacks, such as the specificity of the results to the use of QRS complexes and uncertainty about the neural networks' ability to generalize. Careful selection of training data is crucial for achieving accurate results. This method could be useful for securing computer systems and sensitive applications [17].

### 3.4   The Empirical Mode Decomposition and Hilbert Transform

The authors of this study have introduced an individual identification method based on the analysis of electrocardiograms (ECGs) using Empirical Mode Decomposition (EMD) and the Hilbert transform. They used a database of 20 healthy subjects from the PhysikalischTechnische Bundesanstalt (PTB) to decompose the signals into their fundamental components and calculate the instantaneous frequency of the final component.

By employing the nearest neighbor classifier (1NN), they achieved a high identification rate of 93.22%. The authors suggested enhancing the EMD algorithm by incorporating additional temporal features to produce more independent and orthogonal components, thereby reducing the computational load required for the decomposition process. They concluded that their method applies to various applications such as security and access control. However, a drawback is that it requires a high-dimensional feature space and substantial computational resources to determine the distance between points [3].

### 3.5   Abnormal Cardiac Conditions in Remote Monitoring System

This study presents the use of the Normalize-Convolute-Normalize (NCN) method for identifying individuals using electrocardiogram (ECG) signals with abnormal cardiac conditions in network environments. The abnormal signals are normalized using a technique proposed by the authors. Experiments conducted on 164 subjects from three different databases show that the proposed biometric technique outperforms existing methods in terms of feature extraction efficiency for biometric matching. The method achieved high accuracy rates of 96.7% for the MIT-BIH Arrhythmia Database (MITDB), 96.4% for the MIT-BIH Supraventricular Arrhythmia Database (SVDB), and 99.3% for the Charles Sturt Initiative Diabetic Screening Complications Database (DiSciRi).

To demonstrate the robustness of the proposed NCN technique, four commonly applied classification induction algorithms were used: Bayesian Network (BN), Multilayer Perceptron (MLP), Radial Basis Function (RBF), and k-Nearest Neighbors (kNN). Based on the results from these classifiers, the authors were able to determine the reliability and efficiency of the proposed technique. Finally, the authors suggested that their technique could be further improved by using ECG recordings with a lower sampling frequency and an increased number of ECG samples [4].

### 3.6   The SVM-Based Approach

The researchers of this study developed a biometric identification approach using electrocardiogram (ECG) features. They extracted 21 features from each heartbeat, utilizing temporal and amplitude distances, as well as morphological descriptors. The classifier used was a Support Vector Machine (SVM) with two kernels, Gaussian and polynomial, to determine the best kernel and the appropriate hyperparameter values. The authors evaluated their algorithm on two MIT-BIH databases, demonstrating improved efficiency for normal and arrhythmic signals. The results showed a performance of 98.35% for the MIT-BIH Normal Sinus Rhythm database and 98.10% for the MIT-BIH Arrhythmia database. The authors also proposed using a multimodal biometric method in future work, combining ECG data with other biometric features [18].

# 4  Results and Discussion

In this work, we will apply and test several similarity measures with the K-nearest neighbours (KNN) classifier in biometry. For this purpose, we have selected 15 random person's ECG (records). We have segmented the records every 2.5 s, so each segment contains 900 samples.

## 4.1  Results

Table 1 presents a summary of the obtained results Table 1. Revealing the Results for K = 1.

**Table 1.** Table Revealing the Results for K = 1

| Approach | Accuracy |
|---|---|
| 1NN-SEA | 92% |
| 1NN-DTW | 70% |
| 1NN-QPDTW | 88% |
| 1NN-CIDDTW | 84% |
| 1NN-CID | 70% |
| 1NN-ED | *44%* |

## 4.2  Discussion

According to the results obtained, the 1NN-SEA approach demonstrated the best results with an accuracy of 92% due to its ability to effectively align the quasi-periodic time series.

The DTW method, which is most commonly used in the field of time series, has produced estimated results of 70%.

Figure 3 shows the alignment of two similar but phase-shifted ECG with SEA (Fig. 3a) and DTW (Fig. 3b) traces(a) original traces (b) SEA alignment (c) DTW alignment [6] The QPDTW method has achieved an accuracy of 88%. This is primarily attributed to its effectiveness in resolving significant phase shifts in quasi-periodic time series. Figure 4 shows the alignment of two similar but phase shifted ECG with QPDTW and DTW.

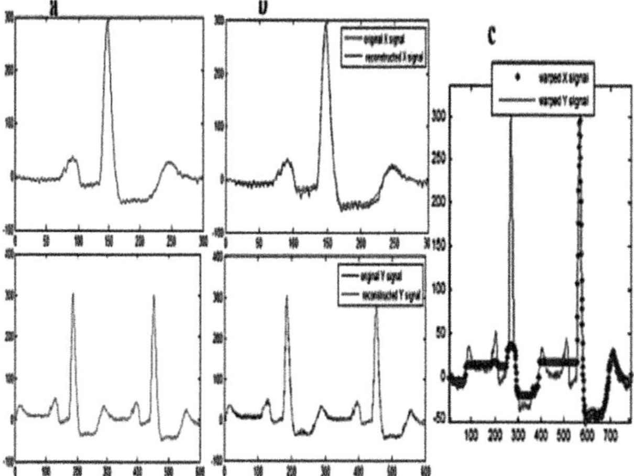

**Fig. 3.** Illustration of SEA and DTW alignment of two phase-shifted quasi-similar.

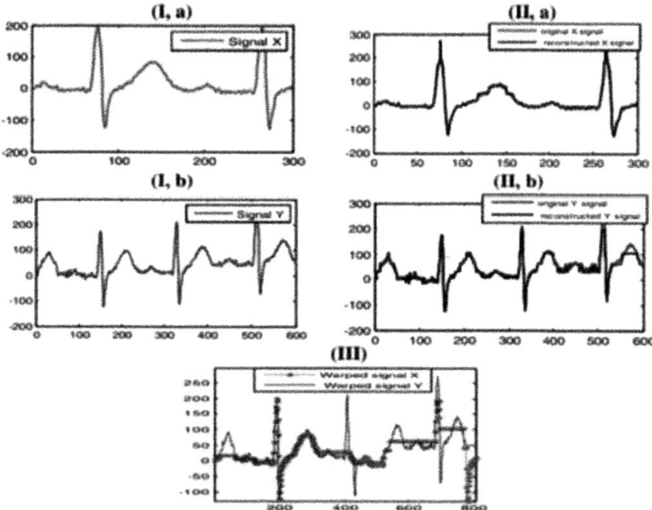

**Fig. 4.** Comparison of Two similar ECGs (from the same person) (I) Original signals.

(II) Original signals vs. reconstructed signals by QPDTW. (III) Original signals vs. Warped signals by DTW The KNN-CIDDTW approach also achieved good results, estimated at 86%. These results demonstrate the effectiveness of this method in the field of biometrics using electrocardiograms (ECG), due to its ability to address the challenges of invariance to morphological changes in complex signals such as ECG.

The CID method achieved estimated results of 70%, making it less effective than other methods in the field of biometrics.

The Euclidean distance achieved the least satisfactory results compared to other approaches, with an accuracy of only 44%. This is because the Euclidean distance cannot effectively compare time series of different lengths or temporally shifted time series.

Figure 5 shows two ECGs compared point by point, using the Euclidean distance. It should be noted that the two peaks of the ECGs are not correctly matched with the Euclidean distance because the ECGs are taken at different time instances (shifted on the time axis). We notice that ED has limitations when the series differ in terms of temporal/amplitude scale or shift.

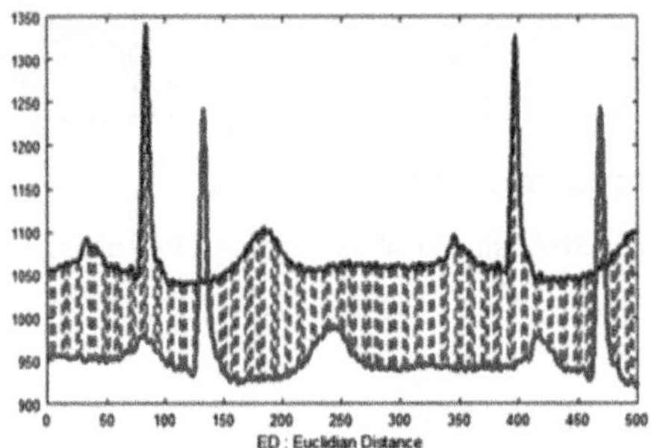


**Fig. 5.** Comparison of two similar but shifted ECGs using ED Distance.

In this work, we have tested and compared similarity distance for ECG-based personal identification. The 1NN-SEA approach yielded the best results, followed by hybrid approaches KNN-QPDTW and KNN-CIDDTW seem to provide good results and perform better than KNN-DTW and KNN-CID approaches. This conclusion is based on the simplicity and recognized effectiveness of the 1NN classifier in the literature, as well as the proven capacity of SEA, QPDTW, and CIDDTW methods to be invariant to morphological changes in complex in quasi-periodic signals as ECG.

## 5   Conclusion

In this work, we tested and compared similarity distances in the field of biometrics. The 1NN-SEA approach yielded the best results, followed by hybrid approaches. KNNQPDTW and KNN-CIDDTW seem to provide relatively good results and perform better than KNN-DTW and KNN-CID approaches. This conclusion is based on the effectiveness of the 1-NN classifier, as well as the proven capability of SEA, QPDTW, and CIDDTW methods to be invariant to morphological changes in complex quasi-periodic signals such as ECG.

## References

1. Biel, L., Pettersson, O., Philipson, L., Wide, P.: «ECG analysis: a new approach in human identification». IEEE Trans. Instrum. Measur. (2001)

2. Wang, Y., Agrafioti, F., Hatzinakos, D., Plataniotis, K.N.: Analysis of human electrocardiogram for biometric recognition». EURASIP J. Adv. Signal Process. 1–11 (2007)
3. Samaneh, K., Dehghani, A., Omranian, S., Boostani, R.: ECG based personal identification using empirical mode decomposition and Hilbert transform.» The 16$^{th}$ CSI International Symposium on Artificial Intelligence and Signal Processing (2012)
4. Sidek, A., Khalil, I., Jelinek, F.: ECG biometric with abnormal cardiac conditions in remote monitoring system. Trans. Syst. Man Cyber. IEEE **44**(111), 1498–1509 (2014)
5. Li, R., Yang, G., Wang, K., Huang, Y., Yuan, F.: Robust ECG biometrics Using GNMF and sparse representation. **129**, 70–76 (2020)
6. Boulnemour, I., Boucheham, B.: I-SEA: improved shape exchange algorithm for quasi-periodic time series alignment. In: IEEE International Conference on Computer Vision and Image Analysis Applications (ICCVIA), Sousse, Tunisia, pp. 1–6 (2015)
7. Batista, G.E., Keogh, E.J., Tataw, O.M.: CID: an efficient complexity-invariant distance for time series. Data Min. Knowl. Disc. **28**, 634–669 (2014)
8. Sakoe, H., Chiba, S.: Dynamic programming algorithm optimization for spoken word recognition. IEEE Trans. Acoust. Speech Signal Process. **26**, 43–49 (1978)
9. Berndt, D.J., Clifford, J.: Using dynamic time warping to find patterns in time series. In: Proceedings of Workshop on Knowledge Discovery in Databases. pp. 359- 370 (1994)
10. Batista, G.E., Wang, X., Keogh, E.J.: A complexity-invariant distance measure. SIAM International Conference on Data Mining. Society for Industrial and Applied Mathematics, pp. 699–710 (2011)
11. Boulnemour, I., Boucheham, B.: QP-DTW: upgrading dynamic time warping to handle quasi periodic time series alignment. J. Inf. Process. **14**, 851–876 (2018)
12. Silva, F., Giusti, R., Keogh, E., Batista, G.: Speeding up similarity search under dynamic time warping by pruning unpromising alignments. Data Min. Knowl. Discovery (2018)
13. Bishop, M.C.: Pattern recognition and machine learning. Springer (2006)
14. Venkatesh, S., Srinivasan, N.: Human electrocardiogram for biometrics using DTW and FLDA. International Conference on Pattern Recognition. IEEE., pp. 3838–3841 (2010)
15. Pratt, K.B., Eugene, F.: Search for patterns in compressed time series. Int. J. Image Graph. (2002)
16. Jun, S., Shu-Di, B.: The PLR-DTW Method for ECG Based Biometric Identification 33rd Annual International Conference of the IEEE EMBS. Massachusetts USA, Boston (2011)
17. Khalil, I., Meli, C.: ECG biometric using multilayer perceptron and radial basisfunction neural networks. Annual International Conference of the IEEE Engineering in Medicine and Biology Society. IEEE (2011)
18. Rezgui, D., Lachiri, L.: ECG biometric recognition using SVM-based approach. IEEJ Trans. Electr. Electron. Eng. (2011)

# Automatic Mapping of UML to EPC Using Graph Transformation for Improving Business Process Modeling

Aissam Belghiat[✉] ⓘ, Houda Rezig, and Wafa Labiad

LaRIA Laboratory, University of Jijel, 18000 Jijel, Algeria
`{aissam.belghiat,houda.rezig,wafa.labiad}@univ-jijel.dz`

**Abstract.** Modeling, analysis, and redesign of business processes are crucial for enhancing the performance of organizations and manage their evolving. UML activity diagrams (ADs) are expressive formalisms broadly utilized for modeling business processes. Conversely, Event-driven Process Chains (EPCs) are commonly adopted flowchart-based representations for business process modeling and can be analyzed using a doted Petri nets formal semantics. This paper presents a method for automatic transformation of ADs into EPC models, facilitating behavior modeling across different formalisms. It consists of using graph transformation for mapping the elements of activity diagrams and their semantics to their corresponding EPC models. This transformation connects two major communities—those utilizing UML and those employing EPC—which allows more flexible and thorough modeling and analysis of business processes.

**Keywords:** Business process · UML Activity Diagram · EPC · Graph Transformation

## 1 Introduction

Modern organizations are relying on efficient and well-structured business processes to stay competitive. Business processes are a collection of related tasks which are ordered in a specific sequence that produces a service or product according to a business goal in an enterprise [1]. Modeling business processes allows their understanding and improvement. There is diversity in organizations and their business processes which results in a variety of communities that utilize different modeling languages.

Among these languages, UML (Unified Modeling Language) activity diagram (AD) [2] is suitable and widely adopted model for representing the behavior of business processes [3]. It offers a comprehensive graphical tool equipped with necessary elements and mechanisms that simplify the modeling task. Conversely, EPC (Event-driven Process Chain) [4, 5] is another modeling language that is used in describing business processes. It is a simple language which is well known especially at information systems community [22].

Mapping UML to EPC helps connect the UML and EPC communities by enabling the transfer of business process models from the widely used UML language to the EPC

© The Author(s), under exclusive license to Springer Nature Switzerland AG 2025
N. Seddari and M. Redjimi (Eds.): ICMSCT 2024, CCIS 2606, pp. 150–160, 2025.
https://doi.org/10.1007/978-3-032-01922-6_13

framework. Moreover, EPC is doted by Petri net-based formal semantics and supported by powerful tools like ProM [6], which provide various verification capabilities. As a result, the mapping offers several advantages and improvements for business processes modeling.

Hence, the current study proposes a transformation of UML ADs to EPC models. It is a part of a project that aims to gathering multiple modeling languages of business processes in one framework and making simple passages between them. This paper presents first reflections and results of the project. We use a graph transformation method to realize the mapping due to their efficiency and since both formalisms is graph. The graph transformation utilizes a combined meta-modeling and graph grammar technique implemented in the $AToM^3$ (A Tool for Multi-formalism Meta-Modeling) tool [7]. Meta-modeling is used for defining two meta-models for UML activity diagrams and EPC models, and consequently generating modeling environments for both formalisms which allow specifying their models. Then, two graph grammars are developed; one for automatically transforming UML activity diagrams, described in our $AToM^3$ integrated tool, into EPC models, and another for converting the resulting EPC models into ".epml" files which can then be directly uploaded into tools like ProM for verification.

The remainder of the paper is as follows. Section 2 presents some related work. Section 3 exposes some preliminaries. Section 4 explains the proposed approach. Section 5 shows an application example. Section 6 ends the paper.

## 2 Related Work

There are a lot of contributions in the literature trying to bridge the gap between activity diagrams and business process modeling languages. In [8, 9], the authors highlighted the convergence points between UML activity diagrams and BPMN. In [10], activity diagrams are transformed to BPEL, and to BPEL4WS in [11]. In [12], a transformation is realized from UML activity diagrams to the YAWL.

The convergence of UML and EPC has been a topic of exploration since the appearance of both languages. In [13, 14], the authors tried to integrate UML diagrams and EPC models. In [15], a comparison experiment has been made about the effectiveness and efficiency of UML activity diagrams and EPCs for requirements engineering. Another similar comparison is done in [16]. In [17], a UML 2 profile for EPCs is developed targeting software developers to view EPC models in a familiar notation.

Some others works have tried to translate activity diagrams to formal methods which are recognized by their suitability for modeling business processes, such as the mapping to the Pi-calculus in [18], to the CSP in [19], to the Petri nets in [20].

In contrast to prior studies, our approach is focused on developing a user-friendly tool specifically designed for mapping UML activity diagrams to EPC models, with a primary aim of modeling and analysis. We have used meta-modeling and graph transformation for this purpose which allows implementing the translation efficiently.

# 3   Preliminaries

ADs [2, 3] are behavioral models of UML used for representing flow and data control in systems. They have been also largely used for modeling business processes thanks to the ability of describing the flow of activities within a process, showing the sequence of actions, decision points, and branching paths. In fact, considering the rich intuitive set of elements ADs offers, they can be an effective tool for communication within organizations.

EPC (Event-driven process) [4, 5] is a flowchart-style language designed for business process modeling. It represents an ordered graph where events and functions are connected through various connectors that support both alternative and parallel process execution. EPCs are commonly adopted in the information systems community because of their simplicity.

Graph transformation is a formal-based method that allows efficient transformation between graphs syntaxes and semantics. It is implemented using different techniques including the Meta-modeling and Graph grammar approach used in this paper.

Meta-modeling is used for specifying the abstract and concrete syntax of any formalism (i.e. a graph), while the graph grammar is used to implement the mapping between formalisms in a Meta level. Both techniques are provided by AToM$^3$ [7].

A graph grammar contains three constituents [21]: an initial action, a set of transformation rules, and a final action. The initial action is for specifying overall information about the transformation. The final action is for making necessary information to end it. Concerning the rules, each rule consists of two parts—a left-hand side (LHS) pattern and a right-hand side (RHS) replacement. During execution, when the system detects a match for the LHS within the source graph, it substitutes it with the RHS structure. A rule carries an execution priority and may specify preconditions that must be fulfilled before application and post-actions that are executed after. The rewriting system continues applying rules iteratively until no transformations exist.

# 4   The Approach

We propose two meta-models: one for UML ADs and the other for EPC models. The meta-models are served for creating two different environments for modeling activity diagrams and EPC models respectively. After that, we develop two graph grammars to carry out the mapping. The first grammar transforms the UML activity diagram meta-model into the EPC meta-model, and the second grammar transforms the EPC meta-model into its XML format.

## 4.1   Activity Diagram Meta-model

The meta-model proposed for ADs is named "AD_META", and is inspired by the OMG specification [2]. It comprises 8 classes and 7 associations. It specifies the essential elements of UML ADs, including InitialNode, FinalNode, Action, DecisionNode, MergeNode, ForkNode, and JoinNode. The abstract syntax of this meta-model is depicted in Fig. 1, which also implicitly contains its concrete syntax.

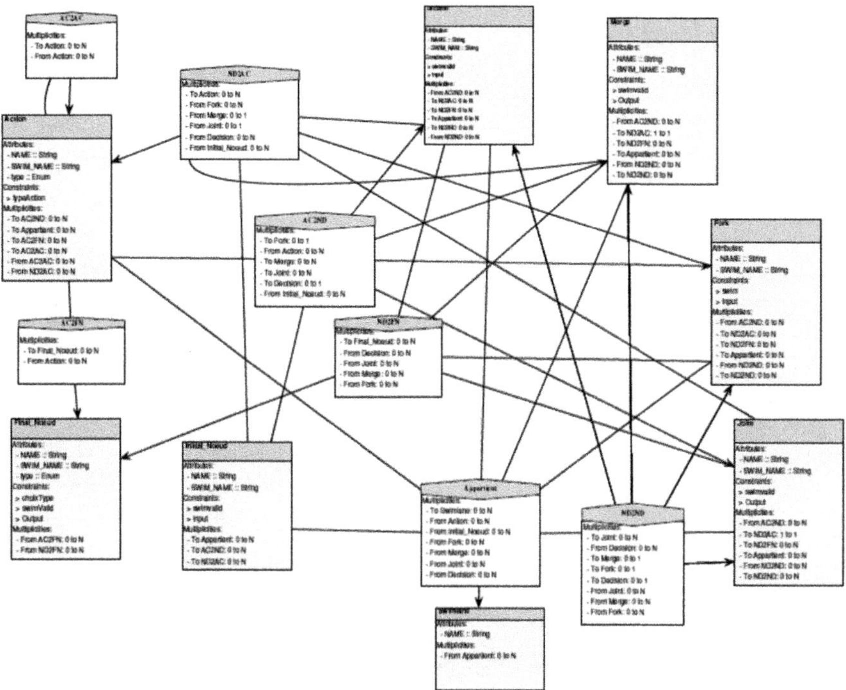

**Fig. 1.** Meta-model of activity diagrams

## 4.2   EPC Meta-model

The meta-model proposed for EPCs is named "EPC_META", and is inspired by [4, 5]. It consists of 9 classes and 24 associations. It represents the core components of Event-driven Process Chains, including Event, Function, and various types of connectors (AND, OR, XOR). The abstract syntax of this meta-model is depicted in Fig. 2, implicitly capturing its concrete syntax.

## 4.3   Rules of the Mapping

To transform UML activity diagrams to EPC models, we propose multiple correspondence rules. This task requires a good understanding of both modeling formalisms. The rules are as follows:

- Initial/Final nodes → Events
- Actions → Functions
- Decision/Merge nodes → XOR operators
- Fork nodes → OR operators
- Join nodes → AND operators

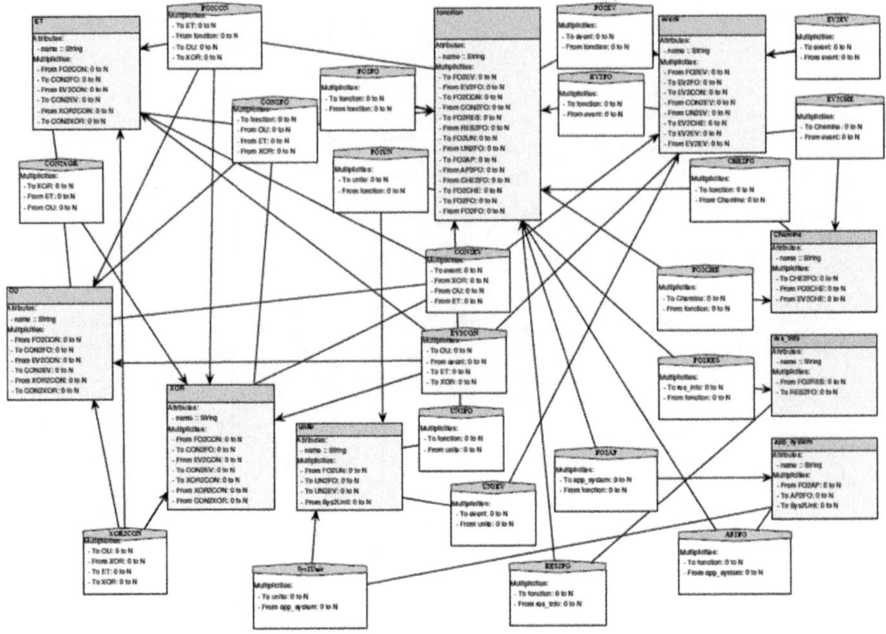

**Fig. 2.** Meta-model of EPC models

## 4.4   The First Graph Grammar, UML-AD to EPC

A graph grammar is used to implement the rules defined previously. It is named *"transfDA2EPC" and* composed of 18 rules. Table 1 shows some details of the implementation; each rule has a name, a priority, a condition, an LHS, an RHS and an action. Due to space constraints, we consider only some rules.

The transformation process executes in three phases:

1.  Transform individual elements (like Rule 1 and 2)
2.  Transform connections between elements (like Rule 3 and 4)
3.  Clean up transformed elements (like Rule 5 and 6)

**Table 1.** The first graph grammar

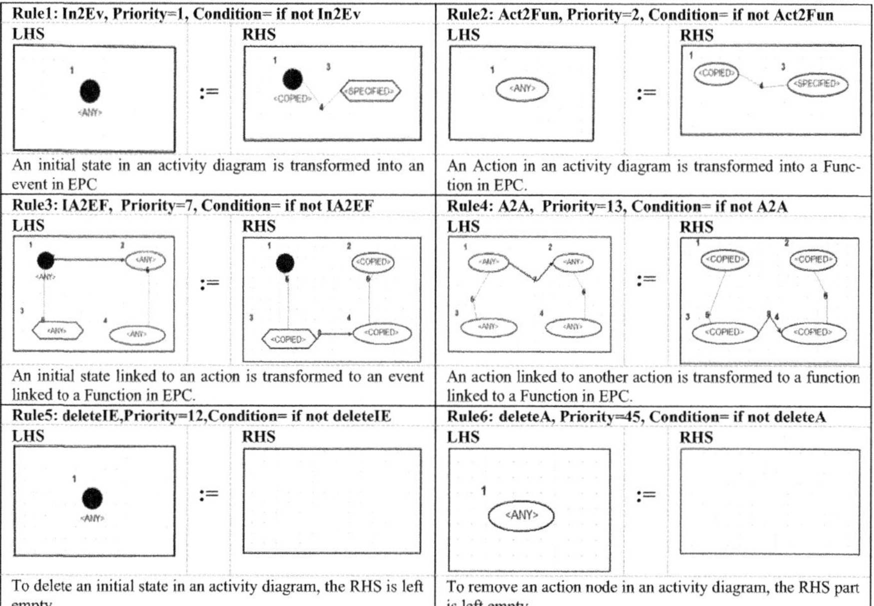

| Rule1: In2Ev, Priority=1, Condition= if not In2Ev | Rule2: Act2Fun, Priority=2, Condition= if not Act2Fun |
|---|---|
| LHS   RHS | LHS   RHS |
| An initial state in an activity diagram is transformed into an event in EPC | An Action in an activity diagram is transformed into a Function in EPC. |
| **Rule3: IA2EF, Priority=7, Condition= if not IA2EF** | **Rule4: A2A, Priority=13, Condition= if not A2A** |
| LHS   RHS | LHS   RHS |
| An initial state linked to an action is transformed to an event linked to a Function in EPC. | An action linked to another action is transformed to a function linked to a Function in EPC. |
| **Rule5: deleteIE,Priority=12,Condition= if not deleteIE** | **Rule6: deleteA, Priority=45, Condition= if not deleteA** |
| LHS   RHS | LHS   RHS |
| To delete an initial state in an activity diagram, the RHS is left empty. | To remove an action node in an activity diagram, the RHS part is left empty. |

## 4.5   The Second Proposed Graph Grammar, EPC to EPML

The second graph grammar is developed to enable mapping EPC models to their textual representation. It is named "*epc2xml_GG_exec*" and it comprises 11 rules. When applying this graph grammar to a resulting EPC model, a file that contains its ".epml" code is created. Only three rules are presented here, the others are similar.

– **The initial action**: In the initial action of the graph grammar, we use Python to create a file named "*EPC.epml*" to store the derived code (see Fig. 3).

```
obFichier = open('EPC.EPML','w')
obFichier.write('\n')
obFichier.write('<?xml version=1.1 encoding ="UTF-8"?>'+'\n')
obFichier.write('<epml:epml xmlns:epml="http:\\www.epml.de"> '+'\n')
obFichier.write('<coordinates '+'\n xOrigin="leftToRight" '+'\n
                            yOrigin="topToBottom" /> '+'\n')
obFichier.write('<directory name="Root"> '+'\n')
obFichier.write('<epc> '+'\n')
obFichier.close()
```

**Fig. 3.** The initial action

- **Set of rules:** the graph grammar is made up of 11 rules. Each rule is characterized by a name, a property, a condition and an action, such as the rule **EvFn2EvFn** bellow (see Fig. 4).

**Rule 1: EvFn2EvFn (priority: 1)**

Condition :

```
node = self.getMatched(graphID, self.LHS.nodeWithLabel(2))
return not hasattr(node, "EV2FCT_executed")
```

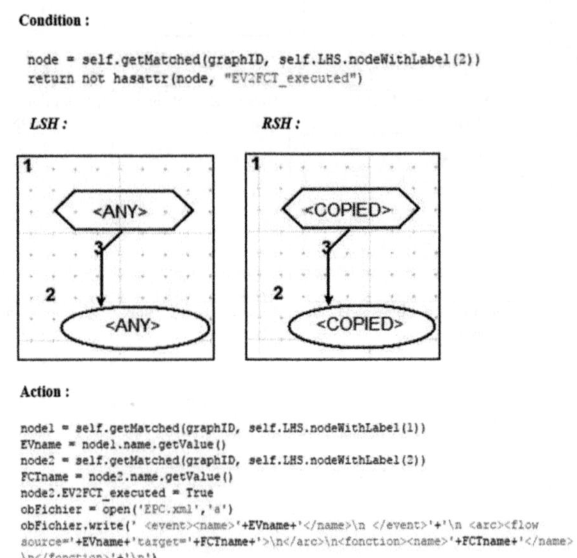


Action :

```
node1 = self.getMatched(graphID, self.LHS.nodeWithLabel(1))
EVname = node1.name.getValue()
node2 = self.getMatched(graphID, self.LHS.nodeWithLabel(2))
FCTname = node2.name.getValue()
node2.EV2FCT_executed = True
obFichier = open('EPC.xml','a')
obFichier.write(' <event><name>'+EVname+'</name>\n </event>'+'\n <arc><flow
source='+EVname+'target='+FCTname+'>\n</arc>\n<fonction><name>'+FCTname+'</name>
\n</fonction>'+'\n')
```

**Fig. 4.** Mapping an EPC event linked to a function to EPML

- **The final action:** used to close the file (see Fig. 5).

```
obFichier = open('EPC.xml','a')
obFichier.write('<\n \n')
obFichier.write('</epc>')
obFichier.write('<\n')
obFichier.write('</directory>')
obFichier.write('<\n')
obFichier.write('</epml>')
obFichier.write('<\n')
obFichier.close()
```

**Fig. 5.** The final action.

## 5   Example

A simple example is shown here to just illustrate the proposed approach. It concerns a client's interaction with a bank ATM.

The process begins when a customer inserts their card into the ATM. The system then verifies the card code. At this point, there is a decision: if the code is incorrect, the customer must re-enter it, but if the code is valid, the process continues to the withdrawal request. When making a withdrawal, there is also a decision. The system either finds the resources to dispense money or declines such operation. In both cases there is a join that gathers both flows and ends the process by returning the card to the client.

According to our approach, this activity diagram must be described in our integrated tool AToM3 as illustrated in Fig. 6.

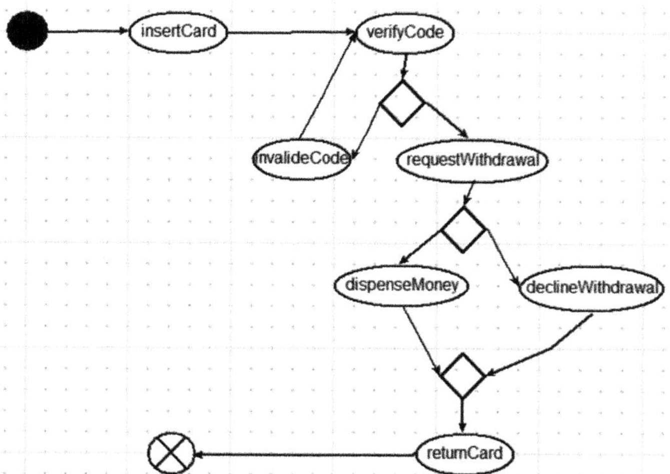


**Fig. 6.** Example of an ATM activity diagram described in our tool.

Applying the first graph grammar, we start transforming the ATM activity diagram to an EPC model. At this stage, the graph is a blind of both formalisms, and as the transformation rules continue executing, the EPC model starts to appear on the canvas as indicated in Fig. 7.

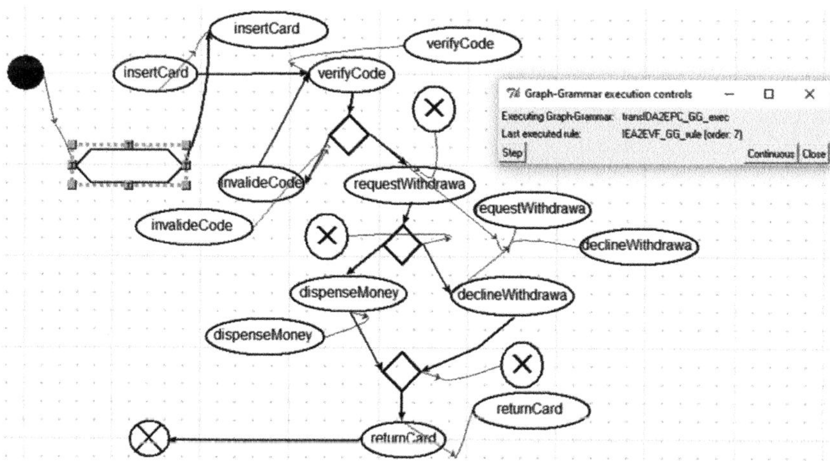


**Fig. 7.** The ATM activity diagram during the execution in our tool.

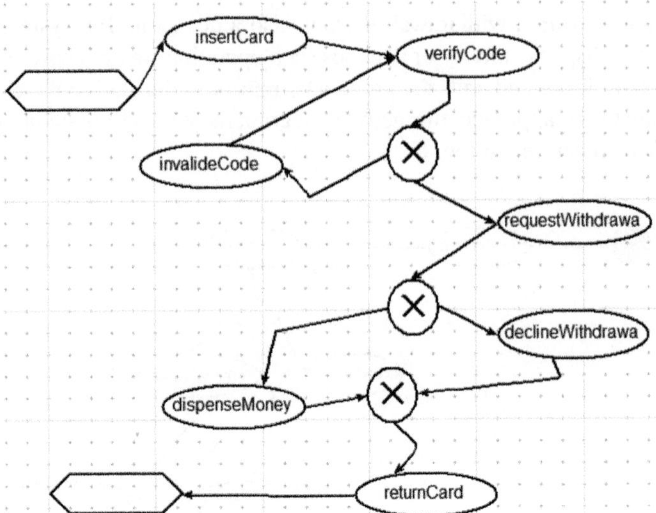


**Fig. 8.** The resulted EPC model.

```
<?xml version="1.0" encoding="UTF-8"?>
<epml:epml xmlns:epml="http://www.epml.de"
  xmlns:xsi="http://www.w3.org/2001/XMLSchema-instance"
  xsi:schemaLocation="http://www.epml.de EPML_111_Draft.XSD">
  <coordinates xOrigin="leftToRight" yOrigin="topToBottom"/>
  <directory name="Root">
    <epc epcId="1" name="swim">
      <event id="1">
        <name>event initial</name>
      </event>
      <arc id="11">
        <flow source="1" target="2"/>
      </arc>
      <function id="2">
        <name>Insert card</name>
      </function>
      <arc id="12">
        <flow source="2" target="3"/>
      </arc>
          .
          .
          .
      <function id="5">
        <name>Choose operation</name>
      </function>
    </epc>
  </directory>
</epml:epml>
```

**Fig. 9.** An extract of the generated ".epml" file

After the execution of the first graph grammar, we obtain the EPC model shown in Fig. 8.

Applying the second graph grammar, an XML file ".epml" is generated by serializing the information existing in the graphical model. This file will be charged in the ProM tool [6] to start the verification task. Figure 9 shows an extract of the file.

## 6  Conclusion

This paper presents an automatic approach for transforming UML activity diagrams to the EPC models. The approach is based on graph transformation. It consists of using a combined meta-modeling and graph grammar technique that is implemented in the $AToM^3$ tool. The meta-modeling is used to build environments for activity diagrams and EPC models. Then, two graph grammars are used. The first one transforms the UML graphical activity diagrams to the EPC graphical models, while the second one is used to generate the textual code of the generated EPC graphical models. This code is uploaded automatically in the ProM tool to start the verification task.

We plan in future work to add other advanced elements and other UML diagrams. Also, we intend make the reverse translation, in order to provide a mapping in the two directions to make each other benefit.

## References


1. Weske, M.: Process management concepts, languages, architectures. ISBN 978-3-662-59431-5. Third Edition, Springer (2019)
2. UML (Unified Modeling Language), Superstructure, v2.5. http://www.omg.org/.
3. Russell, N., Van der Aalst, W., Ter Hofstede, A., Wohed, P.: On the suitability of UML 2.0 activity diagrams for business process modelling. Conceptual modelling 2006: Proceedings of APCCM2006, pp. 95–104 (2006)
4. Hoffmann, W., Kirsch, J., Scheer, A.-W.: Modellierung mit Ereignisgesteuerten Prozeßketten: Methodenhandbuch; Stand, Dezember 1992. In: Scheer, A.-W. (Ed.): Veröffentlichungen des Instituts für Wirtschaftsinformatik, no. 101, Saarbruecken: Universität des Saarlandes (1992). (in German)
5. Keller, G., Nüttgens, M., Scheer, A.-W.: Semantische Prozeßmodellierung auf der Grundlage,"Ereignisgesteuerter Prozeßketten (EPK). In: A.-W. Scheer (ed.): Veröffentlichungen des Instituts für Wirtschaftsinformatik, no. 89, Saarbruecken: Universität des Saarlandes (1992). (in German) http://www.iwi.uni-sb.de/Download/iwihefte/heft89.pdf
6. Dongen, V.B.F., de Medeiros, A.K.A., Verbeek, H.M.W., Weijters, A.J. M. M., van Der Aalst, W.M.: The ProM framework: a new era in process mining tool support. In International conference on application and theory of petri nets (pp. 444–454). Springer, Berlin, Heidelberg (2005)
7. AToM3. http://atom3.cs.mcgill.ca/.
8. Geambaşu, C.V.: BPMN vs UML activity diagram for business process modeling. Account. Manage. Inf. Syst. 11(4), 637–651 (2012)
9. Peixoto, D., Batista, V., Atayde, A., Borges, E., Resende, R., Pádua, C.I.P.S.: A comparison of BPMN and UML 2.0 activity diagrams. VII Simposio Brasileiro de Qualidade de Software 56, 012010 (2008)
10. Ouyang, C., et al.: Translating standard process models to BPEL. Adv. Inf. Syst. Eng. 417–432 (2006)

11. Gardner, T.: UML modelling of automated business processes with a mapping to BPEL4WS (2003)
12. Zhaogang, H., Zhang, L., Ling, J.: Transformation of uml activity diagram to yawl. Enterprise Interoperability IV. Springer, London, pp. 289-299 (2010)
13. Nüttgens, M., Feld, T., Zimmermann, V.: Business process modeling with EPC and UML: transformation or integration?. In: The Unified Modeling Language (pp. 250–261). Physica-Verlag HD (1998)
14. Loos, P., Allweyer, T.: An approach for integrating uml and event-driven process chains (EPC). Publications of the Institut f ur Witschaftsinformatik (1998)
15. Gross, A., Doerr, J.: EPC vs. UML activity diagram-two experiments examining their usefulness for requirements engineering. In: 2009 17th IEEE International Requirements Engineering Conference (pp. 47–56). IEEE (2009)
16. Islam, M.R., Islam, M.R., Alam, M.S., Azam, M.S.: Experiences and comparison study of EPC & UML for business process & IS modeling. Int. J. Comput. Sci. Inf. Secur. (2011)
17. Korherr, B., List, B.: A UML 2 profile for event driven process chains. In Research and Practical Issues of Enterprise Information Systems (pp. 161–172). Springer, Boston, MA (2006)
18. Belghiat, A., Chaoui, A.: A graph transformation of activity diagrams into pi-calculus for verification purpose. In: Proceedings of the 3rd Edition of the International Conference on Advanced Aspects of Software Engineering (ICAASE18), Constantine, Algeria, 1,2-December-2018, published at http://ceur-ws.org
19. Bolton, C., Davies, J.: On giving a behavioral semantics to activity graphs, in UML 2000 Workshop Dynamic Behavior in UML Models: Semantic Questions (2000). http://www.disi.unige.it/person/ReggioG/UMLWORKSHOP/ACCEPTED.html. Accessed 26 July 2006
20. Storrle, H., Hausmann, J.H.: Towards a formal semantics of UML 2.0 activities, in German Software Engineering Conf. 2005 (2005) http://www.pst.ifi.lmu.de/personen/stoerrle/V/AD-11-Limits.pdf. Accessed 26 July 2006
21. Rozenberg, G.: Handbook of Graph Grammars and Computing by Graph Transformation, World Scientific (1999)
22. Scheer, A.-W.: ARIS - business process frameworks (3rd Ed), Springer (1999)

# Adaptive Simulation Framework for IoT Data Processing in Edge-Cloud Environments Using SimPy and Neural Networks


Imane Khoums$^{(\boxtimes)}$ ⓘ, Radouane Nouara ⓘ, and Nabil Belala ⓘ

MISC Laboratory, University of Constantine 2 – Abdelhamid Mehri, Constantine, Algeria
{imane.khoums,redouane.nouara,nabil.belala}@univ-constantine2.dz



**Abstract.** The Internet of Things (IoT) and cloud computing have transformed various industries by facilitating intelligent applications and improving data processing. While IoT devices generate massive volumes of data, cloud computing provides scalable resources for storage and analysis. However, integrating IoT with cloud computing poses challenges, such as latency, bandwidth limitations, and resource allocation. Traditional static processing models fail to adequately improve performance in dynamic IoT environments. This paper presents a novel simulation approach combining SimPy for discrete event simulation with a neural network-based decision-making system. Our method optimises data processing techniques across IoT, edge, and cloud environments. We evaluate the efficiency and scalability of centralised, edge, and hybrid data processing models in cloud-IoT scenarios, determining the most effective technique for each use case. The results demonstrate that edge processing can greatly decrease latency for applications that require real-time responsiveness, while hybrid models are particularly effective in IoT contexts that handle large amounts of data. The high precision of the neural network in predicting optimal methods demonstrates its potential to offer a flexible and adaptive framework for optimising IoT-cloud integration, paving the way for more efficient and responsive IoT systems across various sectors, such as smart cities, industrial IoT, and healthcare applications.

**Keywords:** IoT Simulation · Data Processing · Cloud Computing · Edge Computing · Neural Network


## 1 Introduction

The convergence of cloud computing and the Internet of Things (IoT) is an important improvement in computational efficiency and data management [1]. The infrastructure required for scalable data storage and processing is provided by cloud computing, while IoT contributes continuous streams of data from interconnected devices [2]. This integration enables the efficient management and analysis of large volumes of data. It facilitates the development of sophisticated applications and real-time insights in many fields, such as smart homes, healthcare, and industrial automation [3,4].

However, this convergence also introduces notable challenges, such as concerns about data transmission capacity, the ability to handle increased workload, latency management, and resource use [4].

Furthermore, the integration of IoT with cloud computing for next-generation wireless technology addresses various issues, including real-time data processing, security, and privacy, noting the high latency and limited bandwidth that often accompany the integration of IoT with the cloud [5]. This emphasises the need for energy-efficient algorithms for IoT devices and explores the applications of machine learning and deep learning to improve the services provided by cloud and IoT systems [6].

Despite these advancements, there are still significant challenges in the integration of IoT with the cloud technology. Traditional static models often fail to optimise performances under varying conditions. This indicates the need for flexible strategies to process data that can dynamically adapt to different system and workload conditions [7]. In addition, numerous cloud computing surveys have focused on aspects such as scalability, resource management, and energy efficiency [8,9]. They often neglect the different challenges posed by IoT systems. These include real-time responsiveness, device heterogeneity, and frequent data transmission interruptions [10]. Simulation provides a valuable approach to address these issues by enabling accelerated and risk-free testing.

Most simulation frameworks concentrate on edge computing, which operates at the edge of the network with a small calculation scale, or cloud processing, which operates at a large computing scale. However, these frameworks lack dynamic, hybrid approaches and adaptive decision-making mechanisms for real-time optimal processing strategies in changing IoT environments [11,12].

Our study evaluates the effectiveness of centralised, edge, and hybrid data processing models in cloud-IoT scenarios. We focus on latency, bandwidth usage, and the total amount of data after processing. We use simulation-based methods and neural networks to evaluate system scalability and adaptability, with the potential to improve traffic management in smart cities and optimise resource use in industrial IoT applications.

We review related work on IoT simulation and data processing strategies in Sect. 2, detail the simulation methodology and neural network architecture in Sect. 3, present experimental results and discuss findings in Sect. 4, and conclude with a summary of contributions and future research directions in Sect. 5.

## 2    Related Work

Recent developments in simulation frameworks for IoT, edge computing, cloud systems, and artificial intelligence (AI) have focused on improving performance and efficiency using various novel methods.

Frameworks for edge-focused simulation like EdgeSimPy [13] and PureEdgeSim [14] are Python-based tools for modelling resource management in edge computing environments. However, they have limitations in integration with cloud resources and the lack of dynamic strategy selection mechanisms. Our approach builds on these frameworks by using cloud resources and adaptive processing strategies.

The Lean simulation framework [15] and CloudSim [16] are efficient stress testing tools for cloud-IoT systems, but they have limitations in considering edge computing scenarios and static resource allocation strategies. Our research aims to address these gaps by integrating edge computing capabilities and dynamic resource allocation, built on these two cloud-centric approaches.

IoTSim-Osmosis [17] and EdgeCloudSim [18] frameworks are designed to bridge edge and cloud computing simulations. They offer strengths such as a comprehensive modelling of IoT applications, consideration of user mobility, and efficient network management. However, they have limitations in the adaptability of systems to changing conditions, such as scalability and latency, in addition to the absence of AI-driven decision-making mechanisms. Our approach enhances these hybrid frameworks by adding a neural network for the dynamic selection of strategies that address these limitations.

Recent advances in IoT simulations integrate AI and machine learning. For example, the Digital Twin intelligent system [19] uses reinforcement learning, while EdgeAISim [20] uses deep Q-Networks. These solutions improve efficiency and resource management, but they face issues with high computational overhead and real-time responsiveness. As a solution to this problem, we used a lightweight neural network for quick and adaptive decision-making in diverse IoT scenarios.

In summary, existing frameworks often rely on static rules or simplified models. In contrast, our new simulation framework integrates comprehensive models of edge and cloud resources. It also supports their hybridisation and incorporates dynamic AI-driven strategy selection. Additionally, our framework enables scalable simulation of diverse IoT scenarios. This approach creates a realistic and adaptive simulation environment. It improves resource management and responsiveness, effectively addressing the limitations of prior work. As a result, it advances the field of IoT-cloud integration.

## 3   Methodology

In this section, we detail the methodology employed for the evaluation of the scalability, efficiency, and adaptability of various data processing models, including the implementation of a neural network to predict optimal strategies in dynamic IoT scenarios.

### 3.1   Simulation Framework

The simulation framework was developed with SimPy, a Python package for discrete-event simulation that operates on a process-based model [21]. The decision to choose SimPy was driven by its capacity to effectively simulate intricate, parallel systems with numerous interconnected entities, which is highly compatible with the features of IoT settings [22].

The simulation framework in Fig. 1 combines various components to assess IoT data processing models in edge-cloud scenarios. The procedure starts by creating a simulation environment in which the parameters are defined. The simulation proceeds with the monitoring of key performance metrics. Latency is calculated as the ratio of the time difference between the start and end of the simulation to the total volume of data processed. Bandwidth, which reflects network resource consumption, is measured by totalling the size of data transmitted during the simulation. Additionally, the total data processed is recorded and provided as output. However, addressing the impact of these metrics can be challenging. Latency impacts real-time applications, leading to delayed responses that can affect system performance. Bandwidth limitations can result in increased costs and inefficiencies when managing large data volumes. The resource allocation becomes increasingly complex as more devices are added, requiring dynamic strategies to optimise system performance.

The IoT layer depicted in Fig. 1 is responsible for device initialisation, generation and transmission of data, and the log of events. The Edge layer instructs edge nodes to receive, process, and potentially transmit data to the cloud. The Cloud layer is responsible for receiving and storing data, besides monitoring bandwidth usage. The framework incorporates a static analysis feature. This feature is used to configure simulation parameters, oversee concurrent simulations, gather and consolidate data, and manage problems. An AI component is used to preprocess data, train a neural network, assess its performance, and employ it to forecast the most effective processing strategies. The centralised model processes data directly in the cloud. In contrast, the edge model reduces latency and bandwidth usage through local processing on edge nodes. The hybrid model combines edge and cloud processing for optimal resource utilisation. This setup ensures that the simulation can dynamically adapt to different scenarios.

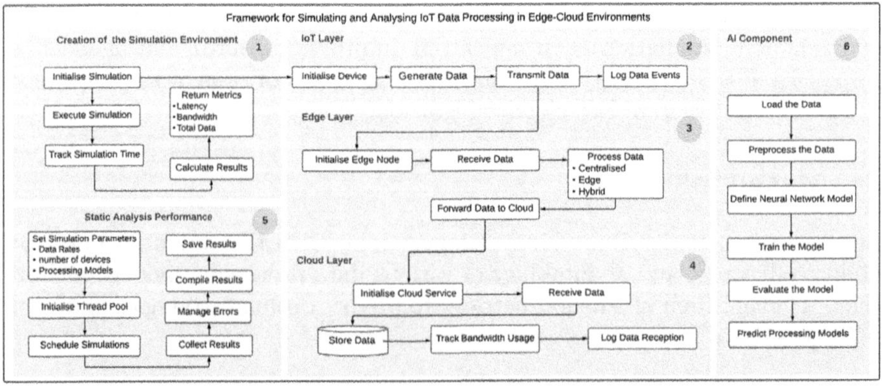


**Fig. 1.** Proposed SimPy Framework for IoT Data Processing Models in Edge-Cloud Environments

## 3.2   Neural Network Architecture

We implemented a feedforward neural network using the *TensorFlow* and *Keras* framework and libraries. The model was chosen for its suitability for tabular data and clear input-output relationships. It is efficient in classification tasks and offers simplicity of implementation. In addition, it demonstrated good performance with moderate computational requirements [23].

We used the dataset produced by our simulation framework with 9000 rows, each representing a scenario. These scenarios encompassed 750 devices with various data rates and processing models. This variety of data rates guarantees that the neural network is trained in an inclusive selection of possible IoT contexts.

The Neural Network architecture depicted in Fig. 2 comprises an input layer, two hidden layers, and an output layer.

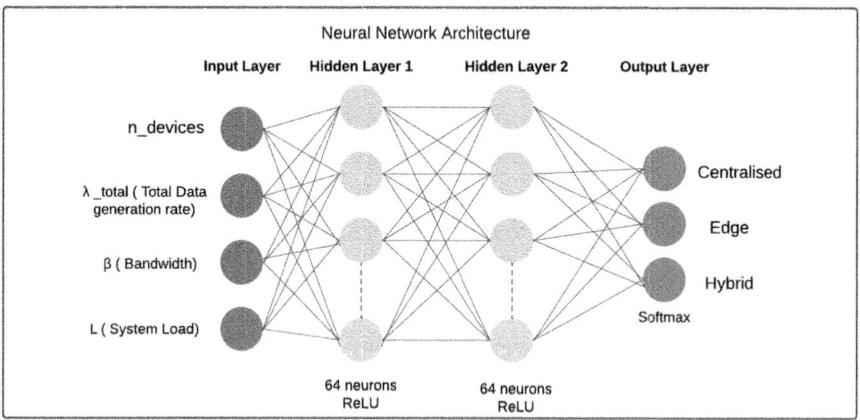


**Fig. 2.** Architecture of Feedforward Neural Network for IoT Strategy Classification.

The dimension of the input layer is determined by the number of features obtained after preprocessing, which guarantees that the model can successfully analyse the tabular data. Every feature corresponds to a distinct characteristic of the data. Preprocessing techniques, such as encoding categorical variables and standardising features, ensure that the input is appropriate for the neural network, resulting in enhanced model performance. The network consists of two hidden layers with 64 neurons, designed to balance complexity and computational efficiency for a dataset with 9000 rows. These layers ensure good generalisation without overfitting. The used *ReLU* (Rectified Linear Unit) activation function introduces non-linearity. It is crucial for learning complex patterns and is computationally efficient, helping to mitigate the problem of vanishing gradients [24]. The size of the output layer is determined by the number of unique classes in the target variable *(Centralised, Edge, Hybrid)*. The *softmax* activation function converts raw output scores into probabilities, facilitating model predictions and assigning the most likely class to each input.

### 3.3   Experimental Setup

The test setup simulated various IoT scenarios, reflecting real-world deployment scalability challenges. We used Python scripts to create virtual IoT devices with unique ID attributes and data generation logic. Scalability was evaluated by systematically varying the number of IoT devices from 1 to 750 in increments of 50, with data rates of 5, 10, 15, and 20 units per time step. For each device count, simulations were conducted in these data rates to observe how performance metrics responded to both increased load and data transmission intensity. Each simulation run spanned 100 time units, during which latency and bandwidth usage were closely monitored. Latency was measured as the time delay between data generation and processing, while bandwidth usage was recorded as the total amount of data transmitted throughout the simulation. By examining trends in varying device counts and data rates, we evaluated the adaptability and scalability of the framework, indicating its potential for large-scale dynamic deployments.

The data packets contained textual data showcasing typical IoT sensor readings or status updates, ensuring relevance to common IoT applications. We conducted the SimPy simulation for a duration of 100 time units, enabling the evaluation of the immediate and long-lasting effects of various processing methods on system performance.

We integrated the neural network model into the simulation framework, allowing it to adjust its processing strategy in response to changing conditions. The model is created using the *Adam* optimiser with a 0.001 learning rate, using a categorical *cross-entropy* loss function [24]. The optimisation is chosen for its ability to manage sparse gradients when most parameters are negligible, which can otherwise slow down the learning process. We chose accuracy as the evaluation metric because it provides a clear and straightforward measure of the model's ability to correctly identify cases. The training procedure was constrained to 100 epochs, with a batch size of 64 and a 20% validation split after preprocessing and filtering noisy data.

The key assumptions of the simulation revolve around the trade-off between the greater processing power of cloud computing and the faster data processing of edge nodes. The simulation uses a controlled environment with consistent bandwidth and no interference sources, allowing for focused analysis of processing models without network variability complexity and assuming that security is ensured. Furthermore, the homogeneity of IoT devices simplifies real-world scenarios, allowing a clearer comparison of processing models.

## 4   Results and Discussion

This section presents key findings from our experimental simulations and neural network model, analysing their implications for IoT data processing strategies. Our model accurately distinguishes between centralised, edge, and hybrid strategies, highlighting its ability to capture nuanced trade-off, especially in complex IoT deployments influenced by data rate, device count, and latency requirements.

## 4.1   Comparative Analysis of Data Processing Models

The comparative analysis of data processing models, depicted in Fig. 3, compares data processing models based on the average total data processed, the average latency, and the average bandwidth usage. The results show that all three models process similar amounts of data but have varying latency and bandwidth usage. The centralised model exhibits the highest latency and the lowest bandwidth, while the edge model has the lowest latency but consumes moderate bandwidth. The hybrid model balances processing strategies but has the highest bandwidth consumption. These results indicate that while the edge model is best for reducing latency, it uses more bandwidth. The hybrid model, despite being balanced, incurs the highest bandwidth costs.

Additional insights into our results highlight significant performance improvements. The framework optimises latency by processing data at the edge, which reduces the time required for data transmission. This is essential for applications that require quick responses. In addition, the hybrid model improves bandwidth efficiency by intelligently managing data flow and resource allocation. These strategies minimise operational costs and ensure efficiency.

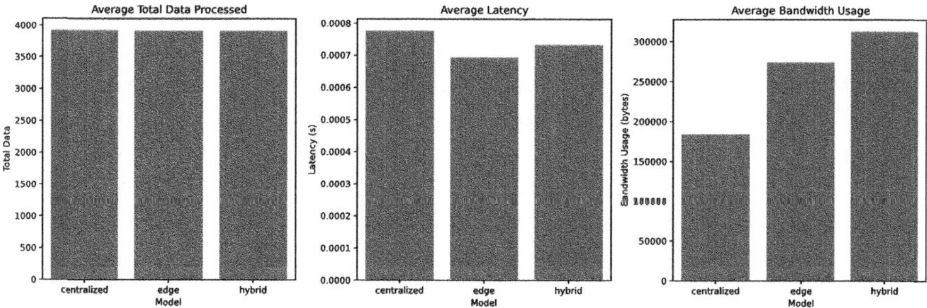


**Fig. 3.** Performance Comparison of Data Processing Models

## 4.2   Scalability Analysis

Figure 4 shows the latency performance of the three data processing models as the number of IoT devices increases from 1 to 750. Initially, all models experience a latency spike, particularly the edge and hybrid models, due to the overhead from excessive resource usage during connection setup. As the number of devices increases, the edge model maintains the lowest latency due to local data processing, making it ideal for real-time applications. In contrast, the centralised model shows stable but higher latency, with bottlenecks as more devices connect. The hybrid model exhibits variability as it dynamically offloads tasks between edge and cloud, depending on system conditions. While not achieving the lowest latency, it scales well by balancing resources. Scalability was also tested across different data rates, revealing that the edge model performs best at lower rates, while the hybrid model handles higher device counts more efficiently.

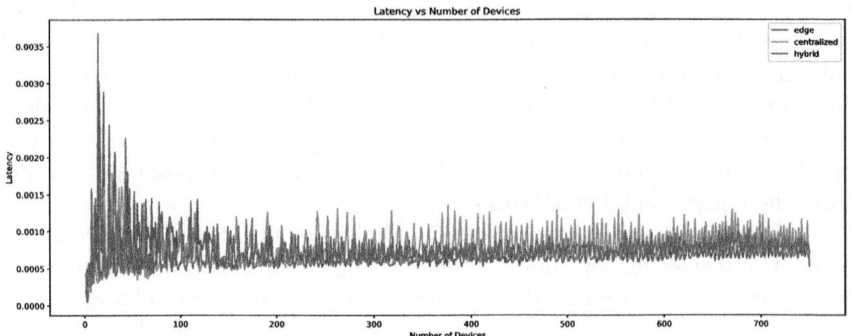

**Fig. 4.** Impact of Increasing IoT Devices on Latency

The analysis of bandwidth usage versus device number reveals different patterns for each processing model. In Fig. 5, the centralised model has the lowest bandwidth consumption, which indicates efficient data aggregation. The edge model has moderate bandwidth usage, while the hybrid model consumes the most due to combined edge-cloud communication. All models show a linear increase in bandwidth usage as the number of devices grows. This highlights scalability challenges, such as increased network congestion and the need for more efficient resource allocation. The centralised model can be suitable for high data transmission costs.

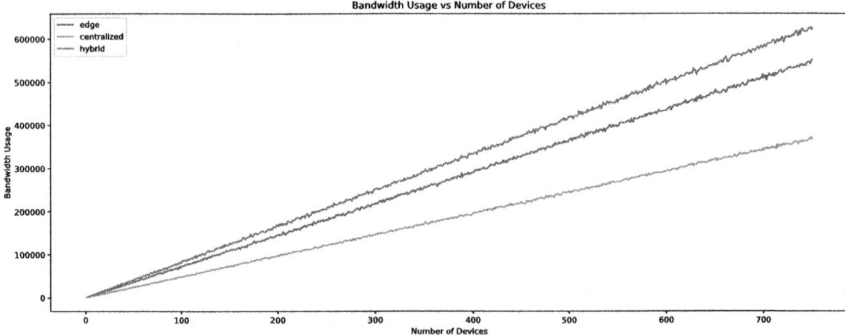

**Fig. 5.** Impact of Increasing IoT Devices on Bandwidth Usage

### 4.3   Impact of Data Rate

The analysis of Fig. 6 reveals that as the data rates increase, the total processed data decrease consistently across all models. The hybrid model processes the most data but experiences the steepest decrease. This decline indicates that, although the hybrid model can initially handle higher data loads, it becomes less efficient at peak data rates due to the combined overhead of edge-cloud communication. Latency trends indicate that the centralised model maintains the highest latency, while the edge model shows increasing latency with higher data rates. Bandwidth usage decreases as data rates increase, with the centralised model using the least bandwidth, the edge model showing moderate usage, and the hybrid model consuming the most.

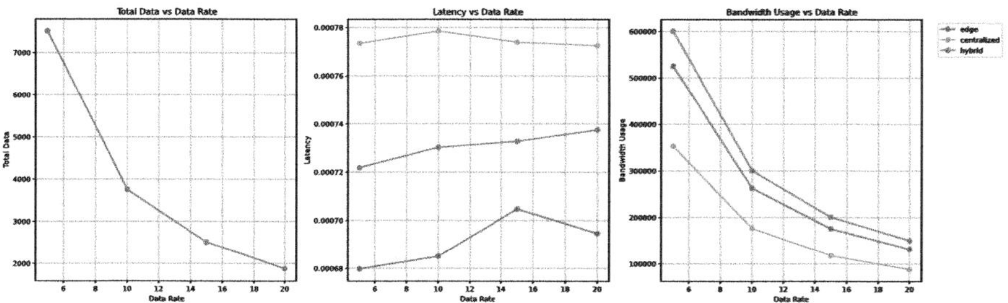

**Fig. 6.** Impact of Data Rate on Processing Models

### 4.4   Statistical Comparison of Processing Models Using ANOVA

The Analysis of Variance (ANOVA) test was applied to systematically compare the performance of different data processing models in various scenarios. It ensures that the observed differences are statistically significant and not due to random chance [25]. The conducted test evaluates the variance of the F-statistic, which measures the differences between group means, and the p-value, which indicates the significance level of those differences, as Table 1 summarises. This statistical approach helps identify which models outperform others in terms of key performance metrics, providing a clearer understanding of the trade-off between each model.

**Table 1.** Analysis of Variance Test Results and Interpretation

| Metric | F-Statistic | p-Value | Interpretation |
| --- | --- | --- | --- |
| Total Data | 0.001029621967808 | 0.9989709080283068 | Not statistically significant ($p > 0.05$) |
| Latency | 74.77462367175406 | 6.2048983053541e-33 | Statistically Significant ($p < 0.05$) |
| Bandwidth usage | 253.56178769852596 | 7.4323892615897e-108 | Statistically Significant ($p < 0.05$) |

The analysis of the three data processing models reveals distinct performance characteristics. Although all models demonstrate almost similar data processing capabilities, they differ significantly in terms of latency and bandwidth usage. The edge model exhibits the best latency performance, with a mean of 0.000691. This is followed by the hybrid model at 0.000731 and the centralised model at 0.000775. ANOVA results confirm these differences as statistically significant $(F = 74.775, p < 0.05)$. In terms of bandwidth usage, the centralized model consumes the least, with a mean of 183,578.03. It is followed by the edge model at 273,387.93 and the hybrid model at 312,425.48. ANOVA results again confirm significant differences $(F = 253.562, p < 0.05)$. These findings highlight the trade-off between latency and bandwidth usage across the different processing models in IoT edge-cloud environments.

## 4.5   Neural Network Performance

Our neural network model demonstrated high accuracy in classifying strategies across various simulated scenarios. The model achieved an overall test accuracy of 98.83%. It indicates a strong ability to predict the optimal data processing strategy.

Table 2 presents a concise overview of the metrics included in the classification report. The precision of each strategy, which quantifies the accuracy of positive predictions. The recall parameter, which indicates the capacity to identify all the positive instances, and the F1-score, which constitutes the harmonic mean of precision and recall. The classification results reveal that the neural network succeed in predicting optimal data processing strategies for IoT scenarios, with precision, recall, and F1-scores between 0.98 and 1.00 across Centralised, Edge, and Hybrid strategies. The Centralised strategy achieves perfect precision, while Edge and Hybrid strategies exhibit slightly lower but still acceptable results. This consistent performance and balanced support (600 instances per strategy) underscore the robustness of the model in real-time decision-making for optimising IoT data processing, with minor variations possibly reflecting the complexities of each strategy.

**Table 2.** Neural Network Classification Report

| Strategy | Precision | Recall | F1-Score | Support |
|---|---|---|---|---|
| Centralised | 1.00 | 0.99 | 0.99 | 600 |
| Edge | 0.98 | 0.98 | 0.98 | 600 |
| Hybrid | 0.98 | 0.99 | 0.99 | 600 |

The training progress of the neural network is illustrated in the graphs depicted in Fig. 7. It displays the accuracy and loss of the model across 100 epochs. The accuracy graph demonstrates an increase, with the model attaining an accuracy of more than 90% during the first 10 epochs, suggesting an effective acquisition of unique characteristics for each strategy. The validation accuracy closely mirrors the training accuracy, indicating the capacity of the model to generalise to unfamiliar data. The loss graph illustrates a consistent decrease, with both the training and validation loss progressively falling. This further indicates the excellent learning process of the model and suggests that it is a well-generalised model. Overall, the final high accuracy of 98.67% and the detailed classification metrics indicate that the neural network has successfully learnt to classify the optimal processing strategy based on the input parameters with high reliability. The periodic adjustments in the learning rate contribute to overcoming potential local optima, resulting in the robust performance of the model.

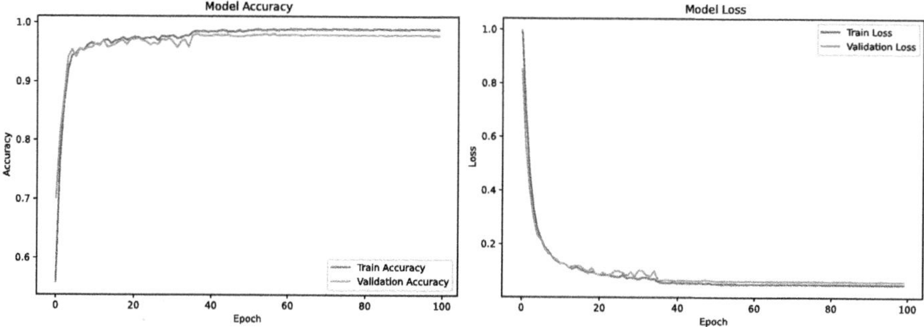


**Fig. 7.** Learning Rate Curve for Neural Network Model.

Before concluding, it is essential to take into account the pragmatic consequences of this research for actual Internet of Things (IoT) implementations. The results highlight the need for organisations to do a thorough evaluation of their processing needs. Choosing between edge, centralised, or hybrid models will significantly affect system performance, especially in terms of latency and bandwidth efficiency. Utilising AI-driven models to determine strategies dynamically helps enhance resource management, guaranteeing that IoT systems function more efficiently and adaptively in different circumstances.

## 5    Conclusion

The convergence of cloud computing and IoT constitutes an advent today. The cloud provide scalable data storage and processing, while IoT contributes continuous streams of data. This combination facilitate the development of applications in different domains.

This study assesses different data processing models (centralised, edge, and hybrid) for cloud-IoT systems, highlighting the importance of selecting suitable strategies based on unique application needs. The results of our study demonstrate that edge models have a substantial impact on reducing latency in applications that require real-time responsiveness. Hybrid models are the most effective in environments dealing with complex and data-intensive tasks. The accuracy of the proposed neural network in forecasting optimal solutions indicates the potential of real-time optimisation in dynamic IoT contexts. Despite limitations such as the lack of real-world tests, future research could explore simulating different types of devices and varying network conditions. We could also study how to incorporate more advanced data analytics and conduct large-scale simulations. Additionally, examining the security implications and validating the results in real-world settings would be valuable areas of study. These technological breakthroughs have the potential to greatly improve the efficiency, scalability, and adaptability of Internet of Things (IoT) systems in a wide range of applications.

# References


1. Haji, L.M., Ahmad, O.M., Zeebaree, S.R.M., Dino, H.I., Zebari, R.R., Shukur, H.M.: Impact of cloud computing and internet of things on the future internet. 62, (2020)
2. Nazari Jahantigh, M., Masoud Rahmani, A., Jafari Navimirour, N., Rezaee, A.: Integration of Internet of Things and cloud computing: a systematic survey. IET Commun. **14**, 165–176 (2020). https://doi.org/10.1049/iet-com.2019.0537
3. Rani, S., Bhambri, P., Kataria, A.: Integration of IOT, big data, and cloud computing technologies: trend of the era. In: Big Data, Cloud Computing and IoT, pp. 1–21. Chapman and Hall/CRC (2023)
4. Mohammed Sadeeq, M., Abdulkareem, N.M., Zeebaree, S.R.M., Mikaeel Ahmed, D., Saifullah Sami, A., Zebari, R.R.: IoT and cloud computing issues, challenges and opportunities: a review. QAJ. 1, 1–7 (2021). https://doi.org/10.48161/qaj.v1n2a36
5. Lahby, M., Saadane, R., Correia, S.D.: Integration of IoT with cloud computing for next generation wireless technology. Ann. Telecommun. **78**, 653–654 (2023). https://doi.org/10.1007/s12243-023-01003-3
6. Chen, Y.: IoT, cloud, big data and AI in interdisciplinary domains. Simul. Model. Pract. Theory **102**, 102070 (2020). https://doi.org/10.1016/j.simpat.2020.102070
7. Wu, Y.: Cloud-edge orchestration for the internet of things: architecture and ai-powered data processing. IEEE Internet Things J. **8**, 12792–12805 (2021). https://doi.org/10.1109/JIOT.2020.3014845
8. Bambrik, I.: A survey on cloud computing simulation and modeling. SN Comput. Sci. **1**(5), 1–34 (2020). https://doi.org/10.1007/s42979-020-00273-1
9. Mansouri, N., Ghafari, R., Zade, B.M.H.: Cloud computing simulators: a comprehensive review. Simul. Model. Pract. Theory **104**, 102144 (2020). https://doi.org/10.1016/j.simpat.2020.102144
10. Internet of things (IoT) fusion with cloud computing: current research and future direction. IJATEE. 9, (2022). https://doi.org/10.19101/IJATEE.2021.876002
11. Shirke, A., Chandane, M.M.: Collaborative offloading decision policy framework in IoT using edge computing. Multimed Tools Appl. (2023). https://doi.org/10.1007/s11042-023-14383-4
12. Cao, K., Liu, Y., Meng, G., Sun, Q.: An overview on edge computing research. IEEE Access. **8**, 85714–85728 (2020). https://doi.org/10.1109/ACCESS.2020.2991734
13. Souza, P.S., Ferreto, T., Calheiros, R.N.: EdgeSimPy: Python-based modeling and simulation of edge computing resource management policies. Futur. Gener. Comput. Syst. **148**, 446–459 (2023). https://doi.org/10.1016/j.future.2023.06.013

14. Mechalikh, C., Taktak, H., Moussa, F.: PureEdgeSim: a simulation toolkit for performance evaluation of cloud, fog, and pure edge computing environments. In: 2019 International Conference on High Performance Computing and Simulation (HPCS), pp. 700–707. IEEE, Dublin, Ireland (2019). https://doi.org/10.1109/HPCS48598.2019.9188059
15. Li, J., Moeini, B., Nejati, S., Sabetzadeh, M., McCallen, M.: A lean simulation framework for stress testing IoT cloud systems. IIEEE Trans. Software Eng. **50**, 1827–1851 (2024). https://doi.org/10.1109/TSE.2024.3402157
16. Sundas, A., Panda, S.N.: An introduction of cloudsim simulation tool for modelling and scheduling. In: 2020 International Conference on Emerging Smart Computing and Informatics (ESCI), pp. 263–268. IEEE, Pune, India (2020). https://doi.org/10.1109/ESCI48226.2020.9167549
17. Alwasel, K., et al.: IoTSim-Osmosis: a framework for modeling and simulating IoT applications over an edge-cloud continuum. J. Syst. Architect. **116**, 101956 (2021). https://doi.org/10.1016/j.sysarc.2020.101956
18. Sonmez, C., Ozgovde, A., Ersoy, C.: EdgeCloudSim: an environment for performance evaluation of edge computing systems. Trans Emerging Tel Tech. **29**, e3493 (2018). https://doi.org/10.1002/ett.3493
19. Stergiou, C.L., Psannis, K.E.: Digital twin intelligent system for industrial internet of things-based big data management and analysis in cloud environments. Virtual Reality Intell. Hardware. **4**, 279–291 (2022). https://doi.org/10.1016/j.vrih.2022.05.003
20. Nandhakumar, A.R., Baranwal, A., Choudhary, P., Golec, M., Gill, S.S.: EdgeAISim: a toolkit for simulation and modelling of ai models in edge computing environments. Measurement: Sensors. 31, 100939 (2024). https://doi.org/10.1016/j.measen.2023.100939
21. Collins, A.J., Sabz Ali Pour, F., Jordan, C.A.: Past challenges and the future of discrete event simulation. J. Defense Model. Simul. 20, 351–369 (2023). https://doi.org/10.1177/15485129211067175
22. SimPy documentation, SimPy 4.1.1. https://simpy.readthedocs.io/, Last Accessed 09 Aug 2024
23. Abdolrasol, M.G.M., et al.: Artificial neural networks based optimization techniques: a review. Electronics **10**, 2689 (2021). https://doi.org/10.3390/electronics10212689
24. Hu, Z., Zhang, J., Ge, Y.: Handling vanishing gradient problem using artificial derivative. IEEE Access. **9**, 22371–22377 (2021). https://doi.org/10.1109/ACCESS.2021.3054915
25. Das, B.K., Jha, D.N., Sahu, S.K., Yadav, A.K., Raman, R.K., Kartikeyan, M.: Concept building in fisheries data analysis. Springer Nature Singapore, Singapore (2023). https://doi.org/10.1007/978-981-19-4411-6_7

# Explore Neuro-Symbolic Approach in Knowledge Inference Based on Learner-Related Factors

Zahoua Amoura[1]([✉]), Farida Bouarab-Dahmani[1], Samia Lazib[1], and Fatiha Tali-Otma[2]

[1] Computer Science Department, Mouloud Mammeri University of Tizi Ouzou, Campus Hasnaoua 2, Tizi Ouzou, Algeria
{zahoua.amoura,bouarabfarida,samia.lazib}@ummto.dz
[2] UMR EFT University Toulouse2, Toulouse, France
fatiha.tali@univ-tlse2.fr

**Abstract.** This article explores the use of a neuro-symbolic approach to support school orientation, based on learner-related factors, Career decision self-efficacy and personality, using the Career Decision Self-Efficacy Scale and Holland's theory. The challenge lies in combining symbolic methods—used for rule-based reasoning, such as assessing self-efficacy levels—with neural models for tasks lacking clear rules, such as predicting suitable university fields. The system provides fields recommendations by integrating self-efficacy profiles and personality and offers personalized advice to enhance autonomy in decision-making. This approach contributes to school orientation by combining neuro-symbolic techniques for more informed and personalised recommendations.

**Keywords:** Neuro-symbolic · Knowledge inference · Learner-related factors · Holland's theory · PGI · Career decision self-efficacy

## 1 Introduction

School orientation is a particularly important subject, as it strongly determines a learner's chances of success in school and at work. However, this process is complex, as it is influenced by multiple factors, such as learner-related factors including self-efficacy [1] and personality [2], When these factors are not adequately considered, learners may experience stress, reduced productivity.

Vocational personalities are widely recognized as the most reliable predictors of both professional choices [3] well as school orientation choices. However, despite the relation between vocational personalities and professional choices, traditional methods still have significant room for improvement in terms of accuracy.

Traditional methods of predicting professional choices based on theoretical rules that describe the relationship between individual vocational personalities and professional aspirations, often leading to their development through symbolic approaches [4]. However, due to the complexity and multidimensionality of these relationships, these methods may lack precision. Consequently, recent researches [5, 6] has focused on utilizing machine learning to refine these predictions by capturing more nuanced interactions enhancing the reliability of the results. Furthermore, our study focuses on Career decision self-efficacy and decision-making autonomy.

In this article, we explore a neuro-symbolic method to generate school orientations recommendations that integrates the strengths of theoretical rules and data-driven approaches, which can improve the overall accuracy of predictions.

## 1.1  Vocational Personality

Vocational personalities are associated with preferences for particular activities and environments [3]. Consequently, vocational personality assessments are designed to evaluate an individual's level of interest in various activities, such as building a birdhouse. Vocational personalities play a crucial role in guiding individuals in their professional and school orientation choices.

### Holland's Theory of Vocational Personalities

Holland's theory of vocational personalities [3] is a well-established framework for understanding careers and vocational choices, developed through extensive research by American psychologist John Holland. This theory, widely utilized in career orientation [7], identifies six distinct personality types, each corresponding to a specific vocational profile known as RIASEC: "Realistic," "Investigative," "Artistic," "Social," "Entrepreneurial," and "Conventional." According to Holland [3], individuals of the same type are often attracted to similar professional activities due to their shared personality traits, goals, and psychological dispositions.

While this assessment was primarily designed to help identify suitable career paths, it is also beneficial for determining the fields of study that learners should follow, as these academic choices can significantly influence their future career opportunities.

## 1.2  Usual Methods of Matching Personality to Aspiration Orientations

One method for matching vocational personalities to aspiration orientations using Holland's RIASEC categories is based on the fact that the person's highest personality code corresponds to the highest code of the profession. In other words, a person whose highest code is Social would have a high match with an occupation that exemplifies social interests.

Another method for evaluating matching is through "profile matching". This approach assesses the extent of alignment between the complete RIASEC score profiles of an individual and a profession by correlating the score sets or calculating the Euclidean distance between the profiles. This particular method has often been used to guide professional choices in practice. For example, the O*NET Interest Profiler [8], which uses the full profile of participant scores, provides a list of the best matching professional choices.

## 1.3  Career Decision Self-efficacy

Career decision self-efficacy is a fundamental concept that emerged from the pioneering research of Taylor and Betz [8]. It plays a crucial role in understanding and addressing career indecision. Taylor and Betz defined self-efficacy as a person's belief in their ability to successfully complete the tasks necessary for making career decisions. They developed the Career Decision Self-Efficacy Scale (CDSeS), which quickly became an essential tool for assessing an individual's confidence in their ability to make decisions related to their career path. The scale has been translated and adapted in various international contexts, including a version for a French sample by [9]. It comprises four factors: selecting long-term goals, changing career paths, job search techniques, and finding and processing information.

Additionally, [10] adapted Gaudron's original career decision self-efficacy scale for learners in the context of school orientation, emphasizing its significance in guiding learners. This adaptation identified four Career decision self-efficacy (CDSE) profiles among high school learners. The first profile, characterized by high career decision self-efficacy, indicates a clear direction and strong confidence in their ability to make informed choices. In contrast, the second profile, while also exhibiting a high level of confidence, demonstrates some uncertainty regarding how to navigate the challenges associated with changing direction. The third and fourth profiles, with medium to low levels of self-efficacy, appear to struggle more with the orientation process, indicating a greater need for support to clarify their career goals and make definitive decisions.

## 2  Related Work

Research directly combining neuro-symbolic approaches with school orientation systems is still rare. However, there is emerging research in the use of neuro-symbolic approaches within education. One example is a framework proposed by [11], which focuses on interpretable AI in education based on neuro-symbolic techniques. This framework is designed to enhance the confidence of educators and learners in data-driven decisions, showing the potential of neuro-symbolic systems to transform school orientation by combining symbolic reasoning with neural networks for deeper insights.

There are notable studies that leverage machine learning techniques, especially those built on Holland's RIASEC theory, which links vocational personality to professional choices. For example, [12] introduced an educational and vocational guidance system in Morocco, which employs a to facilitate career decision-making by aligning user responses with career suggestions through machine learning models.

Another study [6] explores how machine learning can be applied to psychological assessments, aiming to enhance decision-making processes. This research underlines the potential of artificial intelligence (AI) to support and refine the RIASEC model.

While self-efficacy is frequently cited as an important factor in school orientation [10], concrete applications that integrate self-efficacy models into IT-based educational systems are still lacking.

## 3  Proposed Process

### 3.1  Knowledge Inference Module from Learner-Related Factors

This module serves as an expert system that integrates a neural network, forming a neuro-symbolic system. The expert system contains a rule that, when activated, triggers a prediction using a trained neural network model. This prediction generates a fact that subsequently activates other rules within the system. This interconnected framework ensures that the knowledge acquired from the neural network dynamically informs and enriches the reasoning process, thereby enhancing the overall decision-making.

Figure 1 presents the Neuro-symbolic process diagram for knowledge inference from learner-related factors, illustrating how the expert system and neural network interact to facilitate knowledge extraction and decision-making within the school orientation context.

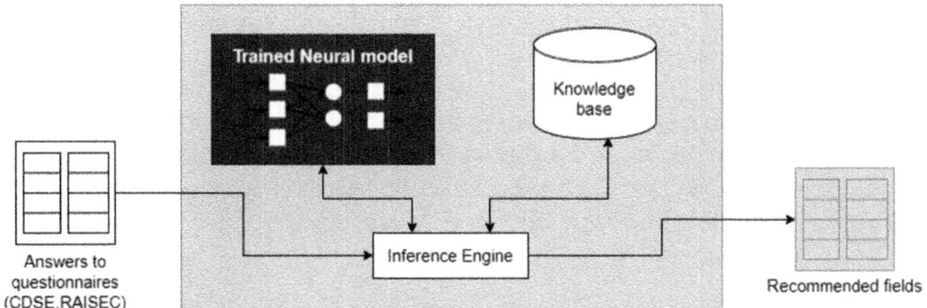


**Fig. 1.** Neuro-symbolic process diagram for knowledge inference from learner-related factors

The expert system relies on a knowledge logic framework, employing propositional rules. The system starts by classifying learners into CDSE categories. This classification aids in generating tailored advice to improve leaner CDSE, fostering greater decision-making autonomy.

In this module, learners complete a questionnaire based on CDSE, measuring their confidence on a scale from 0 to 10 for each question. A rule is activated for each response to calculate the CDSE score for each of the four CDSE factors, previously mentioned in Sect. 1.3 Career decision self-efficacy. Once the questionnaire is completed, other rules determine the learner's CDSE profile.

For instance, consider the case of a learner, Alice, who completes the CDSE questionnaire and receives a score of 5 for the 'changing career paths' factor (f2score). Based on this result and the following rule, the learner will be classified into the 'CDSE_inf' profile:

```
If f2score < 9:
        self.declare(CDSE_Class('CDSE_inf'))
```

Upon classifying the CDSE, advice rules are activated, which encompass four distinct recommendations designed to enhance Career decision self-efficacy. In addition, the rules concerning facts representing the lowest levels of CDSE trigger a personality questionnaire to measure RIASEC codes, and trigger a process of fields recommendations.

In Alice's case, whose profile is 'CDSE_inf', the following rule is triggered, providing her with recommendations to enhance her Career Decision Self-Efficacy and launching the RIASEC questionnaire.

```
if cdse_class.class_name == "CDSE_inf":
  self.advice_recommendations.append(adviceRecommenda-
tion(
      name='Faible CDSE',
      codes=' CDSE_inf',
      description='Students in the CDSE-inf class lack
information about educational programs and rely primarily
on online sources, without direct interactions with
knowledgeable individuals. They hesitate to share their
decisions with close ones and face family pressure re-
garding further studies.',
      conseil=["Diversify your sources of information by
talking to professionals or visiting schools to better
understand careers and educational paths.", "Don't hesi-
tate to discuss your choices with family members and
teachers to get personalized advice and build confidence
in your decisions."]
    ))
  self.launch_PGI_RIASEC_questionnaire()
```

The RIASEC score is calculated using 20 items from the Personal Globe Inventory Mini (PGI-Mini) [13].Participants respond on a scale of 1 to 7, indicating their preferences for various activities.

After completing the questionnaire and obtaining all scores, the inference engine inputs the 20 PGI item scores and the calculated RIASEC scores into a trained neural network to predict aspiration RIASEC scores.

We trained the neural model using a dataset containing individuals' RIASEC vocational personality scores and RIASEC aspiration scores. Using this dataset, the model was trained to predict RIASEC aspiration scores from RIASEC personality scores, which illustrates a supervised learning approach. We used hyperparameter fitting via random grid search and 10-fold cross-validation to optimize the model.

Once the RIASEC codes are predicted, consider Alice's prediction as follows: {'R': 2.63, 'I': 2.33, 'A': 1.56, 'S': 1.6, 'E': 3.5, 'C': 6.89}. The engine calculates the top three Holland codes with the highest scores, a process activated only when the predicted_RIASEC fact exists. This step introduces multiple facts into the engine's fact list, including 1TopCode, 2TopCode, and 3TopCode.Once the final Holland codes are generated, fields recommendation rules are activated.

According to Alice's RIASEC prediction, her 1TopCode is 'C' (Conventional), her 2TopCode is 'CE' (Conventional and Enterprising), and her 3TopCode is 'CER' (Conventional, Enterprising, and Realistic). The rules that will be activated are those where top_codes[0] = = 'C', top_codes[1] = = 'CE', and top_codes[2] = = 'CER'. The following rule is an example:

```
if len(top_codes) > 1 and top_codes[1] == 'CE':
  self.recommendations.append(Recommendation(
    name='Bookkeeping, Accounting, and Auditing Clerks',
    codes=['CE'],
    scores_RIASEC = {'R':2.33, 'I':2.33, 'A':1, 'S':1,
'E':3, 'C':7},
    description='Studying Accounting involves the process
of keeping track of financial accounts, ensuring the flow
of money and handling all money matters.'
    job titles = ["Account Clerk", "Accounting Assis-
tant", "Accounting Associate", "Accounting Clerk", "Ac-
counting Specialist, Accounting Technician, Accounts Pay-
able Clerk", "Accounts Payable Specialist", "Accounts
Payables Clerk", "Accounts Receivable Clerk"])
```

The activated rule generates an instance of the Recommendation class, which includes the field name, corresponding Holland codes, example job titles, field description, and RIASEC scores. These scores facilitate the ranking of recommendations based on their alignment with the learner's profile, determined by the Euclidean distance between each recommendation and the predicted RIASEC.

Some of the recommendations for Alice include Payroll and Timekeeping Clerks with the RIASEC profile {'R': 1.67, 'I': 1, 'A': 1, 'S': 2, 'E': 3.67, 'C': 7}, Bookkeeping, Accounting, and Auditing Clerks with the RIASEC profile {'R': 2.33, 'I': 2.33, 'A': 1, 'S': 1, 'E': 3, 'C': 7}, and Court Clerks with the RIASEC profile {'R': 3.67, 'I': 2, 'A': 1, 'S': 3.33, 'E': 4.33, 'C': 7}. The Euclidean distance is then calculated between these profiles and Alice's predicted profile {'R': 2.63, 'I': 2.33, 'A': 1.56, 'S': 1.6, 'E': 3.5, 'C': 6.89}, and the recommendations are ranked from the smallest distance to the largest. The final ranking according to the Euclidean distance is as follows: **Bookkeeping, Accounting, and Auditing Clerks** (distance of 1.01), **Payroll and Timekeeping Clerks** (distance of 1.79), and **Court Clerks** (distance of 2.28).

## 4   Validation of Our Proposal

### 4.1   Dataset Characteristic

To validate our proposal, we used data from the O*NET database [14]. Providing RIASEC scores for professions and occupations. Each occupation is rated based on its representation across the six RIASEC codes, helping individuals to assess whether a particular orientation matches their own interests and vocational personality.

We also used a publicly available dataset from [15] with N = 84 349 participants. These participants responded to a survey conducted by Time magazine in June 2016, answering questions about their professional preferences and personalities. The personality aspect is assessed using the PGI test, which includes 20 items where each question describes a task (e.g., "Teach people to dance"). Respondents rated their preference on a scale from 1 (Strongly dislike) to 7 (Strongly like).

## 4.2  Results

The following results compare the prediction accuracy of the "profile matching" method, which we will refer to as the 'profile method' in what follows, and which can be implemented with a symbolic approach, against the two neural networks that differ in their input. The first network "Neural Network1" input contains responses to the 20 PGI items and RIASEC scores of vocational personality, while the second "Neural Network2" was trained with only the RIASEC scores of personalities. Both networks aim to predict the RIASEC score of orientation.

The accuracy is evaluated using three methods: high-point hit rate, Euclidean distance, and profile correlation. High-point hit rate describes the match between the RIASEC code with the highest score of a learner's real aspiration and the predicted RIASEC score with the highest score. For example, if a learner's predicted RIASEC profile is {'R': 2.33, 'I': 2.33, 'A': 1, 'S': 1, 'E': 3, 'C': 7}, then the code with the highest score is 'C', with a score of 7. If the real aspiration code with the highest score is also 'C', the prediction is considered a match. However, if the real top aspiration code differs (e.g., it is 'E'), the prediction would not count as a hit. Euclidean distance measures the absolute distance between the RIASEC profile of a learner's aspiration and the predicted aspiration and correlation evaluates how similarly the RIASEC profile of a learner's aspirations aligns with the predicted RIASEC profile.

The results indicate that, in general, the accuracy of both Neural Networks models is identical and represents a better prediction accuracy than the traditional profile method. Table 1 illustrates the results.

**Table 1.** Overall results of the methods evaluated with hit rate, Euclidean distance, and profile correlation results.

|  | Hit Rate | Euclidean | Profile Corr |
|---|---|---|---|
| Profile Method | 0.34 | 4.97 | 0.28 |
| Neural Network1 | 0.36 | 4.28 | 0.35 |
| Neural Network2 | 0.36 | 4.27 | 0.35 |

By examining each category of the predicted RIASEC code, we found that both neural networks tended to provide better prediction accuracy for the RIASEC code categories that were most represented in the dataset (such as Enterprising), while Neural Network2 demonstrated better performance for these categories compared to Neural Network1. In contrast, the profile method tended to have better prediction accuracy for the RIASC code categories that were less represented in the dataset. These results were observed during the evaluation using the high-point hit rate, with results presented in Table 2, and the profile correlation, with results presented in Table 3.

Regarding the evaluation using Euclidean distance, presented in Table 4, the results suggested that both neural networks had superior prediction accuracy compared to the traditional profile method, with Neural Network2 showing similar or higher accuracy.

**Table 2.**  RIASEC code categories evaluation result with high-point hit

|   | Base Rate | Profile Method | Neural Network1 | Neural Network2 |
|---|---|---|---|---|
| R | 0.08 | 0.18 | 0.06 | 0.05 |
| I | 0.21 | 0.30 | 0.29 | 0.24 |
| A | 0.23 | 0.42 | 0.35 | 0.36 |
| S | 0.12 | 0.23 | 0.05 | 0.05 |
| E | 0.29 | 0.42 | 0.70 | 0.73 |
| C | 0.07 | 0.20 | 0.06 | 0.04 |

**Table 3.**  RIASEC code categories evaluation result with Profile correlation

|   | Profile Method | Neural Network1 | Neural Network2 |
|---|---|---|---|
| R | 0.02 | -0.05 | -0.08 |
| I | 0.24 | 0.20 | 0.17 |
| A | 0.31 | 0.43 | 0.45 |
| S | 0.23 | 0.12 | 0.11 |
| E | 0.38 | 0.57 | 0.60 |
| C | 0.28 | 0.38 | 0.38 |

**Table 4.**  RIASEC code category evaluation result with Euclidean distance.

|   | Profile Method | Neural Network1 | Neural Network2 |
|---|---|---|---|
| R | 5.80 | 5.12 | 5.11 |
| I | 5.00 | 4.56 | 4.57 |
| A | 4.76 | 4.06 | 4.02 |
| S | 5.06 | 4.60 | 4.58 |
| E | 4.81 | 3.87 | 3.85 |
| C | 5.08 | 4.37 | 4.37 |

### 4.3  Discussion

School orientation plays an essential role in the lives of learners, as it represents the first decision that significantly influences various outcomes throughout their school journey.

Our work in this article makes two main contributions. Firstly, we proposed that the learner's self-efficacy in making school orientation decisions should be taken into account in order to better understand the learner before supporting them in their school orientation. In addition, we stressed the importance of providing guidance to reinforce this self-efficacy, as an informed decision made consciously by the learner is a better-quality decision.

Secondly, our proposal uses the neuro-symbolic approach, which combines symbolic reasoning techniques with neural learning capabilities. This approach has allowed us to create a unique process where expert knowledge (the classification of learners according to their CDSE) is executed by an expert system, enabling the identification of learners with lower Career decision self-efficacy, i.e., those who are unsure of which path to follow. For these learners, the system triggers the neural process to predict the RIASEC code corresponding to their aspiration (the neural model offers greater accuracy in predicting the RIASEC code). The symbolic system then identifies the fields closest to this aspiration and ranks them, thus providing better support in the school orientation process. Our system generates more accurate recommendations compared to using traditional methods alone.

To assess the accuracy of our process, it is essential not only to examine its performance, but also to compare it with more conventional methods. Some of these methods look for a perfect match between the learner's RIASEC code and that of the profession or field, while others return results randomly. Our process, on the other hand, takes into account the learner's RIASEC code to predict the RIASEC code corresponding to the adequate field for for them, and then generates field recommendations ranked according to how close they are to the predicted adequate field RIASEC code.

## 5  Conclusion

The aim of this article is to propose a neuro-symbolic approach that can be useful in school orientation. We have proposed a method that combines the advantages of expert systems and neural networks. The design we have proposed is an expert system with a traditional knowledge base and trained neural network process. Our method considers career decision self-efficacy (CDSE) and provides advice on how to increase this CDSE. The results showed that, overall, our proposal produced higher predictive accuracy than conventional methods.

For future work, we plan to validate the proposed process real-world applicability by testing it with larger and real-life datasets. This would help ensure the robustness of the proposed approach.

### References


1. Betz, N.E., Rottinghaus, P.J.: Current research on parallel measures of interests and confidence for basic dimensions of vocational activity. J. Career Assess. **14**, 56–76 (2006). https://doi.org/10.1177/1069072705281348

2. Lounsbury, J.W., Smith, R.M., Levy, J.J., Leong, F.T., Gibson, L.W.: Personality characteristics of business majors as defined by the big five and narrow personality traits. J. Educ. Bus. **84**, 200–205 (2009). https://doi.org/10.3200/JOEB.84.4.200-205
3. Holland, J.L.: Making vocational choices: a theory of vocational personalities and work environments. Psychol. Assess. Resour. (1997)
4. Romli, A., Cha, A.P.: An expert system for stress management. In: 2009 International Conference for Internet Technology and Secured Transactions, (ICITST). pp. 1–6. IEEE, London (2009). https://doi.org/10.1109/ICITST.2009.5402579
5. Taylor, K.M., Betz, N.E.: Applications of self-efficacy theory to the understanding and treatment of career indecision. J. Vocat. Behav. **22**, 63–81 (1983). https://doi.org/10.1016/0001-8791(83)90006-4
6. Fokkema, M., Iliescu, D., Greiff, S., Ziegler, M.: Machine learning and prediction in psychological assessment: some promises and pitfalls. Eur. J. Psychol. Assess. **38**, 165–175 (2022). https://doi.org/10.1027/1015-5759/a000714
7. Nauta, M.M.: The development, evolution, and status of Holland's theory of vocational personalities: reflections and future directions for counseling psychology. J. Couns. Psychol. **57**, 11–22 (2010). https://doi.org/10.1037/a0018213
8. Rounds, J., Hoff, K., Lewis, P.: O*NET® Interest Profiler Manua (2021)
9. Gaudron, J.-P., Vautier, S.: Analyzing individual differences in vocational, leisure, and family interests: a multitrait-multimethod approach. J. Vocat. Behav. **70**, 561–573 (2007). https://doi.org/10.1016/j.jvb.2007.01.004
10. Tali, F., Ratinaud, P., Jmel, S., Gaudron, J.-P.: Relations entre les profils d'auto-efficacité vocationnelle et ses sources chez les élèves de terminale au moment du choix d'orientation post-bac. osp. pp. 341–384 (2023). https://doi.org/10.4000/osp.17489
11. Hooshyar, D., Yang, Y.: Neural-symbolic computing: a step toward interpretable AI in Education (2021)
12. Zahour, O., Benlahmar, E.H., Eddaoui, A., Ouchra, H., Hourrane, O.: A system for educational and vocational guidance in Morocco: Chatbot E-Orientation. Procedia Comput. Sci. **175**, 554–559 (2020). https://doi.org/10.1016/j.procs.2020.07.079
13. Tracey, T.J.G.: PGI - personal globe inventory (2021). https://www.psycharchives.org/handle/20.500.12034/4065. https://doi.org/10.23668/PSYCHARCHIVES.4545
14. O*NET: Interests. https://www.onetcenter.org/dictionary/28.3/excel/interests.html.
15. Glosenberg, A., Behrend, T., Tracey, T., Blustein, D., Foster, L.: Vocational interest data from Time magazine. (2022). https://doi.org/10.17605/OSF.IO/BE5JA

# Estimating Performance Costs of Enabling Privacy-Awareness in Data Lifecycles

Lelio Campanile[1]([envelope]) [iD], Mauro Iacono[1] [iD], Michele Mastroianni[2] [iD],
and Bruna Viscardi[3]

[1] DMF, Università degli Studi della Campania "L. Vanvitelli", Caserta, Italy
{lelio.campanile,mauro.iacono}@unicampania.it
[2] DAFNE, Università degli Studi di Foggia, Foggia, Italy
michele.mastroianni@unifg.it
[3] Caserta, Italy

**Abstract.** Privacy-awareness is a mandatory requirement for data management systems. When planning for compliance of existing systems, performance issues have to be considered: in this paper we propose an approach for evaluating performance costs of enabling privacy-awareness in data lifecycles based on risk analysis and queuing networks. We use as running example a case of monitoring for smart cities, starting from an existing case to provide a guideline for practitioners to show how to use simple performance models to evaluate the impact of an intervention to guarantee privacy compliance, exploiting a well-known, state-of-the-art, free, event-based simulator, namely Java Modeling Tools, to avoid practitioners the need for analytical models and related complexity.

**Keywords:** Smart city · Data Lifecycle Model · Simulation · Risk Analysis · Queuing Network

## 1 Introduction

The EU Regulation 2016/679 on General Data Protection Regulation (GDPR) prescribes a new logic for data management, as it enforces privacy protection since the first steps of the design of privacy-sensitive systems. While for new designs the way is indicated by privacy-by-design and privacy-by-default provisions, which may be managed by proper reconfigurations of the design cycles of software engineering [4], existing systems must be mandatorily made compliant, in order to avoid privacy-related risk and related costs.

Existing systems are coherent with their specifications, and privacy compliance result in additional non-functional specifications which must be taken into account in the hardening process, and result in additional derived functional specifications which integrate, and possibly superpose to, the original specifications. As a consequence, interventions on an existing system meant to make it compliant modify its structure, impacting on overall performances. System

B. Viscardi—Independent expert

administrators must ensure that new performance levels are compliant with the overall performance requirements, in order not to introduce potential problems and possible penalties on contracts, due to violations of the agrees Quality of Service (QoS) levels. The situation may be even more complex if human operators are part of the workflow of the system, e.g., to supervise some data processing phase, as, in this case, human operators represent a privacy risk which must be compensated by means of investments in training and assigning additional responsibilities, and GDPR requires in some cases also additional obligations in terms of registries which keep track of any access to privacy related data: this is anyway also a potential security risk, as privacy sensitive information is legitimately accessible to operators.

Tackling this problem requires performing an ex ante performance assessment of the system and a prediction of the effects on performances of the proposed hardening, in order to evaluate the best alternative and to support the design process; this performance assessment should include a differential risk analysis to evaluate the effectiveness towards the desired goals and a cost analysis both to evaluate new regime variable expenditures and to allow make or buy choices. The adoption of a sound data lifecycle approach, which is a good practice in information systems design and operation, facilitates QoS compliance and provides a good basis on which hardening can be designed, as it provides a clear and well organized architecture and a structured data processing framework; moreover, it also supports risk analysis, as organization and data flows are well documented.

In this paper we propose a modeling approach and a simulation based evaluation process to support privacy compliance oriented system hardening starting from the adopted data lifecycle model. Our goal is to provide a possible guidance, based on the analysis of a simple case, in the setup of the hardening process. The analysis of the as-is system is based on understanding of the closest data lifecycle model; the performance analysis of the as-is and to-be organization is based on queuing networks models (and on the Java Modeling Tools [5] simulation framework).

This paper is organized as follows: Sect. 2 presents background, Sect. 3 describes the running example and its hardening process, Sect. 4 describes its performance analysis; conclusions and future research directions close the paper.

## 2   Related Work and Background

### 2.1   Data Lifecycle Models

Data Lifecycle Models (DLM) illustrate how scientific data can be generated, stored, and utilized by academics and practitioners [15]. DLMs have been initially developed in scientific data management by both scholars, government agencies and enterprises with the aim of supporting collaborative working and enhancing standardization in the process of data management, mainly within the big data perspective. Two notable examples of DLM are respectively DataONE [1], developed by National Science Foundation (NSF) with a wide group of stakeholders belonging from both academia and corporation, and USGS [8] by U.S. Geological Survey.

There are some interesting review papers on this topic: in [13], the main issues related to data management and a new DLM particularly suitable for big data and the introduction of security and privacy criteria were presented. In [2] the main security-related issues regarding data have been discussed, with a useful threat map and security model, while in [18] a secure DLM conceived for government big data has been proposed. In [7] some relevant DLMs have been compared aiming to propose a roadmap to help users select the DLM that aligns with their data management goal. In order to do this, the authors determined the pertinent selection cycle criteria and developed a grading system for each of them. The same authors successively proposed a new DLC also including data quality, security and privacy [6].

An interesting survey about DLM is [16], in which a taxonomy is proposed of the various phases of well-known DLMs, and a new DLM is proposed as well.

In [15], the relevance of using a well-structured DLM in the case of data reuse has been highlighted [15]. An intriguing paper [14] compares DLM with Software Development Lifecycle for scientific applications, underlining similarities and differences.


## 2.2    Definition of the DLM Used

As described before, the various DLMs have different number, meaning and names of the phases required to process data accordingly with the particularly chosen. Moreover, some of the phases are to be acted sequentially, other are *transversal* across the whole process.

On the other side, the process we choose to use for this research does not fully adhere to any specific DLM. That said, in the work of Shah et al. [16], authors define a sort of *meta-phases* in order to compare several DLMs to each other. We choose to adopt the phase definition proposed there in terms of name and meaning. The temporal sequence of phases will be represented using UML state machines.

The phases defined by Shah et al. [16] are the following:

1. *Planning*: it produces the data management plan.
2. *Collection*: it defines the moment when the new data or metadata is created in the system
3. *Preparation*: it refers to data pre-processing, filtering and integration While in the pre-processing step noise and errors from collected data are identified and removed, during filtering activities data will be classified in different formats (structured, unstructured, . . . ). Instead, in the integration activity access to data sources has been standardized.
4. *Analysis*: it consist of performing analysis of data to extract knowledge. Different big data analytical tools may be utilized to analyze data.
5. *Visualization*: it has the aim of explaining the meaning of the discovered information through categorization of data visualizations.
6. *Access*: it focuses on communication paths and mechanisms between the data provider and data consumer in a big data ecosystem.

7. *Share*: it highlights which data can and must be made public and how it must be published according to security and integrity measures.
8. *Use, re(use) and feedback*: the use and re-use of data by consumers focuses on the discovery of new information from existing public datasets. In particular, the concept of feedback occurs when data users provide their feedback (e.g. user reactions, comments and suggestions).
9. *Archiving*: it is a process which links a chunk of data with a system through cataloging, indexing, or a related action.
10. *Delete*: in it duplicated data, no longer required data, and useless data is removed from the system.
11. *Storage*: it concerns the operations needed to reliably store data throughout the lifecycle.
12. *Data quality*: the quality phase focuses on ensuring and keeping data quality during the whole data lifecycle.
13. *Security and Protection*: data protection in terms of data integrity, security, access control, and privacy.
14. *Governance*: Ensuring the existence and exploitation of high quality and protected data throughout the entire data lifecycle. It determines policies and procedures to safeguard the preventive and effective management of data assets.

## 3    A Running Example

A smart cities information system has been chosen as running example[1]. The system gathers information from the urban environment close to the road system of a district, including pictures, videos and other information from various IoT sensors.

The whole process is operated by three different kinds of participants (see Fig. 1):

- A **Data Collector**, which is the corporation that collects the data from street cameras and IoT sensors and send data to the Data Broker;
- A **Data Broker**, which manages and process the data acquired by Data Collector and send data to Data Consumers;
- One or more **Data Consumers**, which represent the final recipients of the data.

Acquired information includes privacy-sensitive data, as vehicles move on the road and plates and images of people on board are acquired, and people walk on the sidewalks or cross the road. In the as-is system, the general process consists of several states through which each acquired information job may evolve, represented in the UML state machine in Fig. 2: this also describes the process

---

[1] This running example has been obtained starting from a real world case which cannot be disclosed and adapting it to keep all relevant details of general interest, in order to comply with non-disclosure agreements.

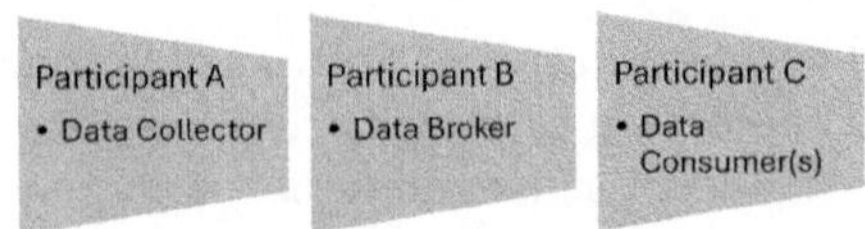

**Fig. 1.** The process data flow among participants.

of interest implemented in the system. As the states also represent effectively the phases a job undergoes during its processing, for the sake of space, abusing notation we will use UML state machines also to describe data management processes in the following, with no loss of expressiveness.

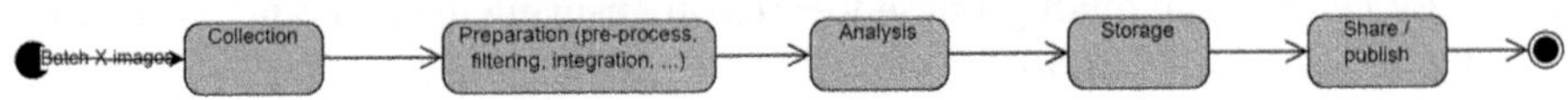

**Fig. 2.** Original data workflow

In our specific case, the *Collection* phase represents the acquisition of a job from the input subsystem, intended as a fixed aggregate volume of heterogeneous data. Each job potentially includes privacy sensitive information, thus this phase is already part of the potential risk exposure; this notwithstanding, the job is automatically processed, with no human intervention. The *Preparation* phase transforms a raw job in usable information set up in a suitable form for the analysis. The *Analysis* phase represents all higher-level information extraction operations, including the generation of semantic information and all what is needed for information indexation and warnings generation; clean data and related metadata are generated in the desired form and made available in the *Storage* phase. After this state, data is available for the customers of the information system.

Data is delivered to users on request according to orders in the *Share* phase. This is a composite state (see Fig. 3) as the use of data is subject to customer acceptance. When data is delivered (state *Delivery*), the customer performs a *Data assessment* to check if the QoS agreement is satisfied; in negative case, human intervention is required in the system, which in turn may produce fixed data sent back to the customer for a new *Data assessment*, may result in a failure in the process or may generate the scheduling of a new, synchronous, on demand *Collection* phase.

### 3.1 Securing Compliance and Hardening of the System

The smart city service above described suffers from several security and privacy issues. In particular, the personal data collected qualifies as i) systematic monitoring of publicly accessible areas; and ii) profiling of the behavior of individuals and, as stated by article 35 of GDPR, both conditions are indicators of high risk for the whole collection/analysis/storage process.

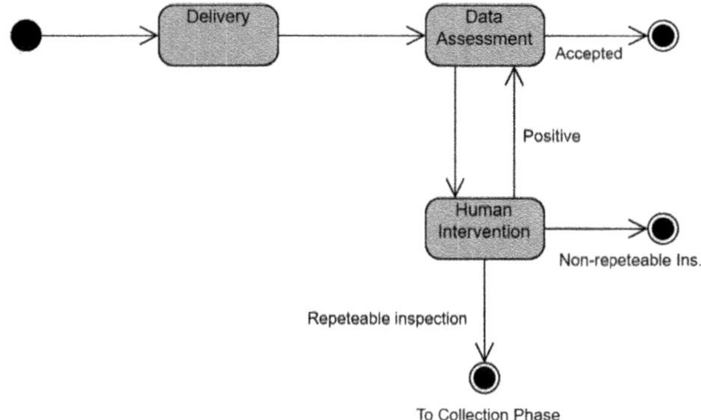


**Fig. 3.** Share composite state

Consequently, a first, trivial solution to the problem is, obviously, to use a regulation shield by turning all involved persons into data processors[2] to get into a defensive position under the legal point of view. The entities involved are all three participant as shown in Fig. 1, so the collector, the broker and consumers (i.e. all the broker's clients). GDPR states also that all those users are to be considered as internal personnel of data controllers or data processors.

Although this is reducing the risk of being liable, this is not really lowering the actual risk, as a security breach is always possible and there is no actual lowering of the risk exposure with this solution beyond purely formal terms. Moreover, even in absence of security breaches, operators can still access privacy sensitive information in disclosed form.

In fact, a cyber-security report published in 2024 by Verizon [17] shows that:

- above 35% of data breaches are caused by internal actors of organizations, and this number has grown in the last years;
- users' errors are growing substantially and there is also a rise in users' carelessness;
- about 50% of privilege misuse is done by internal employees.

Consequently, the data workflow described by Fig. 2 leads to a condition that is not fair for substantial privacy compliance due to the potentially high number of employees involved.

A more effective approach is aiming at hardening the system by introducing a new *Protection* phase devoted to privacy enhancing with the aim of lowering the risk on the basis of the results of the as-is risk analysis, as depicted in Fig. 4. In the case of this running example, exposure to risk both for privacy and security breaches in the system are as lower as earlier in the data management process privacy sensitive information which is not needed for the purposes of the system are made unavailable. This can be done by encryption (to obtain anonymization or pseudonymization as needed [3], or, in extreme cases or when about non

---

[2] As defined in article 28 of GDPR.

necessary information because of spurious data acquisitions, by deletion[3]. Consequently, as the *Analysis* phase is the first phase operating on semantically sound information, it can be extended by including an additional sub-phase in charge of a privacy risk related automated evaluation (e.g., by means of an artificial intelligence based solution e.g. as described in [12] capable of discerning non sensitive cases, which proceed to the actual analysis stage for metadata generation and consolidation, sensitive cases which can be handled automatically, which proceed to the *Protection* phase for encryption or deletion, and cases which require to be handled by *Human inspection*, actually operated by data processors.

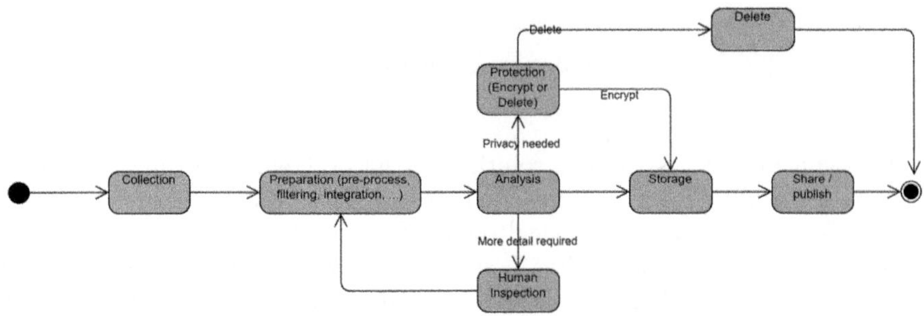


**Fig. 4.** Modified data workflow

This solution ensures that there is no direct[4] privacy risk in case of attacks to the storage subsystem of the system, as the *Storage* phase only operate on encrypted or non-sensitive data, and that there cannot be any exposure to human generated privacy risk in the *Share* phase, as operators cannot in any way access privacy sensitive information (so they do not even need to be data processors). As a secondary effect, consequently, the fact that no privacy sensitive information is stored in the system reduces the attack surface of the system as well, as the system becomes less interesting to attackers and privacy sensitive data breaches may only happen in real time and on an individual job basis during the *Collection, Preparation* and *Analysis* phases, which means mostly raw data rather than information, so with the need for preprocessing and analysis costs for the attacker.

---

[3] We do not include here the cases of deletion imposed by the GDPR, which are mandatory in any system, are related in our case to the *Storage* phase, and are outside the scope of this work.

[4] We cannot exclude offline privacy attacks based on big data techniques in case of massive data theft, which can anyway be mitigated by reducing the period in which data are kept available in the storage subsystem to the minimal needed time to provide the services, similarly to what GDPR prescribes for privacy sensitive information.

## 4 Performance Analysis

In order to analyze the performance cost of the privacy-aware mechanisms, a comparison between the performance indices of the as-is and to-be configurations can be carried on by means of queuing networks models [11].

In order to perform this analysis, two distinct models were constructed: an existing model (as-is model) and a proposed model (to-be model). The proposed model incorporates enhancements pertaining to privacy and security.

The as-is model, depicted in Fig. 5, represents the original workflow, starting from the collection of data from external sources to the sharing stage.

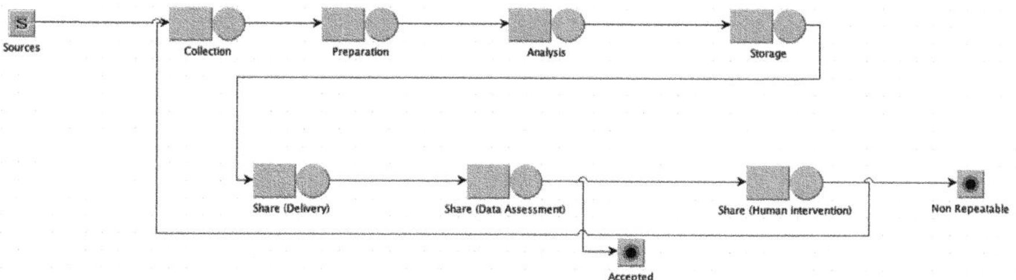


**Fig. 5.** As-is model

In the initial stage, data is collected from external sources, which serves as input to the entire process. Once collected, the data is gathered in the Collection stage, which initiates its propagation through the network. Once collected, the data moves to the Preparation stage, where it is cleaned, structured, and prepared for further processing to ensure consistency and quality. Next, the data is analyzed to provide insights for decision making. After analysis, the data is stored in the Storage stage, providing a repository for future use or reference. The network then moves to the sharing stages, which include Delivery, Data Assessment, and Human Intervention. In the Delivery phase, data is shared with relevant stakeholders who request it. During Data Assessment, the quality and validity of the data are evaluated before further sharing, while Human Intervention involves manual reviews.

Finally, data that passes all checks is either Accepted or marked as Not repeatable, indicating whether it is ready for final use or cannot be processed again.

The to-be model, in Fig. 6, evolves with a stronger emphasis on privacy and data security, incorporating additional steps and decision points that enhance the original network.

An important new step is the Privacy Protection Stage, which is added prior to analysis to ensure that sensitive data is anonymized or otherwise protected. This enhancement directly addresses privacy concerns by protecting data before it is further processed.

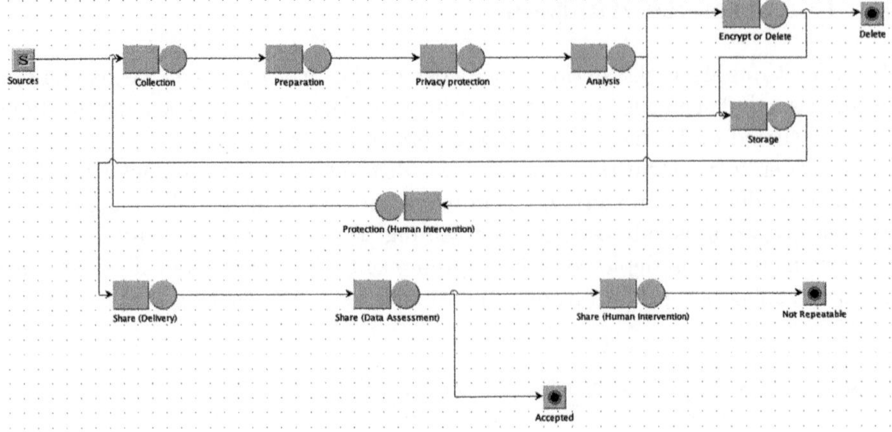



**Fig. 6.** To-be model

After Analysis, the updated model introduces a decision point to either encrypt or delete the data. This additional control over data management allows for the secure retention or removal of sensitive data based on its future use or compliance requirements. If Delete is chosen, the data is permanently removed, reflecting an increased focus on data minimization practices.

The updated network also includes an additional phase called Protection (Human Intervention) after data is stored, where manual review is performed.

Java Modeling Tools (JMT) was used to perform simulation for both model, the performance indicators utilized to evaluate the efficacy of the models are the *system response time* (the total time spent by each processing unit in the system, measured at the *Accepted* sink in the model) and the *system throughput* (the number of processing unit successfully completing their cycle in the system, measured at the *Accepted* sink in the model) as usually defined in queuing networks theory. Performance indicators have been obtained by means of event-based simulation to allow practitioners with little or no background in performance evaluation analytical models and techniques to implement our approach with a mild learning curve.

The system arrival time for entire systems is modeled by an exponential distribution with $\lambda$ set to 0.6. Similarly, all the service times are exponentially distributed according to values in Table 1.

In order to facilitate a comparison between the two models, all values assigned to the common station were set to the same value. This was done in order to evaluate the impact of the introduction of the station related to privacy protection.

Furthermore, a what-if analysis was conducted for the to-be model, varying the service time for the Privacy Protection Station from 1.11 ($\lambda = 0.9$) to 1.72 ($\lambda = 0.58$), with 10 equal intervals between these values, in order to evaluate

**Table 1.** Model parameters

| Stages | as-is | to-be |
|---|---|---|
| Collection | 0.6 | 0.6 |
| Preparation | 0.75 | 0.75 |
| Privacy Protection | / | 0.6 |
| Analysis | 0.7 | 0.7 |
| Encryption or Delete | / | 0.95 |
| Storage | 0.9 | 0.9 |
| Protection (Human Intervention) | / | 0.2 |
| Share (Delivery) | 0.8 | 0.8 |
| Share (Data Assessment) | 0.95 | 0.95 |
| Share (Human Intervention) | 0.1 | 0.1 |

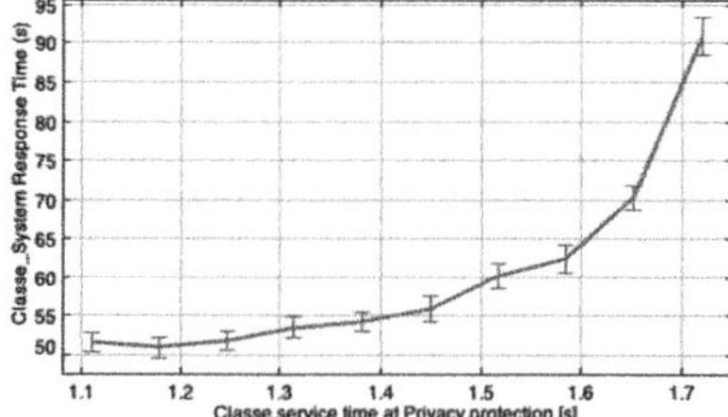

**Fig. 7.** To-be system response time varying Privacy protection service time

**Fig. 8.** To-be system throughput varying Privacy protection service time

the impact of adding a Privacy Protection Station under different service time conditions.

The simulation results demonstrate that in general the performance of the to-be system is comparable to that the original system, as desired. The mean system response time of the as-is model is 43.86, while that of the to-be model is 66.82. A similar trend is observed in system throughput, with both models show an average value above 0.5.

Moreover, the results of the what-if analysis suggest that increasing the service time capacity of the Privacy Protection Station, the mean system response time and the system throughput improve, as depicted in Fig. 7 and Fig. 8.

As a result, the trade-off between the lower performances and the compliance with low risk is advantageous and quantifiable and the provided approach viable.

## 5    Conclusions and Future Work

In this paper we provided a performance evaluation oriented approach for data managers which have to estimate the effects of privacy-related hardening of existing DLM. The approach is based on queuing networks and has been kept as similar as possible to the logic of DLMs, rather than focusing on a refinement of the performance models[5]. We believe that it can be a good tool to bridge practitioners between data management practices and performance evaluation practices.

Future work includes both a multiformalism [10] extension of the approach to account for the actual influence of users' requests and the definition of a domain specific language [9] for the integrated modeling of data lifecycle and performance, and the definition of a complete methodology to support privacy-oriented performance and risk evaluation for DLM: this is due to the will to further lower the learning curve and increase the user friendliness of the approach by providing a specific interface with the simulation formulated in terms of concepts from the domain of data management.

**Acknowledgements.** This research was partially funded by MUR PRIN PNRR 2022 grant number P20227W8ZC.

# References


1. Allard, S.: DataONE: Facilitating eScience through collaboration. J. Sci. Librarianship (2012)
2. Alshboul, Y., Nepali, R.K., Wang, Y.: Big Data LifeCycle: threats and security model. In: AMCIS (2015)
3. Campanile, L., Iacono, M., Levis, A.H., Marulli, F., Mastroianni, M.: Privacy regulations, smart roads, blockchain, and liability insurance: putting technologies to work. IEEE Secur. Priv. **19**(1), 34–43 (2021)
4. Campanile, L., Iacono, M., Mastroianni, M.: Towards privacy-aware software design in small and medium enterprises. In: 2022 IEEE Intelligence Conference on Dependable, Autonomic and Secure Computing, Intelligence Conference on Pervasive Intelligence and Computing, Intelligence Conference on Cloud and Big Data Computing, Intelligence Conference on Cyber Science and Technology Congress (DASC/PiCom/CBDCom/CyberSciTech), pp. 1–8 (2022)
5. Casale, G., Serazzi, G., Zhu, L.: Performance evaluation with java modelling tools: a hands-on introduction. ACM Perf. Ev. Review **45**(3), 246–247 (2018)
6. El Arass, M., Souissi, N.: Data lifecycle: From Big Data to SmartData. In: 2018 IEEE 5th Intelligence Configuration on Information Science and Technology, pp. 80–87 (2018)
7. El Arass, M., Tikito, I., Souissi, N.: Data lifecycles analysis: Towards intelligent cycle. In: 2017 Intelligent Systems and Computer Vision (ISCV), pp. 1–8 (2017)
8. Faundeen, J., et al.: The United States Geological Survey Science Data Lifecycle Model. Tech. rep, US Geological Survey (2014)
9. Gribaudo, M., Iacono, M., Kiran, M.: A performance modeling framework for lambda architecture based applications. Futur. Gener. Comput. Syst. **86**, 1032–1041 (2018)


---

[5] For example, we keep different queues in sequence rather than modeling sequential stages with a compound modeling element, e.g., a single queue with hypoexponentially distributed service times in place of several queues with exponentially distributed service times.

10. Gribaudo, M., Iacono, M.: Theory and application of multi-formalism modeling. IGI Global (2013)
11. Gribaudo, M., Iacono, M.: Places, transitions and queues: new proposals for interconnection semantics. In: Gilly, K., Thomas, N. (eds.) Computer Performance Engineering, pp. 216–230. Springer International Publishing, Cham (2023)
12. Hadi, H.J., Cao, Y., Nisa, K.U., Jamil, A.M., Ni, Q.: A comprehensive survey on security, privacy issues and emerging defence technologies for UAVs. J. Netw. Comput. Appl. **213**, 103607 (2023)
13. Khan, N., et al.: Big Data: Survey, technologies, opportunities, and challenges. Sci. World J. **2014**(1), 712826 (2014)
14. Lenhardt, W.C., Ahalt, S., Blanton, B., Christopherson, L., Idaszak, R.: Data management lifecycle and software lifecycle management in the context of conducting science. J. Open Res. Softw. **2**(1), e15–e15 (2014)
15. Park, J., Cordell, R.: The ripple effect of dataset reuse: Contextualising the data lifecycle for machine learning data sets and social impact. J. Inf. Sci. **0**(0), 01655515231212977 (2023)
16. Shah, S.I.H., Peristeras, V., Magnisalis, I.: DaLiF: a data lifecycle framework for data-driven governments. J. Big Data **8**(1), 1–44 (2021). https://doi.org/10.1186/s40537-021-00481-3
17. Verizon Inc.: 2024 Data Breach Investigations Report (DBIR) (2024). https://www.verizon.com/business/resources/reports/dbir/
18. Zahid, R., et al.: Secure data management life cycle for government Big-Data ecosystem: Design and development perspective. Systems **11**(8) (2023)

# Conciseness Effect on High Utility Itemsets Mining


Ibtissam Boussouf[1,2]([✉]) [iD], Walida Boussouf[3,4], Hamza Haddad[2],
Salah Eddine Mebarek Nacereddine[2], and Mohamed Salah Medjram[1,2] [iD]

[1] Computer Science Department, Ferhat Abbas Setif 1 University, 19000 Setif, Algeria
{mohamed.bensaadi,mohamed.bensaadi,
chafia.kara-mohamed}@univ-setif.dz
[2] LRSD Laboratory, Ferhat Abbas Setif 1 University, 19000 Setif, Algeria
abdelouahab.attia@univ-bba.dz
[3] Computer Science Department, Mohamed El-Bachir El-Ibrahimi University, 34000
Bordj-Bou-Arreridj, Algeria
[4] LMSE Laboratory, Mohamed El-Bachir El-Ibrahimi University, 34000 Bordj-Bou-Arreridj,
Algeria



**Abstract.** High Utility Itemset Mining (HUIM) extends Frequent Itemset Mining. HUIM identifies itemsets with significant impacts on metrics like sales or customer satisfaction. With the advent of Big Data, there is a demand for algorithms to extract these High Utility Itemsets (HUIs) efficiently. While the size of the output can be controlled *via* well-chosen support and confidence, the quality of the results can be altered. To address this challenge, various approaches generate more compact representations (CRs) of high-utility itemsets (HUIs), significantly reducing the output size while maintaining the integrity of the information and ensuring that the derived rules retain their high significance. Among these CRs we find closed, maximal, minimal itemsets, and generators. While CRs-based HUIs are typically compared internally, our study takes a unique approach with an external comparison. We analyze algorithms designed for HUIM without CRs (e.g., FHM, HUI-Miner, UPGrowth) against those incorporating various CRs (e.g., MinFHM, CHUI-Miner, CHUI-Mine, CLS-Miner, GHUI-Miner). Experiments on two large-scale datasets (Cosmetics and Chicago Crimes) evaluate performance based on runtime, memory size, and extracted HUIs. The findings indicate that algorithms utilizing CRs based on closed itemsets surpass those relying on generators; these latter yielding the least efficient results. Specifically, FHM and MinFHM demonstrate efficient extraction of HUIs. Additionally, MinFHM excels in runtime, memory usage, and in the size of extracted HUIs.

**Keywords:** Data mining · Big Data · High utility itemset · Concise representation · Closed itemset · Generators


## 1 Introduction

This paper discusses the data mining evolution, starting from Frequent Itemset Mining (FIM) [1, 2], to High Utility Itemset Mining (HUIM) [3], through Frequent Pattern Mining [4], and its offshoots [5, 6]. The FIM approach, initiated by the Apriori algorithm [7],



N. Seddari and M. Redjimi (Eds.): ICMSCT 2024, CCIS 2606, pp. 196–210, 2025.
https://doi.org/10.1007/978-3-032-01922-6_17

focuses on identifying frequently occurring itemsets but often results in many irrelevant or uninteresting itemsets. To address these FIM limitations, the HUIM approach has been proposed [8]. This latter considers the utility of items (such as profit or relevance) rather than just frequency. HUIM is useful in various applications like marketing and e-commerce. However, HUIM faces challenges like high memory usage due to generating numerous itemsets. Various approaches have been proposed to address this issue. Well-chosen values of support and confidence can reduce the size of the output but there is no guarantee on the quality of this output.

The paper proposes using concise representations (CRs) to reduce the output size without losing information. Different types of CRs—such as closed [9–13], maximal [14, 15], minimal [16], and generators [17]—summarize HUIMs more efficiently. We utilize a classification framework that systematically categorizes HUIM algorithms based on whether they incorporate concise representations (CRs) or operate without them, thereby facilitating a more structured comparative analysis. The paper presents a comparative analysis between two groups. The first group (G1) – HUIM algorithms without CRs – comprises FHM [18], HUI-Miner (High Utility Itemset Miner) [19], UPGrowth [20]. The second group (G2) comprises MinFHM [16], CHUI-Miner [21], GHUI-Miner [17], CLS-Miner [22] and CHUI-Mine [19] as algorithms using different types of CRs to provide an external comparative study. Experiments are conducted using two datasets (Cosmetics and Chicago Crimes), and the algorithms are evaluated based on computational time, memory usage, and the number of high utility itemsets extracted. The structure of the paper covers related work, a description of the algorithms, experiments, and conclusions; each of these in an independent section.

## 2  Related Work

A HUI was defined in [9] as "a set of values that appears in a database and has a high importance to the user, as measured by a utility function". The HUIM technique employs a utility factor to represent the total profit of an itemset, which is then utilized to identify a set of HUIs [23]. Different HUIM algorithms have been developed and are typically categorized based on approach used, into four groups: candidate generation and test-based, pattern growth-based, utility-list-based, and projection-based. In Table 1, we provide some characteristics of G1's algorithms. This category is usually divided into two types of algorithms: one-phase algorithms [24–29] and two-phase algorithms [30–33]. In our comparison, we have chosen from both categories.

**Table 1.** HUIM algorithms with no concise representations.

| Reference | Alg. Name | #Phases | Search Strategy |
| --- | --- | --- | --- |
| Tseng V S et al. [20] | UP-Growth | Two | Depth |
| Liu M et al. [19] | HUI-miner | One | Depth |
| Fournier-Viger P et al. [18] | FHM | One | Depth |

To address the challenges mentioned earlier, the concept of CRs has been introduced. Several algorithms have emerged for extracting concise representations of HUIs [34]. These algorithms are classified as level-wise mining algorithms, tree-based mining algorithms, and utility list-based mining algorithms [9]. Table 2 below, shows for each G2's algorithms the CR generated for HUIM. It also shows the CR used by the algorithm with which the comparison was done in the original paper. From Table 2, we note that no exhaustive external comparison is done.

**Table 2.** HUIM Algorithms with concise representation.

| Reference | CR type | Name | Compared with |
| --- | --- | --- | --- |
| Fournier-Viger et al. [16] | Minimal | MinFHM | CHUD, GHUI-Miner |
| Nguyen et al. [21] | Closed | CHUI-Miner | EFIM-Closed,HMiner-Closed |
| Fournier-Viger P et al. [17] | Generators | GHUI-Miner | CHUI |
| Dam, Tl. Et al. [22] | Closed | CLS-Miner | CHUI-Miner |
| Wu Cw. et al. [19] | Closed and Maximal | CHUI-Mine | CHUD |

In [12], the authors outlined the challenge of mining a lossless, concise, and compact representations (CRs) for HUIs. Consequently, they introduced the closed HUIs (CHUIs) algorithm. This approach helps in reducing the number of discovered patterns, thereby requiring less time for result analysis. Additionally, a novel representation known as the minimal high utility itemsets (MinHUIs) was proposed in [16]. These MinHUIs are defined as the smallest sets of items that yield a significant profit. In [14], the authors introduced two concise representations, defined as two sets. These representations have a smaller cardinality, providing a succinct summary of all frequent high utility sequences. The authors addressed the issue of low minimum utility thresholds and lengthy sequences, which involve computational time and memory. An efficient algorithm named frequent closed high-utility itemset miner (FCHUIM), capable of discovering all frequent and closed high-utility itemsets, was proposed in [13]. Advanced methods such as efficient list structures to reuse memory and quickly access information about items were also investigated in this work. In [15], the authors proposed two algorithms, MaxC-HUIM and C-HUIM, for mining CHUIs and MaxHUIs. In [35], a new approach was introduced for efficiently mining frequent closed HUIs and frequent generators of HUIs using a new weak lower bound utility known as *wlbu*. The authors presented an efficient pruning strategy for early removal of non-closed and/or non-generator high utility branches in the prefix search tree based on *wlbu*. In [18], the authors integrated the concept of producer into HUIM and developed two novel concise representations of HUIs: High Utility Generators (HUGs) and Generator of High Utility Itemsets (GHUIs).

Traditional HUIM algorithms overlook item categorizations, limiting discoveries to the lowest level of abstraction. To address this issue, the MLHMiner (Multiple-Level HMiner) algorithm is designed to incorporate item taxonomies to find HUIs across multiple levels. The approach, enhanced by pruning techniques, is proven effective through experimental evaluations on various datasets. The limitations of the MLHMiner approach include increased computational complexity due to handling multiple abstraction levels, dependency on the availability and accuracy of taxonomy data, potential inefficiencies in pruning irrelevant patterns, and added complexity in algorithm implementation and maintenance. These challenges may affect scalability and general applicability in various domains [36].

While existing vertical algorithms offer scalability and efficiency, they rely on expensive join operations. To address this, the paper presented in [8] introduces a new vertical algorithm, FOTH (Fast sOrted iTemset searcH), which uses the IndexSet structure to eliminate the need for joins, reducing memory and computation costs. Experiments on eight benchmark databases demonstrate that FOTH outperforms state-of-the-art algorithms, particularly in dense databases. Despite the advantages of the FOTH algorithm, several limitations remain. FOTH demonstrates strong performance in dense databases, but its efficiency may be reduced in sparse datasets, where the benefits of its IndexSet structure are less pronounced. Additionally, the complexity of the IndexSet data structure introduces implementation challenges, potentially affecting its ease of use and adaptability. While FOTH reduces computational and memory overhead, scalability issues may still arise when applied to very large databases with diverse patterns. Furthermore, its effectiveness is closely tied to specific database characteristics, limiting its general applicability across all types of datasets.

## 3  Comparative Study

### 3.1  Dataset Description

1) Cosmetics

This quantitative transactional dataset was gathered from a cosmetics store located in the City of Bordj-Bou-Arreridj (Algeria) throughout the year 2023. It encompasses comprehensive information concerning cosmetics, including records pertaining to products, customers, and transactions. The utility of items in this dataset is defined by their respective profits. Based on the ratio (countOfItems/transactionsAverageLength), this dataset is considered as sparse. Details regarding this dataset are outlined in Table 3.

2) Chicago Crimes[1]

The dataset contains records of crimes reported in Chicago from 2001 to 2017. Each transaction is associated with a <month, area> pair, detailing crimes reported in a specific area during a particular month. The items within transactions represent different types of crimes, with the utility of each item indicating the number of occurrences of that particular crime. In Table 4, we give some characteristics of the dataset.

---

[1] https://www.kaggle.com/datasets/chicago/chicago-crime.

**Table 3.**  Cosmetics dataset statistics.

| | |
|---|---|
| The count of items | 16,257 |
| The count of transactions | 170,374 |
| The size of the dataset | 5.95 Mo |
| Total utility | 45,229,212 |
| Average length of transaction | 3.86 |
| Maximum length of transaction | 151 |

**Table 4.**  Chicago Crimes dataset statistics

| | |
|---|---|
| The count of items | 30 |
| The count of transactions | 367,103 |
| The size of the dataset | 4 Mo |
| Total utility | 1,032,793 |
| Average length of transaction | 1.9 |
| Maximum length of transaction | 15 |

## 3.2  Process Steps

Here, we describe the block diagram of our comparative study, between the two groups of algorithms, G1 and G2. Figure 1 illustrates the main three steps of the process.

**Step 1:** Preprocessing of the input data from two datasets cited above: the process includes removal of null value quantities of product and negative quantities.

**Step 2**: Execute G1 and G2 algorithms from the SPMF site [37]. Use these algorithms with different values of minimum utility parameter, *MinUtil*. This execution is done on the two datasets mentioned above.

**Step 3:** Visualization and interpretation of the obtained results.

Pre-processing
- Data cleaning and errors removal.
- Removal of null value and negative quantities.
- Transform transactional database for using SPMF data mining library.

- Choose the Minimum Utility threshold.
- Run G1 algorithms: FHM, HUI-Miner and UPGrowth.
- Run G2 algorithms: CHUI-Miner, GHUI-Miner, CHUI-Mine and CLSMiner, Min-FHM.

Interpretation of algorithms' results.
- Memory usage.
- Number of patterns discovered.
- Runtime.

**Fig. 1.** Diagram for the steps of the proposed comparative study

## 3.3 G1 and G2 Algorithms Description

Further elaboration on the two groups G1and G2 of algorithms is provided below.

1) Group 1
   i. UPGrowth

   The UPGrowth (Utility Pattern Growth algorithm) was introduced in [20]. The authors utilized a data structure known as the UP tree (Utility Pattern tree) to store information about high utility itemsets. This data structure enables the algorithm to traverse the database twice. The results obtained demonstrate that the algorithm reduces the number of candidates and achieves the best execution time. Their study reveals that the method yields favorable outcomes for databases with a significant number of lengthy transactions. UPGrowth is efficient for high-utility itemset mining but can be memory-intensive, struggles with large or dynamic datasets, and may lose patterns due to aggressive pruning. It also relies on sensitive user-defined thresholds and is less suited for multi-dimensional data.

   ii. HUI-Miner

   The HUI-Miner (High Utility Itemsets Miner) utilizes a list containing information about the utility of an itemset along with a pruning heuristic for candidate search [19]. The authors demonstrate in their paper that HUI-Miner surpasses state-of-the-art algorithms in terms of running time and memory consumption. HUI-Miner faces challenges with high memory consumption and computational overhead due to the generation of numerous candidate itemsets. Additionally, it lacks mechanisms to reduce the search space efficiently, leading to potential inefficiencies in large datasets.

   iii. FHM algorithm

   The Fast High-Utility Miner (FHM) algorithm aims to efficiently extract itemsets with HUIs from transactional databases [18]. It employs a technique known as Estimated Utility Co-occurrence Pruning to reduce the number of join operations required during the itemset mining process compared to HUI-Miner. The authors conducted experiments using four datasets from real-life applications. Results demonstrate that FHM is six times faster than FHM. The FHM algorithm, while efficient, can suffer from high memory usage and increased runtime when dealing with dense datasets.

2) Group 2

   i. GHUI-Miner

The concept of generators was used in two concise representations for HUI: High Utility Generators (HUGs) and Generators of High Utility Itemsets (GHUIs) [17]. GHUI-miner was proposed to mine the latter type of itemsets. Experiments were applied on synthetic and real datasets and have shown that the concise representations were 36 times smaller than the HUI sets. GHUI-Miner can encounter performance issues due to the generation of a large number of candidate itemsets, leading to high memory usage and processing time.

   ii. CHUI-Miner

A new HUIM algorithm based on closed itemsets was introduced in [21]. This algorithm is able to work efficiently with sparse and dense datasets. To reduce the mining time, the authors have modified the list structure in HUI-miner. They combined forward and backward cheching method to generate the canddates. To reduce the search space size, pruning strategies have been used. The results obtained highlights the outperformance of this method compared to state-of-the-art algorithms in sparce and dense datasets. However, CHUI-Miner may experience increased computational overhead due to its reliance on generating and managing candidate itemsets, which can lead to higher memory consumption. Additionally, its efficiency can be compromised when applied to dense datasets, resulting in longer processing times.

   iii. CLS-Miner

This algorithm presented in [22] was designed for avoiding the candidate generation step. Itemset utilities are directly calculated from a structure called utility-list. Chain-estimated utility, co-occurrence pruning, lower branch pruning, and pruning by coverage are the four strategies used by authors to tackle the search space issue. CLS-Miner is able to generate closed HUIs. It was executed on six benchmark datasets. The superior performance of the CLS-Miner algorithm compared to current state-of-the-art CHUD and CHUI-Miner algorithms, and its linear scalability have been proved. However, CLS-Miner can face scalability issues due to its dependence on candidate generation, which often results in significant memory usage and longer computation times. Furthermore, its performance may degrade in the presence of large transaction databases with complex item relationships.

   iv. CHUI-Mine

In the CHUI-Mine algorithm [19], the authors were motivated by discovering complete closed HUIs and maximal HUIs without generating candidates. The algorithm has been classified as a one-phase class and used the property called "Pivot Utility Downward Closure" in the process of search space pruning. The algorithm presents an enhancement, in terms of execution time, of several orders of magnitude, compared to other competitors. CHUI-Mine may encounter performance challenges due to the extensive generation of candidate itemsets, leading to high memory consumption and increased processing times.

v. MinFHM algorithm

MinFHM algorithm is an extended version of the FHM [16]. It was designed to handle minimum utility threshold constraints. It was specifically developed for HUIM in transactional databases where a minimum utility threshold is specified. The algorithm efficiently prunes the search space by leveraging the concept of minimal HUIs, allowing it to focus only on itemsets that meet or exceed the specified threshold. The MinFHM algorithm can experience limitations in efficiency due to its reliance on generating a large number of candidate itemsets, which may result in significant memory overhead.

## 4   Results and Discussions

In this section, we report the results of our experiments. All memory estimation was done by the Java™ API. The algorithms were executed on each dataset, while varying the *MinUtil* threshold. The experiments were carried out on a computer with a 64-bit $6^{th}$ generation Core i5 processor, running on Windows 10, and equipped with 16 GB of RAM. Our comparative analysis is based on runtime, memory space usage and the number of HUIs generated.

### 4.1   Runtime

In Fig. 2, the running time comparison is done for all algorithms with Chicago Crimes dataset in (a) and with Cosmetics dataset in Fig. 3 (b).

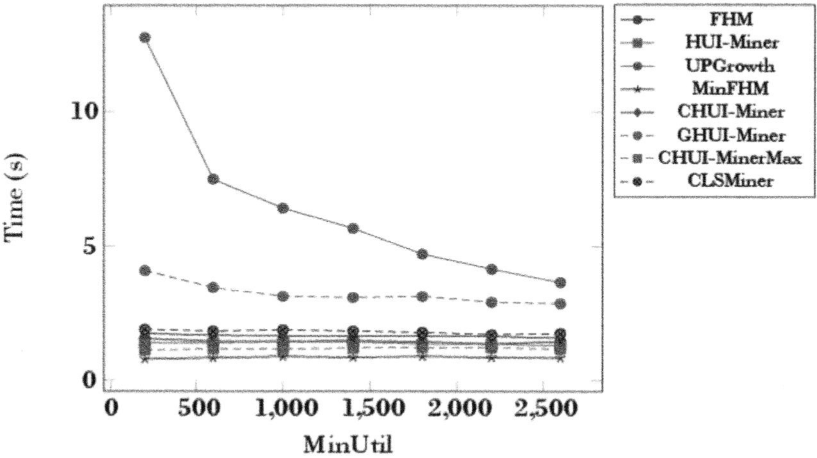


**Fig. 2.** (a) Chicago Crimes

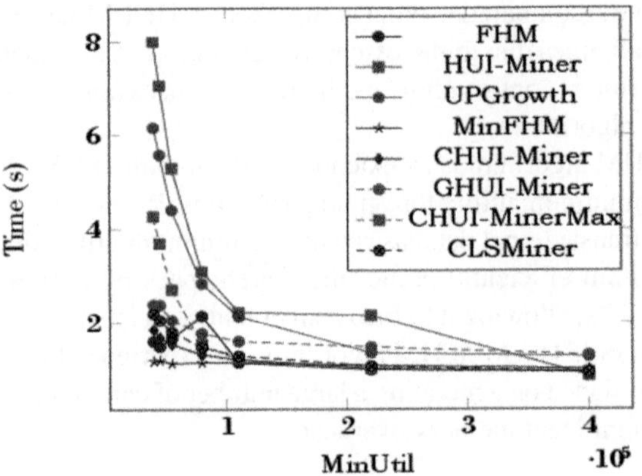



**Fig. 3.** (b) Cosmetics

In the case of Chicago Crimes dataset, the results surpass those obtained with Cosmetics dataset, for almost algorithms. This is due to the difference between the datasets. Hence, from a *MinUtil* value of 2,000 onwards, all algorithms in both groups exhibit running times of less than 5 s, which outperforms the runtimes achieved with Cosmetics dataset. Specifically, GHUI-Miner and UPGrowth are 3 and 8 times slower, respectively. For Cosmetics dataset, we note that the group of HUI algorithms based on CRs outperforms the other group for small values of *MinUtil* except FHM which consumes comparable runtime and is even faster than GHUI-Miner. When the *MInUtil* values increase, the two groups present comparable runtime values, less than 5 s. For both datasets, MinFHM gives best results and all groups consume comparable runtime for values of *MinUtil* above 100,000.

## 4.2  Memory Usage

Figure 4 represents results regarding memory usage comparison for all algorithms. (a) For Chicago Crimes dataset and Fig. 5 (b) for Cosmetics dataset.

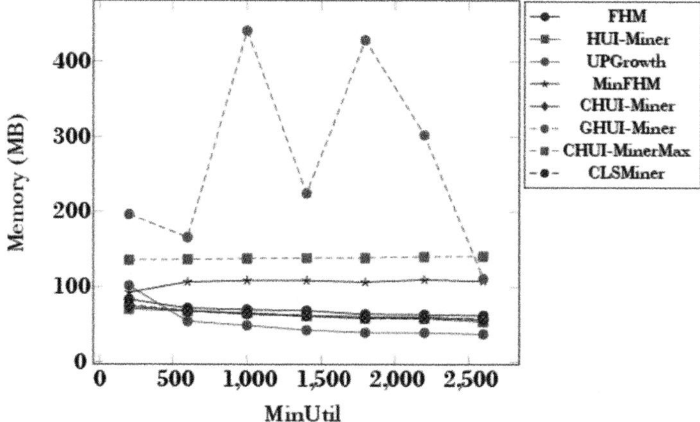

**Fig. 4.** (a) Chicago Crimes

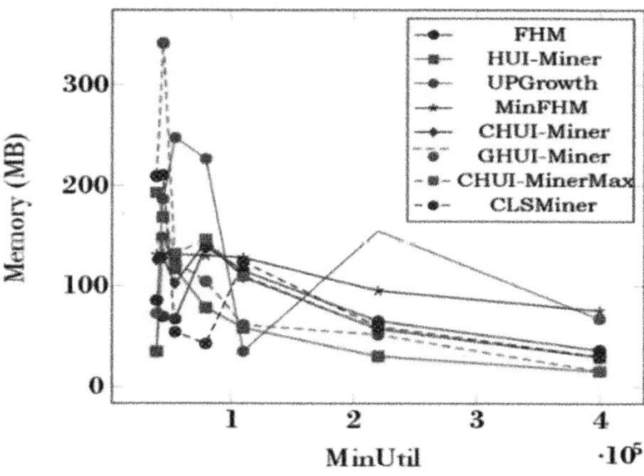

**Fig. 5.** (b) Cosmetics

Memory usage tends to decrease for all algorithms as *MinUtil* values increase. HUI-Miner and MinFHM demonstrate the best results overall. Interestingly, UpGrowth performs well with the Chicago Crimes dataset but not with Cosmetics dataset. Additionally, GHUI-Miner's performance is noteworthy in terms of memory usage, contrasting its position regarding runtime and the size of generated HUIs. In the case of the Chicago Crimes dataset, MinFHM outperforms others for small MinUtil values, while GHUI-Miner shows the least favorable results. However, when considering memory usage overall, no distinct pattern emerges from the results.

### 4.3  The Number of HUIs

Figure 6 shows the comparison for all algorithms based on HUIs number. Figure 6(a) for Chicago Crimes dataset and Fig. 7(b) for Cosmetics dataset.

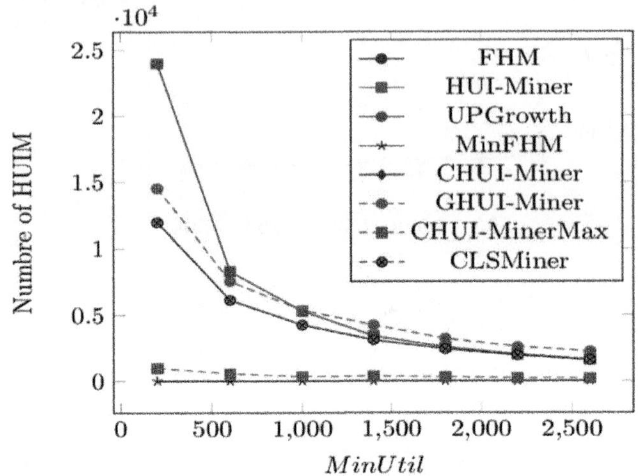

**Fig. 6.**  (a) Chicago Crimes

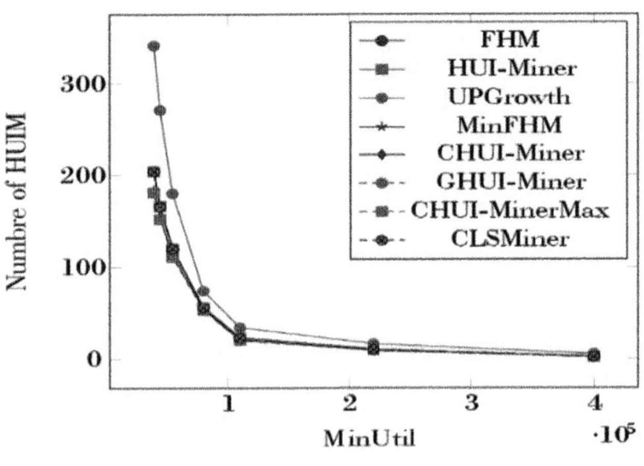

**Fig. 7.**  (b) Cosmetics

When examining the size of HUIs generated by different algorithms, we observe that algorithms utilizing CRs tend to produce smaller sets, which is an expected outcome. For the Cosmetics dataset, we find that all algorithms within both groups exhibit comparable performance for large *MinUtil* values. However, as *MinUtil* values decrease, MinFHM outperforms its counterparts in the same group, while GHUI-Miner delivers the least satisfactory results. In contrast, with the Chicago Crimes dataset, algorithms leveraging CRs notably outperform those in the other group, resulting in sets reduced by 6 to 3 times. Furthermore, MinFHM demonstrates the most favorable outcomes. FHM, HUI-Miner, and UPGrowth yield similar results across all *MinUtil* values.

## 5  Conclusion

In this paper, we reported a comprehensive comparative analysis between two distinct sets of algorithms designed for HUIM. Our study juxtaposes the performance of algorithms belonging to two primary categories: those relying on FIM techniques, such as FHM, HUI-Miner, and UPGrowth, and those harnessing concise itemsets mining strategies, including CHUI-Miner, GHUI-Miner, CLS-Miner, CHUI-MinerMax, and MinFHM, which leverage CRs. To assess the efficacy of these algorithms, we employed two diverse real-world datasets: the Cosmetics dataset and the Chicago Crimes dataset. Our evaluation framework encompasses crucial metrics including memory consumption, runtime efficiency, and the resulting size of HUIs extracted. Our findings highlight a notable performance disparity between the two algorithmic groups. Those coupling CRs as the basis for their operation exhibit superior performance across the evaluated metrics compared to their FIM-based counterparts. Particularly, algorithms relying on CRs demonstrate more efficient utilization of memory and exhibit reduced runtime consumption, ultimately yielding HUIs of more manageable sizes.

Surprisingly, among the standout performers, FHM, which operates outside the CRs-based approaches, emerges as a frontrunner. Despite its departure from CRs, MinFHM shows remarkable efficiency in terms of runtime, memory utilization, and the size of extracted HUIs. This paradoxical observation highlights the importance of considering diverse algorithmic strategies and their nuanced impact on performance metrics. Notably, beyond merely the size of extracted HUIs, the quality of output assumes paramount importance in real-world applications. Therefore, our study emphasizes the significance of not only the quantitative metrics but also the qualitative aspects of algorithmic output.

As a prospective avenue for future research, we aspire to leverage the insights gathered from this comparative analysis in social network analysis. Specifically, we aim to deploy the most promising CRs-based HUIM algorithms to untangle patterns and insights within social network datasets, thereby extending the practical implications of our findings.

## References


1. Luna, J.M., Fournier-Viger, P., Ventura, S.: Frequent itemset mining: a 25 years review. WIREs Data Min. Knowl. **9**(6), e1329 (2019). https://doi.org/10.1002/widm.1329
2. Fournier-Viger, P., Lin, J.C., Vo, B., Chi, T.T., Zhang, J., Le, H.B.: A survey of itemset mining. WIREs Data Min. Knowl. **7**(4), e1207 (2017). https://doi.org/10.1002/widm.1207
3. Fournier-Viger, P., Chun-Wei Lin, J., Truong-Chi, T., Nkambou, R.: A survey of high utility itemset mining. In: Fournier-Viger, P., Lin, J.C.-W., Nkambou, R., Vo, B., Tseng, V.S. (eds.) High-Utility Pattern Mining, vol. 51, in Studies in Big Data, vol. 51, pp. 1–45. Springer International Publishing, Cham (2019). https://doi.org/10.1007/978-3-030-04921-8_1

4. Aggarwal, C.C., Bhuiyan, M.A., Hasan, M.A.: Frequent pattern mining algorithms: a survey. In: Aggarwal, C.C., Han, J. (eds.) Frequent Pattern Mining. Springer International Publishing, Cham, pp. 19–64 (2014). https://doi.org/10.1007/978-3-319-07821-2_2

5. Chen, J., et al.: Incremental high average-utility itemset mining: survey and challenges. Sci. Rep. **14**(1), 9924 (2024). https://doi.org/10.1038/s41598-024-60279-0

6. Kumar, M.J.K., Rana, D.: HARUIM: high average recent utility itemset mining. IJDMMM **16**(1), 66–100. https://doi.org/10.1504/IJDMMM.2024.136217.(2024)

7. Agrawal, R., Srikant, R., et al.: Fast algorithms for mining association rules. In: Proceedings of the 20th International Conference Very Large Data Bases, pp. 487–499 (1994)

8. Yan, Y., Niu, X., Zhang, Z., Fournier-Viger, P., Ye, L., Min, F.: Efficient high utility itemset mining without the join operation. Inform. Sci. **681**, 21218 (2024). https://doi.org/10.1016/j.ins.2024.121218

9. Singh, K., Singh, S.S., Luhach, A.K., Kumar, A., Biswas, B.: Mining of closed high utility itemsets: a survey. RACSC **14**(1), 6–12 (2021). https://doi.org/10.2174/2213275912666190204134822

10. Hidouri, A., Jabbour, S., Raddaoui, B., Ben Yaghlane, B.: Mining closed high utility itemsets based on propositional satisfiability. Data Knowl. Eng. **136**, 101927 (2021). https://doi.org/10.1016/j.datak.2021.101927

11. Pramanik, S., Goswami, A.: Discovery of closed high utility itemsets using a fast nature-inspired ant colony algorithm. Appl. Intell. **52**(8), 8839–8855 (2022). https://doi.org/10.1007/s10489-021-02922-1

12. Nguyen, T.D.D., Nguyen, L.T.T., Vu, L., Vo, B., Pedrycz, W.: Efficient algorithms for mining closed high utility itemsets in dynamic profit databases. Expert Syst. Appl. **186**, 115741 (2021). https://doi.org/10.1016/j.eswa.2021.115741

13. Wei, T., Wang, B., Zhang, Y., Hu, K., Yao, Y., Liu, H.: FCHUIM: efficient frequent and closed high-utility itemsets mining. IEEE Access **8**, 109928–109939 (2020). https://doi.org/10.1109/ACCESS.2020.3001975

14. Truong, T., Duong, H., Le, B., Fournier-Viger, P.: FMaxCloHUSM: an efficient algorithm for mining frequent closed and maximal high utility sequences. Eng. Applic. Artific. Intell. **85**, 1–20 (2019). https://doi.org/10.1016/j.engappai.2019.05.010

15. Duong, H., Hoang, T., Tran, T., Truong, T., Le, B., Fournier-Viger, P.: Efficient algorithms for mining closed and maximal high utility itemsets. Knowl.-Based Syst. **257**, 109921 (2022). https://doi.org/10.1016/j.knosys.2022.109921

16. Fournier-Viger, P., Lin, J.C.-W., Wu, C.-W., Tseng, V.S., Faghihi, U.: Mining minimal high-utility itemsets. In: Hartmann, S., Ma, H. (eds.) Database and Expert Systems Applications, Lecture Notes in Computer Science, vol. 9827, pp. 88–101. Springer International Publishing, Cham (2016). https://doi.org/10.1007/978-3-319-44403-1_6

17. Fournier-Viger, P., Wu, C.-W., Tseng, V.S.: Novel concise representations of high utility itemsets using generator patterns. In: Luo, X., Yu, J.X., Li, Z. (eds.) Advanced Data Mining and Applications, Lecture Notes in Computer Science, pp. 30–43. Springer International Publishing, Cham (2014). https://doi.org/10.1007/978-3-319-14717-8_3

18. Fournier-Viger, P., Wu, C.-W., Zida, S., Tseng, V.S.: FHM: faster high-utility itemset mining using estimated utility co-occurrence pruning. In: Andreasen, T., Christiansen, H., Cubero, J.-C., Raś, Z.W. (eds.) Foundations of Intelligent Systems, Lecture Notes in Computer Science, vol. 8502, pp. 83–92. Springer International Publishing, Cham (2014). https://doi.org/10.1007/978-3-319-08326-1_9

19. Wu, C.-W., Fournier-Viger, P., Gu, J.-Y., Tseng, V.S.: Mining compact high utility itemsets without candidate generation. In: Fournier-Viger, P., Lin, J.C.-W., Nkambou, R., Vo, B., Tseng, V.S. (eds.) High-Utility Pattern Mining, Studies in Big Data, vol. 51, pp. 279–302. Springer International Publishing, Cham (2019). https://doi.org/10.1007/978-3-030-04921-8_11

20. Tseng, V.S., Wu, C.-W., Shie, B.-E., Yu, P.S.: UP-Growth: an efficient algorithm for high utility itemset mining. In: Proceedings of the 16th ACM SIGKDD International Conference on Knowledge Discovery and Data Mining, pp. 253–262. ACM, Washington DC USA (2010). https://doi.org/10.1145/1835804.1835839

21. Nguyen, L.T.T., et al.: An efficient method for mining high utility closed itemsets. Inf. Sci. **495**, 78–99 (2019). https://doi.org/10.1016/j.ins.2019.05.006

22. Dam, L., Li, K., Fournier-Viger, P., Duong, Q.-H.: CLS-Miner: efficient and effective closed high-utility itemset mining. Front. Comput. Sci. **13**(2), 357–381 (2018). https://doi.org/10.1007/s11704-016-6245-4

23. Cheng, Z., Fang, W., Shen, W., Lin, J.C.-W., Yuan, B.: An efficient utility-list based high-utility itemset mining algorithm. Appl. Intell. **53**(6), 6992–7006 (2023). https://doi.org/10.1007/s10489-022-03850-4

24. Krishnamoorthy, S.: Pruning strategies for mining high utility itemsets. Expert Syst. Appl. **42**(5), 2371–2381 (2014). https://doi.org/10.1016/j.eswa.2014.11.001

25. Ryang, H., Yun, U.: Indexed list-based high utility pattern mining with utility upper-bound reduction and pattern combination techniques. Knowl. Inform. Syst. **51**(2), 627–659 (2016). https://doi.org/10.1007/s10115-016-0989-x

26. Peng, Y., Koh, Y.S., Riddle, P.: mHUIMiner: a fast high utility itemset mining algorithm for sparse datasets. In: Kim, J., Shim, K., Cao, L., Lee, J.-G., Lin, X., Moon, Y.-S. (eds.) Advances in Knowledge Discovery and Data Mining, Lecture Notes in Computer Science, vol. 10235, pp. 196–207. Springer International Publishing, Cham (2017). https://doi.org/10.1007/978-3-319-57529-2_16

27. Qu, J.-F., Fournier-Viger, P., Liu, M., Hang, B., Hu, C.: Mining high utility itemsets using prefix trees and utility vectors. IEEE Trans. Knowl. Data Eng. **35**(10), 10224–10236 (2023). https://doi.org/10.1109/TKDE.2023.3256126

28. Qu, J.-F., Fournier-Viger, P., Liu, M., Hang, B., Wang, F.: Mining high utility itemsets using extended chain structure and utility machine. Knowl.-Based Systems **208**, 106457 (2020). https://doi.org/10.1016/j.knosys.2020.106457

29. Wu, P., Niu, X., Fournier-Viger, P., Huang, C., Wang, B.: UBP-Miner: an efficient bit based high utility itemset mining algorithm. Knowl. Based Syst. **248**, 108865 (2022). https://doi.org/10.1016/j.knosys.2022.108865

30. Liu, Y., Liao, W., Choudhary, A.: A two-phase algorithm for fast discovery of high utility itemsets. In: Ho, T.B., Cheung, D., Liu, H. (eds.) Advances in Knowledge Discovery and Data Mining, Lecture Notes in Computer Science, vol. 3518, pp. 689–695. Springer Berlin Heidelberg, Berlin, Heidelberg (2005). https://doi.org/10.1007/11430919_79

31. Tseng, V.S., Shie, B.-E., Wu, C.-W., Yu, P.S.: Efficient algorithms for mining high utility itemsets from transactional databases. IEEE Trans. Knowl. Data Eng. **25**(8), 1772–1786 (2013). https://doi.org/10.1109/TKDE.2012.59

32. Ahmed, C.F., Tanbeer, S.K., Jeong, B.-S., Lee, Y.-K.: Efficient tree structures for high utility pattern mining in incremental databases. IEEE Trans. Knowl. Data Eng. **21**(12), 1708–1721 (2009). https://doi.org/10.1109/TKDE.2009.46

33. Yun, U., Ryang, H., Ryu, K.H.: High utility itemset mining with techniques for reducing overestimated utilities and pruning candidates. Expert Syst. Appl. **41**(8), 3861–3878 (2014). https://doi.org/10.1016/j.eswa.2013.11.038

34. Gan, W., Lin, J.C.-W., Fournier-Viger, P., Chao, H.-C., Tseng, V.S., Yu, P.S.: A survey of utility-oriented pattern mining. IEEE Trans. Knowl. Data Eng. **33**(4), 1306–1327 (2021). https://doi.org/10.1109/TKDE.2019.2942594

35. Tran, T., Duong, H., Truong, T., Le, B.: Efficient mining of concise and informative representations of frequent high utility itemsets. Eng. Appl. Artific. Intell. **126**, 107111 (2023). https://doi.org/10.1016/j.engappai.2023.107111

36. Tung, N.T., Nguyen, L.T.T., Nguyen, T.D.D., Vo, B.: An efficient method for mining multi-level high utility Itemsets. Appl. Intell. **52**(5), 5475–5496 (2021). https://doi.org/10.1007/s10489-021-02681-z
37. Fournier-Viger, P., et al.: The SPMF open-source data mining library version 2. In: Berendt, B., et al. (eds.) Machine Learning and Knowledge Discovery in Databases, Lecture Notes in Computer Science, vol. 9853, pp. 36–40. Springer International Publishing, Cham (2016). https://doi.org/10.1007/978-3-319-46131-1_8

# Selecting an Appropriate Ensemble Machine Learning Algorithm for Intelligent Quality Management in Production


Djahida Belayadi[2,3]($\boxtimes$) [iD], Djamila Bendouda[1,2] [iD], Nesrine Benali[2,3] [iD], and Narimane Hamouche[2,3] [iD]

[1] Laboratory of Innovative Technologies (LTI), Algiers, Algeria
[2] National Higher School of Advanced Technologies (ENSTA), Algiers, Algeria
{djahida.belayadi,djamila.bendouda,n_benali,n_hamouche}@ensta.edu.dz
[3] Department of Industrial Engineering and Maintenance, Algiers, Algeria



**Abstract.** Artificial intelligence (AI) technology has significantly contributed to innovation in manufacturing, particularly in improving quality management processes. One key area within AI is machine learning (ML), which has developed into a distinct field, with ensemble learning gaining increasing attention. This paper explores the concept of ensemble learning and its application in quality management, with a specific focus on quality control and quality assurance. We assess the performance of three widely used tree-based algorithms: Random Forest, Extreme Gradient Boosting (XGBoost), and Adaptive Boosting (AdaBoost). Through a detailed evaluation process, we review models frequently utilized in prior research to highlight their distinctive advantages. Our findings indicate that Random Forest is the most prevalent algorithm, demonstrating superior performance over not only basic ML algorithms but also deep learning models. Its success is largely due to its simplicity, scalability, and robust ability to handle multidimensional data.

**Keywords:** Ensemble learning · RF · XGboost · Adaboost · Quality · Production


## 1 Introduction

In today's competitive manufacturing landscape, production processes optimization is faced to several challenges that are based on three major axes: reducing costs, environmental sustainability, and enhancing quality. In other words, improve efficiency and efficacy. New approaches based on emergent technologies have been developed and used to improve those crucial pillars so that the companies face the market demands and requirements.

As we already mentioned, quality stands as the cornerstone and essential pillar of a company's profitability that should be maintained by quality management system which is defined as a set of procedures followed to guarantee that a service or product meets the customer's expectations. This system is designed and implemented in a company in a way to ensure both effectiveness and efficiency [1]. The goals of quality management include both achieving high standards for goods and services as well as ensuring profitability [2]. In the last century, quality paradigms have seen a continuous evolution. From the first paradigm of quality -*Quality inspection*- until the occurrence of *intelligent quality management*, several tools and methods have been exploited for the optimization and improvement of products, services and processes [3]. With the occurrence of artificial intelligence techniques, new insights have been discussed and recent approaches have been studied that improve the quality in the production process. Among those techniques, machine learning models. Machine learning is a trunk of a tree called artificial intelligence. It focuses on enabling computers to use existing knowledge in order to simulate human work and obtain new skills or knowledge [4]. In order to obtain superior knowledge and attend higher performance, multiple and basic machine learning models are combined, this approach is known as *ensemble learning*. If the ensemble models use the same kind of base machine learning model, the ensemble learning is homogeneous. As for the heterogeneous ensemble learning, different types of base algorithms are combined [5]. Both types are divided into parallel ensembles and sequential ensembles. The parallel ensembles algorithms simultaneously generate diverse base learners for example bagging and random forests models, while sequential ensembles train base models one after the other which is the case in boosting algorithms and stacking models where the result of one model can be used as input to another model [6]. Researches [7–11] have shown that AI, particularly machine learning techniques, offers a promising path to significantly enhance production quality. These advancements enable companies to achieve their goals of producing high-quality products by implementing policies based on ML-powered defect detection [12] and product quality prediction [13]. When basic ML models or individual models show varying performances and limitations, ensemble learning rises to the challenge. By combining the strengths of different models, ensemble methods improve overall accuracy and generalisation which means that ensemble models are mostly able to generate a wider range of relationships and patterns between the variables of the data set [6,14]. From individual to ensemble diving to deep learners, we have a variety of approaches, techniques and methods to improve quality from inspection to total quality management. This paper was written for the purpose of presenting alternatives and efficient methods to enhance quality of products and processes. For this reason we explored the literature of the ensemble learning approach assuming that it surpass the individual machine learning models. In this study, we aim to answer the following questions: How can EL applied to improve quality? Do ensemble models consistently outperform individual models, and if so, which ensemble algorithms show the best performance under different circumstances?

The rest of the paper is organised as follows: Sect. 2 introduces the background needed to understand this research. Section 3 is devoted to the application of machine learning in quality management field. Section 4 presents a study on the application of EL in quality management. In Sect. 5 we present a comparative study between ensemble tree based algorithms. Finally, in Sect. 6 and 7 we discuss the results and the conclusion of this study respectively.

## 2  Background

### 2.1  Quality Management Tools

Quality management tools and techniques are practical methods that can be applied to improve the quality system to achieve positive changes and improvements.

Several new quality tools have emerged, mainly for qualitative data like affinity diagrams, relationship diagrams, tree diagrams, matrix diagrams, arrow diagrams, process decision procedure diagrams (PDPC), and matrix data analysis [15]. From the above tools, those frequently used and appropriate for the quality systems, according to ISO 9001 are presented in Fig. 2.

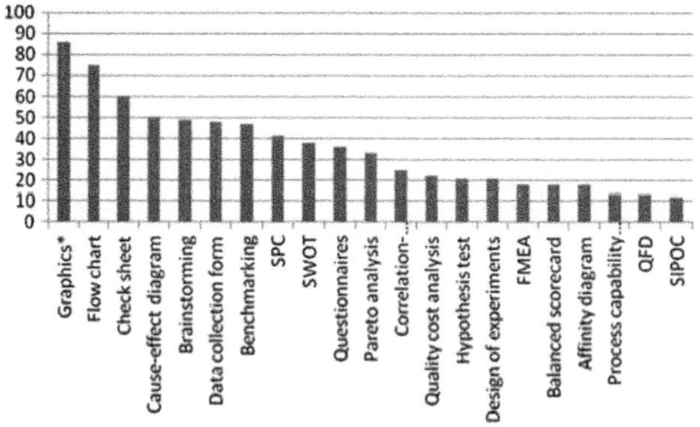

Fig. 1. Level of use of techniques [15].

With the advances of emergent technologies a new term has appeared *Intelligent quality management* which is a system of quality management that goes beyond traditional methods by using advanced data analysis techniques such as machine learning to improve quality [3].

There are several approaches for using machine learning. The ensemble learning approach is covered in the section that follows.

### 2.2  Ensemble Learning

In this section we are going to introduce the most famous and commonly used algorithms of tree based ensemble learning (Fig. 3):

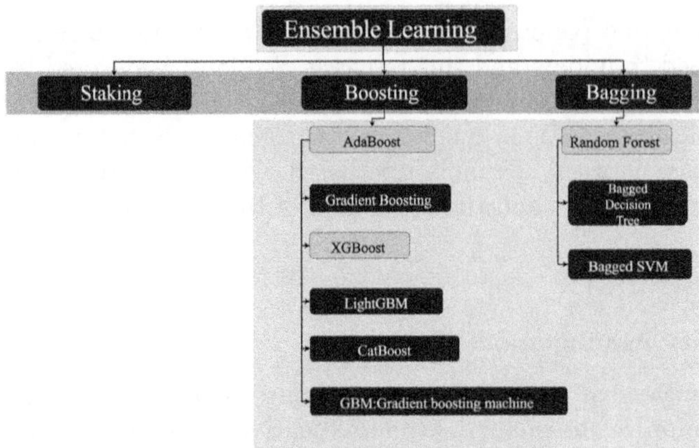

**Fig. 2.** Ensemble learning algorithms.

**Random Forest (RF).** Created by Leo Breiman (statistician and computer scientist), RF is an enhanced version of bagging, known as bagging of CARTs (Classification and Regression Trees) [16]. CARTs represent techniques for dividing the variable space according to a decision tree's embedded set of rules where a decision rule determines how each node divides [16].

**Extreme Gradient Boosting (XGboost).** Another type of boosting algorithms, it is constructed based on an ensemble of gradient boosting decision trees or a sequential ensemble approach known as a sequential decision tree [17]. Adding a regularisation parameter that lessens each regression tree's susceptibility to dataset outliers, which improves the gradient boosting algorithm [18]. With this approach, a weight value is assigned to each data value in the database, defining the likelihood that the value will be chosen for additional examination by a decision tree [17].

**Adaptive Boosting (Ada Boost).** Freund Yoav and Schapire Robert were the first searchers who introduced and built this model in 1995. It is one of the boosting algorithms that was named accordingly because it adapts adaptively to the flaws of the weak hypotheses that WeakLearn returns, in contrast to other algorithms [19]. It may combine any number of base-learners and re-uses the same training set. This means that many classifiers of AdaBoost are trained one after the other. The efficiency of previously taught classifiers is the basis for training each new one [19].

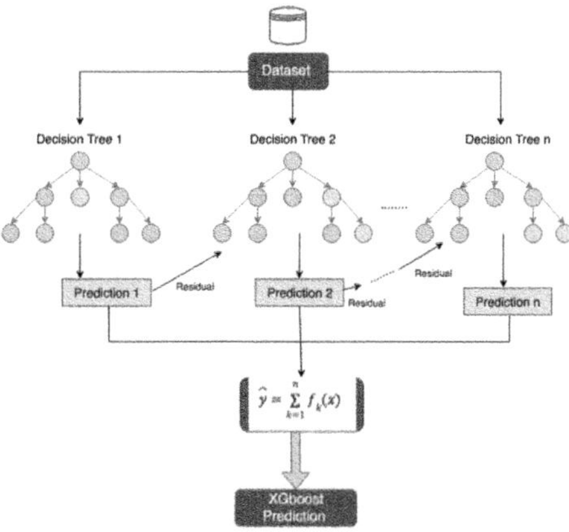



$$\hat{y} = \sum_{k=1}^{n} f_k(x)$$

**Fig. 3.** XGboost algorithm

## 3 Insights of Machine Learning Applications in Quality Management

ML techniques are applied in so many fields and industry domains. For instance computer and electronics manufacturing, chemical manufacturing, metal industry uses those techniques to improve the performance of its discrete and continuous production lines and to solve various problems. Among the key areas where ML shines in order to enhance the performance of the production function, we mention quality optimization to achieve customer satisfaction.

The ML models can optimise quality either by detecting and minimising defects, therefore, achieving better product reliability or by optimising critical product performance parameters [7]. In other words, machine learning can be used indirectly as a diagnostic tool to detect anomalies for processes or products by analysing the underlying causes of quality issues and predicting product quality early on to avoid unfavourable outcomes. As for direct methods, they rely on identifying the best parameters to adjust based on the desired outcome and the characteristics of the product in order to attend high level of quality [8].

Another contribution of machine learning branches shines in the quality improvement tools such as lean management, six sigma, lean six sigma (LSS). In this case, it can be integrated directly for instance in root cause analysis for quality issues. The ML algorithm has shown to be successful in improving competitiveness and anticipatory identifying defects [8].

Abd elnaby et al. [9] integrated machine learning in the analysis phase of the DMAIC approach (Define- Measure-Analyze-Improve-Control) in the context of lean six sigma to improve the quality of plastic bottles.

Another research by Altuğ [10] was maintained to solve defective coating by implementing six sigma which lead to improved quality and reduce waste. After that, deep learning was integrated as an alternative solution. The deep learning model's performance closely matched actual outcomes, validating the effectiveness of the Six Sigma improvements.

As for lean combined with machine learning, Abusaq et al. [11] studied where ML was used with the lean principles to reduce idle time which led to substantial energy savings. Additionally, they developed a machine learning model using various factors to predict energy consumption which helped in optimising job scheduling, further reducing energy consumption and costs.

Even with the remarkable success of machine learning, Kuo-yi et al. [20] aptly point out that many production line challenges demand intelligent management solutions. They argue that ensemble learning (EL) surpasses traditional methods in quality control by eliminating non-conformity factors, ultimately leading to higher quality standards. EL's versatility extends beyond simple detection and isolation: it can pinpoint fault characteristics, predict imminent issues, and even forecast future problems before they manifest.

## 4   Ensemble Learning for Quality

This section provides an introduction to the use of ensemble learning in quality, with a focus on quality assurance and control.

Our study was based on 13 articles in different domains : s

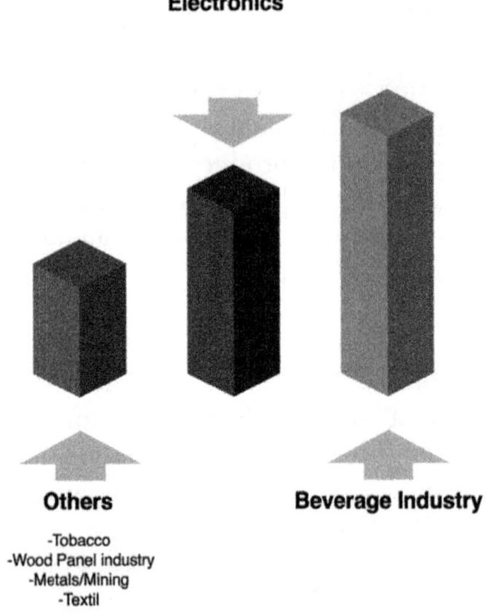


**Fig. 4.** Articles' domains

## 4.1   The Beverage Industry

Jain et al. [21] aimed to predict wine quality using RF, DT, XGboost, Adaboost, Gradient boosting since the acquisition of wine certificates is quite important. For this purpose, they gathered the information for this from several websites. As a result of testing the algorithms in different conditions, the XGboost showed good results with/without feature selection and key variables while RF had the best performance using the key variables. Another key finding of this study showed the effect of feature selection on the performance of the models.

Dahal et al. [22] used ensemble model Gradient Boosting Regressor (GBR) and other machine learning and deep learning models. Similarly to the earlier case study they collected product dataset from a website (UCL repository) unlike Schubert et al. [23], they used check outliers during the data preprocessing aiming to attend higher performance. The models' performance was assessed by the authors using correlation coefficient (R), mean squared error (MSE), and mean absolute percentage error (MAPE). The results showed that the gradient boosting regressor had the best results.

Although the boosting algorithms were found to be effective in detecting and predicting defects in the three earlier studies, Nandan et al. [24] tested the Random Forest classifier in conjunction with machine learners and deep learners, and found that it outperformed the other classifiers with an accuracy of 98.11 per cent.

## 4.2   Electronics

Jabbar et al. [25] used four tree based algorithms (CART, RF, XGboost, Adaboost) in order to make defect detection more efficient and distinguish genuine defects from false alarms. Unlike the case studies where they used SMOTE (Synthetic Minority Oversampling Technique) analysis to handle the imbalanced dataset, in this case,they used data level procedures to modify the data sets by adding or removing entries. They tested the previously mentioned algorithm on the balanced dataset and using the metrics of accuracy, Hamming loss, precision, recall and computation time. They concluded that XGboost outperformed the other algorithms proving that the implementation of this algorithm is a promising approach.

Liu et al. [26] proposed another boosting hybrid technique called random undersampling boosting (RUBoost). The ensemble learner was developed and used to verify electrode quality classification during battery manufacture. This technique outperformed the random forest algorithm in terms of accuracy and forecast the characteristics of the produced electrode.

## 4.3   Tobacco

In this case study, Shi et al. [27] integrated prediction ensemble learning models to anticipate the quality of tobacco products. To resolve the class imbalance in the datasets gathered from the tobacco industry's management information platform, a SMOTE study was performed. The ensemble methods were examined following the preprocessing of the data and feature selection. Recall, F1, and training time were used to test each algorithm. Although it took longer, using the SMOTE analysis enhanced the algorithms' performance. Furthermore, in all scenarios, Xgboost performed better in categorization and product quality prediction than the random forest approach.

### 4.4  Software

While maintaining quality for physical items was the focus of the previous cases we saw, Saheed et al. [28] also examined software quality. Several datasets were collected from the website and analysed. A single method (logistic regression) as well as an ensemble of algorithms were used to train and evaluate the dataset. The models were assessed using the following metrics: F measure, AUC (area under the curve), accuracy, recall, precision, and MCC (Matthew correlation coefficient). Overall, the ensemble models fared better than the individual models with the various datasets. The catboost algorithm performed the best among the ensemble models.

### 4.5  Wood Fibre Industry

Schubert et al. [23] used machine learning to predict pertinent wood fibre board qualities in real time for better quality control. They began by gathering and preparing data. The models used were assessed using the correlation coefficient R, the coefficient of determination R2, the mean absolute percentage error (MAPE), and the root-mean-square error (RMSE). Out of all the models that were used, the RF performed the best. It's also important to note that ensemble learners performed well despite the fact that just data generalisation was employed during data preparation rather than data cleaning. The scientists viewed this as a benefit of ensemble learning over statistical methods, as ensemble learning is more resilient to outliers and noisy data.

### 4.6  Plastic Industry

Jung et al. [29] used machine learning techniques in an injection moulding company to address quality prediction issues. The injection moulding manufacturer provided the data utilised in this investigation, which were subsequently balanced using the SMOTE approach. Because of this, the autoencoder performed best in terms of accuracy, precision, and f1 score. Its recall score of 1 indicates that it can identify every flaw.

### 4.7  Metallurgy

Li et al. [30] aimed to propose an alternative solution for quality prediction using stacking -ensemble learning approach- with ensemble base models. They first gathered the required data,then tested and evaluated the based algorithms with R2, RMSE and percentage of error. The boosting algorithms had better performance than the other ensemble. Then the based ensemble models were used with stacking and averaging ensemble models. The last two models showed better performance and were more robust predicting steel quality control.

## 4.8  Textile Industry

In the paper of Demirel et al. [31], a novel concept was covered with Regressor chain algorithms. In order to assess the quality of the textiles, data was gathered from the textile production process. The data was then separated into ten clusters, one of which had odd patterns. Before assessing the ensemble regressor chain performance, they examined the individual models to see which was the best. It was a really impressive performance by the random forest. The usage of ensemble regressor chains produced superior results when it came to odd data segments, according to statistical analysis. These results seem promising to further applications to automated quality control in the context of industry 4.0.

## 4.9  Manufacturing

In the work of Sankhye et al. [18], efficient quality control was achieved by the application of categorization machine learning in quality inspection. Following the manufacturing unit's data gathering, the dataset was processed and cleaned using feature engineering and SMOTE analysis. The algorithms were assessed using accuracy and a different statistic known as Cohen Kappa after they were put into practice in order to gather further information about the algorithms' overall performance. XGboost and the RF were put through four tests. When the features were chosen, the performance of both models improved and was still rather decent. However, based on Cohen's kappa metric and confusion matrix, the XGboost had the best performance, indicating a significant degree of ability to forecast minority classifications.

## 5  Comparative Study of EL Algorithms

As shown in Fig. 7 we have found that the tree based ensemble learners are the most used. As a result, we have focused only on three models (Random forest, XGboost, Adaboost) in this part.

This approach was motivated by the article [28] in which the authors contrasted the ensemble algorithms' performance study in comparison to other papers.We assessed the performance of these models based on several metrics:

***Accuracy in General:*** calculates a model's effectiveness as a percentage of actual outcomes over the total count of the instances [25, 32].

***Parametrisation:*** The required knobs that configure an algorithm are called parameters. These figures influence the behaviour of the algorithm in terms of duration and accuracy [25, 32].

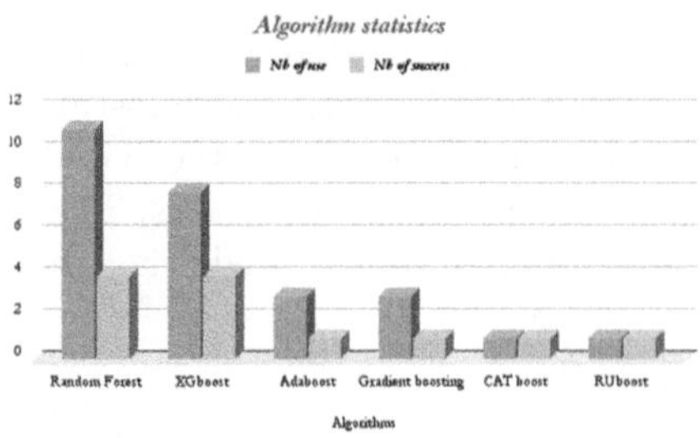



**Fig. 5.** Algorithm statistics

***Overfitting Tendency:*** We refer to this as the overfitting problem while learning a noisy database. An excessively complicated model with too many parameters leads to overfitting [32].

***Learning time (speed):*** The amount of time needed to train the dataset is called the "learning time," and it varies based on the size of the dataset and the technique being used [25,32].

***Prediction time (speed):*** The amount of time needed to test the dataset is called prediction time. The magnitude of the data and the technique we employ determine this [25,32].

***Flexibility:*** A network's flexibility is defined as its capacity to adjust to the patterns found in the database [32].

***Interpretability:*** is the degree to which a human can consistently predict the model's result. This means that a model is more interpretable when it is easier for people to reason about and trace why the model made its predictions [15].

***Robustness:*** refers to an algorithm's ability to uphold consistent performance, even in the presence of new samples that belong to the same subset as the testing samples. it signifies that when confronted with comparable samples, the algorithm's performance remains stable or predictable [33].

The evaluation of each method is indicated by the number of stars:
In general : *** High **Average *Low
For parametrisation: ***small number of parameters **Average number of parameters *Large number of parameters
For Time: *** Slow **Average * Fast
For Algorithm type : ** means that the algorithm is used for both classification and regression [32] (Table 1).

## 6    Discussion

Similar to how scientific experiments may yield different results in various conditions, machine learning algorithms are also subject to this phenomenon. The

**Table 1.** Evaluation of algorithm according to selected criteria

| Criteria | RF | XGboost | Adaboost |
| --- | --- | --- | --- |
| Accuracy | *** | *** | * |
| Parametrisation | ** | * | *** |
| Robust to Overfitting | *** | ** | ** |
| Learning time | ** | * | *** |
| Prediction time | ** | *** | ** |
| Flexibility | *** | *** | *** |
| Interpretability | ** | * | *** |
| Algorithm Type | ** | ** | ** |
| Robustness | *** | ** | ** |
| Score | 22 | 18 | 21 |

selection of an appropriate algorithm is contingent upon a number of factors, including the data, the nature of the problem under investigation, and the requirements of the study. As we saw in the case studies presented previously, multiple algorithms have been employed and compared in the examined articles to determine which is the most appropriate. Among the different models and approaches applied, ensemble learning algorithms especially the tree based algorithms outperformed the other algorithms.

This leads us to conclude that using ensemble learning for both regression and classification issues appears to be a viable strategy in the quality sector since it does not only surpass the basic machine learning, it also outperformed the deep learning models in cases. [22, 23]. Although the authors Dahal et al. [22] showed preservation when it comes to neural networks assuming that ANN may have the best performance if the training set was larger or much more complex. Beside bagging and boosting algorithms stacking can also lead to very efficient results [30].

Despite the good performance of stacking algorithms, we focused on bagging and boosting techniques. Our evaluation of three commonly used tree-based algorithms (Random Forest, XGBoost, and AdaBoost) within quality systems suggests that Random Forest and AdaBoost are potentially the best models for this specific case. Even thought XGBoost may be faster and more flexible in generating predictions.

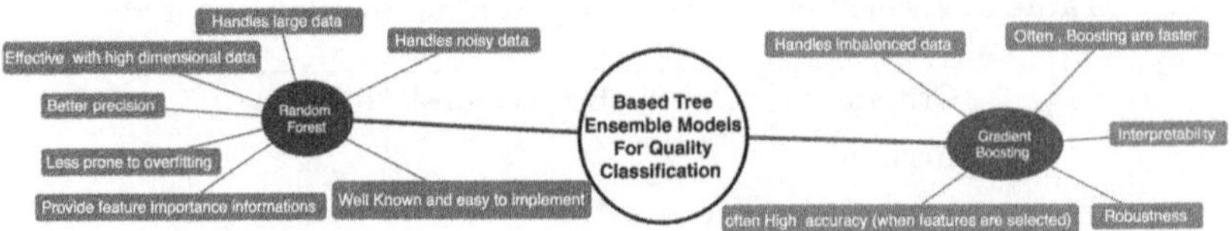


**Fig. 6.** Criteria of selecting the algorithms

# 7    Conclusion

This review confirms that ensemble learning can be used as a technique to predict quality metrics or classify the defects components to improve quality. In this research, we introduced the most common boosting and bagging algorithms, then we collected articles to enrich our study. We found that tree based algorithms are well known and most used in quality prediction. Accordingly, we assessed the performance of the most used algorithms (RF, XGboost, Adaboost) based on several metrics. We found that RF and Adaboost are suggested due to their high score in the multicriteria analysis. RF with their simplicity, capacity and ability to handle multidimensional data is a good solution for quality regression and classification problems, while Gradient Boosting sequential learners are a suitable choice when precision is targeted.

Despite the results discussed, selecting an appropriate algorithm should be based on a number of principles, for instance the minimum description length (MDL) principle, which states that the optimal model is the one that minimises the total description length, which includes the complexity of the model and the encoded representation of the data.

In addition, further investigation and case studies are required in this work to deeply understand and highly assess the performance of the ensemble algorithms.

# References


1. Abdulridha, M. S., et faithi, N.A.: Enhancing organizational performance in oil and gas industry: a comprehensive review of quality management practices. J. Appl. Eng. Des. Simul. (JAEDS), 4(1):1–12, (2024)
2. Helmold, M.: Virtual and innovative quality management across the value chain. 2023
3. Werner, T., Weckenmann, A., Akkasoglu, G.: Quality management–history and trends-. The TQM J. 27(3):281–293, 2015
4. Wang, H., Ma, C., Zhou, L.: A brief review of machine learning and its application. In: International Conference on Information Engineering and Computer Science, pp. 1–4, 2009
5. Mienye, Y., Domor, I., Sun.: Ensemble learning approaches. Rule Based Systems For Big Data: A Machine Learning Approach. 13, 2016
6. Weckenmann, A., Akkasoglu, G., Werner, T.: A survey of ensemble learning: concepts, algorithms, applications, and prospects. IEEE Access **10**(3), 99129–99149 (2022)

7. Kang, Z., Catal, C., Tekinerdogan, B.: Machine learning applications in production lines: a systematic literature review. Comput. Ind. Eng. **149**, 106773 (2020)
8. Weichert, D., Link, P., Stoll, A., Rüping, S., Ihlenfeldt, S.: and Stefan Wrobel. a review of machine learning for the optimization of production processes. Int. J. Adv. Manufact. Technol. 104(5-8):1889–1902, (2019)
9. Abd Elnaby, Z., Zaher, A., Abdel-Magied, R. K.: Improving plastic manufacturing processes with the integration of six sigma and machine learning techniques: a case study. J. Ind. Prod. Eng. 41(1):1–18, (2024)
10. AltĞ, M.: Application of six sigma through deep learning in the production of fasteners. Int. J. Lean Six Sigma, 14(7):1376–1402, (2023)
11. Abusaq, Z., Zahoor, S., Habib, M.S.: Improving energy performance in flexographic printing process through lean and AI techniques: a case study. Energies, 16(4):1972, (2023)
12. Czimmermann, T., Ciuti, G., Milazzo, M.: Visual-based defect detection and classification approaches for industrial applications—a survey. Sensors **20**(5), 1459 (2020)
13. Sankhye, S., Hu, G.: Machine learning methods for quality prediction in production. Logistics, 4(4):35, (2020)
14. Li, N., Yu, Y., Zhou, Z.: Diversity regularized ensemble pruning. In: Peter A. Flach, Tijl De Bie, and Nello Cristianini, editors, Machine Learning and Knowledge Discovery in Databases. ECML PKDD 2012, volume 7524 of Lecture Notes in Computer Science, pp. 330–345. Springer, (2012)
15. Ismyrlis, V., Moschidis, O.: The use of quality management systems, tools, and techniques in ISO 9001: 2008 certified companies with multidimensional statistics: the greek case. Total Qual. Manag. Bus. Excellence **26**(5–6), 497–514 (2015)
16. Tyralis, H., Papacharalampous, G., Langousis, A.: A brief review of random forests for water scientists and practitioners and their recent history in water resources. Water **11**(5), 910 (2019)
17. Bhati, B.S., Chugh, G., Al-Turjman, F., Bhati, N.S.: An improved ensemble based intrusion detection technique using xgboost. Trans. Emerg. Telecommun. Technol. 32(6):e4076, (2021)
18. Sankhye, S., Guiping, H.: Machine learning methods for quality prediction in production. Logistics **4**(4), 35 (2020)
19. Freund, Y., Schapire, R.E.: A decision-theoretic generalization of on-line learning and an application to boosting. J. Comput. Syst. Sci. 55(1):119–139, (1997)
20. Lin, K.Y., Huang, C.: Ensemble learning applications in multiple industries: A review. Inf. Dyn. Appl **1**(1), 44–58 (2022)
21. Jain, K., Kaushik, K., Gupta, S.K., Mahajan, S., Kadry, S.: Machine learning-based predictive modelling for the enhancement of wine quality. Sci. Rep. 13(1):17042, (2023)
22. Dahal, K.R., Dahal, J.N., Banjade, H., Gaire, S.: Prediction of wine quality using machine learning algorithms. Open J. Stat. **11**(2), 278–289 (2021)
23. Schubert, M., Luković, M., Christen, H.: Prediction of mechanical properties of wood fiber insulation boards as a function of machine and process parameters by random forest. Wood Sci. Technol. **54**(3), 703–713 (2020). https://doi.org/10.1007/s00226-020-01184-3
24. Nandan, M., Gupta, H.R., Mondal, M.: Building a classification model based on feature engineering for the prediction of wine quality by employing supervised machine learning and ensemble learning techniques. pp. 1–7, (2023)
25. Jabbar, E., Besse, P., Loubes, J.M., Roa, N.B., Merle, C., Dettai, R.: Supervised learning approach for surface-mount device production. pp. 254–263, (2019)

26. Liu, K., Hu, X., Meng, J., Guerrero, J.M., Teodorescu, R.: Ruboost-based ensemble machine learning for electrode quality classification in li-ion battery manufacturing. IEEE/ASME transactions on mechatronics, 27(5):2474–2483, 2021

27. Shi, Q., Wang, H., Xiaoshuang, X., Hailong, L., Zhang, Z.: The application of tobacco product quality prediction using ensemble learning method. 1, 1780–1784 (2019)

28. Saheed, Y.K., Longe, O., Baba, U.A., Rakshit, S., Vajjhala, N.R.: An Ensemble Learning Approach for Software Defect Prediction in Developing Quality Software Product. In: Singh, M., Tyagi, V., Gupta, P.K., Flusser, J., Ören, T., Sonawane, V.R. (eds.) ICACDS 2021. CCIS, vol. 1440, pp. 317–326. Springer, Cham (2021). https://doi.org/10.1007/978-3-030-81462-5_29

29. Jung, H., Jeon, J., Choi, D., Park, J.-Y.: Application of machine learning techniques in injection molding quality prediction: Implications on sustainable manufacturing industry. Sustainability 13(8), 4120 (2021)

30. Li, F., Wu, J., Dong, F., Lin, J., Sun, G., Chen, H., Shen, J.: Ensemble machine learning systems for the estimation of steel quality control. pp. 2245–2252, 2018

31. Demirel, K. C., Sahin, A., Albey, E.: Ensemble learning based on regressor chains: a case on quality prediction, pp. 267–274, 2019

32. Ouadah, A., Zemmouchi-Ghomari, L., Salhi, N.: Selecting an appropriate supervised machine learning algorithm for predictive maintenance. Appl. Artif. Intell. 119(7), 4277–4301 (2022)

33. Huan, X., Mannor, S.: Robustness and generalization. Machine learning 86, 391–423 (2012)

34. Eason, G., Noble, B., Sneddon, I.N.: On certain integrals of lipschitz-hankel type involving products of bessel functions. Philos. Trans. R. Soc. London. Series A, Mathematical and Physical Sciences, 247(935):529–551, 1955

35. Maxwell, J.M.: A treatise on electricity and magnetism, volume 1. Clarendon press, 1881

36. Breiman, L.: Random forests. Machine learning 45, 5–32 (2001)

37. Ali Shebl, Dávid Abriha, Amr S Fahil, Hanna A El-Dokouny, Abdelmajeed A Elrasheed, and Árpád Csámer. Prisma hyperspectral data for lithological mapping in the egyptian eastern desert: Evaluating the support vector machine, random forest, and xg boost machine learning algorithms. *Ore Geology Reviews*, page 105652, 2023

38. Biau, G., Scornet, E.: A random forest guided tour. TEST 25(2), 197–227 (2016). https://doi.org/10.1007/s11749-016-0481-7

39. Ramón Díaz-Uriarte and Sara Alvarez de Andrés: Gene selection and classification of microarray data using random forest. BMC Bioinformatics 7, 1–13 (2006)

40. Ziegler, A., König, I.R: Mining data with random forests: current options for real-world applications. Wiley Interdisciplinary Reviews: Data Mining and Knowledge Discovery, 4(1):55–63, 2014

41. Wager, S., Athey, S.: Estimation and inference of heterogeneous treatment effects using random forests. J. Am. Stat. Assoc. 113(523), 1228–1242 (2018)

42. Wu, J., Li, Y., Ma, Y.: Comparison of xgboost and the neural network model on the class-balanced datasets, pp. 457–461, 2021

43. Chen, T., et al.: Xgboost: extreme gradient boosting. R package version 0.4-2, 1(4):1–4, 2015

44. Kalcheva, N., Todorova, M., Marinova, G.: Naive bayes classifier, decision tree and adaboost ensemble algorithm–advantages and disadvantages. pp. 153–157, 2020

45. Ying, C., Qi-Guang, M., Jia-Chen, L., Lin, G.: Advance and prospects of adaboost algorithm. Acta Automatica Sinica 39(6), 745–758 (2013)

46. Schapire, R.E.: Explaining adaboost, pp. 37–52, 2013
47. Shahraki, A., Abbasi, M., Haugen, Ø.: Boosting algorithms for network intrusion detection: A comparative evaluation of real adaboost, gentle adaboost and modest adaboost. Eng. Appl. Artif. Intell. **94**, 103770 (2020)
48. Freund, Y., Schapire, R., Abe, N.: A short introduction to boosting. Journal-Japanese Society For Artificial Intelligence **14**(771–780), 1612 (1999)
49. Randhawa, K., Loo, C.K., Seera, M., Limand, C.P., Nandi, A.K.: Credit card fraud detection using adaboost and majority voting. IEEE access, 6:14277–14284, 2018
50. Bhardwaj, P., Tiwari, P., Olejar Jr, K., Parr, W., Kulasiri, D.: A machine learning application in wine quality prediction. Mach. Learn. Appl. 8:100261, 2022
51. Fotopoulos, C., Psomas, E.: The use of quality management tools and techniques in ISO 9001: 2000 certified companies: the greek case. Int. J. Product. Perform. Manag. **58**(6), 564–580 (2009)
52. Mohamed Abdel-Hamid and Hanaa Mohamed Abdelhaleem: Improving the construction industry quality using the seven basic quality control tools. J. Miner. Mater. Charact. Eng. **7**(6), 412–420 (2019)
53. HANSEN, B., Mark, H., et YU.: Model selection and the principle of minimum description length. J. Am. Stat. Ass. 96(454):746–774, 2001

# NutriSafe: Anti-Allergy Food Recommendation System Based on Content and Word Embedding


Didoune Nadia[1]($\boxtimes$), Sad-Houari Nawal[2], Reguieg Hicham[1], and Hassani Djihad[3]

[1] Department of Mathematics and Computer Science Computer Science Specialist, University of Science of Technology Oran Mohammed Boudiaf Oran, Oran, Algeria
nadia.didoune@univ-usto.dz
[2] Oran Computer Science Laboratory, Department of Living and Environment, Faculty of Natural and Life Sciences, University of Science of Technology Oran Mohammed Boudiaf Oran, Oran, Algeria
[3] Department of Science of Nature and Life Specialist Nutritionist Dieteticienne a EPH of Mascara Meslem Taib Mascara, Mascara, Algeria



**Abstract.** Food recommendation systems have become increasingly popular in recent years. With the growth of online commerce and the availability of a wide variety of food products, it becomes difficult for consumers to make informed choices. This is where food recommendation systems come into play. The objective of this work is to propose a food recommendation system to help consumers make choices that are tailored to their individual needs. The proposed system provides personalized nutritional recommendations taking into account preferences, allergies, and the health status of the patients. To do this, we have exploited dishes and their ingredients using the Word Embedding technique, notably GloVe and binary representation. Then, we measured distances using three different methods, namely, Cosine similarity, Euclidean distance, and Manhattan distance. The results of various experiments show that our system was able to recommend relevant foods taking into account the allergies and health issues of each user.

**Keywords:** Recommendation system · Foods · allergies · Allergy · content-based filtering · similarity measurement · Word-Embedding · GloVe · binary representation


## 1 Introduction

Food choices have a considerable impact on individual health. Maintaining good health requires a balanced diet that provides essential nutrients such as proteins, carbohydrates, fats, vitamins, and minerals. Nutritional deficiencies can lead to health problems, as our diet directly influences aspects such as weight, energy, and vitality. Selecting specific foods can contribute to preventing chronic diseases, while consuming unhealthy products can lead to serious issues. Therefore, our food choices play a central role in our overall health [1,2]. It is essential to make wise food decisions considering our nutritional needs, preferences, and health goals in order to maintain a balanced and healthy lifestyle.

The development of a food recommendation system based on allergies is crucial for allergic individuals who need information about allergenic ingredients. Traditional recommendation systems often overlook the importance of considering specific ingredients, which can result in potentially dangerous recommendations. Recognizing the need for personalized recommendations, a food recommendation system based on allergies accurately takes into account allergens. This system also integrates user ratings, timing information, and changes in preferences over time to provide accurate and relevant recommendations [3,4].

The aim of this work is to create a food recommendation system specifically designed for people with food allergies to offer them healthy meals taking into account their allergies, individual needs, and nutritional preferences. To promote a varied and balanced diet, it is essential to include a diversity of nutrient sources, such as fruits, vegetables, whole grains, lean proteins, and dairy products or alternatives. It is also crucial to consider cultural differences in dietary recommendations, integrating traditional foods while respecting the principles of a balanced diet. For people with allergies, intolerances, or specific medical conditions, recommendations must be adapted by avoiding allergens and adjusting portions according to medical restrictions. By taking into account these various issues, we can develop precise and personalized dietary recommendations.

Our article is structured as follows: in the following section, we present a state-of-the-art by summarizing several existing works. Then, we detail our proposed approach, followed by an analysis and discussion of the obtained results. Finally, we conclude with a synthesis and some perspectives for future research.

## 2   Related Work

Recommendation systems have become an essential element of our daily lives, allowing us to personalize the suggestions we receive during our interactions. Their creation relies on a combination of advanced techniques from the fields of machine learning, artificial intelligence, and data analysis. Here, we present the most relevant works from various recommendation systems.

Elsweiler et al. have leveraged biases in food choices to recommend healthier recipes using machine learning techniques. They have employed several machine learning techniques, such as logistic regression, decision trees with cosine similarity as a measure of similarity [5].

Still in the food domain, Yera et al. have presented a general framework for daily meal plan recommendations for individuals with chronic diseases, including diabetes, integrating simultaneous management of nutritional information and preferences. They compared between different existing food recommendation systems. The results revealed four main groups of food recommendation systems for diabetic patients: (a) Semantic-based approaches, (b) Optimization-based approaches, (c) Rule-based and classification-based approaches, and (d) Interaction-based approaches [6].

In the medical field, El Bouhissi et al. presented a recommendation system in the health domain for smart cities. The proposed system helps tourists find the appropriate doctor by using collaborative filtering and the K-means algorithm. The results showed that the developed recommendation system holds great promise in improving smart city tourism services [7].

The article presented in (Singh et al., 2023) [2] deals with the development of a food product recommendation system based on collaborative filtering and content-based techniques. The aim is to help users choose the best food products by providing personalized recommendations. The issues related to manual search are discussed, justifying the use of recommendation techniques. The authors used the K-nearest neighbors algorithm to implement this system. The latter recommends foods based on the food name, food identifier, cooking type, and dietary regimen (vegetarian or non-vegetarian). To recommend foods using collaborative filtering, the authors utilized user identifier, food identifier, and rating as attributes. Results showed that the proposed method improved the accuracy of food product recommendations compared to other methods proposed in the literature.

The article [A. Cardozo et al., 2022] [8] presents a meal recommendation system to enhance health and prevent non-communicable diseases caused by unhealthy eating habits. The authors propose a general framework for daily meal plan options that combine nutritional knowledge with user's physiological and personal data. The system employs machine learning methods and algorithms such as Random Forest classification and K-means clustering to recommend specific foods to each user based on their data. Results indicate that this system can assist users in maintaining and improving their health.

The article by [Chhipa et al., 2022] [9] presents a method for recipe recommendation using content-based filtering, specifically the TF-IDF algorithm and cosine similarity. Their goal is to create a mobile application where users can search for recipes based on ingredients available to them, including vegetables. The authors utilized a dataset consisting of Indian recipes and applied content-based recommendation using TF-IDF and cosine similarity to provide recommendations of Indian recipes based on user-available ingredients. They developed an API for their model using the Flask web framework and hosted it on Pythonanywhere. Results demonstrate the effectiveness of their method with respect to user-provided ingredients.

## 3   Our Contribution

Our user-centered contribution enables the creation of accurate databases and the development of more effective solutions for preventing food allergies, thereby enhancing the safety and quality of life for affected individuals. To understand the impact of allergenic ingredients, we collected food data from real users, validated by nutritionists. This collection was carried out through online questionnaires where users recorded their daily meals and reported any allergic reactions encountered. The collected data included detailed information about the ingredients of each consumed dish, allowing for the precise identification of potential allergens. The system provides personalized recommendations and eliminates dangerous ingredients.

We have created a robust and relevant database, facilitating the analysis of links between certain ingredients and allergic reactions, based on the experiences of real users and validations from nutritionists. In this study, we present an innovative method based on content, leveraging artificial intelligence with advanced techniques such as word embeddings with GloVe, and compare this approach to binary representation. Our research details the process of data collection and processing by nutritionists, as well as the customization of recommendations based on the specific needs of users.

## 4   Proposed Approach

To overcome the challenges related to dietary issues caused by the lack of nutritional knowledge among users, we have developed a personalized nutritional recommendation system based on content and knowledge. This system aims to provide relevant dietary recommendations for individuals with food allergies. We will explore the different steps necessary to create such a recommendation system, focusing on personalization and adaptation to the preferences and individual health status of each user.

Figure 1 presents the proposed architecture of our content-based food recommendation system.

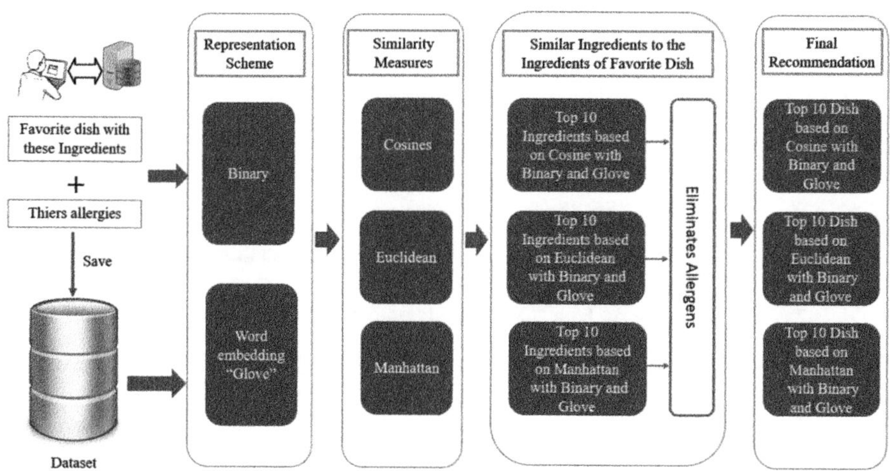


**Fig. 1.** Proposed architecture of the content-based food recommendation system.

To create our system, we followed the following steps:

## 4.1   Data Collection

Data collection is a crucial step in developing our system to offer a personalized experience to users with food allergies. For this purpose, we used an original and unique dataset developed in collaboration with nutrition specialists and specialized biologists. With the help of a nutritionist, we identified and classified all possible foods and their potential allergens.

Our dataset not only lists the allergens present in each recipe but also includes those that may occur when combining multiple foods, a common occurrence in cooking.

Our database includes 929 recipes (see Fig. 2), 1145 allergens, and 73 allergies. In our case, it is essential for the user to report their allergy in order to exclude dishes from recommendations that contain these allergens. However, we encountered difficulties when implementing this system due to data issues: most users did not mention their allergies or did not have any. This is why we adopted a constraint-based system. Thus, if a user reports a specific allergy to a food, all dishes containing that allergen will be automatically excluded from the recommendations.

## 4.2   Data Preprocessing

After collecting data on recipes and allergies, we preprocessed our database of dishes before training our models. First, we translated our database into English, as the Glove corpus in English is 2.4 GB, compared to a smaller size in French. Then, we performed thorough cleaning and preprocessing, including extracting all the ingredients from each dish and removing stop words. We also normalized words by converting uppercase to lowercase and replacing plurals with singulars. These steps allowed us to obtain a clean database ready for further operations.

|   | Aliment | ingrediants (mains-sucre-gras-epice) |
|---|---|---|
| 0 | Abricots Séchés | abricots, sulfites, sucre. |
| 1 | Aich | berkoukes,legumes,epice,piment,viande,poulet,a... |
| 2 | Ailes De Poulet Au Four | Ailes de poulet,Huile d'olive,Assaisonnements |
| 3 | Ailes De Poulet Buffalo | Ailes de poulet,Huile végétale,Sauce buffalo |
| 4 | Aliments En Pâte Frits | huile, œuf, farine de blé,lait,sel |

Fig. 2. An excerpt from our food database with these ingredients.

## 4.3   Data Representation

After cleaning, we proposed two methods of data representation for generating recommendations, namely:

– Binary representation.
– Semantic representation based on Glove Word Embedding.

**Binary Representation.** In our study, we adopted an approach based on the use of a binary matrix. This method was chosen for its ability to efficiently process data and provide accurate results within the scope of our research.

This method relies on a dataset consisting of a list of ingredients for each recipe, where each recipe is represented by a binary vector (see Fig. 3). A binary vector assigns the value 1 to an ingredient if it is present in the recipe, and 0 otherwise. This simplified representation effectively captures information about the composition of recipes while reducing computational complexity.

|   | aliment | abricot | sulfite | sucre | berkouke | legume | epice | piment | viande | poulet |
|---|---------|---------|---------|-------|----------|--------|-------|--------|--------|--------|
| 0 | Abricots Séchés | 1 | 1 | 1 | 0 | 0 | 0 | 0 | 0 | 0 |
| 1 | Aich | 0 | 0 | 0 | 1 | 1 | 1 | 1 | 1 | 1 |
| 2 | Poulet Au Four | 0 | 1 | 1 | 0 | 0 | 1 | 0 | 0 | 1 |
| 3 | Zlabia | 0 | 0 | 1 | 0 | 0 | 1 | 0 | 0 | 0 |

**Fig. 3.** Binary matrix of recipes.

**Semantic Representation Based on Glove Word Embedding.** In our approach, we chose to use Word Embedding, a technique that represents words or phrases as real-number vectors in a vector space model. Recently, two major methods have emerged in this field: Word2Vec and GloVe, which compete for prominence in literature and semantic similarity projects. Studies have shown that GloVe is more effective than Word2Vec, mainly due to the limitation of vector dimensionality in Word2Vec, which hinders the integration of all corpus information [10]. For this reason, we have favored GloVe, one of the most renowned methods for creating this vector representation of words.

GloVe, developed by Pennington and his colleagues at Stanford University, combines both local and global information in the vector representation of words.

It is an unsupervised word embedding that uses word co-occurrence statistics in a corpus. GloVe is widely used for similarity and entity identification [11].

To capitalize on the semantic richness of natural language, we have chosen to exploit word embedding, particularly the GloVe technique with a dimension of 300, to highlight the relationships between different recipes and choose the dimension that offers the best results. For ingredients that do not exist in the GloVe vector, we have proposed two solutions, which are:

- The first is to filter the words to keep only those with vector representations in the GloVe files.
- The second is to replace absent words with null vectors.

After retrieving the GloVe vector for each ingredient, we calculated the average of the vectors for ingredients composed of compound words, such as "olive oil." (see Fig. 4).

| Aliment | ingredient_0 | ingredient_1 | ingredient_2 | ingredient_3 |
|---|---|---|---|---|
| Abricots Séchés | [0.0, 0.0, 0.0, 0.0, 0.0, 0.0, 0.0, 0.0, 0.0, ... | [-0.28033000230789185, 0.21202999353408813, -0... | [0.8223000168800354, -0.16590000689029694, -0.... | [0.012001000344753265, 0.20750999450683594, -0... |
| Aich | [0.0, 0.0, 0.0, 0.0, 0.0, 0.0, 0.0, 0.0, 0.0, ... | [-0.5960800051689148, -0.038371000438928604, -... | [0.0, 0.0, 0.0, 0.0, 0.0, 0.0, 0.0, 0.0, 0.0, ... | [0.2046400010585785, 0.08120299875736237, -0.2... |
| Ailes De Poulet Au Four | [0.18578000366687775, -0.1621199995279312, -0.... | [-0.1770399957895279, 0.28492000699043274, -0.... | [-0.4401099979877472, 0.639739990234375, 0.564... | [0.01900400035083294, -0.21548999845981598, 0.... |
| Ailes De Poulet Buffalo | [0.18578000366687775, -0.1621199995279312, -0.... | [-0.1770399957895279, 0.28492000699043274, -0.... | [-0.4401099979877472, 0.639739990234375, 0.564... | [0.05847899988293648, -0.16144999861717224, 0.... |
| Aliments En Pâte Frits | [-0.4401099979877472, 0.639739990234375, 0.564... | [0.0, 0.0, 0.0, 0.0, 0.0, 0.0, 0.0, 0.0, 0.0, ... | [0.3743700087070465, 0.19147999584674835, 0.14... | [-0.08179199695587158, 0.17901000380516052, -0... |

Fig. 4. Representation by the GloVe model.

## 4.4  Similarity Calculation

After obtaining the binary matrices and GloVe associated with the dishes, we extracted the user preferences and also converted them into binary matrices and GloVe. We then used several similarity measures to calculate the distance between the user's preference ingredients and those of the dishes in our database. This allows us to determine which dishes are most similar by calculating the distance between the ingredients. To do this, we used three different similarity measures, namely:

1. **Cosine similarity** This is a measure of the degree of similarity between two vectors. This measure is often used in applications where the importance of vectors lies more in their orientation than in their magnitude, such as natural language processing (NLP). In the context of recipes and their ingredients, it allows us to evaluate the similarity between two ingredients. The values obtained range from 0 to 1: a value of 0 means that there is no similarity between the ingredients, while a value of 1 indicates that the ingredients are identical. [12]

$$Sim(x_\beta, y_\beta) = \frac{\sum_{\beta=1}^{n} |x_\beta y_\beta|}{\sqrt{\sum_{\beta=1}^{n} x_\beta^2}\sqrt{\sum_{\beta=1}^{n} y_\beta^2}} \tag{1}$$

2. **Euclidean similarity** Euclidean distance is the measure of the distance between two points in a multidimensional space. It is calculated by taking the square root of the sum of the squares of the differences between the corresponding coordinates of the two points. The advantage of this measure is its intuitive nature and its efficiency in spaces where the differences in magnitude between vectors are significant. [13]

$$DistEucl(x_\beta, y_\beta) = \sqrt{\sum_{\beta=1}^{n} (x_\beta - y_\beta)^2} \tag{2}$$

3. **Manhattan similarity** Manhattan distance is the sum of the absolute values of the differences between the corresponding coordinates of two points. In the context of recipes and calculating similarity between ingredients, this measure is useful for evaluating differences between ingredients when direct paths are not relevant. It allows for comparing ingredients by considering variations in their characteristics cumulatively. [13].

$$DistManh(x_\beta, y_\beta) = \sum_{\beta=1}^{n} |x_\beta - y_\beta| \tag{3}$$

### 4.5   Recommendation

After calculating the similarity between the user's favorite dish ingredients and those of all the dishes in our database, we ranked these ingredients in descending order of similarity. We then selected the five most similar ingredients. Before recommending dishes, we checked that these ingredients do not contain any potential allergens for the user, to avoid any dangerous recommendations. This verification involves eliminating dishes containing allergens or substituting these ingredients with safer alternatives. Then, we searched our database for dishes containing one or more of the selected allergen-free ingredients and classified them based on the number of corresponding ingredients. Finally, we recommended the top ten dishes from this list.

## 5   Experiments

To carry out this work, we used a machine described in Table 1.

**Table 1.** Machine Specification

| Machine | CPU | RAM | Storage | OS |
|---|---|---|---|---|
| Dell | Intel(R)Core(TM) | 16.00 | 512 | Windows11 |
| | i7-8665U | GB | GB | Professional 64bit |
| | CPU | | | x64-based |
| | @ 1.90GHz | | | processor |
| | 2.11 GHz | | | |

To evaluate our food recommendation system, we conducted an online evaluation with users. A group of 12 people was gathered to test the performance and effectiveness of our system.

First, we explained the concept and operation of our model to each participant. Then, the users provided their personal information, after which the system generated personalized recommendations for each of them. The participants then evaluated each recommendation by rating it on a scale of 1 to 5, where 1 means they do not like the dish, 3 means they moderately like it, and 5 means they like it a lot.

To guide the users in their ratings, we defined several criteria:

- **Health** Is this dish healthy according to your preferences?
- **Available ingredients** Were the ingredients easy to find?
- **Consumption frequency** Is this a dish you would eat frequently?
- **Cost** Is the dish economical?

The information and evaluations of each participant were recorded in a database. Once all the evaluations were completed, we conducted an overall evaluation by calculating the average ratings for each dish. Here is an excerpt from the evaluation conducted by user number 1 (See Fig. 5):

| Utilisateur 1: | | | | |
|---|---|---|---|---|
| Plat préféré: ['pizzas'] | | | | |
| Ingrédients : ['tomato sauce', 'flour', 'meat', 'cheese'] | | | | |
| Ingrédients plus proche: ['tomato sauce', 'creamy tomato sauce', 'tomato', 'sauce', 'pesto sauce', 'flour', 'wheat flour', 'barley flour', 'cereal flour', 'rye flour', 'meat', 'ground meat', 'salted meat', 'dried meat', 'meat or vegetable', 'cheese', 'mozzarella cheese', 'feta cheese', 'parmesan cheese', 'brie cheese'] | | | | |
| User 0 Preferences: ['pizzas'] | | | | |
| N | plat similaire | 8,00 | 1,00 | 1,00 |
| 1 | Annaba Brik: 4 | 5 | | |
| 2 | Algerian Coca: 3 | 5 | | |
| 3 | Eggplant Gratin: 3 | | 3 | |
| 4 | Pickle Hamburgers: 3 | 5 | | |
| 5 | Stuffed Kesra: 3 | 5 | | |
| 6 | Mchahma: 3 | 5 | | |
| 7 | Mhadjeb: 3 | 5 | | |
| 8 | Pizzas: 3 | 5 | | |
| 9 | Dough-Coated and Fried Foods: 2 | 5 | | |
| 10 | Antipasto: 2 | | | 1 |

**Fig. 5.** Excerpt from an evaluation by the user.

## 5.1  Results Obtained with the Binary Matrix

The evaluation results of user satisfaction with the results provided by the binary matrix are presented in Table 2. The recommendations were also evaluated using three different distance measures: Cosine, Euclidean, and Manhattan.

**Table 2.** Content-based filtering results with the binary matrix

| | Note 5 | Note 1 | Note 3 |
|---|---|---|---|
| Cosines | 5 | 2,833333333 | 2,166666667 |
| Euclidean | 3,333333333 | 1,916666667 | 4,75 |
| Manhattan | 4,25 | 1,583333333 | 4,166666667 |

– **Cosine** With binary representation, the cosine measure averages 5 for ratings 5, 2.83 for ratings 3, and 2.17 for ratings 1.

- **Euclidean** The Euclidean measure averages 3.33 for ratings 5, 1.92 for ratings 3, and 4.75 for ratings 1, showing poor performance for quality recommendations.
- **Manhattan** The Manhattan measure averages 4.25 for ratings 5, 1.58 for ratings 3, and 4.17 for ratings 1.

Recommendations based on binary representation show lower performance, especially for Euclidean and Manhattan measures, where ratings 1 are very high, indicating less relevant recommendations.

## 5.2   Results Obtained with GloVe Word Embedding

The evaluation results of user satisfaction using GloVe Word embedding are presented in Table 4.

**Table 3.** Content-Based Filtering Results Using GloVe Vectors

|           | Note 5 | Note 1 | Note 3 |
| --------- | ------ | ------ | ------ |
| Cosines   | 6,67   | 1,92   | 1,42   |
| Euclidean | 6,83   | 1,75   | 1,42   |
| Manhattan | 6,25   | 2,08   | 1,67   |

The recommendations were also evaluated using the same three distance measures, but this time with GloVe Word embedding.

- **Cosine** The cosine measure averages 6.67 for ratings 5, 1.92 for ratings 3, and 1.42 for ratings 1.
- **Euclidean** The Euclidean measure performs best with an average of 6.83 for ratings 5, 1.75 for ratings 3, and 1.42 for ratings 1.
- **Manhattan** The Manhattan measure averages 6.25 for ratings 5, 2.08 for ratings 3, and 1.67 for ratings 1.

Word embedding-based recommendations are generally better rated overall. Ratings for Rating 5 are significantly higher for all distance measures compared to binary representation.

## 5.3   Comparative Analysis

1. **Embedding vs Binary Representation**
   - **Word Embedding** Recommendations based on word embedding are generally better rated. The Euclidean distance performs slightly better for Rating 5, followed closely by cosine distance and Manhattan distance. This indicates that users find the recommendations more relevant when based on word embeddings and measured by Euclidean distance.

- **Binary Representation** Recommendations based on binary representation show less satisfactory performance. The scores for Euclidean and Manhattan distances are particularly low, with high scores of 1, suggesting that these recommendations are often deemed irrelevant.

2. **Most Effective Similarity Measure**
   - **For Word Embedding** Euclidean distance achieves the highest scores with ratings of 5, making it the most effective measure in this context.
   - **For Binary Representation** Cosine distance is the best measure for ratings of 5, although its scores are lower than those for word embedding-based recommendations.

Comparative analysis reveals that word embedding-based recommendations outperform those based on binary representation in terms of effectiveness and user evaluation. Specifically, Euclidean distance measures applied to word embeddings stand out as the most effective, especially for recommendations rated 5 by users, indicating increased relevance. In contrast, binary representation, although cosine distance is slightly better, remains overall less satisfactory, with notably lower relevance scores. These results highlight the importance of the representation method and similarity measurement in the quality of recommendations, placing word embedding and Euclidean distance at the top of the analyzed techniques.

## 6  Conclusion

In this article, we have presented the different stages of development of a food recommendation system. The experiments conducted have shown promising performances, thus demonstrating the effectiveness of our approach in recommending content tailored to individual user preferences. Throughout this study, we paid particular attention to the detailed description of the hardware configurations used, as well as the rigorous evaluation of the models developed through a series of experiments. These elements have proven essential in enlightening our understanding of the system's performances and guiding our perspectives on its potential improvements.

However, despite the successes achieved, we have identified aspects to delve deeper into. We thus plan to test other natural language processing techniques and integrate machine learning and/or deep learning algorithms to enhance the performances of our food recommendation system.

## References


1. Abdulmalek, S., et al.: IoT-based healthcare-monitoring system towards improving quality of life: a review, healthcare, vol. 10, no 10, p. 1993 (2022). https://doi.org/10.3390/healthcare10101993.
2. Singh et , R., Dwivedi, P.: Food recommendation systems based on content-based and collaborative filtering techniques (2023)

3. Ross, G.M.S., Bremer, M.G.E.G., Nielen, M.W.F.: Consumer-friendly food allergen detection: moving towards smartphone-based immunoassays. Anal. Bioanal. Chem. **410**(22), 5353–5371 (2018). https://doi.org/10.1007/s00216-018-0989-7

4. Rostami, M., Oussalah, M., Farrahi, V.: A Novel Time-Aware Food Recommender-System Based on Deep Learning and Graph Clustering , IEEE Access, vol. 10, pp. 52508-52524 (2022). https://doi.org/10.1109/ACCESS.2022.3175317.

5. Elsweiler, D., Trattner, C., Harvey, M.: Exploiting food choice biases for healthier recipe recommendation , In: Proceedings of the 40th International ACM SIGIR Conference on Research and Development in Information Retrieval, Shinjuku Tokyo Japan: ACM, août 2017, pp. 575-584. https://doi.org/10.1145/3077136.3080826.

6. Yera, R., Alzahrani, A. A., Martínez, L., Rodríguez, R. M.: A systematic review on food recommender systems for diabetic patients , Int. J. Environ. Res. Public. Health, vol. 20, no 5, p. 4248, févr. (2023). https://doi.org/10.3390/ijerph20054248.

7. Bouhissi, H. E., Tagzirt, D., Bouredjioua, F., Pavlova, O.: Health recommender system for smart cities

8. Cardozo, A., Poojari, M., Joseph, J., Yadav, O.: Food recommendation system , J. Emerg. Technol. Innov. Res. JETIR Wwwjetirorg, vol. 9, no 5 (2022)

9. Chhipa, S., Berwal, V., Hirapure, T., Banerjee, S.: Recipe recommendation system using TF-IDF , ITM Web Conf., vol. 44, p. 02006 (2022). https://doi.org/10.1051/itmconf/20224402006.

10. Mohammed, S. M., Jacksi, K., Zeebaree, S. R. M.: Glove word embedding and DBSCAN algorithms for Semantic Document Clustering , In: 2020 International Conference on Advanced Science and Engineering (ICOASE), Duhok, Iraq: IEEE, déc. 2020, pp. 1-6. https://doi.org/10.1109/ICOASE51841.2020.9436540.

11. Guohua, W., Yutian, G.: Using density peaks sentence clustering for update summary generation ', In: 2016 IEEE Canadian Conference on Electrical and Computer Engineering (CCECE), Vancouver, BC, Canada: IEEE, mai 2016, pp. 1-5. https://doi.org/10.1109/CCECE.2016.7726719.

12. Park, K., Hong, J. S., Kim, W.: A methodology combining cosine similarity with classifierfor text classification , févr. 2020. https://doi.org/10.1080/08839514.2020.1723868.

13. Walsh, T.: Strategy proof mechanisms for facility location in euclidean and manhattan space. arXiv, 2020. https://doi.org/10.48550/ARXIV.2009.07983.

# Addressing Imbalanced Datasets in Cyberbullying Detection on Social Media


Sihem Nouas[✉][iD], Fatima Boumahdi, Lharti Abdesatar,
and Oussama Rekrouk

LRDSI Laboratory, Department of Computer Science, Faculty of Sciences,
University of Blida 1, Blida, Algeria
`nouas.sihem.b3@gmail.com`



**Abstract.** Social media platforms play a crucial role in the dissemination of cyberbullying, making automated detection challenging due to the varied forms of content. This task is further complicated by class imbalance, which leads to biased machine learning models. We propose a hybrid solution for detecting cyberbullying using imbalanced datasets. First, to address data imbalance, we employ techniques such as Random Oversampling, SMOTE (Synthetic Minority Oversampling Technique), and Borderline-SMOTE. Second, to capture essential features and patterns while reducing noise, we introduce a Bi-LSTM-based Auto-encoder (AE). Experimental results using standard evaluation metrics demonstrate the efficacy of our proposed method for cyberbullying detection, achieving F1-scores of 0.71, 0.89, and 0.72 across three benchmark datasets. These results represent a significant improvement over existing approaches, particularly in addressing class imbalance.

**Keywords:** Cyberbullying · Imbalanced datasets · Auto-encoder · Oversampling


## 1 Introduction

The advent of digital and social media has revolutionized communication, making it instantaneous and far-reaching. However, this technological advancement has also facilitated the misuse of these platforms for spreading hatred and engaging in cyberbullying. Smith et al. [19] described cyberbullying as aggressive acts or behaviors executed through electronic means by an individual or a group, repeatedly and over time, targeting a victim who is unable to easily defend themselves. This phenomenon has been reported to be most prevalent among school-aged children and adolescents, who are particularly vulnerable [13]. The impact of cyberbullying can be severe, leading to significant psychological consequences for victims, including increased stress, anxiety, depression, and diminished self-esteem [21].



N. Seddari and M. Redjimi (Eds.): ICMSCT 2024, CCIS 2606, pp. 239–248, 2025.
https://doi.org/10.1007/978-3-032-01922-6_20

Data imbalance arises when a dataset exhibits an unequal distribution of class instances, resulting in one class typically the majority class being overrepresented compared to the minority class. This imbalance presents a significant challenge in machine learning, as models are prone to favor the majority class, impairing their ability to accurately predict instances of the minority class. In the context of cyberbullying detection, datasets frequently display this imbalance, with a higher number of non-cyberbullying instances compared to actual cases of cyberbullying.

To address the issue of imbalanced datasets, we employ three commonly used oversampling techniques: SMOTE [5], Borderline-SMOTE [11], and random oversampling. These methods aim to enhance the representation of the minority class by generating synthetic data points. Following this, we investigate the use of auto-encoders, specifically a Bi-LSTM-based model, to extract meaningful features and mitigate the noise introduced by the oversampling methods. By leveraging auto-encoders, we aim to improve the quality of the features extracted from the data, thereby enhancing the model's ability to accurately identify cyberbullying instances and reducing the potential distortions caused by synthetic data generation.

The remainder of the paper is organized as follows: Sect. 2 reviews related work. Section 3 presents our proposed approach, including details on preprocessing, the auto-encoder and oversampling framework, datasets, and evaluation metrics. Section 4 discusses the experimental results and analysis. Finally, Sect. 5 summarizes our findings and suggests directions for future research.

## 2   State of the Art

Most of the previous work in cyberbulying detection mainly focused on supervised based approaches using LR classifier (Logistic Regression) [4,7,23], SVM classifier (Support Vector Machine) [1,12] and Random Forest classifier (RF) [20]. Davidson et al. [7] used an LR classifier to differentiate between hate speech, offensive language, and neutral content. They preprocessed tweets by lowercasing, stemming, and creating TF-IDF-weighted n-gram features. They also used NLTK (Natural Language Toolkit) for POS tagging (Part Of Speech tagging), sentiment scores from a social media lexicon, and features for hashtags, mentions, retweets, URLs, and tweet length. Atoum [1] employed sentiment analysis for cyberbullying detection. The preprocessing steps included normalization, tokenization, named entity recognition (NER), removal of stop words, stemming, and application of N-grams. Experiments were conducted using Support Vector Machine (SVM) and Naive Bayes (NB) classifiers. Among various n-gram models, the 4-gram language model demonstrated superior performance in terms of accuracy, precision, recall, F-measure, and ROC, with the best results achieved using SVM. Talpur and O'Sullivan [20] assessed various algorithms for classifying cyberbullying severity on Twitter, using Word2Vec embeddings, PMI-sentiment orientation, and lexicon features, along with SMOTE for data balancing. The Random Forest (RF) classifier achieved the best performance in binary classification.

Deep learning models such as Word2Vec, Convolutional Neural Networks (CNN), LSTM (Long Short-Term Memory), BiLSTM (Bidirectional LSTM), BERT (Bidirectional Encoder Representations from Transformers) [10], GloVe (Global vectors for word representation) and FastText [3] are commonly used for cyberbullying detection. For instance, De Gibert et al. [9] implemented a CNN based model, an LSTM based model, and SVM for classifying sentences from a white supremacy forum. Among these models, the LSTM classifier performed the best. Saleh et al. [18] developed a domain-specific word embedding model to capture the meanings of hate speech terminology, acronyms, and misspelled words. Their evaluation showed that a BiLSTM-based model with these embeddings improved performance over existing state-of-the-art techniques. Additionally, the BERT language model demonstrated higher performance compared to the domain-specific embedding model in binary hate speech classification. Lindenmayr et al. [15] employed a CNN based model to identify hateful tweets related to the 2020 Vienna terror attack. They compared three techniques for hate speech detection: a lexicon-based approach, an SVM classifier using Bag of Words (BOW) and TF-IDF for feature extraction, and a deep learning approach with CNN and GloVe embeddings [17]. The CNN model outperformed the other methods. They also observed that hateful tweets were more frequently retweeted and often came from anonymous accounts.

In the context of imbalanced datasets in cyberbullying detection, various methods are employed to address the issue. These include oversampling techniques such as SMOTE to create new synthetic data [20], and undersampling methods. For example, Verdikha et al. [22] tackle the problem of imbalanced data in hate speech classification using the Instance Hardness Threshold (IHT) method, which involves undersampling by removing frequently misclassified data. They compare various estimators for the IHT method, finding that Logistic Regression (IHT(LR)) achieves the fastest process time, balanced class ratios, and the highest Index of Balanced Accuracy (IBA) scores. Additionally, a combination of both oversampling and undersampling is often employed [6]. For instance, Khairy et al. [14] combined SMOTE with Tomek links for cyberbullying detection. Cost-sensitive methods are also commonly utilized [12,16]. For example, Hee et al. [12] employed a cost-sensitive SVM approach, wherein misclassifications of the minority positive class are assigned a higher penalty compared to classification errors associated with the majority negative class.

While commonly employed oversampling techniques like SMOTE and its variants effectively address class imbalance, they may introduce noise by generating synthetic instances that do not fully represent the minority class distribution. Additionally, traditional machine learning models, such as LR, SVM, and NB, which are frequently used in cyberbullying detection, rely on handcrafted features and struggle to adapt to complex social media data. To mitigate these issues, we propose a hybrid framework that combines SMOTE-based oversampling with a Bi-LSTM-based Auto-encoder (AE). This approach could not only reduces the noise introduced by oversampling but also captures essential patterns in the data, offering a more robust solution for detecting cyberbullying in imbalanced datasets compared to existing techniques.

# 3   The Proposed Approach

In this section, we will outline our proposed approach for addressing cyber-bullying detection in imbalanced datasets. First, in Sect. 3.1, we discuss the preprocessing steps necessary to prepare the data for analysis. Following that, in Sect. 3.2, we introduce our auto-encoder and Oversampling Framework. In Sect. 3.3, we provide details about the datasets used in our experiments.

## 3.1   Preprocessing and Textual Representation

We perform several data preprocessing tasks. Text is cleaned by standardizing format, removing irrelevant elements such as URLs, mentions, usernames, hash-tags, and unnecessary punctuation, and converting to lowercase. Additionally, stop words, which are common and less informative are removed to reduce noise. An example of a tweet before and after preprocessing is illustrated in Fig. 1. After preprocessing, we adopt GloVe embeddings for textual representation.

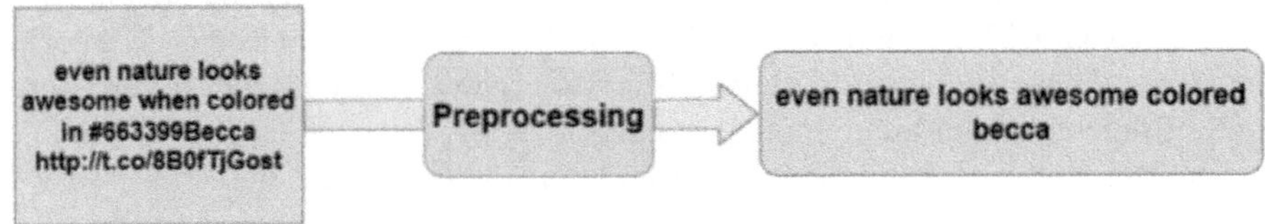


**Fig. 1.** Example of Preprocessed Tweets.

## 3.2   Auto-encoder and Oversampling Framework

As illustrated in Fig. 2. Our framework consists of the following steps:

- **Oversampling Techniques:** To improve the representation of the minority class, we generate synthetic data points using three methods: random over-sampling, borderline-SMOTE, and SMOTE. Random oversampling helps to increase the number of minority instances, providing the classifier with more examples to learn from. SMOTE and its variant, borderline-SMOTE, create synthetic examples by interpolating between existing minority instances. These generic methods are widely used and regarded as simple and effective for mitigating the effects of imbalanced classes.
- **Feature Extraction and Noise reduction:** We apply a Bidirectional Long Short-Term Memory (Bi-LSTM) based auto-encoder to extract meaningful patterns and reduce noise introduced by the oversampling. The first layer processes these sequences using a Bidirectional LSTM, which captures dependencies from both past and future contexts. This is followed by a dense layer that reduces the dimensionality of the data. An attention mechanism is then applied to focus on important parts of the sequence. The attention output is combined through a dot product operation. The resulting representation is fed into another Bidirectional LSTM layer. Finally, a dense layer outputs the refined feature set, matching the original input shape.

– **Final Classification:** We use a Support Vector Machine (SVM) as the final classifier, as SVMs are well-established for their effectiveness in textual classification [1,12].

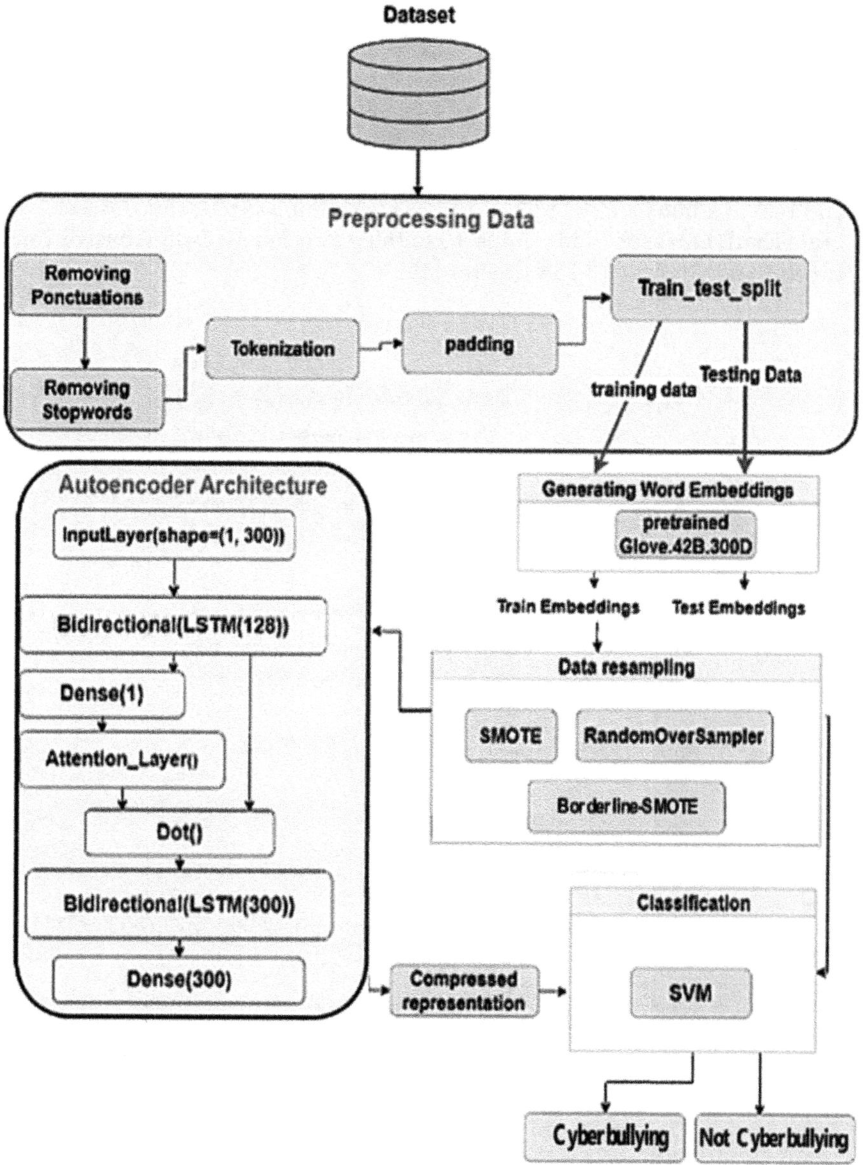


Fig. 2. Overview of the Proposed Method.

### 3.3  Datasets

In this paper, we present a binary classification study conducted on the following three imabalanced datasets:

- **Hate Speech Dataset**: This dataset was first used by De Gibert's [9]. It contains it contains 10,703 English sentences have been extracted from Stormfront and has two classes, Not hate (9,507 posts) and Hate (1,196 posts).
- **Davidson Combined Dataset**: The original dataset [7] consists of 24,783 English tweets distributed across three classes: Hate (1,430 tweets), Offensive (19,190 tweets), and Neutral (4,163 tweets). For the following experiment, the Hate and Offensive classes were combined into a single 'Hate' class. As a result, the dataset now categorizes tweets into 'Hate' and 'No Hate' categories and remains imbalanced. Figure 3 presents examples from this dataset.
- **HateEval Dataset**: This dataset contains two labels: Non-Abusive (5,217 tweets) and Abusive (3,783 tweets) [2].

| | class | tweet |
|---|---|---|
| 0 | 0 | !!! rt @mayasolovely: as a woman you shouldn't... |
| 1 | 0 | !!!!! rt @mleew17: boy dats cold...tyga dwn ba... |
| 2 | 0 | !!!!!!! rt @urkindofbrand dawg!!!! rt @80sbaby... |
| 3 | 0 | !!!!!!!! rt @c_g_anderson: @viva_based she lo... |
| 4 | 0 | !!!!!!!!!!! rt @shenikaroberts: the shit you... |
| ... | ... | ... |
| 993 | 1 | @JRubinBlogger @SenFeinstein I wonder what's i... |
| 994 | 1 | @Lauren_Southern You're just a skank and every... |
| 996 | 1 | @DanReynolds STFU BITCH! AND YOU GO MAKE SOME ... |

**Fig. 3.** Hate Speech Dataset with two classes 1: hate, 0: Not Hate).

### 3.4  Evaluation Metrics

In this paper, we evaluated the proposed method using Accuracy, Precision, Recall, and F-score.

$$\text{Accuracy} = \frac{TP + TN}{TP + TN + FP + FN} \tag{1}$$

$$\text{Precision} = \frac{TP}{TP + FP} \tag{2}$$

$$\text{Recall} = \frac{TP}{TP + FN} \tag{3}$$

$$F\text{-score} = \frac{2 \cdot Precision \cdot Recall}{Precision + Recall} \tag{4}$$

# 4   Experimental Results and Discussion

In this experiment, we evaluated the proposed method using several combinations to address the challenge of imbalanced datasets in cyberbullying detection. The combinations tested include:

- **SVM**
- **AE + SVM**
- **SMOTE + SVM**
- **SMOTE + AE + SVM**
- **Oversampling + AE + SVM**
- **Borderline-SMOTE + AE + SVM**

**Table 1.** Comparison of State of the Art Results and Proposed Approach

| Model | Dataset | Precision | Recall | F1-score (Macro) | Accuracy |
|---|---|---|---|---|---|
| [8] | Hate Speech | / | / | / | 78.00 |
| SVM | | 69.50 | 50.00 | 47.50 | 89.00 |
| SMOTE+SVM | | 67.40 | 73.00 | 69.55 | 86.00 |
| AE+SVM | | **88.80** | 57.10 | 60.00 | **90.25** |
| SMOTE+AE+SVM | | 68.30 | 72.50 | 70.00 | 87.00 |
| RandomOverSampler +AE + SVM | | 67.65 | **79.70** | 70.85 | 87.90 |
| **Borderline-SMOTE+AE+SVM** | | 67.76 | 78.81 | **71.00** | 88.00 |
| SVM | Davidson Combined | **89.89** | 89.09 | 89.48 | **94.16** |
| SVM + SMOTE | | 86.54 | 92.33 | 89.04 | 93.36 |
| **AE + SVM** | | 88.22 | 91.52 | **89.75** | 94.00 |
| SMOTE +AE + SVM | | 87.14 | 92.00 | 89.30 | 93.40 |
| RandomOverSampler +AE + SVM | | 86.99 | **93.00** | 89.61 | 93.68 |
| Borderline-SMOTE +AE + SVM | | 88.22 | 91.52 | 89.16 | 94.00 |
| SVM | HateEval | 70.74 | 66.67 | 66.55 | 66.55 |
| SMOTE+SVM | | 70.80 | 68.20 | 68.30 | 70.45 |
| AE+SVM | | 72.00 | 70.60 | 70.90 | 72.20 |
| SMOTE+AE+SVM | | 71.30 | **72.00** | 71.50 | 72.50 |
| RandomOverSampler+AE+SVM | | **72.45** | 71.60 | 71.90 | 72.00 |
| **Borderline-SMOTE+AE+SVM** | | 71.35 | 71.90 | **72.55** | **72.40** |

Table 1 provides a comparative analysis of various combinations of the proposed model. For the Hate Speech dataset, the Borderline-SMOTE + AE + SVM model achieved the highest F1-score (71%), closely followed by the RandomOverSampler + AE + SVM model with an F1-score of 70.85%, and the SMOTE + AE + SVM model with an F1-score of 70%, showcasing the best balance between precision and recall. For the combined Davidson dataset, the AE + SVM model achieved the highest F1-score (89.75%), closely followed by the RandomOver-Sampler + AE + SVM model with an F1-score of 89.61%. And finally, in HateEval, the best F-score was achieved using Borderline-SMOTE +AE+SVM followed by RandomOverSampler+AE+SVM (71.90%) and SMOTE+AE+SVM (71.50%). Compared to the baseline results, the Hatespeech and HateEval datasets showed the most significant improvements, with rates of 49.47% and 9.02%, respectively. The Davidson dataset showed a more modest improvement of 0.82%. The experimental results further demonstrate that the hybrid approach, incorporating oversampling, auto-encoder, and SVM, consistently outperforms the standalone SMOTE + SVM model across all datasets. The comparative analysis of SMOTE+SVM and the full hybrid approach (incorporating SMOTE) reveals F-score improvements of 0.65%, 0.80%, and 4.83% for the Hate Speech, Davidson, and HateEval datasets, respectively.

The consistent performance of the oversampling-AE-SVM model across all datasets suggests the effectiveness of a combined approach using oversampling, auto-encoders, and SVM classifier for handling imbalanced datasets. Additionally, the results from the AE+SVM model show that while auto-encoders improves performance compared to SVM alone, including oversampling in the combined model can yield additional improvements.

The observed increase in F1-scores across all datasets can be attributed to this enhanced feature extraction process. The auto-encoder not only mitigates the noise from the synthetic data but also strengthens the representation of relevant patterns, allowing the final classifier to make better-informed decisions. Consequently, the integration of the auto-encoder demonstrated its efficacy in improving classification performance in the context of cyberbullying detection using imbalanced datasets.

It is important to note that most of the parameters in the proposed method were kept at their default settings. A study on parameter sensitivity could potentially provide more insights into how variations in these parameters impact the final results.

## 5   Conclusion

This paper proposes a hybrid oversampling-autoencoder framework to address the common class imbalance challenge in social media cyberbullying detection. The proposed approach improves classification performance by combining oversampling techniques with an autoencoder. The autoencoder serves two functions: improving feature representation and reducing potential noise caused by oversampling. Empirical evaluation demonstrates that our hybrid model outperforms conventional methods, including those based solely on oversampling, by achieving a more equitable balance between precision and recall. Our findings suggest that the proposed hybrid framework is a promising strategy for effectively addressing class imbalance in social media data. Future research may investigate the incorporation of additional multidimensional features, such as user profile information, to improve the model's ability to detect cyberbullying.

# References


1. Atoum, J.O.: Cyberbullying detection through sentiment analysis. In: 2020 International Conference on Computational Science and Computational Intelligence (CSCI), pp. 292–297 (2020)
2. Basile, V., et al.: Semeval-2019 task 5: multilingual detection of hate speech against immigrants and women in twitter (2019)
3. Bojanowski, P., Grave, E., Joulin, A., Mikolov, T.: Enriching word vectors with subword information (2016)
4. Ginting, P.S.B., Irawan, B., Setianingsih, C.: Hate speech detection on twitter using multinomial logistic regression classification method. In: 2019 IEEE International Conference on Internet of Things and Intelligence System (IoTaIS), pp. 105–111 (2019)
5. Chawla, N.V., Bowyer, K.W., Hall, L.O., Kegelmeyer, W.P.: Smote: synthetic minority over-sampling technique. J. Artif. Intell. Res. **16**, 321–357 (2002)
6. Colton, D., Hofmann, M.: Sampling techniques to overcome class imbalance in a cyberbullying context (2019)
7. Davidson, T., Warmsley, D., Macy, M., Weber, I.: Automated hate speech detection and the problem of offensive language (2017)
8. de Gibert, O., Perez, N., García-Pablos, A., Cuadros, M.: Hate speech dataset from a white supremacy forum. In: Proceedings of the 2nd Workshop on Abusive Language Online (ALW2), Brussels, Belgium, pp. 11–20. Association for Computational Linguistics (2018)
9. de Gibert, O., Pérez, N., Pablos, A.G., Cuadros, M.: Hate speech dataset from a white supremacy forum. CoRR, abs/1809.04444 (2018)
10. Devlin, J., Chang, M.-W., Lee, K., Toutanova, K.: BERT: pre-training of deep bidirectional transformers for language understanding. CoRR, abs/1810.04805 (2018)
11. Han, H., Wang, W., Mao, B.: Borderline-smote: a new over-sampling method in imbalanced data sets learning. In: International Conference on Intelligent Computing (2005)
12. Van Hee, C., et al.: Automatic detection of cyberbullying in social media text. PLoS ONE **13** (2018)
13. Jin, X., et al.: Cyberbullying among college students in a Chinese population: prevalence and associated clinical correlates (2023)
14. Khairy, M., Mahmoud, T.M., Abd-El-Hafeez, T.: The effect of rebalancing techniques on the classification performance in cyberbullying datasets (2023)
15. Lindenmayr, M., Kusen, E., Strembeck, M.: "Stronger than hate": on the dissemination of hate speech during the 2020 Vienna terrorist attack. In: 2021 Eighth International Conference on Social Network Analysis, Management and Security (SNAMS), pp. 01–08 (2021)
16. Nouas, S., Boumahdi, F., Madani, A., Berrhail, F.: Cyberbullying: a BERT Bi-LSTM solution for hate speech detection. In: Zantout, H., Hassen, H.R. (eds.) Proceedings of the International Conference on Applied Cybersecurity (ACS), pp. 57–65. Springer, Cham (2023)
17. Pennington, J., Socher, R., Manning, C.D.: Glove: global vectors for word representation (2014)
18. Saleh, H., Alhothali, A., Moria, K.: Detection of hate speech using BERT and hate speech word embedding with deep model (2021)
19. Smith, P.K., et al.: Cyberbullying: its nature and impact in secondary school pupils. J. Child Psychol. Psychiatry Allied Disciplines **49**(4), 376–385 (2008)

20. Talpur, B.A., O'Sullivan, D.: Cyberbullying severity detection: a machine learning approach. PLoS ONE **15** (2020)
21. Geel, M., Goemans, A., Toprak, F., Vedder, P.: Which personality traits are related to traditional bullying and cyberbullying? A study with the big five, dark triad and sadism. Personality Individ. Differ. **106**, 231–235 (2017)
22. Verdikha, N.A., Adji, T.B., Permanasari, A.E.: Study of undersampling method: instance hardness threshold with various estimators for hate speech classification (2018)
23. Waseem, Z., Hovy, D.: Hateful symbols or hateful people? predictive features for hate speech detection on twitter. In: NAACL (2016)

# Detecting Aggressive Driving Behaviors on Highways Using 1D Convolutional Neural Networks


Mohamed Aissaoui[1], Fadia Bouchenafa[1], Oussama Annad[2(✉)], and Sadek el Amine Boughaleb[2]

[1] National Higher School of Advanced Technologies (ENSTA), Algiers, Algeria
{m_aissaoui,f_bouchenafa}@ensta.edu.dz
[2] Laboratory of Innovative Technologies (LTI), National Higher School of Advanced Technologies (ENSTA), Algiers, Algeria
{oussama.annad,sadekelamine.boughaleb}@ensta.edu.dz



**Abstract.** Road traffic accidents pose a significant global public health challenge, resulting in millions of deaths and injuries each year. This paper investigates the use of artificial intelligence (AI) and machine learning (ML) techniques to analyze and classify driver behaviors, with a particular focus on detecting aggressive driving style on highways. A one-dimensional convolutional neural network (1D CNN) was employed to process and analyze driving data from the UAH-DriveSet dataset and independently collected real datasets for safe driving, with aggressive driving data simulated from this safe driving dataset. The model developed in this research demonstrated good generalization capabilities across different drivers. The integration of this model into real-time driver monitoring systems has the potential to significantly enhance road safety by alerting drivers to dangerous behaviors and encouraging safer driving practices.

**Keywords:** Road safety · driver behavior · artificial intelligence · machine learning · aggressive driving detection · Real-time monitoring


## 1 Introduction

Road traffic accidents, a major global public health burden, result in millions of deaths and injuries annually. World Health Organization (WHO) data indicate that approximately 1.19 million people die on roads each year, with an additional 20 to 50 million suffering non-fatal injuries [8]. Driver behavior monitoring plays a crucial role in promoting road safety, as aggressive driving behaviors, in particular, have often been identified as key factors contributing to serious accidents [8]. In light of this alarming trend, developing and implementing innovative solutions to enhance road safety is crucial. This article explores the application of artificial intelligence (AI) techniques to classify driver behaviors, with a particular focus on detecting aggressive driving on highways.


© The Author(s), under exclusive license to Springer Nature Switzerland AG 2025
N. Seddari and M. Redjimi (Eds.): ICMSCT 2024, CCIS 2606, pp. 249–260, 2025.
https://doi.org/10.1007/978-3-032-01922-6_21

Researchers have made significant advancements in leveraging machine learning techniques to identify aggressive driving behaviors, paving the way for safer roads. These developments have been driven by the application of cutting-edge algorithms, including convolutional neural networks (CNNs), support vector machines (SVMs), and recurrent neural networks (RNNs), to analyze a wide range of driving data.

CNNs have proven particularly adept at decoding visual cues such as signs, lane markings, and traffic patterns, enabling them to accurately classify driving styles based on adherence to traffic rules and road etiquette. Karaduman and Eren (2017) [4] demonstrate the effectiveness of CNNs in achieving an accuracy of 88.02% in distinguishing between safe and aggressive driving behaviors.

SVMs, on the other hand, demonstrate strength in pattern recognition and classification tasks, making them well-suited for identifying aggressive driving patterns from sensor data. Wang et al. (2017) [10] showcase the potential of SVMs by presenting a system that uses a semi-supervised learning approach on simulated data to classify aggressive drivers with 86.6% accuracy.

RNNs, especially long short-term memory (LSTM) networks, have also emerged as powerful techniques for analyzing sequential data, such as the temporal patterns exhibited in driving behavior. Mumcuoglu et al. (2019) [7] take advantage of this ability by combining FCNs and LSTMs to achieve an F1 score of 95.88% in classifying normal and aggressive driving styles on the UAH-DriveSet dataset.

Moreover, researchers have explored motion-based features to identify aggressive driving patterns. For instance, Matousek et al. (2019) [5] utilized a random forest model to extract these features, achieving an impressive area under the curve (AUC) of 97.10% in predicting aggressive driving behavior.

In addition to analyzing individual driving behaviors, researchers have also investigated how situational factors influence driving styles. Zheng et al. (2017) [11] delve into this area by examining how driving behavior in online car-hailing services varies depending on the task at hand, using k-means clustering to identify aggressive, normal, and cautious driving patterns.

Chen et al. (2021) [1] present a supervised approach leveraging Labeled Latent Dirichlet Allocation (LLDA) to classify driver behavior. This method categorizes drivers into three distinct styles: aggressive, moderate, and careful driving. Notably, the LLDA model achieves an average accuracy of 60.5%, surpassing the performance of traditional classifiers like Support Vector Machines (SVM), Naive Bayes (NB), and K-Nearest Neighbors (KNN).

Several studies have focused on classifying driver aggressiveness. Works by Jahangiri et al. (2017) [3] and Moukafih et al. (2019) [6] employed Support Vector Machines (SVM) and Long Short-Term Memory (LSTM) algorithms, respectively, to categorize driver behavior with scores reflecting aggressiveness levels. These approaches achieved impressive accuracy, reaching 86.67% and 92.8%. Beyond classification, other research has aimed to continuously track driving styles. For instance, the study by Wang et al. (2021) [9] utilized the Next Generation Simulation (NGSIM) dataset and a genetic algorithm (GA) to analyze car-following behaviors under various driving scenarios.

These advances in detecting aggressive driver behavior hold immense potential for improving road safety. By integrating these technologies into advanced driver assistance systems and in-vehicle monitoring solutions, we can effectively alert drivers to potential dangers and encourage safer driving practices. As research continues to improve these methods and explore new applications, we can anticipate a future in which aggressive driving behaviors become increasingly rare, paving the way for a safer and more harmonious driving experience for everyone.

## 2 Methodology

To develop an effective model for detecting aggressive driving behaviors, a systematic approach was adopted. This methodology section outlines the processes of data collection, data preprocessing, model design, and training strategy. The objective is to ensure the robustness and reliability of the model through careful selection and processing of data, followed by rigorous model training and evaluation.

### 2.1 Data Collection

The data collection process involved two primary sources: the publicly available UAH-DriveSet dataset and an independent data collection effort. The UAH-DriveSet dataset provides comprehensive data on driver behavior, including over 500 min of driving data from six different drivers on highways. This dataset encompasses raw sensor data such as GPS coordinates, accelerometer readings, gyroscope data, and detailed video recordings.

In addition to the UAH-DriveSet, we conducted an independent data collection to supplement and diversify our dataset. This independent data collection involved an experienced driver navigating the highway, with data gathered using an Android application. The collected data consisted of 25 min of safe driving only. Aggressive driving data was then simulated from this safe driving session by manipulating parameters such as speed, safety distance, and lane changes, among other parameters.

### 2.2 Data Preprocessing

To ensure the data was suitable for model training and evaluation, several crucial preprocessing steps were undertaken. The base variables used, along with their meanings, are detailed in Table 1.

**Table 1.** Base variables and their meanings

| Feature | Description |
| --- | --- |
| Acceleration in Z (Gs) | Direct acceleration for speeding and braking. Key indicator of aggressive driving behaviors |
| Acceleration in Y (Gs) | Lateral acceleration. Identifies lane changes and swerving |
| Filtered Acceleration (Z & Y) (Gs) | Acceleration filtered by Kalman filter. Provides cleaner signals for detecting driving maneuvers |
| Distance to ahead vehicle (meters) | Distance to the vehicle ahead. Assesses tailgating behavior |
| Time of impact to ahead vehicle (seconds) | Time to collision based on speed and distance. Detects risky driving |
| Number of detected vehicles | Traffic density. Context for driving behavior |
| GPS speed (km/h) | Vehicle speed. Identifies speeding and sudden changes |
| Estimated current lane | Lane position. Understands lane-changing behavior |
| Video footage | Real-time driving sessions. Extracts safety distance, time of impact, and vehicle count using YOLO object detection |

*Signal Processing.* The raw sensor data were processed using Fast Fourier Transform (FFT) to convert the signals from the time domain to the frequency domain. This conversion enabled the filtering of noise and the extraction of relevant features. FFT was applied four times with different component thresholds (80, 300, 600, 900), each chosen for its specific importance:

- 80 components: Captured low-frequency information, important for identifying steady and long-term driving patterns.
- 300 components: Balanced between low and mid frequencies, useful for detecting moderate changes in driving behavior.
- 600 components: Focused on mid-frequency information, which helps in identifying more dynamic driving actions.
- 900 components: Captured high-frequency details, crucial for detecting sharp and rapid driving maneuvers.

*Additional Data Processing for Self-collected Dataset.* For the independently collected dataset, the phone camera was utilized to capture real-time video during the driving sessions. YOLO (You Only Look Once) object detection algorithm was employed to analyze the video frames:

- Safety distance calculation: YOLO was used to detect vehicles ahead by drawing bounding boxes around the vehicles. The focal length method was used to calculate the distance between the subject vehicle and the detected vehicles.
- Vehicle count: Additionally, YOLO provided the count of vehicles ahead, contributing additional context to the driving behavior data (Figs. 1 and 2).

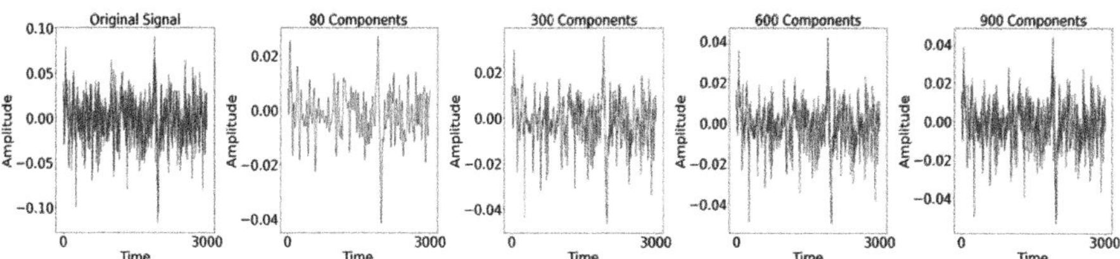

**Fig. 1.** Comparison of a signal filtered four times with different FFT component thresholds: 80, 300, 600, and 900. The figure demonstrates how varying thresholds affect the signal's frequency content and smoothness.

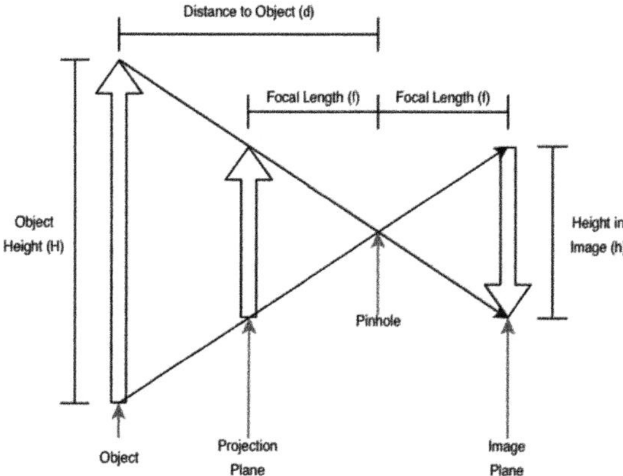


**Fig. 2.** The pinhole method for calculating distances in the context of object detection. [2].

*Data Splitting.* The dataset was divided into training and testing sets to facilitate robust model evaluation. For the data splitting, cross-validation was used with the UAH-DriveSet to ensure robust model training and validation. Specifically, a leave-one-subject-out cross-validation approach was implemented, where each time, data from one driver was set aside as the test set while the data from the remaining drivers were used for training the model. This process was repeated until each driver's data had been used once as the test set. The independently collected data, including the simulated aggressive driving data, was used exclusively as an additional test set. This division ensured that the model was evaluated on unseen data, providing a reliable measure of its generalization capabilities.

*Sequence Generation and Data Augmentation.* To capture the temporal dynamics of driving behaviors, fixed-length sequences of 3000 steps (approximately 100 s) were generated from the continuous data streams. Random start indices were employed to ensure that the sequences covered a broad spectrum of driving behaviors and conditions. This method also served as a form of data augmentation, enhancing the diversity of the training data.

Additional data augmentation was performed by manipulating driving parameters to simulate aggressive and normal driving behaviors. For aggressive driving, parameters such as speed were increased, and safety distance was decreased. Conversely, for normal driving, speed was moderated, and safety distance was increased. This approach helped in creating a more comprehensive dataset, allowing the model to learn from varied driving scenarios.

### 2.3   Model Design

Given the complexity and real-time application requirements, a one-dimensional convolutional neural network (1D CNN) architecture was chosen. This model was selected for its efficiency in processing time-series data and its ability to effectively capture temporal patterns.

- Convolutional Layers: These layers extract spatial features from the input data, identifying patterns related to driving maneuvers.
- Dense Layers: Fully connected layers were used for classification, outputting probabilities for normal and aggressive driving styles.

### 2.4   Training Strategy

The model training involved careful optimization of hyperparameters and the implementation of specific techniques to enhance performance:

- Learning rate adjustment: The ReduceLROnPlateau callback was used to monitor validation loss and reduce the learning rate by a factor of 0.2 after two epochs without improvement, continuing until a minimum learning rate threshold was reached (Fig. 3).

One-dimensional convolutional neural network

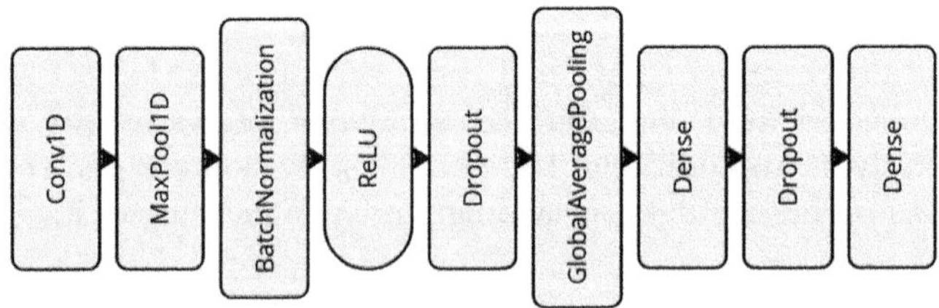


**Fig. 3.** Architecture of the 1D CNN model used in this study. Source: Authors.

- Early stopping: To prevent overfitting, the EarlyStopping callback halted training after five consecutive epochs without improvement in validation loss.

- Optimizer and loss function: The model was compiled using the 'adam' optimizer, known for its adaptive learning rate capabilities, and trained with the 'sparse categorical cross-entropy' loss function.
- Validation split: A validation split of 30% was used to ensure the model's generalization. The model was trained for 20 epochs with a batch size of 32.

By implementing these training strategies, we aimed to optimize the model's performance and generalization capabilities, ensuring its effectiveness in real-time applications.

## 3   Results

### 3.1   UAH-DriveSet Drivers Scores

Before evaluating the model performance, it is essential to understand the real scores of each driver provided in the UAH-DriveSet. For each trip, the driver has an aggressiveness score and a normal score, which serve as a baseline for validating the model's predictions. Table 2 provides more details about these scores.

### 3.2   Model Performance

The one-dimensional convolutional neural network (1D CNN) was evaluated using cross-validation on the UAH-DriveSet dataset and the independently collected dataset. The following metrics were used to assess the model's performance: accuracy, precision, recall and F1-score. These metrics were calculated for both the cross-validation folds and the collected dataset to provide a comprehensive evaluation of the model.

**Table 2.** Driver scores: Driving scores, on a scale of 10, of drivers instructed to drive aggressively on the highway. Source: Romera, Bergasa, & Arroyo, (2016) [12].

| Driver | Normal behavior score | Aggressive behavior score |
|---|---|---|
| Driver 1 | 5.1 | 4.0 |
| Driver 2 | 1.2 | 6.1 |
| Driver 3 | 5.4 | 3.4 |
| Driver 4 | 3.7 | 5.3 |
| Driver 5 | 1.3 | 7.3 |
| Driver 6 | 4.8 | 4.6 |

**Cross-Validation Results**   The leave-one-subject-out cross-validation approach applied to the UAH-DriveSet provided detailed performance metrics for each driver. The performance metrics across all cross-validation folds are summarized in Table 3, and confusion matrices for each driver are provided in Fig. 4.

**Table 3.** Driver performance metrics: Precision, Recall, and F1 Score for normal and aggressive driving styles

| Style | Driver 3, 4, 5 | | | Driver 1 | | | Driver 2 | | | Driver 6 | | |
|---|---|---|---|---|---|---|---|---|---|---|---|---|
| | Pre | Recall | F1 Score | Pre | Recall | F1 Score | Pre | Recall | F1 Score | Pre | Recall | F1 Score |
| Aggressive | 1.00 | 1.00 | 1.00 | 0.00 | 0.00 | 0.00 | 1.00 | 0.60 | 0.75 | 1.00 | 0.73 | 0.84 |
| Normal | 1.00 | 1.00 | 1.00 | 0.50 | 1.00 | 0.67 | 0.71 | 1.00 | 0.83 | 0.79 | 1.00 | 0.88 |
| Average | 1.00 | 1.00 | 1.00 | 0.25 | 0.50 | 0.33 | 0.86 | 0.80 | 0.79 | 0.89 | 0.86 | 0.86 |

**Collected Dataset Results.** The results for the independently collected dataset are as follows:

- Accuracy: 99%
- Precision: 99%
- Recall: 99%
- F1-score: 99%

**Model Efficiency and Lightweight Characteristics.** An essential aspect of the 1D CNN model is its lightweight architecture, making it suitable for real-time applications on devices with limited computational resources. The model's size, number of parameters, and inference time were evaluated to illustrate its efficiency. The 1D CNN model contains approximately 15,474 parameters, resulting in a total model size of 60.45 KB. The inference time was measured on Google Colaboratory CPU (Intel(R) Xeon(R) Platinum 8259CL CPU @ 2.50 GHz) at approximately 0.12 seconds per prediction, highlighting the model's capability for real-time performance.

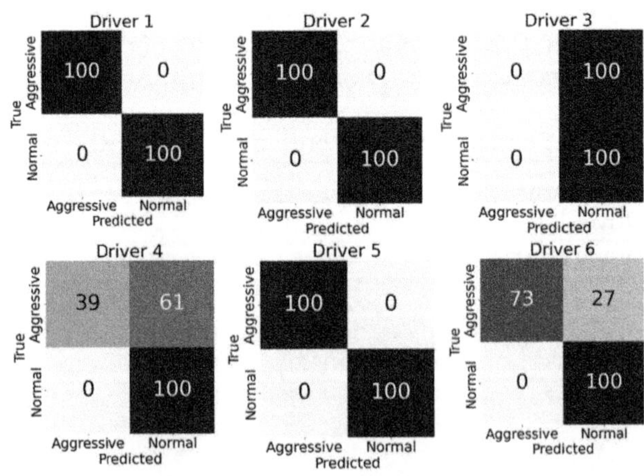


**Fig. 4.** Confusion matrices for each driver. Source: Authors.

# 4   Discussion

## 4.1   Comparison with UAH-DriveSet Real Scores

The performance of the 1D CNN model was thoroughly evaluated against the real scores provided in the UAH-DriveSet. Each driver's aggressiveness and normal driving scores were compared with the model's confusion matrices to validate the accuracy of the model's predictions. The comparison is summarized in Table 4.

**Table 4.** Comparison of UAH-DriveSet real scores and our model Predictions

| Driver | UAH-DriveSet real scores | | Our model predictions | | Alignment |
|---|---|---|---|---|---|
| | Normal | Aggressive | Normal | Aggressive | |
| Driver 1 | 5.1 | 4.0 | 0 | 100 | Good |
| Driver 2 | 1.2 | 6.1 | 0 | 100 | Perfect |
| Driver 3 | 5.4 | 3.4 | 100 | 0 | Good |
| Driver 4 | 3.7 | 5.3 | 61 | 39 | Not bad |
| Driver 5 | 1.3 | 7.3 | 0 | 100 | Perfect |
| Driver 6 | 4.8 | 4.6 | 27 | 73 | Good |

The alignment between the model's predictions and the UAH-DriveSet real scores demonstrates that the 1D CNN effectively identifies aggressive and normal driving behaviors. This validation confirms the model's robustness and reliability in practical scenarios. Moreover, the consistent performance across various drivers indicates the model's strong generalization capability. This ability to accurately predict driving behaviors across different individuals showcases the model's potential for widespread application in real-world settings, ensuring that it can adapt to diverse driving patterns and styles.

Another essential characteristic of the this model is its lightweight nature. With a total of 15,474 parameters and a model size of 60.45 KB, the model is compact enough for real-time applications. The inference time of approximately 0.12 s further supports its suitability for deployment in resource-constrained environments, such as mobile devices and in-vehicle systems.

## 4.2   Classification of Self-collected and Simulated Driving Data

The model's ability to classify both collected normal data and simulated aggressive data demonstrates its sensitivity to the modified variables. This evaluation was crucial in confirming that the 1D CNN model not only performs well on standard datasets but also adapts effectively to variations introduced in the driving behavior.

The aggressive data was simulated by altering several key driving variables, including:

- Speed: Increased speeds to mimic aggressive driving behavior.
- Lane Changing: More frequent and abrupt lane changes.
- Traffic Volume: Increased traffic volume to simulate more complex driving scenarios.
- Safety Distance: Reduced safety distances to ahead vehicles, representing riskier driving patterns.
- Time of Impact: Decreased time of impact to the vehicle ahead, indicating more aggressive following behavior.

The results showed that the model effectively recognized these changes and classified the driving behavior accurately. This demonstrates that the 1D CNN model takes into account the modified variables to simulate aggressive driving, confirming its robustness and sensitivity to different driving conditions. This sensitivity is essential for real-world applications where driving behaviors can vary significantly, ensuring that the model remains reliable and effective in diverse environments.

### 4.3   Limitations and Future Work

While the model shows promising results, several limitations were identified:

- Variability in driving behavior: The natural variability among different drivers can pose challenges for model generalization. Further training with a more diverse driver population is recommended.
- Simulated data limitations: Although the simulated aggressive driving data provided useful testing scenarios, it may not fully capture the complexity of real-world aggressive driving behaviors. Future research should focus on collecting more authentic aggressive driving data.
- Real-time constraints: Although the model itself is fast, the data preprocessing steps, particularly the safety distance measuring and vehicle count using YOLO, need to be optimized for real-time application in systems with limited computational resources.

Future work should aim to address these limitations by expanding the dataset with more diverse and realistic driving behaviors, optimizing the model for various hardware platforms, and exploring additional features that could further enhance the model's performance.

## 5   Conclusion

This study aimed to develop and evaluate a lightweight one-dimensional convolutional neural network (1D CNN) model for detecting aggressive driving behaviors using both the publicly available dataset UAH-DriveSet and an independently collected dataset. A key objective of the study was to ensure the model's generalizability across different drivers, thereby enhancing its applicability in diverse real-world scenarios. The model's performance was validated against real scores from the UAH-DriveSet, demonstrating high accuracy and robustness in identifying normal and aggressive driving patterns.

Key findings include:

- High performance and generalizability: The alignment between the model's confusion matrices and the UAH-DriveSet real scores for each driver confirms its reliability and highlights its strong capability to generalize across different drivers. This ability to consistently perform well across a diverse set of drivers is the main goal of the study, showcasing the model's potential for widespread application in real-world settings.
- Efficiency: The model's lightweight architecture makes it suitable for real-time applications. However, optimization of the preprocessing steps, particularly the safety distance measuring and vehicle count using YOLO, is necessary for deployment in resource-constrained environments.

In conclusion, the 1D CNN model developed in this study demonstrates significant promise for real-time driver behavior monitoring systems, providing a valuable tool for enhancing road safety and reducing the incidence of aggressive driving.

**Disclosure of Interests.** The authors have no competing interests to declare that are relevant to the content of this article.

# References


1. Chen, D., Chen, Z., Zhang, Y., Qu, X., Zhang, M., Wu, C.: Driving style recognition under connected circumstance using a supervised hierarchical Bayesian model. J. Adv. Transp. **2021**(1), 6687378 (2021)
2. Güldenstein, J.: Comparison of measurement systems for kinematic calibration of a humanoid robot. Bachelor thesis at the University of Hamburg (2019)
3. Jahangiri, A., Berardi, V.J., Machiani, S.G.: Application of real field connected vehicle data for aggressive driving identification on horizontal curves. IEEE Trans. Intell. Transp. Syst. **19**(7), 2316–2324 (2017)
4. Karaduman, M., Eren, H.: Deep learning based traffic direction sign detection and determining driving style. In: 2017 International Conference on Computer Science and Engineering (UBMK), pp. 1046–1050 (2017). https://doi.org/10.1109/UBMK.2017.8093453
5. Matousek, M., EL-Zohairy, M., Al-Momani, A., Kargl, F., Bösch, C.: Detecting anomalous driving behavior using neural networks. In: 2019 IEEE Intelligent Vehicles Symposium (IV), pp. 2229–2235 (2019). https://doi.org/10.1109/IVS.2019.8814246
6. Moukafih, Y., Hafidi, H., Ghogho, M.: Aggressive driving detection using deep learning-based time series classification. In: 2019 IEEE International Symposium on INnovations in Intelligent SysTems and Applications (INISTA), pp. 1–5. IEEE (2019)
7. Mumcuoglu, M.E., et al.: Driving behavior classification using long short term memory networks. In: 2019 AEIT International Conference of Electrical and Electronic Technologies for Automotive (AEIT AUTOMOTIVE), pp. 1–6 (2019). https://doi.org/10.23919/EETA.2019.8804534

8. World Health Organization: Global status report on road safety 2023: summary. World Health Organization (2023)
9. Wang, J., Zhang, Z., Liu, F., Lu, G.: Investigating heterogeneous car-following behaviors of different vehicle types, traffic densities and road types. Transp. Res. Interdisc. Perspect. **9**, 100315 (2021)
10. Wang, W., Xi, J., Chong, A., Li, L.: Driving style classification using a semisupervised support vector machine. IEEE Trans. Hum.-Mach. Syst. **47**(5), 650–660 (2017). https://doi.org/10.1109/THMS.2017.2736948
11. Zheng, Y., Chase, R.T., Elefteriadou, L., Sisiopiku, V., Schroeder, B.: Driver types and their behaviors within a high level of pedestrian activity environment. Transp. Lett. **9**(1), 1–11 (2017). https://doi.org/10.1080/19427867.2015.1131943

# DeMoFs: Multi-objective Feature Selection Using Binary Differential Evolution (BDE) Optimization on Breast Cancer Gene Expression Data


Mohamed Djellal Serandi[1(✉)], Amina Houari[1], Fatma Boufera[1], and Farid Flitti[2]

[1] LISYS Laboratory, Mustapha Stambouli University, Mascara, Algeria
{mohamed.djellalserandi,amina.houari,fboufera}@univ-mascara.dz
[2] Higher Colleges of Technology Campus, Dubai, UAE
fflitti@hct.ac.ae



**Abstract.** The DNA microarray allows for the simultaneous measurement of thousands of genes in one experiment. Microarray-derived gene expression data is highly dimensional, making its analysis difficult due to the well-known curse of dimensionality [16]. Because many genes in these datasets are unrelated to illness, efficient feature selection is necessary to enhance the precision and readability of predictive models. Binary Differential Evolution (BDE) with a newly developed mutation operator has been recently used for biclustering [3]. This work proposes a novel feature selection strategy based on multi-objective optimization to identify a reduced subset of significant genes from breast cancer microarray data. Extensive experiments on a breast cancer dataset demonstrated the effectiveness of the proposed technique in terms of classification accuracy.

**Keywords:** Microarray data · feature selection · multi-objective · Differential evolution · Classification · Mutation · Selection · Optimization


## 1 Introduction

A challenge presented by the explosion of microarray gene expression data is its high dimensionality (thousands of genes) combined with limited sample number (tens or hundreds), leading to the well-known "curse of dimensionality": as the dimensionality increases, the number of samples required for inference grows rapidly. This problem is addressed by feature selection (FS), by selecting a pertinent subset of genes to enhance model performance and interpretability. The development of high-throughput genomics, such as microarrays, has indeed transformed our knowledge of diseases such as breast cancer. However, although microarray data provides a wealth of information, finding the genes that significantly contribute to disease categorization requires appropriate feature selection. The emphasis on correctness in traditional FS approaches frequently results in enormous gene sets that can be computationally costly and challenging to analyze.



N. Seddari and M. Redjimi (Eds.): ICMSCT 2024, CCIS 2606, pp. 261–271, 2025.
https://doi.org/10.1007/978-3-032-01922-6_22

This study proposes a novel multi-objective FS method based on Differential Evolution (DE) for breast cancer microarray data. In order to improve the model's performance and interpretability, the Mean Squared Residual (MSR) is optimized simultaneously with limiting the number of selected gene. Additionally, a new mutation operator called Biclustering Binary Differential Evolution Mutation (BBDE) [3] is integrated to adapt the optimization algorithm to features selection.

This paper is organized as follows: Sect. 2 presents a literature review of FS methods. Section 3 exhibits related work and optimization methods, and Sect. 4 details our proposed multi-objective DE-based FS approach. Section 5 is reserved for the experimental setup and results. Section 6 concludes the paper and outlines future research directions.

## 2    Review on Feature Selection Methods

Feature selection [2,17] is a common machine learning methodology to deal with the problem of dimensionality. It aims at reducing the number of original features by selecting reduced set of features pertinent to the classification task. Feature selection can be divided into three classes according to whether or not they are independent of the classification algorithms: filter, wrapper, and hybrid [8,9].

A DNA microarray dataset is presented in as a genomic data matrix, where genes are represented by rows, conditions are represented by columns, and gene expression levels are represented by cells. The formal definition of this data matrix is as follows: $G = \{1, 2, ..., n\}$ a set of indices of $n$ genes, $C = \{1, 2,..., m\}$ a set of indices of $m$ conditions, and $M(G, C)$ the data matrix associated with all genes and conditions.

### 2.1    Feature Selection Definition

Feature selection refers to the process of identifying and selecting a subset of relevant features (variables, predictors) from a larger set of available features in a dataset. Feature selection can be either supervised or unsupervised.

### 2.2    Types of Feature Selection

**Supervised Feature Selection.** Using the labeled data, supervised gene selection chooses the pertinent features from the gene expression data.

- **Wrapper Methods:** Selection of features is formulated as a search problem, in which different feature subsets are created and evaluated using a machine learning algorithm. The performances obtained with these subsets are compared iteratively to select the one yielding best training performance [15].
- **Filter Methods :** Features are selected on the basis of statistics measures. This method does not depend on the learning algorithm and chooses the features as a preprocessing step [15].

- **Embedded Methods :** Embedded methods combined the advantages of both filter and wrapper methods by considering the interaction of features along with low computational cost [15].

**Unsupervised Feature Selection.** Unsupervised feature selection techniques do not require labeled data. These include Principal and independent Component Analysis, PCA and ICA, t-distributed Stochastic Neighbor Embedding, t-SNE, autoencoders, and bio-inspired algorithms. In particular, two bio-inspired algorithms are studied for feature selection, i.e., Swarm-based Algorithms (SWA) and Evolutionary Algorithms (EA) [15].

- SWA methods are inspired by the natural and artificial systems composed of many individuals that coordinate using decentralized control and self-organization, e.g., colonies of ants and termites, schools of fish, flocks of birds, etc.
- EA uses mechanisms inspired by biological evolution, which include reproduction, mutation, recombination, and selection. Candidate solutions to the optimization determine the quality of the solutions.

## 2.3   MSR (Mean Squared Residue) Metric

The MSR is introduced in this section due to its importance in maintaining data fidelity after dimensionality reduction.

Minimizing the Mean Squared Residual (MSR) ensures that the new dataset obtained after feature selection preserves its informational value and retains all relevant data without any loss.

MSR for a solution $S$ that consists of $I$ rows and $J$ columns is defined as follows:

$$MSR(S) = \frac{1}{|I||J|} \sum_{i=1}^{|I|} \sum_{j=1}^{|J|} \left( x_{ij} - x_{iJ} - x_{Ij} + x_{IJ} \right)^2 \tag{1}$$

**Where:**

- $x_{ij}$: represents the expression value of gene $i$ under condition $j$ within the solution $S$.
- $x_{iJ}$: represents the mean expression value of gene $i$ across all conditions in the solution $S$. It is calculated as follows:

$$x_{iJ} = \frac{1}{|J|} \sum_{j \in J} x_{ij}$$

- $x_{Ij}$: represents the mean expression value of all genes under condition $j$ in the solution $S$. It is calculated as follows:

$$x_{Ij} = \frac{1}{|I|} \sum_{i \in I} x_{ij}$$

- $x_{IJ}$: denotes the overall mean expression value of the solution $S$. It is calculated as follows:

$$x_{IJ} = \frac{1}{|I| \cdot |J|} \sum_{i \in I, j \in J} x_{ij}$$

## 3   Related Work

Many FS algorithms use metaheuristic search methods in high dimensionality space to reduce the dimension. A summary of the most recent methods is presented.

- The authors in [14] use Fisher score to reduce the search space's dimensions. The smaller search space is searched using operators from genetic algorithms.
- The authors in [12] use three algorithms, i.e., IGR, Relief-F, and Chi-squared to filter out less significant features. Thus, the search space of the FPA algorithm is first made smaller before the search is launched.
- The authors in [11] reduce the dimensions of the search space of the CSO before starting the search process by converting the search space into several tasks and filtering the less important features determined by three algorithms, i.e., PCC, Relief-F, and TV.
- The authors in [1] search a large space of high-dimensional data with the SFE algorithm, finding important and relevant features in an initial search phase that removes unimportant features from the search space. Then, the PSO algorithm is used to search the reduced search space.
- The authors in [5] utilize PCC, SU, K-Means, and other algorithms to group data and use one bit to display several grouped features in order to reduce the size of the search space. Genetic Algorithm Operators refine gradually the solution of the FS.
- The authors in [18] introduce a novel feature filter and group evolution hybrid feature selection algorithm (FG-HFS) aimed at addressing the limitations of existing methods. It employs spectral clustering to group features, filters redundant features using symmetric uncertainty, and utilizes a group evolution multi-objective genetic algorithm for optimal feature selection. The effectiveness of the FG-HFS algorithm is validated through experiments on five datasets from different fields, demonstrating its practical application value in gene expression data analysis.
- The authors in [10] The paper introduces a hybrid gene selection model that combines filter and wrapper methods for feature selection. This approach enhances the identification of relevant genes crucial for breast cancer classification, thereby improving the accuracy of the model. It demonstrates the application of deep neural networks in the classification of molecular subtypes of breast cancer.
- The authors in [13] The study successfully addresses the difficulties caused by the absence of label information in clustering tasks by presenting a unique

unsupervised feature selection technique based on an enhanced binary differential evolution algorithm. Furthermore, it combines feature selection with clustering optimization

**Our Contribution and Technical Gap:** Unlike previous methods using standard mutation operators, our approach introduces the Biclustering Binary Differential Evolution Mutation (BBDE) operator to adapt the optimization algorithm for feature selection. This allows for better solutions with fewer iterations, addressing a technical gap in optimizing feature selection tasks (Fig. 1).

# 4   DeMoFs Algorithm

## 4.1   Review of Differential Evolution (DE)

DE is a competitive optimization technique used to solve numerical optimization problems with actual parameters. The following is a clear explanation of the main algorithmic steps of DE:

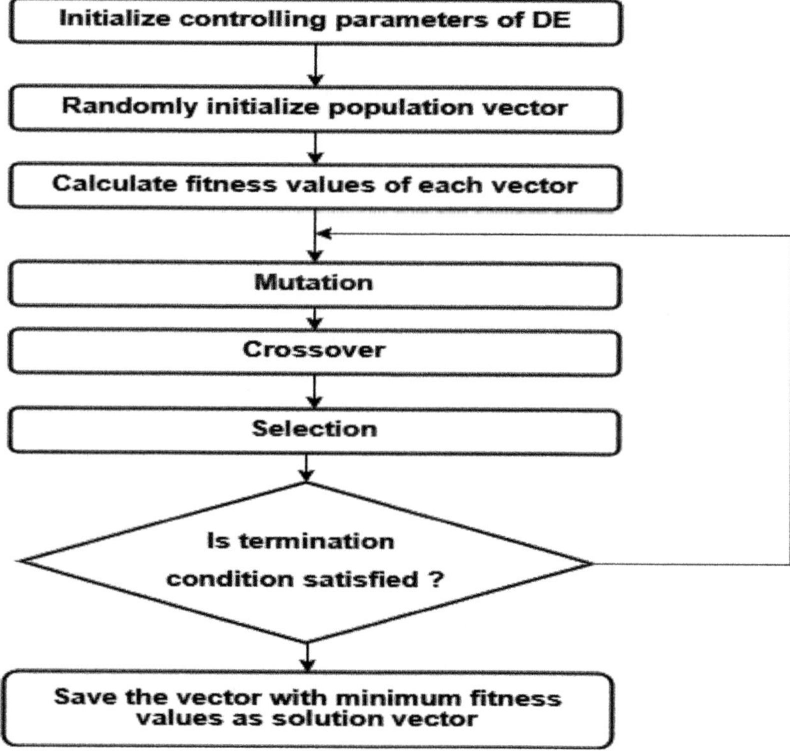


**Fig. 1.** Phases of DE.

– Initialization: DE starts by generating a population of potential solutions.
– Mutation: this step involves the creation of new individuals by perturbing existing ones. This is done by computing the difference between two randomly selected individuals.
– Crossover: this involves randomly choosing components from the mutated solution and mixing them with the remaining components from the original solution to create what is called a trial solution.
– Selection: DE then compares the trial solution created by crossover to the original solution. If a trial solution proves to be more successful than the corresponding original solution, it replaces it in the population.
– Termination: The algorithm repeats these steps for a certain number of generations or until a particular convergence condition is fulfilled.

## 4.2   Illustration of DeMoFs Algorithm

This work proposes a Multi-objective Differential Evolution for the FS method called DeMoFs.

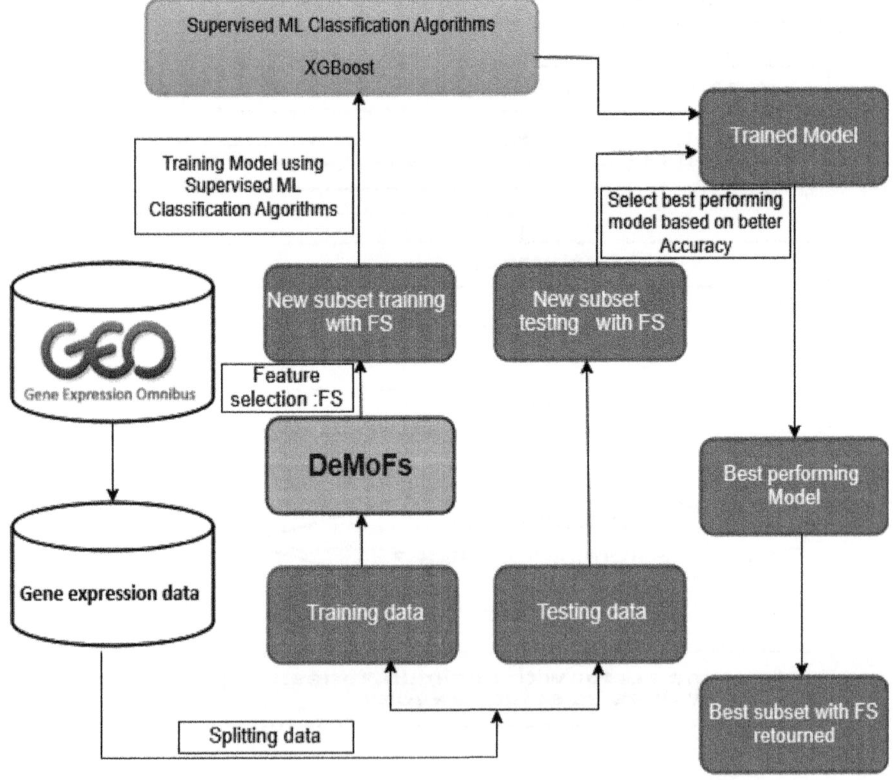


**Fig. 2.** Illustration of DeMoFs scheme.

The proposed solution leverages a binary variant of the differential evolution algorithm (BDE). Given the binary nature of the feature selection (FS) process, the population members in BDE are initialized as binary vectors. Assuming N features or dimensions (equivalent to the number of genes), each solution in the DE algorithm comprises N components, with each component corresponding to a feature. As illustrated in Fig. 3, the values 1 and 0 denote whether a feature is selected or not.

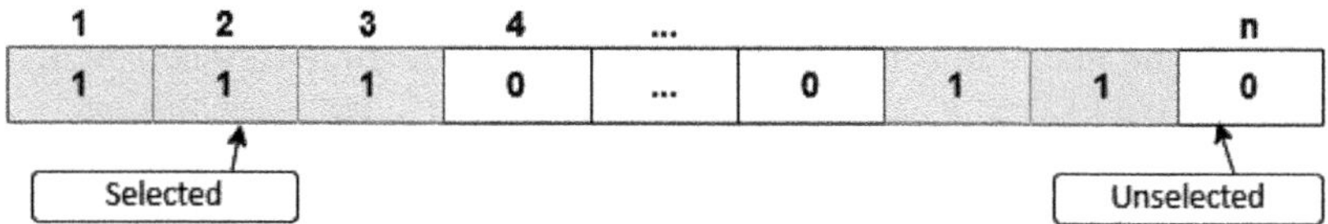


**Fig. 3.** Representation of binary string encoding for FS solution.

## Illustrative Example of the Binary Differential Evolution (BDE) Algorithm

The following is an example of how the Feature Selection Using Binary Differential Evolution (BDE) works:

Consider the data matrix X with 5 genes (features): g1, g2, g3, g4 and g5, and 4 samples: s1, s2, s3 and s4.

Suppose we have a binary solution vector v= [1, 0, 1, 0, 0], which indicates which features are selected:

g1 and g3 are included (1), while g2, g4, and g5 are excluded (0) (Fig. 4).

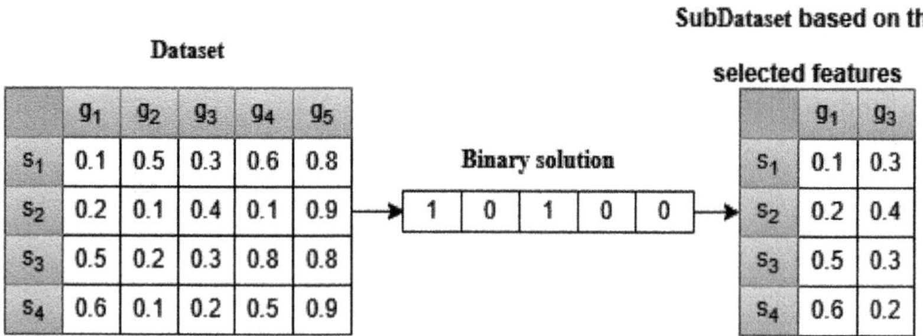


|  | $g_1$ | $g_2$ | $g_3$ | $g_4$ | $g_5$ |
|---|---|---|---|---|---|
| $s_1$ | 0.1 | 0.5 | 0.3 | 0.6 | 0.8 |
| $s_2$ | 0.2 | 0.1 | 0.4 | 0.1 | 0.9 |
| $s_3$ | 0.5 | 0.2 | 0.3 | 0.8 | 0.8 |
| $s_4$ | 0.6 | 0.1 | 0.2 | 0.5 | 0.9 |

| 1 | 0 | 1 | 0 | 0 |
|---|---|---|---|---|

|  | $g_1$ | $g_3$ |
|---|---|---|
| $s_1$ | 0.1 | 0.3 |
| $s_2$ | 0.2 | 0.4 |
| $s_3$ | 0.5 | 0.3 |
| $s_4$ | 0.6 | 0.2 |

**Fig. 4.** Feature Selection Using Binary Differential Evolution: A case study with 5 genes and 4 samples

## 4.3   Algorithm Description

Differential evolution is used in the proposed solution to guide the search of a population of FS solutions toward better solutions. Initially, it uses a predefined binary string size (n = genes) to represent the FS and initializes the parent population by randomly selecting FS. The genes are represented by a one bit string in this binary string. A bit is reset to zero if the relevant gene is not present in the FS, and set to one if it is (cf. Fig. 3).

Our approach utilize multi-objective differential evolution. It simultaneously finds a variety of suitable feature subsets in a binary search space by applying a non-dominated sorting genetic algorithm (NSGA-II) [6].

The DE algorithm uses the mutation operator to modify an existing individual in order to create the offspring population after initializing the population and it uses the multi-objective fitness functions to calculate the score of all solutions.

The fitness functions are selected based on the specific nature of the problem. In this study, selected features' size (f1) and Mean Squared Residual (MSR) (f2) were utilized to guide the algorithm towards achieving the best accuracy. These adjustments may follow a random. After the mutation step, the resulting mutant vector is used in the crossover operation with a probability initialized at 0.75. After that, the algorithm performs the selection process. The algorithm completes by returning the final population after completing a number of predefined iterations.

We split our dataset into two subsets, training and testing, and then apply the DeMoFs method to the training set. We take the identical genes that the algorithm returned on the test set and classify them without running the algorithm again.

As shown in the (cf. Fig. 2), supervised classification can be achieved using several techniques. We chose the XGBoost [4] classifier for its reported high performance and speed in training.

To test the effectiveness of our algorithm, we compared the classification using the original dataset and using the features selected by the proposed method.

Create mutant using BBDE mutation operator

Function BBDE(x: vector, y: vector, z: vector, F: real): vector w

$w = \text{nodeAddition}\,(x, \text{nodeDeletion}(y, z, F), F)$

$$\text{nodeAddition}(x, y, F) = \begin{cases} 1, & \text{if } x_i = 1 \text{ or } (y_i = 1 \text{ and randi} < F) \\ 0, & \text{otherwise} \end{cases} \qquad (2)$$

$$\text{nodeDeletion}(x, y, F) = \begin{cases} 0, & \text{if } x_i = 0 \text{ or } (y_i = 1 \text{ and randi} < F) \\ 1, & \text{otherwise} \end{cases} \qquad (3)$$

## 5   Experimental Results

The DeMoFs algorithm was evaluated using real DNA microarray data by comparing accuracies of supervised classifications of the new dataset, generated through feature selection (FS), and the original dataset.

The well-known dataset GSE31192 [7] were used in this study. This dataset features gene expression profiles for 54675 genes across 33 samples. The data is split into 70% for training and 30% for testing.

After applying our algorithm, we observed that most of the solutions reduced the dataset by approximately 50%. Instead of having 54675 features, the proposed method reduced the features to only 27337. The XGBoost classification method was used to evaluate the quality of the selected features.

The population was fixed at 20 and several experiments were conducted by changing the number of iterations to observe the convergence. Table 1 and Fig. 5 summarize the experiments.

**Table 1.** Comprehensive Table of Experimental Results

| Experiment Number | Number of Iterations | Accuracy Train | Accuracy Test |
|---|---|---|---|
| 1 | 0 (original dataset) | 0.82 | 0.80 |
| 2 | 50 | 0.95 | 0.90 |

The results clearly validate the pertinence of the selected features as both training and testing accuracies obtained with the selected features exceed those obtained with the original data.

Further tests are currently conducted to compare the proposed method with alternative approaches by expanding the population size to 100 and progressively increasing the number of iterations.

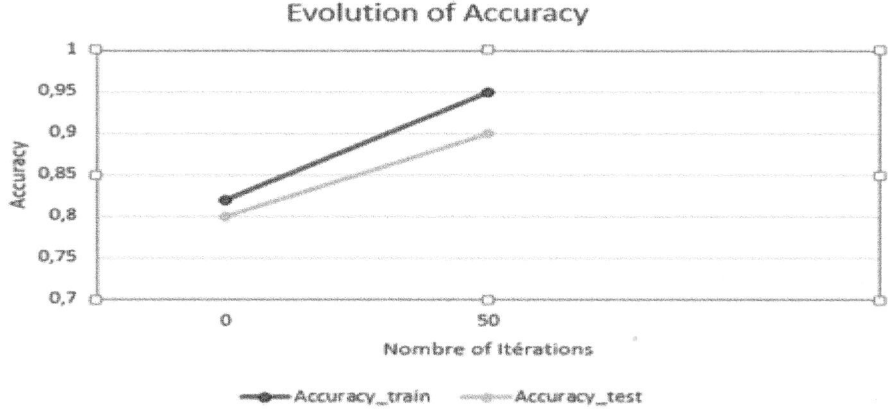


**Fig. 5.** Accuracy Plot for Experimental Results

## 6 Conclusion

This paper introduces a new and effective feature selection mechanism called DeMoFs: Multi-objective Feature Selection using binary Differential Evolution (BDE) that uses Multi-Objective Differential Evolution to find high quality feature subsets. The outcomes shown in Sect. 5 demonstrate very good performance of the proposed FS algorithm compared to using the original data set.

For a more complete evaluation of the proposed algorithm DeMoFs, a comparison with various alternative feature selection methods, including Genetic Algorithms (GA) and Particle Swarm Optimization PSO-based techniques will be conducted using the validation used in this paper in addition to other performance metrics such as F1 score and Matthew's Correlation Coefficient (MCC). Besides, the proposed algorithm will be applied to other datasets to validate its generalizability across different genomic domains.

## References


1. Ahadzadeh, B., Abdar, M., Safara, F., Khosravi, A., Menhaj, M.B., Suganthan, P.N.: SFE: a simple, fast, and efficient feature selection algorithm for high-dimensional data. IEEE Trans. Evol. Comput. **27**(6), 1896–1911 (2023)
2. Bolón-Canedo, V., Alonso-Betanzos, A.: Ensembles for feature selection: a review and future trends. Inf. Fusion **52**, 1–12 (2019)
3. Charfaoui, Y., Houari, A., Boufera, F.: Amodebic: an adaptive multi-objective differential evolution biclustering algorithm of microarray data using a biclustering binary mutation operator. Expert Syst. Appl. **238**, 121863 (2024)
4. Chen, T., Guestrin, C.: XGBoost: a scalable tree boosting system. In: Proceedings of the 22nd ACM SIGKDD International Conference on Knowledge Discovery and Data Mining, pp. 785–794 (2016)
5. Cheng, F., Cui, J., Wang, Q., Zhang, L.: A variable granularity search-based multi-objective feature selection algorithm for high-dimensional data classification. IEEE Trans. Evol. Comput. **27**(2), 266–280 (2022)
6. Deb, K., Pratap, A., Agarwal, S., Meyarivan, T.: A fast and elitist multiobjective genetic algorithm: NSGA-II. IEEE Trans. Evol. Comput. **6**(2), 182–197 (2002)
7. Harvell, D.M.: Genomic signatures of pregnancy-associated breast cancer epithelia and stroma and their regulation by estrogens and progesterone. Hormones and Cancer **4**, 140–153 (2013)
8. Kannan, S.S., Ramaraj, N.: A novel hybrid feature selection via symmetrical uncertainty ranking based local memetic search algorithm. Knowl. Based Syst. **23**(6), 580–585 (2010)
9. Khammassi, C., Krichen, S.: A NSGA2-LR wrapper approach for feature selection in network intrusion detection. Comput. Netw. **172**, 107183 (2020)
10. Lamba, M., Munjal, G., Gigras, Y.: A hybrid gene selection model for molecular breast cancer classification using a deep neural network. Int. J. Appl. Pattern Recogn. **6**(3), 195–216 (2021)
11. Li, L.: An evolutionary multitasking algorithm with multiple filtering for high-dimensional feature selection. IEEE Trans. Evol. Comput. **27**(4), 802–816 (2023)
12. Li, M., Ke, L., Wang, L., Deng, S., Yu, X.: A novel hybrid gene selection for tumor identification by combining multifilter integration and a recursive flower pollination search algorithm. Knowl. Based Syst. **262**, 110250 (2023)

13. Li, T., Dong, H.: Unsupervised feature selection and clustering optimization based on improved differential evolution. IEEE Access **7**, 140438–140450 (2019)
14. Li, T., Zhan, Z.H., Xu, J.C., Yang, Q., Ma, Y.Y.: A binary individual search strategy-based bi-objective evolutionary algorithm for high-dimensional feature selection. Inf. Sci. **610**, 651–673 (2022)
15. Pham, T.H., Raahemi, B.: Bio-inspired feature selection algorithms with their applications: a systematic literature review. IEEE Access **11**, 43733–43758 (2023)
16. Trunk, G.V.: A problem of dimensionality: a simple example. IEEE Trans. Pattern Anal. Mach. Intell. **3**, 306–307 (1979)
17. Tubishat, M., Alswaitti, M., Mirjalili, S., Al-Garadi, M.A., Rana, T.A., et al.: Dynamic butterfly optimization algorithm for feature selection. IEEE Access **8**, 194303–194314 (2020)
18. Xu, Z., et al.: FG-HFS: a feature filter and group evolution hybrid feature selection algorithm for high-dimensional gene expression data. Expert Syst. Appl. **245**, 123069 (2024)

# DCGAN-Based Synthetic Data for Enhancing MRI Brain Tumors Dataset


Abdelhamid Mounis[1]($\boxtimes$), Boudjelal Meftah[1], and Samia Benyahia[2]

[1] University of Mustapha Stambouli, Mascara, Algeria
`abdelhamid.mounis@univ-mascara.dz`

[2] Higher School of Management and Digital Economy (HSMDE/ESGEN), Kolea, Tipaza, Algeria



**Abstract.** Civilization is witnessing a continuous development in human life and society, the medical field is constantly improving. However, modern medicine still faces many limitations, including challenging and previously unsolvable problems. One of the most common challenge in the medical field is the scarce of data. With the Evolution of Artificial Intelligence and its impact on various fields including medical diagnosis, it's important to benefit from AI characteristics to improve the effectiveness of medical diagnosis, We Present Generative adversarial network (GAN) as one of most relevant and effective method for data augmentation that can help us to generate realistic data that resemble the original data. We use an improved GAN architecture that involves Convolutional Neural Network (CNN) to improve the quality of generated images called Deep Convolutional Generative Adversarial Network (DCGAN). The model can learn the distribution of data and generate realistic synthetic data with a similar distribution of the original data allowing the ability to extend the dataset with entirely new data. For evaluation we use Inception score (IS), Higher IS values indicate a high quality and diversity. Our model shows a promising results and generate realistic data that we can use to train machine learning and deep learning models.

**Keywords:** data augmentation · medical image processing · generative adversarial network · deep learning · Magnetic Resonance Imaging


## 1 Introduction

The expansion of generative adversarial networks (GANs) as novel solution to synthetic data generation for training convolutional neural networks (CNN) led to scoring remarkable results and attracting the interest of the research community [1]. In healthcare, synthetic data can be applied for various purposes, such as data augmentation, anonymization, and model prediction [2]. The generation of synthetic diverse data is particularly important in medical imaging, where a vast amount data is needed. The large availability of quality labeled data is a requirement for the development of highly accurate deep learning models, which is crucial in healthcare. However, medical datasets are scarce, due mainly to issues of patient privacy and the high human costs of annotation procedures [3].




N. Seddari and M. Redjimi (Eds.): ICMSCT 2024, CCIS 2606, pp. 272–281, 2025.
https://doi.org/10.1007/978-3-032-01922-6_23

In Magnetic Resonance Imaging (MRI), in particular, the annotation of brain tumors is not a straightforward task due to both the time and complexity associated with manual annotation. This is particularly true for datasets involving distributions that require the separation of dimensions into several endophenotypes, including categories such as large, small, single, or multiple tumors [4].

## 1.1  Research Objectives

The objectives of this research are to develop a DCGAN-based synthetic data generation method for enhancing the MRI brain tumors dataset. And we summarize it as below:

***Extends MRI Brian Tumors dataset.*** Since GAN's models can generate realistic synthetic data that similar to the original data, it's possible to use this generated data to solve issues like scarce and imbalanced data.

***Demonstrate the effectiveness of the synthetic data.*** The GAN models have the potential to generate unlimited realistic data. Deep learning models require a substantial amount of data, which is rare and scarce in the medical field. However, with the unlimited data source provided by GAN models, we can train very powerful deep learning models for medical purposes.

**Table 1.** Enhancing Datasets with GANs: Practical Examples.

| Paper | Published | Main Results |
| --- | --- | --- |
| [5] | 20 May 2023 | Improvement of accuracy over a DCGAN-based synthetic data training for the MNIST and CIFAR 10 datasets |
| [6] | 14 November 2023 | Achieve 99.33% accuracy, without DCGAN data augmentation it was 86.27% |
| [7] | 19 December 2019 | This paper mention data augmentation as one of GAN's applications |

The Table 1 highlights significant improvement of accuracy for image classification tasks using DCGAN-based synthetic data training on the MNIST and CIFAR-10 datasets. [5] discuss the overall improvement in accuracy depending on extended dataset using synthetic data generated by DCGAN. [6] achieve a remarkable accuracy result on MNIST dataset by using DCGAN's generated data. [7] made a significant review of the GAN's application and mention the data augmentation as one of the key uses of GANs.

## 2  Related Works

It is important to mention that training a model with scarce data, imbalanced data, or even corrupted data leads to poor performance or underfitting problems. A common issue appears in medical diagnosis which is data scarcity. Data augmentation has been a valuable solution for this type of issues. Traditional methods of data augmentation depend on geometric transformations, such as rotation, lighting variation, zoom in/out, and flipping. However, these methods have limitations, including limited data creation and potential loss of information. Therefore, it is important to explore more sophisticated methods to maximize the effectiveness of data augmentation for the collected data. Dewi et al. demonstrated that the utilization of DCGAN to generate synthetic data proved effective in enhancing the original dataset, leading to a notable enhancement in the accuracy of the traffic sign recognition model. They conducted a comparative analysis of various CNN models, such as ResNet50 and DenseNet. The findings indicated that DenseNet outperformed other models in terms of accuracy and computational efficiency, suggesting its potential as a valuable resource for data augmentation in diverse computer vision tasks [9]. Haque et al. focused on improving brain tumors detection using DCGAN-based data augmentation and Vision Transformer, the augmented dataset generated by DCGANs improved the accuracy of ViT model, leading to enhance the brain tumor detection accuracy [6]. Mourad et Oseni. Explored the use of DCGAN to generate high-quality MRI images, by employing adversarial training, the generator improves its capability to produce MRI slices that closely resemble real data, showcasing the potential of DCGAN in medical imaging research, finely they suggest an improved DCGAN architectures such as using Wasserstein GAN or Least Squares GAN Structures [10]. Liu et al. proposed Dual-Branch GAN (DBGAN) for under sampled MRI reconstruction, they devised a modified two-stage cascaded U-Net based networks as generator to reconstruct the aliased images from coarse to fine, introducing Cross-Stage Skip Connection to preserve more reconstruction details [11]. Tariq et al. focused on Generative models as a successful method for synthetic data generation, it uses Adaptive Discriminator Augmentation (ADA) to enhance the GAN training process by modifying the discriminator's augmentation policies based on current state of training, leading for better convergence and high quality synthetic data [12]. Altun Güven et Talu. Utilized a Generative Adversarial Network (GAN) architecture known as Progressive Growing GAN (PGGAN) for creating high-resolution brain MRI images and segmentation, achieving high quality images. PGGAN starts with producing low quality images then progressive refines them to a higher resolution. This architecture reduces the computing costs and generate a detailed and realistic images [13]. Zhou et al. introduced a 3D-VQGAN model to generate realistic and diverse 3D brain tumor in order to address imbalance issues in two popular datasets the Multimodal Brain Tumor Segmentation Challenge (BraTS) 2019 and the internal pediatric LGG (pLGG) dataset tumor ROIs [14].

## 3   Methodology

Generative Adversarial Networks (GANs) are a type of deep learning model that can generate realistic images from data, such as enhancing low-quality images, transforming images from one domain to another, or creating new images from scratch [7]. In this section, we will have a comprehensive review of DCGAN architectures, dataset, data preprocessing, training process and evaluation metrics.

Generative Adversarial Networks (GANs) have shown promising results in various fields, including data augmentation. Introduced by Goodfellow et al. GANs consist of two networks: a generator and a discriminator. The generator receives random noise (z) and generates data samples that are similar to the real data. The discriminator, on the other hand, receives both the real data and the fake data and attempts to distinguish between them. They are trained in an adversarial manner: the generator tries to produce indistinguishable samples to fool the discriminator, while the discriminator tries its best to separate the real samples from the fake ones, like in a game [8].

## 3.1 Deep Convolutional GAN

Radford and his colleagues made a big breakthrough in 2015 when they came up with Deep Convolutional Generative Adversarial network, or DCGAN for short. DCGAN is a special kind of GAN that uses CNN and GAN techniques to create amazing, realistic images with lots of details. DCGANs are smart enough to learn and generate images by themselves, without needing any extra help, which is great for unsupervised learning problems. DCGANs are also easy to train, because they use some cool features like strided convolutions, batch normalization, and leaky Rectified Linear Unit (ReLU) activation functions [15]. DCGANs have done really well on big image datasets like CIFAR-10 and ImageNet. But there are some challenges too. DCGANs need a lot of computing power, they are sensitive to how you set the parameters, and they can run into problems like not having enough variety in the images they generate or getting stuck in a bad mode [16]. Still, DCGANs have been very useful for different things like making new images, changing the style of images, and making images sharper. DCGANs have changed the game of generative modeling and inspired a lot of new ideas [17].

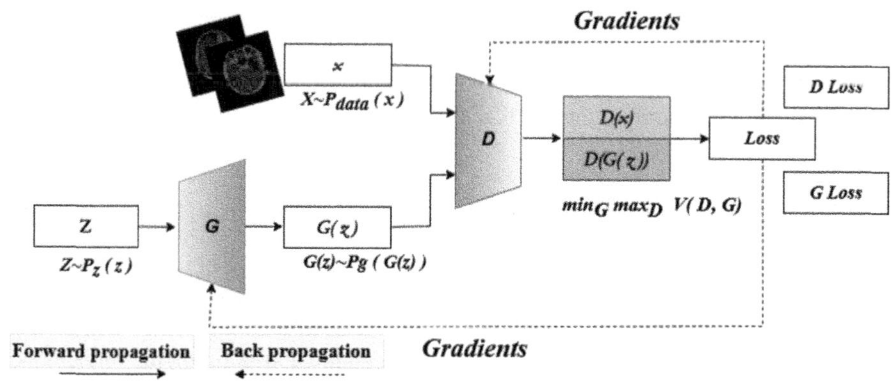


**Fig. 1.** GAN architecture [20]

## 3.2 Data Collection and Preprocessing

In our experiment, we utilize a dataset of MRI brain tumors comprising 7023 MRI images [18] (Fig. 2).

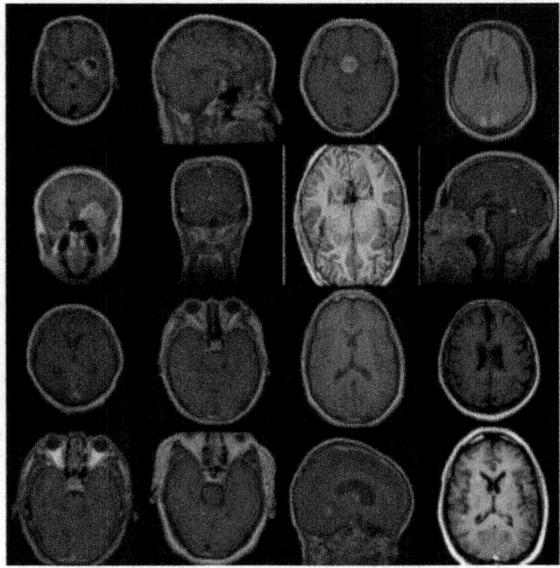

**Fig. 2.** Sample images in the brain tumor dataset

**Preprocessing.** To prepare the MRI Brain Images for training, we resized these images to a standardized resolution of $256 \times 256$ pixels to ensure consistency, in addition, we converted them to Pytorch tensors which is a numerical format for training deep learning models. To accelerate the training process, we normalized the pixel intensity to a specific range (between $-1$ and 1). Finally, in order to use Mini-Batch Gradient Descent we divided our dataset into batches of 128 images.

### 3.3 Deep Convolutional GANs (DCGANs) Architecture

**Generator.** Our enhanced generator's architecture based on the original DCGAN architecture [15], the goal is to generate high-quality MRI Brian images. We increased the resolution of generated images allowing for more detailed images, by adding Batch Normalization layers we made the training more stable and improve the model's performance leading to improve the quality of generated samples. The ReLU activation function applied to prevent vanishing gradient problem enabling the model to learn complex features and produce synthetic images with a lot of details. Finally, in the output layer we applied the tanh activation function to ensure get pixel values within the range $[-1, 1]$ (Fig. 3).

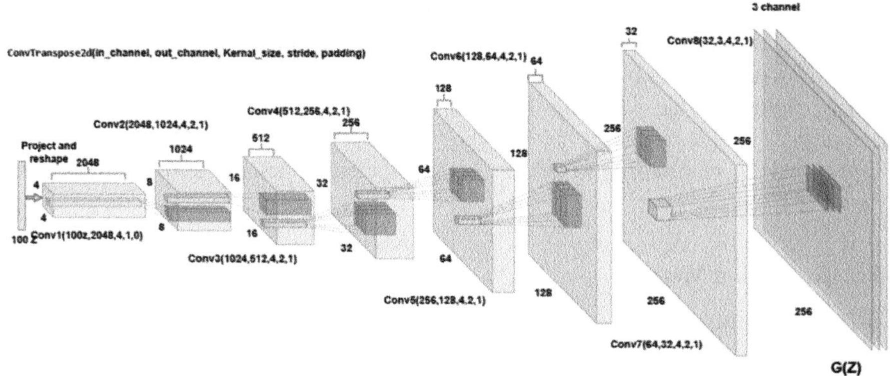

**Fig. 3.** The generator's architecture

**Discriminator.** Similarly, the design of the Discriminator follows the idea of original DCGAN [15]. However, given the large image size($256 \times 256$), our Discriminator employed seven convolutional layers, each followed by Batch Normalization and ReLU activation function except the last one, the last one used a sigmoid activation function. The input is a tree channels images from the original dataset and the images generated by G (G(z)) with dimensions of $256 \times 256$, while its output dimension is 1, in order to classify the input image as fake or real. Batch Normalization is to stabilize the training while ReLU to prevent vanishing gradient (Fig. 4).

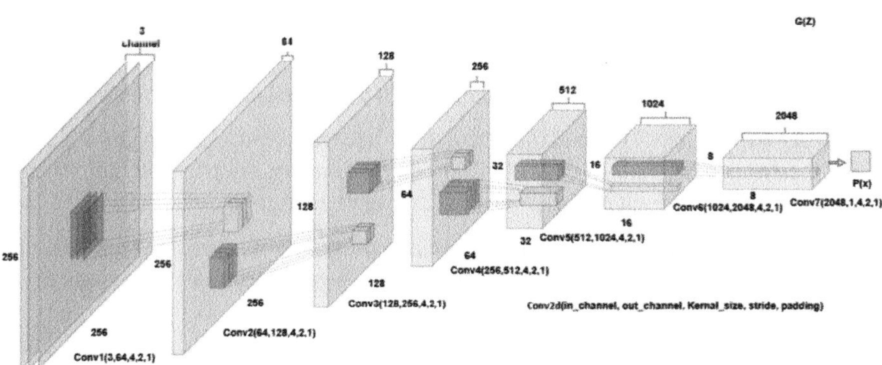

**Fig. 4.** The discriminator's architecture.

## 3.4 The Training Process

In the training process, the generator aims to generate synthetic fake data that is as the real data as possible to deceive the discriminator, while the discriminator aims to classify the fake data as fake and the real data as real. Therefore, the discriminator is fed with both synthetic and real data, the discriminator's loss calculated and used to update the discriminator's weights through backpropagation. The generator starts generate synthetic data to fool the discriminator, this data is fed to the discriminator, using discriminator's output we calculate the generator's loss and use it to update the weights of it. Training involved an adversarial process, the generator's goal is to minimize its loss by generate a high resolution images that similar to the real one. Conversely, the discriminator tries to minimize its loss by accurately classifying real and fake images. This minimax game improves the performance of both generator and discriminator. The training process is shown in Fig. 1.

### 3.5  Evaluation Metrics

Evaluating the quality of the generated images is critical for assessing the performance of generative models. Inception Score(IS) is a metric used to evaluate the quality of synthetic images [19], particularly these generated using Generative Adversarial Network (GAN). IS uses an Inception v3 network pre-trained on ImageNet then calculates a statistic of the network's outputs when applied to the generated images.

$$IS(G) \ = \ \exp\left(E_{x \sim P_g} D_{KL}(p(y|x) \,\|\, p(y))\right), \tag{1}$$

where $x \sim p_g$ represents that x is an image from $p_g$. $D_{KL}\,(p \,\|\, q)$ is the KL-divergence between two distributions p and q. $p(y|x)$ is the conditional probability of a class label given image x, $p(y)$ is the marginal probability of y class label across all synthetic images.

## 4  Experimental Results

### 4.1  Experimental Setting

In the context of Generative adversarial networks (GANs), there are several hyper parameters impact the model performance. These parameters made affect the optimization process and influence the diversity and quality of generated samples.

**Learning Rate (Lr).** This parameter controls the step size during weight updates. A small Learning rate (like Lr = 0,0002 in our case) means more stable convergence. Preventing drastic weight updates that could destabilize training.

**Number of Epochs.** The number of epochs represent how many times the training proceeds the entire dataset in our case we achieve convergence after 80 epochs and we train our model until 144 epochs.

**Latent Vector Size (Z).** The latent vector (often called Z) represent the generator's input, it's random vector from a normal distribution (usually 100 dimensions). A 100-dimensional latent vector from a normal distribution offers enough complexity for the generator to create diverse and realistic synthetic images, without being overly complicated.

**Checkpoint Frequency.** These checkpoints capture the model's progress allowing resume training from a check point or use the saved models to generate samples. This operation helps to avoid losing progress due unexpected interruptions.

**Kernel Size.** We use $4 \times 4$ for both generator and discriminator. The kernel size represents the dimensions of the filter in a convolutional layer.

## 4.2 Synthetic Data Visualization

In Fig. 5, we observe that DCGAN can generate realistic images suitable for training deep learning and machine learning models. Using synthetic data generated by DCGAN can address many challenges including data scarcity, imbalanced dataset, data privacy.

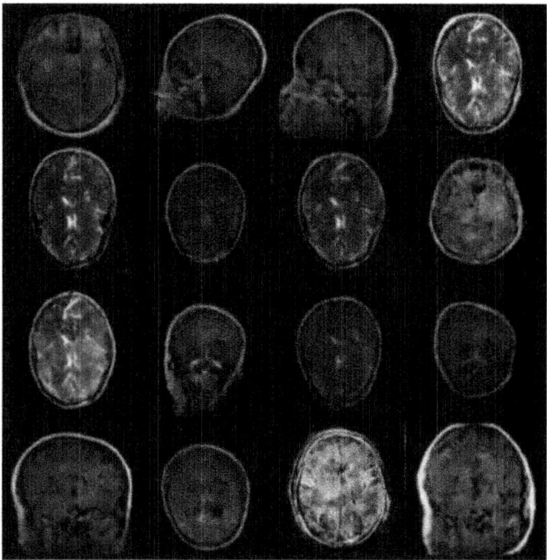

**Fig. 5.** Synthetic images

In Fig. 6, a higher IS value indicate high quality and realism in the generated images. Conversely, a lower standard deviation (SD) indicate that the inception scores for different generated images are not widely spread, so the models are consistently producing images from similar quality.

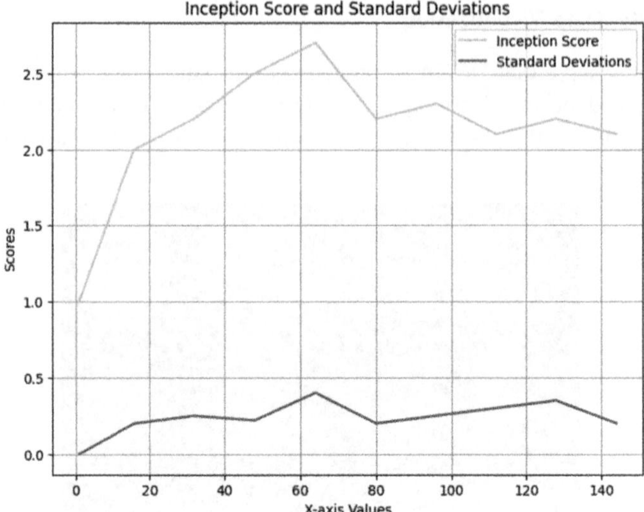

Fig. 6. Inception Score

## 5 Conclusion and Future Directions

The medical imaging faces significant challenges including data availability, data privacy and imbalance dataset. In our work, a common medical problem has been solved using a relevant data augmentation method based on DCGAN. Data scarcity is one of the most relevant challenges that effect the performance of Brain Tumors detection models. It poses a formidable obstacle in developing accurate and effective machine learning and deep learning models. Compared to the traditional methods used for data augmentation like geometric transformations, DCGAN show a promising result by generate unlimited realistic synthetic images which is similar to the real distribution data. With the generated images we can extends the dataset addressing data scarcity challenge and improve the model's performance leading to more stable training and accurate models.

Moving forward, exploring improved GAN architectures, such as Wasserstein GAN, Progressive Growing GAN, Least Squares GAN, Conditional GAN could make further advancements in medical image processing. Combining GANs with other deep learning architectures can produce new hybrid GANs for more accurate models. Expanding GANs applications in medical image processing can be a powerful insight for addressing critical challenges such as data privacy, data scarcity and imbalance, by generating synthetic X-ray, Computed Tomography(CT), positron emission tomography (PET) and other medical images. GANs can offer a significant contribution to augment datasets enabling the development of more accurate machine learning and deep learning models.

## References

1. Murtaza, H., Ahmed, M., Khan, N.F., Murtaza, G., Zafar, S., Bano, A.: Synthetic data generation: state of the art in health care domain. Comput. Sci. Rev. **48**, 100546 (2023)

2. Pezoulas, V.C., et al.: Synthetic data generation methods in healthcare: a review on open-source tools and methods. Comput. Struct. Biotechnol. J. **23**, 2892–2910 (2024)
3. Arora, A., Arora, A.: Generative adversarial networks and synthetic patient data: current challenges and future perspectives. Future Healthc. J. **9**(2), 190–193 (2022)
4. Saeedi, S., Rezayi, S., Keshavarz, H., Niakan Kalhori, R.S.: MRI-based brain tumor detection using convolutional deep learning methods and chosen machine learning techniques. BMC Med. Inform. Decis. Mak. **23**(1), 16 (2023)
5. Rather, I.H., Kumar, S.: Generative adversarial network based synthetic data training model for lightweight convolutional neural networks. Multimed. Tools Applic. **83**(2), 6249–6271 (2024)
6. Haque, M.M., Paul, S.K., Paul, R.R., Islam, N., Rashidul Hasan, M.A.F.M., Hamid, M.E.: Improving performance of a brain tumor detection on MRI images using DCGAN-based data augmentation and vision transformer (ViT) approach. In: Solanki, A., Naved, M. (eds.) GANs for Data Augmentation in Healthcare, pp. 157–186. Springer International Publishing (2023)
7. Alqahtani, H., Kavakli-Thorne, M., Kumar, G.: Applications of generative adversarial networks (GANs): an updated review. Archiv. Comput. Methods Eng. **28**(2), 525–552 (2021)
8. Goodfellow, I.J., ET AL.: Generative Adversarial Networks (2014). arXiv:1406.2661
9. Dewi, C., Chen, R.-C., Liu, Y.-T., Tai, S.-K.: Synthetic data generation using DCGAN for improved traffic sign recognition. Neural Comput. Appl. **34**(24), 21465–21480 (2022)
10. Mourad, D., Oseni, K.O.: Synthetic Brain Images: Bridging the Gap in Brain Mapping With Generative Adversarial Model (2024). arXiv:2404.08703
11. Liu, X., Du, H., Xu, J., Qiu, B.: DBGAN: a dual-branch generative adversarial network for undersampled MRI reconstruction. Magn. Reson. Imaging **89**, 77–91 (2022)
12. Tariq, U., et al.: Brain tumor synthetic data generation with adaptive StyleGANs. In: Longo, L., O'Reilly, R. (eds.) Artificial Intelligence and Cognitive Science, pp. 147–159. Springer Nature Switzerland (2023)
13. Altun Güven, S., Talu, M.F.: Brain MRI high resolution image creation and segmentation with the new GAN method. Biomed. Signal Process. Control **80**, 104246 (2023)
14. Zhou, M., Wagner, M.W., Tabori, U., Hawkins, C., Ertl-Wagner, B.B., Khalvati, F.: Generating 3D Brain Tumor Regions in MRI using Vector-Quantization Generative Adversarial Networks (2023). arXiv:2310.01251
15. Radford, A., Metz, L., & Chintala, S. Unsupervised Representation Learning with Deep Convolutional Generative Adversarial Networks. arXiv.Org. (2015, November 19)
16. Lucic, M., Kurach, K., Michalski, M., Gelly, S., Bousquet, O.: Are GANs Created Equal? A Large-Scale Study (2018). arXiv:1711.10337
17. Chakraborty, T., S, U.R.K., Naik, S.M., Panja, M., Manvitha, B.: Ten Years of Generative Adversarial Nets (GANs): A survey of the state-of-the-art (2023). arXiv:2308.16316
18. Brain Tumor MRI Dataset (n.d.). https://www.kaggle.com/datasets/masoudnickparvar/brain-tumor-mri-dataset
19. Salimans, T., Goodfellow, I., Zaremba, W., Cheung, V., Radford, A., Chen, X.: Improved Techniques for Training GANs (2016). arXiv:1606.03498
20. Cai, L., Chen, Y., Cai, N., Cheng, W., Wang, H.: Utilizing amari-alpha divergence to stabilize the training of generative adversarial networks. Entropy (2020)

# A New Approach for Detecting Communities of Interest in Social Networks


Noureddine Seddari[1,2]([✉]), Adlen Kerboua[3], Sohaib Hamioud[4,5], Chaoui Fayçal[1], Imen Boulnemour[1], Fouzia Krim[1], and Kenza Redjimi[1]

[1] LICUS Laboratory, Department of Computer Science, Université, 20 Août, 1955 Skikda, Algeria
{n.seddari,i.boulnemour1}@univ-skikda.dz
[2] Future Technology Lab, University of Parma, Parco Area Delle Scienze 181/A, Parma, Italy
[3] LGMM Laboratory, Université, 20 Août, 1955 Skikda, Algeria
ad.kerboua@univ-skikda.dz
[4] LISCO Lab, Department of Computer Science, Badji Mokhtar University, Annaba, Algeria
[5] Earth Sciences Department, Kasdi-Merbah University, Ouargla, Algeria
sohaib.hamioud@univ-ouargla.dz



**Abstract.** A social network, such as Facebook or Twitter, is a collection of social actors—like individuals or organizations—connected through relationships that represent social interactions. It describes a dynamic social structure made up of nodes and edges. The ability to analyze and identify communities within the network can aid in understanding and visualizing the structure of these networks. Social entities interact in various ways, forming groups, sending messages, sharing articles, joining discussion groups, and so on. Numerous approaches have been proposed to discover community structures within networks. While some methods have produced acceptable results, no algorithm has yet been able to deliver completely accurate outcomes. In this work, we present a novel approach to community detection in social networks. Our method is based on the Edge-Betweenness algorithm, which identifies communities by analyzing the exchange of information between nodes about their community memberships. We implemented an evaluation of our approach and compared the results with those found in the literature. The results obtained are considered satisfactory and demonstrate the potential of our method in accurately identifying community structures in social networks.

**Keywords:** Social networks · social network analysis · community detection · Edge-Betweenness algorithm


## 1 Introduction

In today's world, social media platforms such as Facebook, Twitter, and LinkedIn have emerged as significant new forms of communication. A social network, as represented by these platforms, consists of social actors—either individuals or organizations—connected by various forms of social interaction. This can be visualized as a dynamic social structure, which is typically modeled using graphs composed of nodes (the social actors) and edges (the interactions or relationships between them) [1]. Social network analysis, primarily based on graph theory and sociological analysis, aims to explore various facets of these networks. Among the most crucial are community detection, identifying influential actors, and studying and predicting network evolution. In this paper, the focus is specifically on community detection within social networks.

The concept of networks exists in numerous fields of research, including computer science and other disciplines such as biology, sociology, and even economics. By modeling these networks as graphs, researchers can better study and understand their structures, making use of graph theory for deeper analysis.

Graphs are composed of nodes and edges, where edges may have directions and be labeled with weights or attributes. For instance, in biology, metabolic networks are often represented as graphs, where nodes are proteins and edges represent chemical interactions between these proteins (e.g., metabolic pathways) [2]. Similarly, in sociology, nodes can represent individuals or social entities such as organizations, companies, or nations. The nature of the edges connecting these nodes varies according to the type of network. In a social network, for instance, two individuals are connected if they know each other. In a collaboration network, two individuals are connected if they work together on a common project [3]. In communication networks, two nodes are connected whenever they exchange an email [2].

Generally, the density of the edges between nodes can vary across different regions of a network, which suggests the existence of clusters of highly connected nodes, known as communities. These communities can be defined as groups of nodes that are strongly interconnected, but only weakly connected to nodes outside of their group. Identifying such communities is critical because they represent real-world social phenomena. In addition, understanding these communities enables us to determine the roles of various actors both within their communities and across the entire network.

### 1.1 Problem Statement

Community detection is an important task with applications in many fields and real-world scenarios [4–7]. For example, social networks often reveal communities of individuals who share common interests or exhibit frequent interactions [8]. By identifying these communities, we can make predictions about consumer behavior by analyzing the purchases and activities of other members of the same community. Additionally, community detection helps us understand the role of different actors within both the community and the broader network. A node with a central position in its community—such as one that shares many edges with other nodes in the same community—can exert significant control and contribute to the network's stability. Furthermore, nodes situated on the boundaries between communities often play a crucial role in mediating communication between different groups [2, 9].

### 1.2 Objective

There are several methods for community detection, which can be broadly classified into three categories [2, 10]:

- Hierarchical clustering methods: These allow for choosing a community structure from multiple hierarchical levels, representing different possible community structures.
- Optimization-based methods: These methods focus on maximizing a quality function (such as modularity) to identify optimal community structures.
- Model-based methods: These use predefined formalisms applied iteratively to the nodes in the network until a stable community structure is obtained.

The primary objective of this work is to develop a stable, accurate, and efficient approach to community detection, particularly for large real-world networks. To achieve this, we have implemented a community detection method based primarily on the edge-betweenness algorithm [11]. While this algorithm already exists, our contribution involves introducing modifications and improvements to enhance its performance and accuracy.

This paper is organized as follows: Sect. 2 provides an overview of the relevant background and related work. Section 3 introduces the proposed approach, detailing the methodology, centrality calculation, and the proposed algorithm's structure. In Sect. 4, we present and discuss our experiments and results, analyzing performance in terms of execution and complexity. Finally, Sect. 5 concludes the paper, summarizing our findings and outlining future directions for this research.

## 2   Overview

Community detection is a critical task in network analysis, aimed at identifying groups of nodes, or "communities," that are more densely connected to each other than to the rest of the network. The identification of these structures helps in uncovering the underlying organization of networks, which is applicable in various fields, such as social networks, biological networks, and information networks. Methods for community detection can be broadly categorized into three major groups: hierarchical clustering methods, optimization-based methods, and model-based methods [12]. Each of these approaches brings distinct techniques and focuses for uncovering the structure of a network.

Hierarchical clustering methods are among the most versatile for community detection [1]. These methods allow for the exploration of multiple hierarchical levels within a network, providing the flexibility to represent different possible community structures. The process typically begins with either all nodes assigned to a single cluster or each node assigned to its own cluster. Through iterative merging or splitting of clusters, a hierarchy of community structures is generated, which can be represented in a dendrogram. The user can select the most appropriate level of hierarchy to define the community structure. This method is particularly useful in networks where communities may exist at multiple scales, and the flexibility to choose the most relevant level is advantageous for various applications, such as analyzing social dynamics at both macro and micro levels.

Optimization-based methods focus on the maximization of a quality function, with modularity being one of the most widely used metrics [13]. These methods aim to partition the network in a way that maximizes the modularity score, which quantifies the density of edges within communities compared to random connections between nodes. By iterating through potential partitions and calculating the modularity for each configuration, the method seeks to find the optimal community structure that reflects the most meaningful divisions in the network. This approach has proven effective in detecting well-separated communities and is especially suited for networks where the goal is to maximize the internal cohesion of groups while minimizing the external connections. However, the reliance on modularity as a quality function has certain limitations, such as the resolution limit, which may cause smaller communities to go undetected.

Model-based methods apply predefined probabilistic or deterministic models to identify communities within the network [14]. These methods are based on formal assumptions about the structure of the network, which are iteratively applied to the nodes and edges until a stable community structure is reached. For example, Stochastic Block Models (SBMs) are a popular approach within this category [15]. SBMs assume that the probability of a connection between two nodes depends solely on their community memberships, and by fitting the model to the observed network data, the algorithm can infer the most likely community assignments for each node. Model-based methods are powerful because they incorporate domain-specific knowledge or hypotheses about the network, which can improve detection accuracy, particularly in complex or noisy networks. However, they also require the careful selection of models and parameters to ensure robust results.

Social Network Analysis (SNA) is a structural method used to describe and model the relational structure of a group of actors without prior assumptions [16]. It starts by observing the presence or absence of relationships between members of a group and reconstructing a system of connections based on these observations. SNA provides a framework for visualizing and modeling social relations as nodes (individuals, organizations) and ties (the relationships between nodes). It allows for the observation and quantification of degrees, strengths, or density of connections among actors within a network. By doing so, SNA takes a structural approach to analyzing relationships within an organized social environment. SNA utilizes mathematical tools drawn directly from graph theory, as well as specific techniques unique to the field. Thus, SNA often employs graph-theoretical terms such as degree, strength, or weight of a tie. According to Lazega, SNA enables the description of interdependencies between actors, providing a "simplified representation of a complex social system" [17]. Studies by Divjak and Peharda further define social network analysis as the mapping and measurement of relationships and flows between individuals, groups, organizations, or any other entity that processes information [18]. In this context, the nodes in the network represent people or groups, while the ties depict the relationships or flows between these nodes. SNA offers both a visual and mathematical analysis of human relationships.

## 3 Proposed Approach

### 3.1 Methodology

In our approach, we employ an algorithm based on the following idea: if a link frequently appears in the shortest paths between nodes in a graph, it is likely not within a specific community, but rather acts as a bridge between distant parts of the graph, connecting distinct communities. By progressively removing the link with the highest "centrality", we can decompose the network into distinct blocks. Here's how the algorithm works:

- **Step 1**: Calculate the betweenness centrality of each link.
- **Step 2**: Remove the link with the highest centrality.
- **Step 3**: Repeat the process until no links remain.
- **Step 4**: The remaining connected components represent the communities.
- **Step 5**: This results in a hierarchical decomposition of the network.

The algorithm operates as follows:

**Data:** A graph $G$

**Result:** The hierarchy of nested partitions of $V(G)$

- **Initial graph**: $G_0 := G$
- **Initial partition**: $P_0 := \{V(G)\}$
- **Iteration**: i $= 0$
- **Component count**: c $= 0$

While there are still edges in $G_i$:

1. Compute the centrality $S_i = \{B(e): e \in E(G_i)\}$ for each edge.
2. Find the maximum centrality $\max_S$ and the corresponding edges $E_i := \{e \in E(G_i): B(e) = \max_S\}$.
3. Remove the edge(s) from the graph $G_{i+1} := G_i \setminus E_i$.
4. If the connectedness of the graph decreases, increment c and update the partition $P_c$ with the connected components of $G_{i+1}$.
5. Increment $i$ and continue until no edges remain.

The connected components from each iteration give the final decomposition into communities.

### 3.2  Centrality Calculation

An intermediate node is a node through which multiple geodesic (shortest) paths pass, linking various node pairs in a network. The betweenness centrality of an edge eie_iei is calculated as follows:

$$EBC(e_i) = \sum_{j<k} \frac{NP_{vjvk}(e_i)}{NP_{vjvk}} \tag{1}$$

where:

- $NP_{vjvk}$ is the total number of shortest paths between nodes $v_j$ and $v_k$.
- $NP_{vjvk}(e_i)$ is the number of shortest paths between $v_j$ and $v_k$ that pass through the edge $e_i$.

An edge with a high betweenness centrality is crucial for information flow in the network, as it lies between distinct communities. Calculating this metric is computationally expensive, as it requires evaluating all possible paths between every pair of nodes for each edge. However, a faster method with optimization justifications was presented by Newman [11]. Newman introduced a more detailed algorithm to efficiently calculate betweenness centrality.

### 3.3   Proposed Algorithm Description

This approach is based on the idea that edges connecting different communities are often part of numerous geodesic paths between nodes, resulting in a high betweenness centrality value. Subsequent community-separating edges have lower centrality values. Therefore, the algorithm iteratively removes the edge with the highest EBC value and repeats this process until all edges are removed. The result is a hierarchical breakdown of the network into communities. Figure 1 shows the measurements of betweenness centrality for a network.

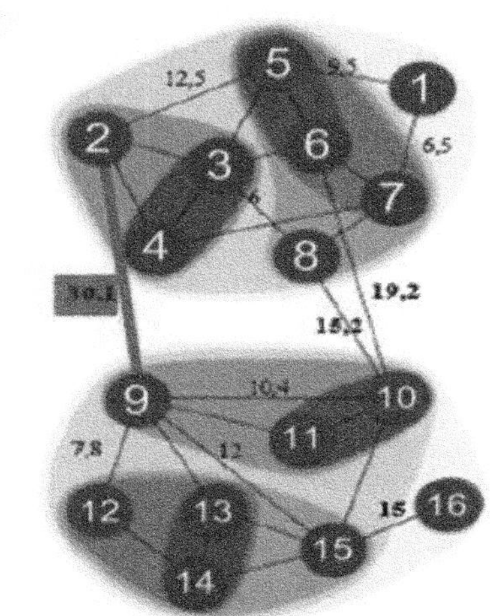

**Fig. 1.** Betweenness centrality measure of a network

## 4   Experiments and Results

This section provides a comprehensive evaluation of the proposed approach, showcasing the experiments, tools, and platforms used to assess its performance and effectiveness. The experiments were conducted utilizing Python 3.8.8 as the primary programming language, chosen for its versatility and widespread use in data science, machine learning, and network analysis. For network-related operations, we employed the NetworkX package, which is specifically designed for the creation, manipulation, and study of the structure, dynamics, and functions of complex networks.

To visualize the results, we utilized the Matplotlib library, a powerful plotting library in Python that allowed us to generate detailed graphs and visual representations of the data, which helped in better interpreting the outcomes of the experiments.

The entire workflow was executed on Google Chrome, ensuring a modern browser environment for web-based operations. Moreover, we made extensive use of Google Colab, a cloud-based platform that facilitates the execution of Python code in a collaborative manner. Colab provided an efficient computational environment by offering powerful hardware resources such as GPU and TPU acceleration, which played a crucial role in speeding up the experiments, especially when dealing with larger datasets and more complex models.

The experiments and tests were conducted on a physical machine (a personal computer) with the following specifications:

- Processor: Intel® Core™ i7-3687U CPU running at 2.10 GHz
- Installed Memory (RAM): 8.00 GB (7.87 GB usable)
- System Type: 64-bit operating system
- Operating System Edition: Windows 10 Professional

These hardware and software configurations were used for all installations and tests. By integrating these tools and platforms, we were able to conduct a series of experiments that provided valuable insights into the behavior and performance of the proposed approach. The results are presented in a detailed manner in the following (Figs. 2 and 3):

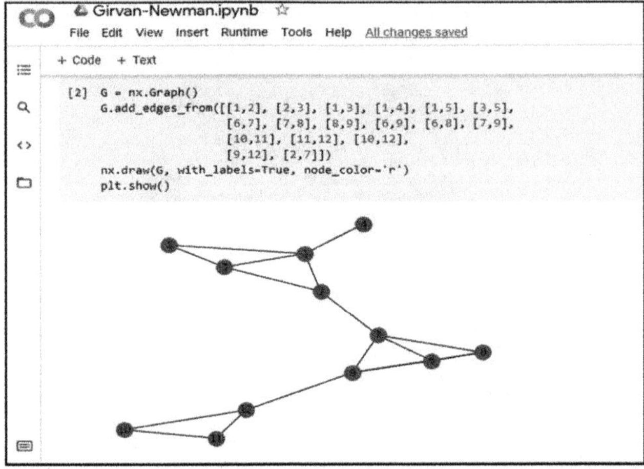


**Fig. 2.** The input of a network (graph)

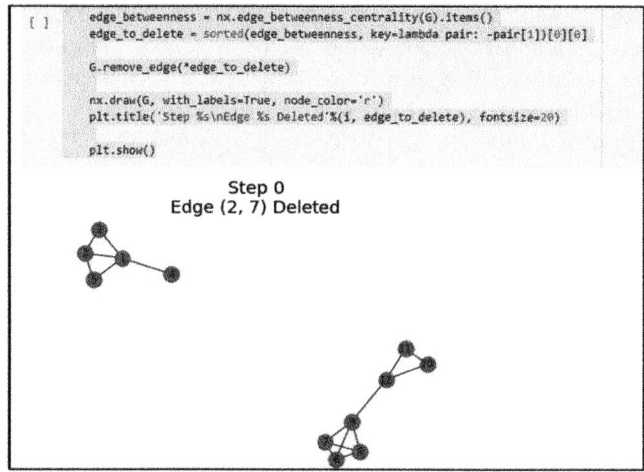

Fig. 3. The initial step in the execution of our method

In this phase, the program removes the links that have the highest EBC values.

Edge Betweenness Centrality is a measure used to identify the importance of links in a network. It helps to determine which connections play a critical role in the flow of information or interaction across the network. By eliminating the links with the highest EBC, the program is likely focusing on simplifying the network or identifying clusters, as the removal of such links can break the network into smaller, more manageable communities.

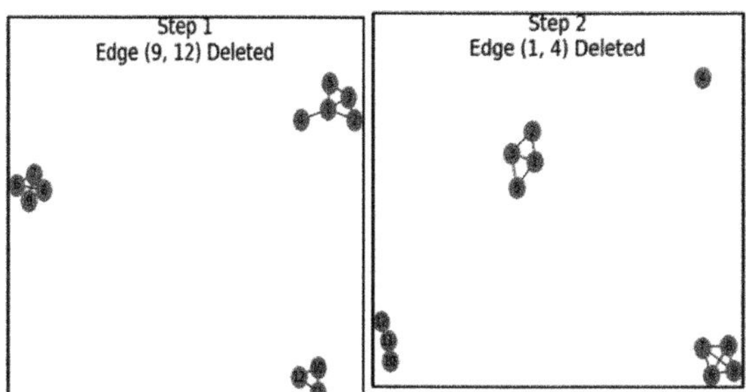

Fig. 4. The sequence of execution steps

## 4.1 Discussion

For a community detection algorithm to be considered valuable, the communities it identifies must be meaningful and relevant. However, this is often challenging to demonstrate, as there is no universally accepted definition of what constitutes a "good" community. The criteria for determining the quality of detected communities can vary, and without a clear standard, evaluating an algorithm's effectiveness becomes complex.

In our study, we sought to address this issue by testing our algorithm on real-world networks where the number of communities is already known. This allowed us to have a benchmark for comparison. Specifically, we compared the performance of our algorithm against well-established community detection algorithms, namely Label Propagation [19], the Louvain algorithm [20], and the Walktrap algorithm [21]. These algorithms are widely recognized for their efficiency and accuracy in detecting communities in various types of networks.

By comparing the communities identified by our algorithm with those produced by Label Propagation and Louvain, we aimed to assess the relevance and quality of the communities discovered. This comparative approach provides insights into how well our algorithm performs in relation to existing methods, offering a clearer understanding of its strengths and potential areas for improvement.

Through this process, we ensured that our evaluation was grounded in practical, real-world examples, rather than relying solely on theoretical metrics. This allowed us to more accurately determine the effectiveness of our algorithm in identifying meaningful communities within complex networks.

The comparison was conducted using Zachary's Karate Club dataset [22], a well-known and widely used network in community detection studies. This network is highly popular because it serves as a standard benchmark for testing community detection algorithms. It consists of 34 nodes and 78 edges, representing the social relationships within a karate club.

One of the reasons this network is frequently chosen for evaluation is that it naturally splits into two distinct communities, based on a real-life division that occurred within the club. This division makes it an ideal test case for assessing the accuracy and effectiveness of community detection algorithms. By applying our algorithm to this dataset, we were able to evaluate its ability to accurately detect these two known communities and compare the results to those produced by other established methods.

The simplicity of the network, combined with its well-defined community structure, provides a clear reference for measuring the performance of different algorithms. This makes the Zachary Karate Club dataset a crucial tool in validating the effectiveness of community detection techniques (Fig. 5 and Tables 1 and 2).

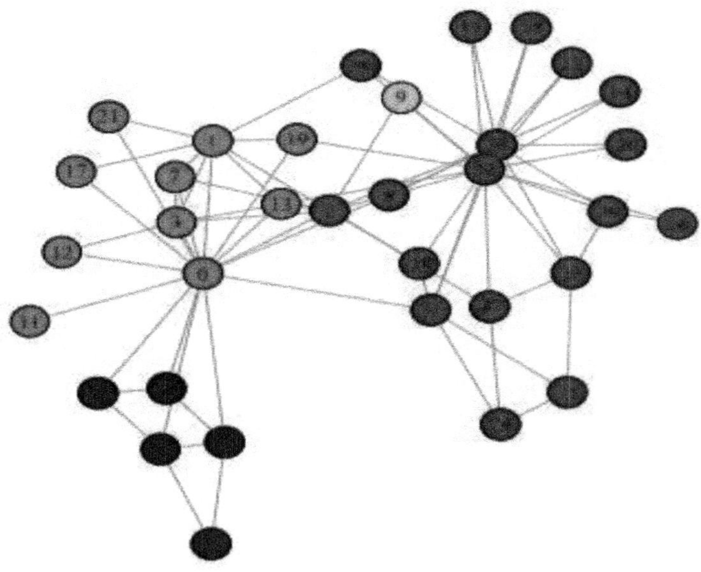

**Fig. 5.** The community structure identified by our algorithm in the Zachary network

**Table 1.** Execution results of algorithms on the Zachary network.

| Method | Number of communities |
| --- | --- |
| Proposed algorithm | 5 |
| Louvain | 4 |
| Label Propagation | 3 |
| Walktrap | 3 |

## 4.2  Execution Time

In terms of execution time, the Girvan-Newman method has consistently demonstrated its efficiency. It generally outperforms other algorithms, particularly as network complexity increases. For smaller networks with fewer than 500 nodes, the algorithm tends to be slower compared to alternatives like Louvain and Label Propagation. However, as both the number of nodes (n) and the number of links (m) grow, the Girvan-Newman algorithm begins to exhibit superior performance. This scalability makes it more efficient for larger and more complex networks, surpassing the speed of the Louvain and Label Propagation methods.

Additionally, across various types of networks, our algorithm consistently maintains its edge in terms of speed, proving itself as the fastest option for detecting communities. This adaptability and ability to handle an increasing number of nodes and connections further solidify its status as a highly effective algorithm for community detection in larger-scale networks, where computational efficiency becomes a critical factor.

This combination of scalability and consistent performance across different network structures makes the Girvan-Newman method particularly suitable for analyzing large and intricate networks where other algorithms may struggle to keep up with growing complexity.

**Table 2.** Execution time in seconds of the three algorithms with different networks.

|  | (n = 100, m = 450) | (n = 500, m = 12 450) | (n = 1 000, m = 49 950) | (n = 5 000, m = 1 250 000) |
|---|---|---|---|---|
| Proposed algorithm | 1.5 | 3.35 | 3.50 | 9 |
| Lauvain | 0.21 | 3.30 | 5.33 | 63 |
| Label Propagation | 0.10 | 2.70 | 3.50 | 10.90 |
| Walktrap | 0.20 | 3.75 | 5.50 | 147 |

### 4.3  Complexity

Calculating the complexity of our algorithm is not a straightforward task because, during the second phase, the number of iterations involved in the clustering process depends on the number of communities identified in the first phase, as well as the relationships between these communities. This dynamic interaction between the two phases adds a layer of variability to the complexity estimation.

However, based on our analysis, the overall estimated complexity of our algorithm can be approximated as $O(m^2n)$, where:

- n represents the number of nodes in the network, and
- m represents the number of edges.

This quadratic dependence on the number of edges and linear dependence on the number of nodes suggest that the algorithm's performance is strongly influenced by the density of the network. Denser networks, with more edges relative to nodes, will incur higher computational costs due to the squared factor of m. In the Table 3, we will provide a comparative analysis of the complexity of our algorithm against other well-known community detection algorithms. This comparison will help highlight the computational efficiency and scalability of our approach in different network structures.

**Table 3.** Comparison of algorithm complexities.

| Method | Complexity |
|---|---|
| Proposed algorithm | $O(m^2\,n)$ |
| Lauvain | $O(mn^3)$ |

(continued)

**Table 3.** (*continued*)

| Method | Complexity |
|---|---|
| Label Propagation | O(m) |
| Walktrap | $O(n4)$ |

## 5 Conclusion

In this paper, we propose a method for community detection in social networks by iteratively removing links between nodes and using Newman's modularity function. Covariance is employed as a network-specific metric for the links. The advantage of this metric is that it reduces the number of iterations needed for link removal and improves the quality of the detected communities. To validate our approach, we compared it with three well-known community detection algorithms: Walktrap, Louvain, and Label Propagation. These algorithms were chosen because they represent different types of community detection methods—agglomerative, divisive, model-based, and heuristic—and are considered benchmarks in the field for evaluating both the accuracy of community boundary identification and execution times.

We also conducted tests on real networks of various sizes, which confirmed the stability and performance of our approach. In many cases, the quality of partitioning achieved by our method surpassed that of the other algorithms, for both synthetic and real networks. Our method proved to be relatively stable, consistently performing on par with the best algorithms across several scenarios. Modularity measures on real networks also supported these findings. As for execution time, our method was the fastest after Label Propagation, making it a competitive option compared to well-established methods in the literature. It is particularly effective in scenarios where mixing parameters are high, which correspond to extreme situations where community boundaries are not easily distinguishable. Additionally, it performs well with large and/or densely connected networks.

For future work, our goals are to optimize the algorithm and reduce execution time to achieve equal or better results. We also plan to add a method for evaluating the resulting communities and to integrate our approach with artificial intelligence to adjust the probability of node membership within a community dynamically.

## References


1. Fortunato, S.: Community detection in graphs. Phys. Rep. **486**(3–5), 75–174 (2010)
2. Barabási, A.: Network Science. Cambridge University Press, Cambridge (2016)
3. Freeman, L.C.: The development of social network analysis: a study in the sociology of science. BookSurge Publishing, Charleston (2004)
4. Fumanal-Idocin, J., Alonso-Betanzos, A., Cordón, O., Bustince, H., Minárová, M.: Title. Futur. Gener. Comput. Syst. **113**, 25–40 (2020)
5. Azaouzi, M., Rhouma, D., Ben Romdhane, L.: Community detection in large-scale social networks: state-of-the-art and future directions. Soc. Netw. Anal. Min. **9**(1), 23 (2019)

6. Wickramasinghe, A.N., Muthukumarana, S.: Social network analysis and community detection on spread of COVID-19. Model. Assist. Stat. Appl. **16**(1), 37–52 (2021)

7. Moosavi, S.A., Jalali, M.: Community detection in online social networks using actions of users. In: 2014 Iranian Conference on Intelligent Systems (ICIS), pp. 1–7. IEEE (2014)

8. Khatoon, M., Banu, W.A.: A survey on community detection methods in social networks. Int. J. Educ. Manage. Eng. **5**(1), 16–23 (2015)

9. Negara, E.S., Andryani, R.: A review on overlapping and non-overlapping community detection algorithms for social network analytics. Far East J. Electron. Commun. **18**(1), 1–27 (2018)

10. Choudhury, D., Paul, A.: Community detection in social networks: an overview. Int. J. Res. Eng. Technol. **2**(14), 83–88 (2013)

11. Newman, M.E.J., Girvan, M.: Finding and evaluating community structure in networks. Phys. Rev. E **69**(2), 026113 (2004)

12. Iliho Saritha, S.K.: Community detection methods in social network analysis. In: Emerging Technologies in Data Mining and Information Security: Proceedings of IEMIS 2018, vol. 2, pp. 849–858. Springer, Singapore (2019)

13. Fortunato, S., Hric, D.: Community detection in networks: a user guide. Phys. Rep. **659**, 1–44 (2016)

14. Zhu, Y., Leung, H.F., Gong, M.: A survey of community detection methods based on statistical models and optimization models. IEEE Trans. Knowl. Data Eng. **34**(11), 5353–5370 (2022)

15. Zhang, P., Moore, C.: A survey of community detection in complex networks using statistical methods. Stat. Sci. **36**(4), 492–515 (2021)

16. Rossetti, G., Cazabet, R.: Community discovery in dynamic networks: a survey. ACM Comput. Surv. (CSUR) **51**(2), 1–37 (2018)

17. Lazega, E.: Analyse de réseaux et sociologie des organisations. Rev. Fr. Sociol. **35**(2), 293–320 (1994)

18. Divjak, B., Peharda, P.: Relation between student academic performance and their position in a social network. In: Central European Conference on Information and Intelligent Systems, pp. 1–10 (2010)

19. Raghavan, U.N., Albert, R., Kumara, S.: Near linear time algorithm to detect community structures in large-scale networks. Phys. Rev. E. **76**(3), 036106 (2007)

20. Blondel, V.D., Guillaume, J.L., Lambiotte, R., Lefebvre, E.: Fast unfolding of communities in large networks. J. Stat. Mech: Theory Exp. **2008**(10), P10008 (2008)

21. Pons, P., Latapy, M.: Computing communities in large networks using random walks. In: 20th International Symposium on Computer and Information Sciences (ISCIS 2005). Istanbul, Turkey, October 26–28, 2005, pp. 284–293. Springer, Berlin, Heidelberg (2005)

22. Zachary, W.W.: An information flow model for conflict and fission in small groups. J. Anthropol. Res. **33**(4), 452–473 (1977)

# MBTI Personality Type Prediction: A Machine Learning Analysis

Samia Bousalem[1]([⊠])  , Naila Marir[1,2], and Fouzia Benchikha[1]

[1] LIRE Laboratory, University of Constantine 2 - Abdelhamid Mehri, Constantine, Algeria
{samia.bousalem,naila.marir,fouzia.benchikha}@univ-constantine2.dz
[2] Effat College of Engineering, Effat University, Jeddah, Saudi Arabia
namarir@effat.edu.sa

**Abstract.** Personalized learning is a foundational element of effective education, aiming to tailor teaching methods to individual learners' needs. Central to this approach is understanding learners' personalities, as it can significantly influence their learning preferences and outcomes. In this paper, we propose a novel method for predicting learner personality. Our method leverages the Linear Support Vector (Linear SVC) classifier technique and the Myers-Briggs Type Indicator (MBTI) model to predict learner personalities. We utilize the PersonalityCafe dataset for this purpose, employing natural language processing (NLP) preprocessing techniques and the TfidfVectorizer for feature extraction. The experimental results reveal the effectiveness of the linear SVC, achieving an impressive accuracy of 84% in predicting MBTI personality types among learners. By accurately predicting learner personality traits, our method enhances the personalization of learning experiences, leading to improved learning outcomes and engagement.

**Keywords:** Learner Personality · MBTI · Machine Learning · Linear SVC · Prediction

## 1 Introduction

In the evolving education landscape, acknowledging individual differences and customizing teaching approaches has become essential for fostering compelling learning experiences. One aspect influencing the dynamics of the learning process is the unique personality traits each learner exhibits. In the context of this study, personality refers to the unique and individual characteristics, preferences, and traits that influence how a person learns, engages with educational materials, and interacts with the learning environment.

Researchers have developed numerous models to conceptualize and categorize personality traits, each offering a unique lens to understand individual differences. Among these, the Big Five personality traits [1], encompassing openness, conscientiousness, extraversion, agreeableness, and neuroticism, provide a widely recognized framework for personality assessment. Another prominent model is the Myers-Briggs Type Indicator (MBTI) [2], which classifies individuals into distinct personality types based on preferences in four dichotomies: extraversion/introversion, sensing/intuition, thinking/feeling, and judging/perceiving. While the Big Five offers a comprehensive view of personality dimensions, the MBTI, which emphasizes preferences and cognitive styles, is a valuable tool for understanding individuals in educational contexts. This study focuses on the MBTI model, leveraging its nuanced approach to predict learner personalities and inform tailored teaching strategies.

This study introduces a novel approach for predicting learner personality by integrating sentiment analysis within the Myers-Briggs Type Indicator (MBTI) framework. Leveraging machine learning techniques, including the Linear SVC algorithm, we analyze the sentiment expressed in learners' interactions to infer their MBTI personality types. By incorporating sentiment as a key feature, we aim to capture nuanced aspects of individual preferences and cognitive styles. Our methodology goes beyond traditional personality assessments, offering a dynamic and context-aware model for personality prediction in educational settings.

The remainder of this paper is organized as follows: Sect. 2 reviews related works, providing a backdrop for our study. Section 3 offers a brief appraisal of the MBTI model. Section 4 outlines our methodology, emphasizing the application of sentiment analysis and machine learning techniques. Section 5 presents the results of our study. Finally, Sect. 6 concludes the paper by summarizing key insights and suggesting directions for future research.

## 2 Literature Review

This section synthesizes current research on predicting personalities, with an emphasis on the MBTI model and machine learning. Two fundamental approaches are distinguished: the binary approach and the multi-class approach. Through focused exploration, we aim to uncover insights, methodologies, and gaps in the landscape of personality prediction using the MBTI.

In a study by Zumma et al. [3], the authors focused on forecasting personality traits from Twitter data by applying machine learning techniques, specifically utilizing the Myers-Briggs Type Indicator (MBTI) dataset. The model underwent training with six state-of-the-art classifiers, with the SVM classifier surpassing the others, exhibiting superior accuracy in personality prediction.

Another strategy for predicting personality based on the MBTI personality model has been presented in three studies [4–6]. These works proposed a comprehensive multi-class approach, employing a range of machine learning and deep learning techniques, including Random Forest (RF), XGBoost (XGB), Logistic Regression (LR), K-Nearest Neighbors (KNN), Support Vector Machine (SVM), BERT, One vs Rest, Multilayer Perceptron (MLP), Naive Bayes (NB), and Convolutional Neural Network (CNN). The application of CNN achieved the most promising results, attaining an accuracy of 81%, while RF was the least effective with a 21% accuracy rate.

Most research in the domain of predicting MBTI personality traits through a binary approach has explored a variety of classifiers, including Random Forest (RF), XGBoost (XGB), Logistic Regression (LR), Support Vector Machine (SVM), BERT, and Naive Bayes (NB) [7–14]. The most impressive results came from Vasquez et al. [11]. They got 88% correct for Introversion/Extraversion (I/E), 92% accurate for Sensing/Intuition (S/N), 88% valid for Feeling/Thinking (F/T), and 84% correct for Judging/Perceiving (J/P).

The current research draws inspiration from prominent researchers in machine learning and deep learning, particularly in predicting MBTI personality types. The authors' central objective in this study is to explore learner personality types and attain a deeper understanding of their psychological preferences.

## 3   Background

This section introduces the Myers-Briggs Type Indicator (MBTI), a widely-used framework that categorizes individuals into 16 distinct personality types based on preferences in four dichotomies.

The MBTI is based on the work of psychologist Carl Jung and was developed by Katharine Cook Briggs and her daughter, Isabel Briggs Myers, in the 1940s. It is an introspective self-report questionnaire indicating differing psychological preferences in how people perceive the world and make decisions. It is divided into four scales: energy, perceiving, judging, and orientation [15].

### 3.1   MBTI Scales

The Myers-Briggs Type Indicator (MBTI) focuses on four scales to classify individuals into one of 16 personality types (see Fig. 1). These scales evaluate an individual's inclinations in perceiving the world and making decisions, ultimately culminating in identifying their personality type.

### 3.2   Personality Types in the Myers-Briggs Type Indicator

This project employs the MBTI framework to determine learners' personalities based on key dimensions, resulting in sixteen distinct personality types, such as "INFJ" or "ENFP". For example, "INFJ" stands for Introversion, Intuition, Feeling, and Judging. Figure 2 illustrates these sixteen personality types.

## 4   Proposed Methodology

Figure 3 illustrates the research methodology employed in this study. In this section, we will detail each stage of the methodology to provide a comprehensive understanding of our approach.

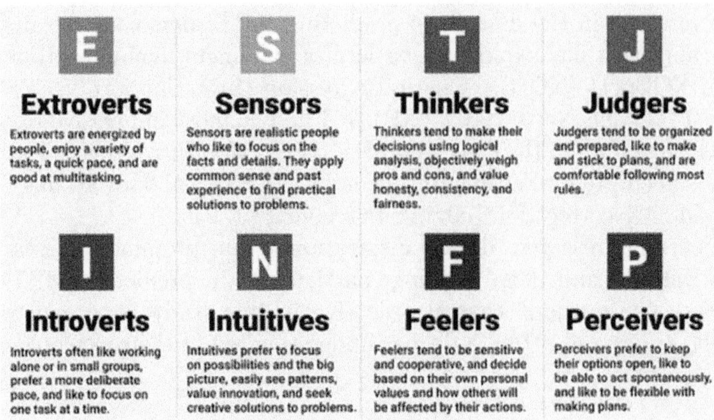



**Fig. 1.** MBTI Personality Scales [16].

### 4.1    Data Collection

The dataset, sourced from PersonalityCafe, comprises 8675 instances and is a fundamental component of our study. Collected through interactions within the PersonalityCafe community, this dataset captures a rich array of responses to the MBTI questionnaire. The diversity within the dataset ensures a comprehensive representation of individuals with varying personality types, laying a robust foundation for the training and evaluation of our machine learning model.

This dataset contains over 8600 rows of data, with each row representing a person's (see Fig. 4):

- Type (the person's 4-letter MBTI code/type)
- A section of each of the last 50 things they have posted (each entry separated by "|||" (3 pipe characters))

Data visualization techniques, including histograms, are used to gain insights into the distribution of the MBTI personality dataset. Our objective is to provide a nuanced overview of the prevalence of various personality types. These visualizations illuminate the distribution landscape, enhancing our understanding of the dataset and setting the stage for subsequent analytical steps. Figure 5 depicts a histogram illustrating the distribution of the MBTI personality dataset.

### 4.2    Data Preprocessing

In the data preprocessing section, text data undergoes a series of essential steps to enhance its quality and suitability for analysis. To ensure thorough refinement and preparation of the data for subsequent analysis, we diligently follow these steps:

- **Lowercasing**: Convert the text to lowercase for uniformity.
- **Link Removal**: Eliminate web links to focus on the core text content.

**Fig. 2.** MBTI Personality Types [16].

- **Stop words Removal**: Optionally remove common stop words to streamline the text.
- **Punctuation Removal**: Optionally eliminate punctuation to focus on essential text elements.
- **Numerical Values Removal**: Optionally remove numerical values to enhance the textual analysis.

For instance, within the MBTI Dataset, the post data is delimited by the pipeline notation "|||". To streamline the information, we need to eliminate all occurrences of these separators. Python offers a built-in function called "replace ()" that effectively removes these pipeline symbols, ensuring a cohesive and unified dataset. In addition to these steps, the dataset undergoes a division phase to facilitate practical model training and evaluation. Following standard practice, the dataset is split into two subsets: 70% for training and 30% for testing.

## 4.3   Model Learning: Training for Prediction

In this phase, we construct and train a predictive model to associate textual features with corresponding personality types. We employ the TfidfVectorizer ('Term Frequency-Inverse Document Frequency Vectorizer) to transform the text data into numerical representations. This method converts raw text into vectors

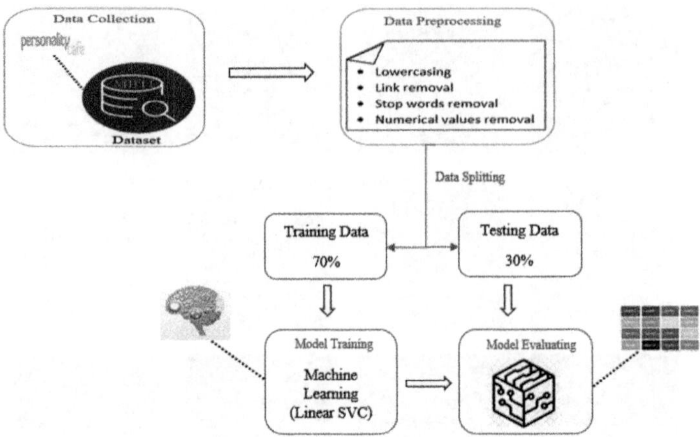


**Fig. 3.** Proposed Methodological Approach.

that reflect word frequencies, adjusted for the rarity of words across the corpus. By highlighting more informative words that are significant within a document but not common across all documents, TfidfVectorizer effectively captures the essence of each individual's written content.

After vectorizing the textual data, we apply a Linear Support Vector Classifier (Linear SVC). This algorithm is well-suited for high-dimensional spaces, making it ideal for text classification tasks. It works by finding the hyperplane that best separates the classes (in this case, personality types), maximizing the margin between data points from different classes. Linear SVC was particularly chosen for its efficiency in handling large datasets and strong generalization performance.

Once the model is trained on labeled data, we test it on unseen, unlabeled data to predict the personality types of learners. This approach enables us to translate written text into personality predictions, enhancing learner profiles for integration into adaptive learning systems. We expanded the previously established profile ontology described in prior work [17] by incorporating learner personality as instances associated with the personality class.

## 4.4   Model Evaluating

This section focuses on testing the performance of a trained model using new, unseen data. This helps to determine how well the model can make predictions on data it hasn't encountered before. By using this fresh data, we can assess the model's accuracy and generalization, ensuring it performs well not just on the training set but also in real-world scenarios. This process is crucial for understanding the model's effectiveness and reliability.

| type | posts |
| --- | --- |
| **0**   INFJ | 'http://www.youtube.com/watch?v=qsXHcwe3krw\|\|\|... |
| **1**   ENTP | 'I'm finding the lack of me in these posts ver... |
| **2**   INTP | 'Good one _____ https://www.youtube.com/wat... |
| **3**   INTJ | 'Dear INTP, I enjoyed our conversation the o... |
| **4**   ENTJ | 'You're fired.\|\|\|That's another silly misconce... |

**Fig. 4.** Kaggle Dataset on MBTI Personality Types.

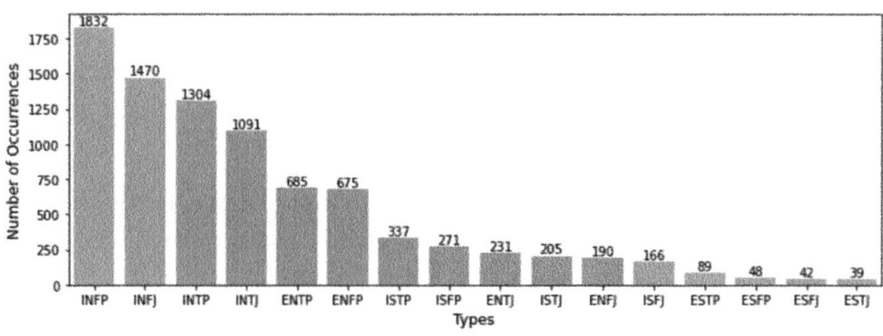


**Fig. 5.** Distribution of Personality Types According to the MBTI Scale.

The following section will present and discuss the results obtained from the study.

## 5   Results and Discussion

This section presents the outcomes of our model's learning and application efforts, shedding light on the model's predictive capabilities in determining learner personalities. We delve into the performance metrics obtained during testing and discuss the implications and significance of the results comprehensively. Figure 6 represents the classification report of the Linear SVC classifier.

The linear SVC classifier's classification report highlights its performance across different personality types, showing variations in precision, recall, and F1-score. Most personality types demonstrate high prediction accuracy, indicating the model's robustness in capturing distinctive features. Macro-average scores provide an overview of overall performance, while weighted averages show effectiveness in handling class imbalance. The classifier's ability to maintain accuracy across diverse learner profiles underscores its efficacy in predicting personalities.

```
              precision    recall  f1-score   support

       ENFJ       0.86      0.66      0.75        56
       ENFP       0.83      0.83      0.83       196
       ENTJ       0.91      0.69      0.78        58
       ENTP       0.82      0.80      0.81       201
       ESFJ       1.00      0.40      0.57        10
       ESFP       1.00      0.33      0.50         9
       ESTJ       0.93      0.76      0.84        17
       ESTP       1.00      0.72      0.84        25
       INFJ       0.83      0.83      0.83       434
       INFP       0.83      0.90      0.86       566
       INTJ       0.79      0.89      0.84       311
       INTP       0.84      0.87      0.86       404
       ISFJ       0.96      0.79      0.86        61
       ISFP       0.85      0.74      0.79        84
       ISTJ       0.86      0.71      0.78        70
       ISTP       0.88      0.76      0.81       101

   accuracy                           0.84      2603
  macro avg       0.89      0.73      0.78      2603
weighted avg       0.84      0.84      0.83      2603
```

**Fig. 6.** Classification Report for Linear SVC Classifier.

Other machine learning algorithms have been tested, and Table 1 summarizes the results for each, noting that Linear SVC achieved the highest accuracy.

**Table 1.** Accuracy Of Each Model

| Algorithm | KNN | DT | RF | MLP | AdaBoost | LGBM | RIDGE |
|---|---|---|---|---|---|---|---|
| Accuracy | 26 % | 45 % | 21 % | 56 % | 35 % | 66 % | 65 % |

We used the confusion matrix to evaluate the model better. This matrix provides a granular view of the model's performance by breaking down predictions into four categories: true positive (TP), true negative (TN), false positive (FP), and false negative (FN). The confusion matrix for our Linear SVC classifier reveals the distribution of these outcomes across different personality types. Figure 7 displays the confusion matrix of our model.

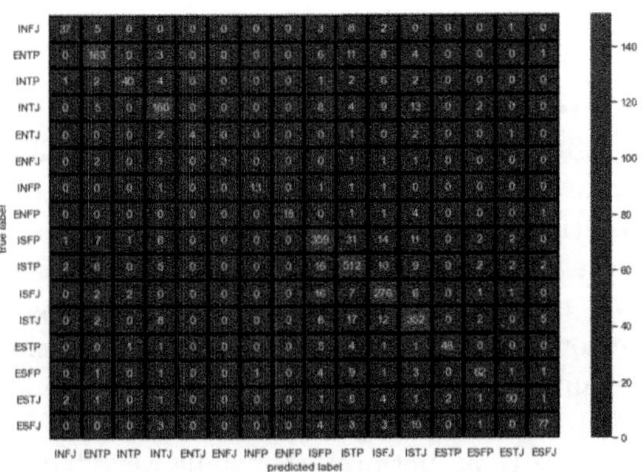

**Fig. 7.** Confusion Matrix For Linear SVC Classifier.

The comparative study presented below critically analyzes various MBTI personality prediction models as detailed in references [3–8], and [11]. Table 2 encapsulates the disparities between these seven distinct approaches and our proposed method, focusing mainly on the accuracy metric. This comparative study table outlines machine and deep learning methods for predicting MBTI personality types. It indicates that binary classification approaches employ diverse classifiers and achieve varied accuracy levels, with the BERT model consistently performing well, especially for the Sensing/Intuition dimension. The results stress the importance of considering specific MBTI dimensions, dataset nature, computational resources, and model interpretability. Our multi-class approach using a Linear SVC achieves competitive accuracy (84%), notable given the complexity of multi-class classification.

**Table 2.** Comparative Study With Previous Work.

| Paper | Approach | Classifier | Best result |
|---|---|---|---|
| [3] | Binary | RF, XGB, SGD, LR, KNN, SVM | I/E = 78%<br>S/N = 86%<br>F/T = 73%<br>J/P = 66% |
| [7] | Binary | LR, NB, XGB, SVM | I/E = 80%<br>S/N = 81%<br>F/T = 81%<br>J/P = 79% |
| [8] | Binary | BERT | I/E = 84%<br>S/N = 89%<br>F/T = 84%<br>J/P = 78% |
| [11] | Binary | BERT | I/E = 88%<br>S/N = 92%<br>F/T = 88%<br>J/P = 84% |
| [4] | Multi-class | RF, XGB, LR, KNN, SVM | 76.06% |
| [5] | Multi-class | RF, KNN, One vs Rest, MLP | MLP 31% |
| [6] | Multi-class | NB, RF, SVM, CNN | CNN 81% |
| Our approach | Multi-class | Linear SVC | 84% |

Our methodology demonstrates robustness compared to various algorithms, including BERT, which is known for its high accuracy in text classification.

Despite BERT's reputation, our approach achieves relative success due to tailored machine learning model selection, particularly suitable for limited dataset sizes. Machine learning is preferable over BERT in such scenarios, where smaller datasets are standard. Our decision reflects the need to extract insights effectively from modest datasets, where machine learning techniques show superior adaptability and performance. Our approach highlights the significance of selecting models tailored to the specific demands of MBTI personality prediction.

## 6    Conclusion

In conclusion, our study explored predicting learner personalities using machine learning techniques, specifically employing the MBTI model. The application of Linear SVC yielded compelling outcomes with high precision, recall, and F1-score values across various personality types. A thorough analysis of the classification report and confusion matrix highlighted the model's robustness and ability to navigate class imbalances within the dataset. Our model's predictive power provides valuable insights for improving learner profiles in educational systems, resulting in more individualized and adaptive learning experiences.

There are numerous opportunities for future research and development in this domain. Expanding the dataset to encompass more diverse and extensive learner profiles could enhance the model's generalization capabilities while also addressing the challenge of imbalanced data. By employing techniques such as oversampling, undersampling, or synthetic data generation, we aim to mitigate the effects of class imbalance, thereby improving the model's accuracy. Additionally, exploring deeper models such as BERT in this context could significantly enhance the model's predictive power and impact. Future work will delve into these advanced architectures to better capture the complex relationships within the data. Exploring the integration of other personality models, such as the Big Five, might provide a more comprehensive understanding of learner traits. Moreover, investigating the real-time application of personality prediction in adaptive learning systems could pave the way for transformative advancements in personalized education.

## References


1. Goldberg, L.R.: An alternative "description of personality": the Big-Five factor structure. In: Personality and Personality Disorders, pp. 34–47. Routledge (2013)
2. Myers, I.B.: The Myers-Briggs Type Indicator: Manual (1962). Consulting Psychologists Press (1962)
3. Zumma, M.T., Munia, J.A., Halder, D., Rahman, M.S.: Personality prediction from twitter dataset using machine learning. In: 2022 13th International Conference on Computing Communication and Networking Technologies (ICCCNT), pp. 1–5. IEEE (2022)
4. Bruno, A., Singh, G.: Personality traits prediction from text via machine learning. In: 2022 IEEE World Conference on Applied Intelligence and Computing (AIC), pp. 588–594. IEEE (2022)

5. Kuchhal, M., Jangid, P., Saini, M., Jindal, R.: Personality prediction and group detection using social media posts. In: 2022 IEEE 7th International Conference for Convergence in Technology (I2CT), pp. 1–5. IEEE (2022)
6. Pradhan, T., Bhansali, R., Chandnani, D., Pangaonkar, A.: Analysis of personality traits using natural language processing and deep learning. In: 2020 Second International Conference on Inventive Research in Computing Applications (ICIRCA), pp. 457–461. IEEE (2020)
7. Dagha, R., Dhaord, R., Lakhani, V., Hirlekar, V.: Personality prediction based on users' tweets. In: 2022 2nd International Conference on Intelligent Technologies (CONIT), pp. 1–5. IEEE (2022)
8. Majima, S., Markov, K.: Personality prediction from social media posts using text embedding and statistical features. In: 2022 17th Conference on Computer Science and Intelligence Systems (FedCSIS), pp. 235–240. IEEE (2022)
9. Pansare, A., Panwar, P., Kosamkar, P.: Personality prediction with natural language processing using questionnaire responses. In: 2022 IEEE Pune Section International Conference (PuneCon), pp. 1–6. IEEE (2022)
10. Orynbekova, K., Talasbek, A., Omar, A., Bogdanchikov, A., Kadyrov, S.: MBTI personality classification using apache spark. In: 2021 16th International Conference on Electronics Computer and Computation (ICECCO), pp. 1–4. IEEE (2021)
11. Vásquez, R.L., Ochoa-Luna, J.: Transformer-based approaches for personality detection using the MBTI model. In: 2021 XLVII Latin American Computing Conference (CLEI), pp. 1–7. IEEE (2021)
12. Mushtaq, Z., Ashraf, S., Sabahat, N.: Predicting MBTI personality type with k-means clustering and gradient boosting. In: 2020 IEEE 23rd International Multitopic Conference (INMIC), pp. 1–5. IEEE (2020)
13. Bhamare, M., Ashokkumar, K.: Personality prediction through social media posts. Int. J. Performability Eng. **18**(11) (2022)
14. Nisha, K.A., Kulsum, U., Rahman, S., Hossain, M.F., Chakraborty, P., Choudhury, T.: A comparative analysis of machine learning approaches in personality prediction using MBTI. In: Computational Intelligence in Pattern Recognition: Proceedings of CIPR 2021, pp. 13–23. Springer, Cham (2022)
15. Allen, J.: Using the Myers Briggs type indicator–part of the solution. Br. J. Nurs. **3**(9), 473–477 (1994)
16. Tieger, P.D., Barron, B., Tieger, K.: Do what you are: discover the perfect career for you through the secrets of personality type. Hachette, UK (2014)
17. Bousalem, S., Benchikha, F., Chelghoum, M.: Modeling learner profiles using ontologies and machine learning. In: 2022 2nd International Conference on New Technologies of Information and Communication (NTIC), pp. 1–6. IEEE (2022)

# Bi-Embedding and Metadata Enhancement for Collaborative Prediction


Sahraoui Kharroubi[1(✉)], Maatoug Abdelfettah[2], Lalia Benathmane[1], and Omar Nouali[3]

[1] Research Laboratory of Artificial Intelligence and Systems, University of Tiaret, 14000 Tiaret, Algeria
{sahraoui.kharroubi,lalia.benathmane}@univ-tiaret.dz
[2] Faculty of Applied Sciences and Technology, University of Tiaret, 14000 Tiaret, Algeria
abdelfettah.maatoug@univ-tiaret.dz
[3] Basic Software Laboratory, C.E.R.I.S.T., 16000 Algiers, Algeria
o.nouali@cerist.dz



**Abstract.** Users are overwhelmed by the gigantic amount of information generated instantaneously by shopping sites, social networks, web services and so on. Having a relevant item (document, film, news, business, etc.) on time and without difficulty is a real challenge for developers. Recommendation systems are effective and powerful tools to address this problem. This research Intituled "Bi-embedding and metadata enhancement for collaborative prediction" (Bi-MeCp) proposes a recommendation method based on bipartite graph topology. To recommend an item to a user, we study the evolution of the graph to find the weights of non-existent links. The link weight between an item node and a user node measures the degree of importance for such a recommendation based on score similarity. The idea is to extract and aggregate the implicit and asymmetric connectivity "Bi-embedding" to build vector embeddings. Then, we structure and enrich the content of the item node with metadata (DC, MPEG, IEEE-lom, etc.) for personalized classification. The use of metadata allows for a variety of recommendations and goes beyond the tradition of being interested in recommending a single type of item. In this way, we have taken into account purely negative user judgments to refine and boost the recommendation process. This study has been tested on real datasets widely used in the field of recommendation and has produced encouraging and promising results.

**Keywords:** Graph · Embedding vector · Prediction · Classification · Similarity


## 1 Introduction

There are a high proportion of daily activities that take place online. On the other hand, there is a huge amount of data that exists and occurs instantly. Timely access to the relevant resource becomes difficult, despite its availability, it is lost in the mass (resource overload) [1]. Recommender systems have been widely used by companies to recommend relevant resources to users using a similarity process. Large companies and websites such as Netflix, Amazon, Facebook, YouTube, Twitter, etc. integrate recommendation techniques into their servers [2]. The methods of recommendation systems can be mainly classified into three approaches [3, 4]: the content-based approach which compares the content of the user profile with the content of the item; the collaborative approach which depends on the feedback of neighbour's ratings; and the hybrid approach which aims at objectively combining the two methods. The collaborative approach is the most widely used and effective in many areas. However, finding the nearest neighbours is the critical phase of the collaborative approach. In addition, providing high-quality recommendations to users with a minimum of common feedback is a major challenge for recommendation systems [5, 6]. Currently, several in-depth studies are focused on modelling using bipartite graph [7, 8]. The structure of bipartite graph is perfectly adapted as a theoretical and practical model for several systems in the real world. Recent algorithms in the field of recommendation are based on a multidimensional graph representation. A knowledge graph consists of nodes and links (edges), where each node is represented by a feature vector and the link represents the interaction between the nodes. The graph structure facilitates the application of machine learning and deep learning to the recommendation task. In our study, we modelled the recommendation by a bipartite heterogeneous graph, which takes into account the weights of existing links to predict new links. This study is organised as follows: it begins with a description of the context and prerequisites in Sect. 2, followed by the presentation of our methodology in Sect. 3. The experimentation and discussion of the results are presented in Sect. 4, Finally, we conclude this work.

## 2   Context

Recommender systems (RS) are increasingly useful in online services such as e-commerce and streaming. The goal is to offer users the products (items) that best match their tastes and preferences. RS automatically identify user's preferences through their interactions (implicit and/or explicit feedback) to build their profiles. Recommender systems are based on information filtering (IF) approaches [9]. IF is divided into three main techniques: content-based filtering (CBF), collaborative filtering (CF) and hybrid filtering. CBF compares the content of products (items) with the content of the user profile and calculates a similarity correlation between the profile vector and the product vectors. If the correlation is positive, the product is recommended to the user. CF compares users or items based on ratings, knowing that similar ratings form a community. If the community rates an item positively, it will be recommended to a user (who has not yet rated it) by the community. This type of filtering is the most popular [10] and is used by recognised platforms such as NetFlix, Movielens, Amazon, etc. In fact, a graphical modelling is a typical case to represent the user and the items as nodes and the relations between them as edges. In particular, a bipartite graph consists of two different sets: one set being dedicated to users and the other to items, and the relations between the nodes of the first set and the nodes of the second set. Formally, a bipartite graph $BG(U, I, E)$, where U and I are the set of users and the set of items, respectively. $E \subseteq U \times I, E = \{(u, i) : u \in U, i \in I\}$, the edge $(u, i)$ is created between user u and item i whose weight is $w_{u,i}$. The value of the weight is $\{0, 1\}$ in the case of a binary rating matrix and $[0-5]$ in the case of real ratings (MovieLens, Amazon, etc.). Graphical modelling is very popular in terms of interpretability and implementation [11, 12]. A first example is to build a graph from explicit user ratings of items (rating matrix) and then apply a random walk algorithm such as PageRank to propose recommendations. In this way, Graph neural networks (GNNs) are an extension of classical Deep Learning neural networks, which are used to process structured data and are also capable of manipulating data in a graph. In GNN, a graph uses feature vectors (embeddings) to encode all the information about a graph, such as its structure and relationships. Thus, embeddings build a digital representation of a graph that can later be used to train machine learning algorithms [13–15]. There are two types of embeddings: node embeddings and edge embeddings.

## 3  The Proposal

In order to improve the quality of recommendations, our proposal (Bi-MeCp) is to enrich the nodes and edges of a graph by integrating metadata. The proposal (Fig. 1) favours the analysis of links between nodes and the assignment of weights to missing links. This modelling faithfully reflects the architecture of a collaborative recommendation system. The weight of the link measures the importance of an item and influences the prediction function.

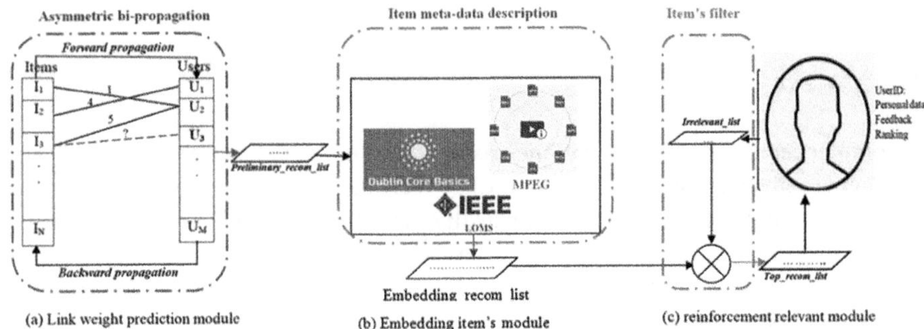


**Fig. 1.** (a) Link weight prediction module. (b) Embedding item's module. (c) reinforcement relevant module

In order to recommend the top items to a user (top recommended list), we structured our model into three sub-modules:

*Link weight prediction module:* This module is used to study network evolution and predict the weight of a link through a transitive forward-backward progression of neighbouring links. It produces a preliminary recommendation list (*preliminary_recom_list*).

*Meta-data standard for item embedding:* For each item in the recommendation list, we search for similar items using metadata such as (Dublin core, IEEE-Lom, MEPG, etc.) [16, 17]. This module ensures recommendation diversity and to offer the user a variety of items (book, movie, news, etc.).

*Negative judgment reinforcement module:* This module exploits user's rating feedback and utilises purely negative judgments to further filter the final recommendation list (*Top_recom_list*).

## 3.1  Link Weight Prediction

### 3.1.1  Link Weight Prediction in a Bipartite Graph

Several complex systems are modelled using mono-graph [18], which include a single type of nodes (links between atoms, web pages or social networks) or bi-graph, which include two types of nodes. A bipartite graph is a common model for recommendation systems [7, 8]. In a bipartite graph BG, there are two sets: set **I** of $N$ items (a first type of node) and set **U** of $M$ users (a second type of node). A link weight $w(t, u)$ represents a value of interest of an item $t$ to a user $u$. There is no link between nodes in the same set. The objective is to determine the missing link weights based on the existing ones. Additionally, it is worth mentioning that item represents a physical or abstract web resource (phone, vehicle, software, travelling, etc.) (Fig. 2).

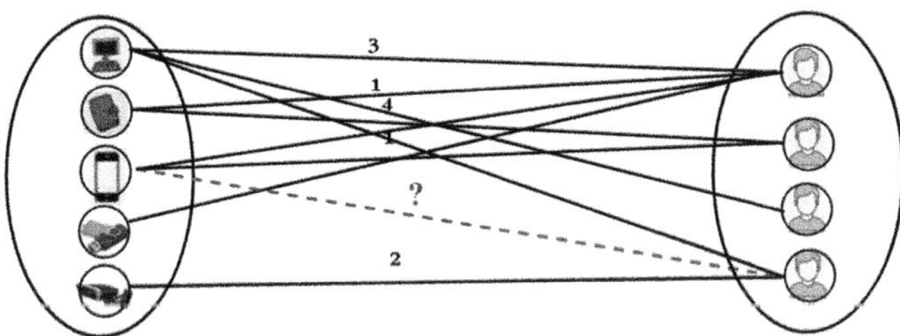


**Fig. 2.**  Weight item-user bipartite graph

Thus, the graph is composed of a set of items $I$, a set of users $U$ and a set of links $L$. Formally, this network $BG(I, U, L)$ is characterised by:

$$\begin{cases} L = \left\{ (t, u, w(t, u)), t \in I, u \in U, w(t, u) \in \mathbb{R}^+ \right\} \\ L = L^R \cup L^P \end{cases}$$

$L^R$: The set of links rated by users.
$L^P$ : The set of links predicted by the system.

### 3.1.2  Item Weight Link Prediction

In the field of collaborative recommendation, two major prediction approaches exist. The first, user-based [5], relies on the neighbourhood (similar ratings) of the active user. The second, item-based, depends on the neighbourhood of the active item. Numerous recommendation studies address the prediction of links on the user side. We have already demonstrated the advantage of an improvement provided by the item-based approach [4]. However, most research only predicts link existence (like, dislike, purchase, positive, negative, etc.) and overlooks link weight, which determines the degree of admiration.

The link weight prediction module focuses on a new approach to predicting link weights. We first explore network forward propagation, followed by backward propagation. The bi-embedding of the graph involves determining shared link weights on both the item and user sides and extracting hidden information. This idea is rooted in the network's inter-connectivity at time $t$. The question is: can we deduce the weights of probable links at $t + 1$?

Forward propagation phase: at a time $t$, each item $i$ has a shared proportion of its own links that indicates a cumulative importance distributed to users. Since the system includes $N$ items, $M$ users and weighted links $w(i, j)$, the shared proportion of the item i is given by the relationship:

$$SP(i) = \sum_{j=1}^{M} w(i, j) \tag{1}$$

The proportion dedicated to a user $j$ is expressed by the relationship:

$$UP(j) = \sum_{i=1}^{M} \frac{w(i, j)}{SP_i} \tag{2}$$

*Backword propagationphase:* This phase focuses on inverse propagation; it includes the cumulative importance of user weights and the proportion of each item. The cumulative importance of the weights of a user $j$ is expressed by:

$$WC(j) = \sum_{i=1}^{N} w(i, j) \tag{3}$$

The proportion dedicated to each item is represented by:

$$IP(i) = \sum_{j=1}^{M} \frac{w(i, j)}{WC(j)} \tag{4}$$

Using dual propagation, we can calculate the prediction of a link weight for item $t$ to user $u$ using the following equation:

$$P(t, u) = \sum_{i=1}^{N} \left[ \frac{1}{\sum_{j=1}^{M} w(i, j)} * \sum_{j=1}^{M} \frac{w(i, j) * w(t, j)}{\sum_{i=1}^{N} w(i, j)} \right] * w(i, u) \tag{5}$$

By replacing the shared proportion of item $i$ (1) and the cumulative weights of users (3), we obtain,

$$P(t, u) = \sum_{i=1}^{N} \left[ \frac{1}{SP(i)} * \sum_{j=1}^{M} *w(i, u) \frac{w(i, j) * w(t, j)}{WC(j)} \right] \tag{6}$$

Exploiting the bipartite graph using the bi-propagation weight method enabled us to calculate predictions of missing weight links at time $t + 1$. This process allowed us to extract hidden information between nodes. The following example is applied to a bipartite network with 4 items and 5 users. (Table 1) describes the link weights (ratings) with some missing weights.

To obtain predictions of the missing link weights, we apply the formulas established previously to find:

**Table 1.** Weight links in bipartite graph

| | $u_1$ | $u_2$ | $u_3$ | $u_4$ | $u_5$ | $SP(i)$ |
|---|---|---|---|---|---|---|
| $i_1$ | 2 | ? | 3 | 4 | 1 | 10 |
| $i_2$ | ? | 5 | 4 | 5 | 2 | 16 |
| $i_3$ | 3 | 4 | 2 | 3 | ? | 12 |
| $i_4$ | 1 | 4 | ? | ? | 3 | 8 |
| $WC(j)$ | 6 | 13 | 9 | 12 | 6 | |

Pred $(1,2) = 2.3704$; Pred $(2,1) = 1.9334$; Pred $(3,5) = 1.2208$; Pred $(4,3) = 1.0931$;

---

**Algorithm1 prediction missing weight link**

---

*Input* $[N, M]$ ←*size of [items, users];*
    *W* ← *Matrix of link weights w(i,j);*
*Output* *P  #Matrix prediction with missing link weights*
*# t item*
*# u user*
*P # initialisation*
*for t=1:N*
  *for u=1:M*
    *pi=0; # item prediction*
   *for i=1:N*
*sj=0;*
   *for j=1:M*
*sj=sj+W(i,j)*W(t,j)/sum(W(:,j)); #sharing propagation*
   *end*
 *pi=pi+1/sum(W(i,:))*sj*W(i,u);*

  *end*
*P(t,u)=pi*
 *end*
*end*
*return P*

---

This link prediction module takes into account the bi-propagation of quotas shared between graph nodes, which greatly reduces the sparsity rate.

## 3.2 Meta-Data for Item Embedding

Most recommendation systems deal with single types of items, such as Last.fm, Movie-Lens, Netflix andCDnow. However, users require a variety of items (books, films, news, blogs, etc.) in their daily lives [19, 20]. To measure the degree of similarity between items, we chose standard metadata to describe the items.

### 3.2.1  Elements Enhancement

Metadata is structured data (elements or attributes) used to describe an item [21, 22]. For each type of item, there is an associated metadata that includes a certain number of elements.

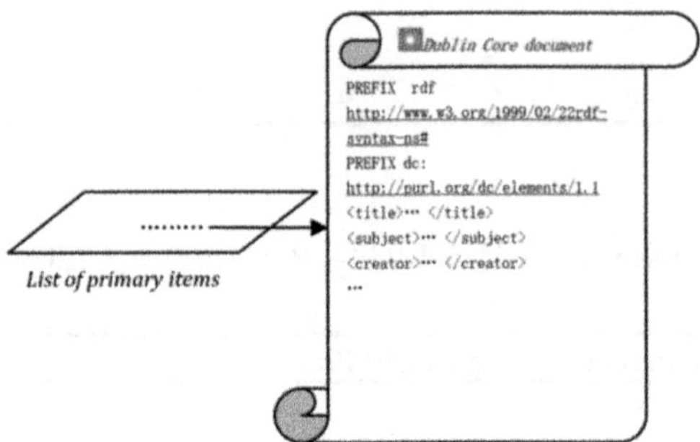


**Fig. 3.** Dublin Core item description

*MARC (Machine Readable Cataloging)* is the first set of metadata in XML format; it was developed by Congress to encode newspapers and facilitate their exchange.

*Dublin Core DC* (http://dublincore.org/documents/dces/) – a metadata for describing digital documents – has been adopted by several international agencies. In its extended version DC includes 22 elements (title, subject, creator, editor, contributor, language, cover, right, etc.). Figure 3 represents some of these elements (title, subject, creator, etc.) for describing an item in standard DC metadata.

*IEEE-LOM* is an XML metadata for describing educational and learning-based items.

*MPEG* – Moving Picture Experts Group – is used to describe audio, image and video resources using elements pertaining to compression, encoding, sampling, etc.

### 3.2.2  Item's Similarity

To calculate the similarity between two items $t_1$, $t_2$ when given a metadata standard MS composed of $K$ elements through this standard, we refer to the cosine measure:

$$SIM^{MS}(t_1, t_2) = \frac{\sum_{i=1}^{k} t_{1i}^{MS} \cdot t_{2i}^{MS}}{\sqrt{\sum_{i=1}^{k} (t_{1i}^{MS})^2} \cdot \sqrt{\sum_{i=1}^{k} (t_{2i}^{MS})^2}}$$
$$\begin{cases} if \quad SIM^{MS}(t_1, t_2) \geq \delta^{MS} \quad t_1 \quad similar\ to\ t_2 \\ else \quad t_1 \quad not\ similar\ to\ t_2 \end{cases} \tag{7}$$

where, $i$ represents the index of all the elements of the MS.

$\delta^{MS}$ is a system parameter which represents the similarity threshold of the MS. We note that normalisation of the elements of an MS is necessary to calculate the similarity between items. Practically, we are only interested in the relevant and normalisable attributes in the comparison, and assign 'null' values to superfluous elements. The diversity module is used to take each item from the preliminary list resulting from the first module and search for items that are similar to it by different MS.

### 3.3  Negative Ranking Reinforcement Module

In a recommendation system, each user possesses a profile that gradually develops based on their reactions and preferences. The profile may include a unique identifier ($ID_{user}$), individual information, interests, ranking feedback, etc. For each undesirable item, the algorithm seeks the most similar items to eliminate. The pseudo algo2 illustrates the steps for constructing the final recommendation list.

---

**Algorithm2:reinforcement recommendation**

---

*Input Irrelev_list IRL◇;*
*Diversity_recom_list DRL◇;*
*Output Top_recom_list TRL◇;*
*#ti item ∈ IRL*
*#sj item ∈ DRL*
*for each ti*
*norm(ti, ek) ; # elements normalized MS*
  *for each sj*
*norm(sj, ek);*

*while(MS(ti)==MS(sj))*
   *if (SIM(ti, sj)>= $\delta_{ir}$ # irrelevant threshold*
   *drop (sj)*
   *else*
   *add (sj) to TRL*
 *end*
 *end*
 *end*
 *return TRL*

---

## 4  Experiments

The aim of this section is to have clear answers to the questions below:

- How to explain the influence of neighbouring nodes on the interaction of an active node?
- Does the enrichment of the graph nodes with metadata diversify the recommendation?
- Does including negative judgments improve predicting quality?

### 4.1  Datasets

*Netflix Prize dataset:*  Is a chronological and multivariate dataset used in the competition for the best collaborative recommendation algorithm. The dataset consists of 480,000 customers, 17,000 films and around 100 million ratings.

*Million song dataset:*  It is a collection of metadata for one million music tracks, annotated in XML format for easy integration. The dataset is the result of a collaborative project between Echo Nest and LabROSA for research testing.

*Social Network Influencer:* This is a dataset based on Twitter activity, containing a pairwise preference machine learning task (the most influential between two individuals). The archive contains three csv files: *sample_predictions, train* and *test*.

*MovieLens:* It has several versions of different sizes, depending on the details of the files and their ratings. The full set includes 330,975 users, 86,000 movies and 33,000,000 ratings.

*Goawalla:* Is a dataset created using a mobile networking application that records places (nature, monuments, archaeological areas, etc.) visited by users. Users mark and share their favourite places using a public API provided by the company Goawalla. The dataset consists of 196 591 nodes, 950 327 edges and 6442890 records collected over 20 months.

## 4.2   Metrics

### 4.2.1   MAE (Mean Absolute Error)

This measures the prediction error, i.e., the difference between the system's prediction and the user's rating.

$$\text{MAE} = \frac{1}{n} \sum_{i=1}^{n} \left| r_{ij} - p_{ij} \right| \tag{8}$$

### 4.2.2   Recall

This measures the coverage rate of relevant items by measuring the ratio of relevant items selected by the system and comparing it to existing relevant items.

$$recall = \frac{SR}{R} \tag{9}$$

### 4.2.3   AUC (Area Under Curve)

Measures the overall performance of the model and takes into account the trade-off between sensitivity and specificity. This metric indicates the rate at which positive examples are correctly classified over negative examples.

## 4.3   Benchmark Methods

### 4.3.1   Benchmark1 (LinkFND)

In the study [21] the authors propose a framework based on the detection and minimisation of False Negatives for the task of link prediction in recommendation systems. The relationship between two nodes in the same group is more positive as long as there are as many neighbours in common, and in interaction, the relationship is more negative when there is a minimum of neighbours in common. This study creates a support view to calculate a similarity score SimGCL [22], which measures the degree of overlap between neighbouring nodes (positive or negative). The algorithm is applied to contrastive graph learning and tested on several datasets such as Gowalla, Yelp2018, Amazon-book, Alibaba.

### 4.3.2  Benchmark2 (AP_CF)

The study [23] is based on the incorporation of significant information on the nodes of the graph. In particular, focuses on the relationships and opinions in the user-item graph and adapts an attention process to differentiate the heterogeneous forces of social interactions. The study is based on a matrix factorisation technique that depends on a preference relation and exploits a preference aggregation strategy. It has been tested on MovieLens and Netflix dataset.

### 4.3.3  Benchmark3 (RS_GEm)

The research [24] relies on exploiting the user classification graph. The first step is to generate classes by ensemble learning in a heterogeneous information graph and integrate vector features. The most similar neighbours to the target user were ranked using ensemble learning, fuzzy rules and decision tree.

### 4.3.4  Benchmark4 (GN_SR)

This study [25] is done in two steps: the first uses matrix factorisation of preferences to obtain the top recommendation list, the second is based on graphical aggregation to maintain the regularity of preferences among similar users.

### 4.4  Results and Discussion

To measure the performance of our proposal, we divided the datasets into two parts: training (80%) and test (20%). Table 2 shows the performance of the different algorithms (Link_FND, AP_CF, RS_GMe, GN_SR) and that of our proposal (Bi-MeCp). In this experiment, we show the advantage of the "link weight prediction" module for the classification of predicted items based on the AUC metric.

**Table 2.**  Rate of AUC

| Datasets | LinkFND | AP_CF | RS_GEm | GN_SR | Bi − MeCp |
|---|---|---|---|---|---|
| NetFlixPrize | 0.7112 | 0.6842 | 0.6881 | 0.7343 | 0.7403 |
| MillionSong | 0.6602 | 0.6001 | 0.6723 | 0.7077 | 0.6981 |
| SocialNetwork | 0.7204 | 0.6701 | 0.6706 | 0.7461 | 0.7603 |
| MovieLens1M | 07423 | 0.6891 | 0.7182 | 0.7572 | 0.7810 |
| Goawalla | 0.7176 | 0.6809 | 0.7034 | 0.7350 | 0.7397 |
| **Average** | **0,71034** | **0,66488** | **0,69052** | **0,73606** | **0,74388** |

The results obtained show an advantage in favour of the proposal (Bi-MeCp) in terms of the classification rate of the predicted items, where the AUC reached 74.388%.

The MAE and RMSE are measures to test the performance of an algorithm. Figure 4(a) shows the evolution of the prediction error as a function of the size of similar neighbours in the social network dataset. The experiment shows the advantage of our Bi-MeCp method over the GN_SR method. The ideal size of similar neighbours is in the range of [35–50] neighbours. From [0–35] there is an insufficiency of similar neighbours, beyond 50 neighbours there is a slight degradation of neighbour similarity. For 40 neighbours, the MAE reached 0.75 for Bi-MeCp against 0.8199 for GN_SR.

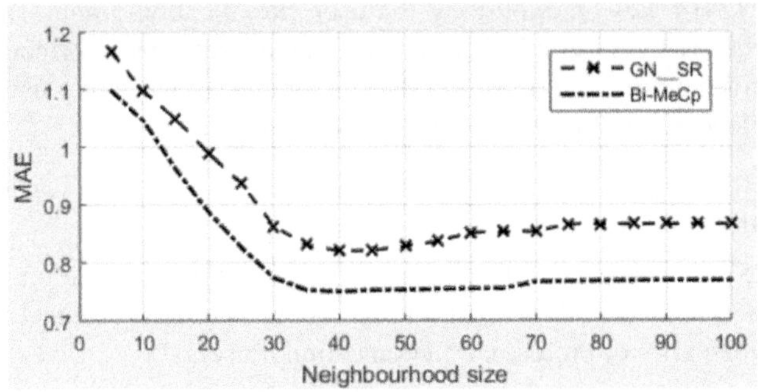


**Fig. 4.** **(a).** On Social Networks dataset

Figure 5(a) and (b) show the performance of the AP_CF, RS8GEm and Bi-MeCp methods using the MAE metric according to the MovieLens datasets 100K and 1M. For MovieLens 100K, the recorded RMSEs are 0.8776, 0.8092 and 0.7805 respectively.For a neighbourhood size of 60, AP_CF records a value of 0.869, while RS_GEm and Bi-MeCp record values of 0.8 and 0.7805, respectively, for a neighbourhood size of 55.Note that the 100K set has an advantage over 1M due to the sparsity rate, i.e. 100K is less sparse than 1M.

The experiments shown in Figs. 6 and 7 measure the recall and precision rates for the three algorithms. We can see that the Bi-MeCp algorithm achieves the best recall and accuracy rates (76.5%, 84.53%). The RS_GEm algorithm comes in second place, with recall and precision rates of 71.175% and 79.8%, respectively. AP_CF recorded rates (66.5%, 76.75%) for recall and precision. These results justify the proposal's theoretical and practical foundations which are based on propagation and metadata standards.

The latest experience is to visualise the variety of recommendations through the Amazon product reviews dataset. We focused on the different metadata in order to use the attributes of the items to diversify the predictions. The table below (Table 3) illustrates the percentage of recommendation for each type of item by our method (Bi-MeCp). Similarity calculations are based on the attributes of each metadata standard depending on the type of item. We filter the percentage of each category (movies, song, books and others) using the category field of the dataset and measure the recommendation accuracy by RMSE for different neighbourhood sizes.

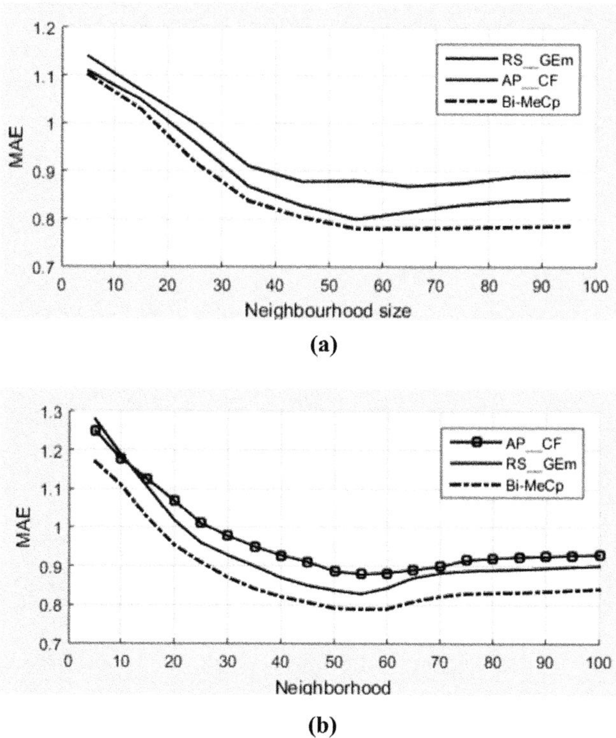

**(a)**

**(b)**

**Fig. 5.** **(a).** MovieLens 100K. **(b).** MovieLens 1M.

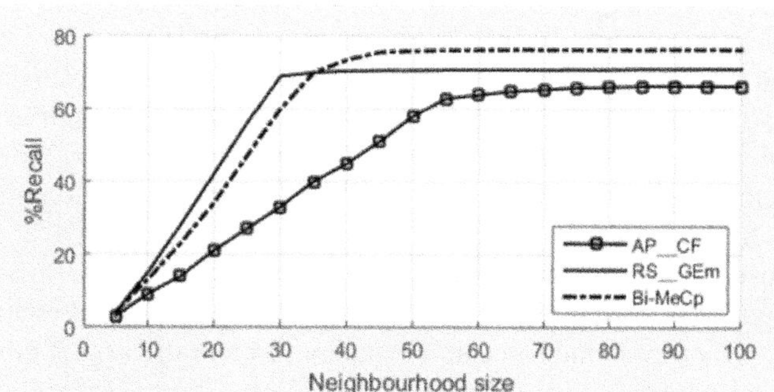

**Fig. 6.** Recall rate

The results of this experiment demonstrate the adaptability of our method to multivariate recommendation in response to frequent user requests. Users can choose between similar items based on criteria like quality and price.

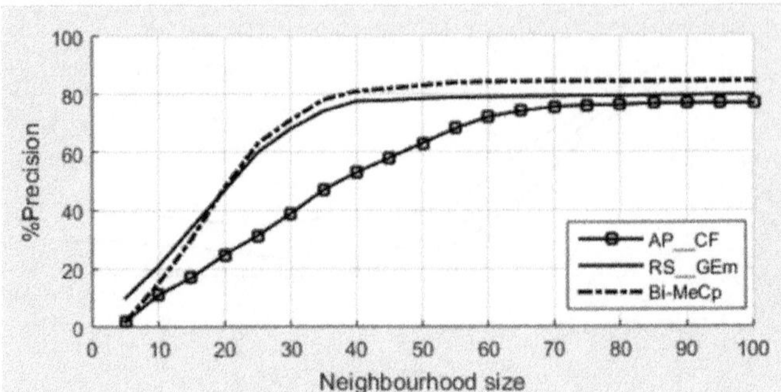


**Fig. 7.** Precision rate

**Table 3.** Rate of item types

| Neighborhood size | *Movies* | *Song* | *Books* | *Others* | *RMSE* |
|---|---|---|---|---|---|
| #20 | 19.248% | 1.465% | 7.074% | 72.213% | **0.987** |
| #30 | 19.107% | 1.628% | 7.988% | 71.277% | **0.902** |
| #40 | 19.093% | 1.631% | 7.991% | 71.285% | **0.851** |
| #50 | 19.095% | 1.630% | 7.991% | 71.284% | **0.854** |
| #60 | 19.097% | 1.630% | 7.992% | 71.281% | **0.870** |

# 5   Conclusion

Improving the performance of a recommendation system is essential for a large community of Internet users. In this work, we focused on a collaborative recommendation system using a bipartite and weighted graph topology, facilitating implementation and interpretation. Finding the weight of a non-existent link depends on the previous state of the graph, and this same link is exploited for a future state. In the first part of the solution, we discriminated the asymmetric interest of an item, noted by a set of users, and the interest of a user towards items noted by transitive propagation. Solution avoids the arduous and complex task of calculating similarity by Pearson correlation or Cosine. Additionally, we propose multi-recommendation through the exploitation of standard and normalised metadata for the description of items, particularly films, books and music. This architecture is flexible and standardised, easily applicable to any type of item and ensures varied recommendations. Although the bipartite graph representation is a faithful model of a collaborative filtering system, the integration of item/user features allows aggregation and transfer of messages between nodes. The integration of item/user features allows the aggregation and transmission of messages between nodes. However, the prediction of links is related to the quality and quantity of data of each node or stop. In a more sparse graph, some links may be missing, substituted or incorrect (noisy networks). On the other hand, most of the experiments in the studies are limited to two layers. To benefit from good deep learning, the model needs to be trained in several layers.

# References


1. Arnold, M., Goldschmitt, M., Rigotti, T.: Dealing with information overload: a comprehensive review. Front. Psychol. (2023). https://doi.org/10.3389/fpsyg.2023.1122200
2. Liu, Q., et al.: Multimodal recommender systems: a survey. ACM Comput. Surv. (2024). https://doi.org/10.1145/3695461
3. A survey on accuracy-oriented neural recommendation: from collaborative filtering to information-rich recommendation. https://ieeexplore.ieee.org/abstract/document/9693280.
4. Su, X., Khoshgoftaar, T.M.: A survey of collaborative filtering techniques. Adv. Artif. Intell. **2009**, 1–19 (2009). https://doi.org/10.1155/2009/421425
5. Kluver, D., Ekstrand, M.D., Konstan, J.A.: Rating-based collaborative filtering: algorithms and evaluation. In: Lecture Notes in Computer Science. pp. 344–390 (2018). https://doi.org/10.1007/978-3-319-90092-6_10
6. Najafabadi, M.K., Mahrin, M.N.: A systematic literature review on the state of research and practice of collaborative filtering technique and implicit feedback. Artif. Intell. Rev. **45**, 167–201 (2015). https://doi.org/10.1007/s10462-015-9443-9
7. Wu, Z., Song, C., Chen, Y., Li, L.: A review of recommendation system research based on bipartite graph. MATEC Web Conf. **336**, 05010 (2021). https://doi.org/10.1051/matecconf/202133605010
8. Giamphy, E., Guillaume, J.-L., Doucet, A., Sanchis, K.: A survey on bipartite graphs embedding. Soc. Netw. Anal. Min. (2023). https://doi.org/10.1007/s13278-023-01058-z
9. Torkashvand, A., Jameii, S.M., Reza, A.: Deep learning-based collaborative filtering recommender systems: a comprehensive and systematic review. Neural Comput. Appl. **35**, 24783–24827 (2023). https://doi.org/10.1007/s00521-023-08958-3
10. Papadakis, H., Papagrigoriou, A., Panagiotakis, C., Kosmas, E., Fragopoulou, P.: Collaborative filtering recommender systems taxonomy. Knowl. Inf. Syst. **64**, 35–74 (2022). https://doi.org/10.1007/s10115-021-01628-7
11. Gao, C., et al.: A survey of graph neural networks for recommender systems: challenges, methods, and directions. ACM Trans. Recommender Syst. **1**, 1–51 (2023). https://doi.org/10.1145/3568022
12. Wu, S., Sun, F., Zhang, W., Xie, X., Cui, B.: Graph neural networks in recommender systems: a survey. ACM Comput. Surv. **55**, 1–37 (2022). https://doi.org/10.1145/3535101
13. Wang, Y., et al.: Enhancing recommender systems with large language model reasoning graphs. https://arxiv.org/abs/2308.10835.
14. Khan, Z.Y., Niu, Z., Yousif, A.: Joint deep recommendation model exploiting reviews and metadata information. Neurocomputing **402**, 256–265 (2020). https://doi.org/10.1016/j.neucom.2020.03.075
15. Reer, A., Wiebe, A., Wang, X., Rieger, J.W.: FAIR human neuroscientific data sharing to advance AI driven research and applications: legal frameworks and missing metadata standards. Front. Genet. (2023). https://doi.org/10.3389/fgene.2023.1086802
16. Wierling, A., et al.: FAIR metadata standards for low carbon energy research—a review of practices and how to advance. Energies **14**, 6692 (2021). https://doi.org/10.3390/en14206692
17. Sarah, C., Jane, B., Rónán, O., Ben, R.: Quality assurance for digital learning object repositories: issues for the metadata creation process. ALT-J. **12**, 5–20 (2004). https://doi.org/10.1080/0968776042000211494

18. Fan, W., et al.: A Graph Neural network framework for social recommendations. IEEE Trans. Knowl. Data Eng. **34**, 2033–2047 (2020). https://doi.org/10.1109/tkde.2020.3008732
19. Mitropoulou, K., Kokkinos, P., Soumplis, P., Varvarigos, E.: Anomaly detection in cloud computing using knowledge graph embedding and machine learning mechanisms. J. Grid Comput. (2023). https://doi.org/10.1007/s10723-023-09727-1
20. Hammi, S., Hammami, S.M., Belguith, L.H.: Advancing aspect-based sentiment analysis with a novel architecture combining deep learning models CNN and bi-RNN with the machine learning model SVM. Soc. Netw. Anal. Min. (2023). https://doi.org/10.1007/s13278-023-011 26-4
21. Sanghun, K., Hyeryung, J.: Simple framework for false negative detection in recommendation tasks with graph Contrastive Learning. https://ieeexplore.ieee.org/document/10368024.
22. Yu, J., Yin, H., Xia, X., Chen, T., Cui, L., Nguyen, Q.V.H.: Are graph augmentations necessary: Simple graph contrastive learning for recommendation. In: Proc. 45th Int. ACM SIGIR Conf. Res. Develop. Inf. Retr., pp. 1294–1303 (July 2022)
23. Pujahari, A., Sisodia, D.S.: Aggregation of preference relations to enhance the ranking quality of collaborative filtering based group recommender system. Expert Syst. Appl. **156**, 113476 (2020). https://doi.org/10.1016/j.eswa.2020.113476
24. Forouzandeh, S., Berahmand, K., Rostami, M.: Presentation of a recommender system with ensemble learning and graph embedding: a case on MovieLens. Multimedia Tools Appl. **80**, 7805–7832 (2020). https://doi.org/10.1007/s11042-020-09949-5
25. Fan, W., et al.: Graph neural networks for social recommendation. https://arxiv.org/abs/1902. 07243.

# Apple Leaf Disease Classification Using Deep Learning Approach on Lab-Built and Realistic Datasets


Mohamed Reda Lakehal[(✉)] and Youcef Ferdi

Bioengineering Laboratory, National Higher School of Biotechnology, Constantine, Algeria
{m.lekhal,y.ferdi}@ensbiotech.edu.dz



**Abstract.** To study diseases that can affect apple leaves, such as cedar rust and scab, some datasets are available in the literature. These datasets have been annotated and labeled with the assistance of experts in the field. Two main categories of datasets can be distinguished. The first category involves Lab-built Datasets which are constructed under controlled conditions, often featuring images taken against a compact background that does not accurately reflect real-world environments. Models trained on these datasets may perform well in controlled settings but often face challenges in practical applications. The second category includes datasets collected in realistic field conditions, capturing the variability and complexity of natural environments. The performance of deep learning algorithms can be significantly affected by the differences in image quality and context between these two types of datasets. In this study, two specific datasets were utilized: the Plant Village dataset and the Plant Pathology dataset, focusing on apple leaves categorized into three classes: healthy, rust, and scab. The study employed Convolutional Neural Networks (CNNs) for classification, experimenting with both models trained from scratch and transfer learning models, including VGG16, InceptionV3, MobileNetV2, and Xception. The experimental results indicated that the MobileNetV2 model performed exceptionally well when applied to the Plant Village dataset, which is likely due to its efficiency and ability to generalize from the lab-built images. Conversely, the custom CNN model trained from scratch showed superior performance on the Plant Pathology dataset, which consists of more realistic images, suggesting that it was better suited to handle the complexities of real-world data.

**Keywords:** Deep learning · convolutional neural network · transfer learning · plant disease classification · apple leave disease


## 1 Introduction

The plant diseases can affect in drastic way the quantity and the quality of the yield, leading to huge losses of the corps, with a negative financial impact on the farmers and government incomes. Therefore, recently the plant disease research attracts more attention and become a research hotspot [1]. There are mainly two approaches for detection


© The Author(s), under exclusive license to Springer Nature Switzerland AG 2025
N. Seddari and M. Redjimi (Eds.): ICMSCT 2024, CCIS 2606, pp. 321–331, 2025.
https://doi.org/10.1007/978-3-032-01922-6_27

and classification of plant diseases. The first one is considered as traditional approach that lay on agriculture experts who inspect visually or microscopically the field crops, but this identification is laborious, high-cost expenses and time consuming. The second approach is based on the use of imaging technology and artificial intelligence, which is considered more effective and efficient.

Many works based on the second approach have been proposed in literature to address the problem of plant disease recognition and classification. Some studies have used Machine Learning methods such as Support Vector Machine [2, 3]. An automatic detection and classification of plant leaf diseases is carried out based on Color Co-occurrence Matrix (CCM) and neural network classifier [4]. Alternatively, for the same purpose, it is proposed in [5] the use of K-Means based segmentation and neural networks for classification. An embedded image recognition technology is proposed in [6] for wheat leaf rust disease detection. An integral intelligent system is proposed in [7] for the diagnosis and control of tomato diseases and pests in hydroponis greenhouses. Based on Deep Learning (DL) architectures [8], many works can also be found in the literature for the purpose of identification and classification of plants' disease. In [9], the recognition of plant diseases is implemented via a convolutional neural network, which was able to distinguish plant leaves from their surroundings. In [10], based on AlexNet and GoogLeNet's inception networks, a novel architecture is proposed to classify four apple leaf diseases using a dataset of synthetic images. In the same perspective, an improved version of VGG-16 network is proposed in [11], where the model was trained on lab-built dataset. Also, in [12], where the identification of the tomato leaf diseases is implemented via basic convolutional neural network (CNN) architecture like AlexNet, GoogLeNet and ResNet. LeNet architecture was implemented in order to detect the diseases in banana leaf, the evaluation of the model in Color and Gray Scale modes used the classification accuracy and F1-score [13]. According to [14], the detection of cassava disease is implemented using Inception-v3 DL model. In [15], the two basic versions of CNN were used to classify the plant diseases in cucumber. AlexNet and SqueezeNet v1.1 models were used for the classification of tomato plant disease [16]. Moreover, in [17], AlexNet and VGG-16 DL architectures were used to classify six tomato plant diseases.

This paper outlines a comparative analysis regarding the classification of apple leaf diseases (healthy, rust, and scab) using two types of datasets: lab-built (Plant Village) and realistic (Plant Pathology), using the most recent Convolutional Neural Network (CNN) architectures, based on transfer learning, namely VGG16, InceptionV3, MobileNetV2 and Xception. While these models perform well on lab-built datasets, they show significant performance degradation when applied to realistic images from the Plant Pathology dataset [18]. This highlights a common challenge in deep learning where models trained on idealized data struggle with variability in real-world data. To address the performance issues observed with realistic datasets, we have proposed a new model developed from scratch. This model is inspired by the VGG16 architecture but features lighter architecture blocks and lower computational time. This adaptation aims to enhance performance specifically when dealing with realistic images from the field.

The remaining sections of the paper are structured as follows: Sect. 2 provides a brief description of the datasets and the different deep learning architecture used. In Sect. 3, we explore the application of the different deep learning architectures on a lab-built and realistic datasets. Lastly, the conclusion will be presented in Sect. 4.

## 2   Materials and Methods

### 2.1   Datasets

In this study, we have used two datasets, namely Plant Village dataset and Plant pathology dataset.

- **Plant Village Dataset**

The main characteristic of Plant Village dataset is that all leaf images background is uncluttered (black or gray). The camera used for the acquisition task is a Sony DSC—Rx100/13, with 20.2 megapixels using the automatic mode. The images of the dataset include healthy and diseased leaves of 14 crop species: tomato, strawberry, cherry, corn, grape, orange, peach, bell pepper, potato, raspberry, soybean, squash, blueberry and apple.

- **Plant Pathology Dataset**

The Plant Pathology dataset [19] is composed of 3651 RGB images of multiple apple foliar disease symptoms with realistic background. To shot the photos of the dataset, a camera Canon Rebel T5i DSLR and smartphones are used under various noise conditions. The annotations of the dataset are driven manually and confirmed by an expert plant pathologist.

### 2.2   Deep Learning Networks

To conduct our comparative study, we have adopted four well-known convolutional neural network architectures, namely VGG-16, InceptionV3, MobileNetV2 and Xception, which are well explained in [20, 21]. As for the proposed CNN model from the scratch, it was inspired by VGG16 architecture with lighter compounding blocks, containing a stack of 6 convolutional layers, each of which is followed by a Max Pooling layer. The final Max Pooling layer is followed by flatten layer and two fully connected (FC) layers as depicted in Fig. 1.

| Layer (type) | Output Shape | Param # |
|---|---|---|
| sequential (Sequential) | (16, 224, 224, 3) | 0 |
| conv2d (Conv2D) | (16, 222, 222, 32) | 896 |
| max_pooling2d (MaxPooling2D) | (16, 111, 111, 32) | 0 |
| conv2d_1 (Conv2D) | (16, 109, 109, 64) | 18,496 |
| max_pooling2d_1 (MaxPooling2D) | (16, 54, 54, 64) | 0 |
| conv2d_2 (Conv2D) | (16, 52, 52, 64) | 36,928 |
| max_pooling2d_2 (MaxPooling2D) | (16, 26, 26, 64) | 0 |
| conv2d_3 (Conv2D) | (16, 24, 24, 64) | 36,928 |
| max_pooling2d_3 (MaxPooling2D) | (16, 12, 12, 64) | 0 |
| conv2d_4 (Conv2D) | (16, 10, 10, 64) | 36,928 |
| max_pooling2d_4 (MaxPooling2D) | (16, 5, 5, 64) | 0 |
| conv2d_5 (Conv2D) | (16, 3, 3, 64) | 36,928 |
| max_pooling2d_5 (MaxPooling2D) | (16, 1, 1, 64) | 0 |
| flatten (Flatten) | (16, 64) | 0 |
| dense (Dense) | (16, 64) | 4,160 |
| dense_1 (Dense) | (16, 3) | 195 |

**Fig. 1.** CNN Model from the scratch architecture.

## 3 Results and Discussions

To assess the performance of the CNN deep learning models on apple leaf datasets taken from both Plant village (1890 apple leaf images) and from plant pathology databases (1548 apple leaf images), we developed Python scripts using Keras 2.15.0 library built on the TensorFlow 2.15.0 framework. All experiments were conducted on a laptop equipped with 3.0 GHz Intel I9, and 16 GB of RAM.

Three ground truths are involved for the classification task: "healthy", "scab" and "rust". The dataset was divided into training, validation and test datasets with a ratio of 8:1:1, respectively. All images were resized to $224 \times 224 \times 3$ that complies with the shape expected by the network and also all of them are scaled to have values between 0 and 1. During the training phase, the data augmentation technique was applied, including horizontal and vertical flip for the purpose to manage the unbalance of the classes and improve the model robustness.

Both CNN models trained from scratch and those developed using transfer learning were applied. In the transfer learning approach, pre-trained CNN models originally trained on the ImageNet dataset [22] were adapted during the training phase. The hyperparameters configurations used are presented in Table 1. Once the model weights are adjusted to achieve the lowest validation loss, the RMSprop optimizer is used to record the obtained model's performance in terms of accuracy, loss and training/test time. In the first task, we start by applying the proposed four transfer learning models namely, VGG16, InceptionV3, MobileNetV2, Xception along with the proposed CNN from the scratch on Plant village dataset. Based on Table 2, VGG16, InceptionV3, MobileNetV2, Xception, the obtained accuracy results for the test set are 98.75%, 97.27%, 100% and 99.80% respectively, while the proposed CNN model achieved an accuracy of 99.81%. The comparison of the scores obtained between the training, validation and test sets, show that the aforementioned models exhibit high accuracy accompanied with low loss function values. Slight advantage is recorded in favor of MobileNetV2. The corresponding accuracy value for this architecture related to training after the 40th epoch reaches 99.32% as can be observed in Fig. 2. a, while the results of the Loss values are presented in Fig. 2. b, getting values close to zero, with some oscillation values. The computational performance is an important factor in models ranking. As shown in Table 3, the details of the training/test time required for each model are provided, where it appears that MobileNetV2 required shorter time frames, followed by InceptionV3, the proposed CNN model, Xception, and on the last position came the VGG-16 with its big architecture. From the results we can see that the MobileNetV2 outperforms the other models, by presenting a trade-off between computational time and performance in classifying apple leaf images with an uncluttered background from plant village dataset. In the second task, we applied the same models' architectures but this time on plant pathology dataset with a realistic background image of apple leaf. According to Table 4, VGG16, InceptionV3, MobileNetV2, Xception obtained accuracy results for the test set of 74.74%, 90.49%, 89.49% and 89.39% respectively, while the proposed CNN model performed better with an accuracy of 97.16%. The obtained results in this second experiments, either on training, validation or test sets, show that all the proposed models showed lower accurate results, compared to the proposed CNN from the scratch, which accuracy values during training phase after the 40th epoch reaches 97.21% as observed in Fig. 3. a, while the results of the Loss values going downward to zero with less oscillations are presented in Fig. 3. b. For the computational performance evaluation of the models, the top-ranking position is attributed to the CNN from the scratch with lower test times, as shown in Table 5.

Table 1. Models hyperparameters.

| Parameters | Values |
| --- | --- |
| Optimizer | RMSprop |
| Loss function | Cross-entropy |
| Learning rate | **2e-5** |
| Batch size | 16 |

(continued)

Table 1. (continued)

| Parameters | Values |
| --- | --- |
| epochs | 40 |

**Table 2.** Accuracies and Loss obtained with in the Plant village dataset (training/validation/test set).

| | Train Data Set | | Val Data Set | | Test Data Set | |
| --- | --- | --- | --- | --- | --- | --- |
| | Loss | Accuracy | Loss | Accuracy | Loss | Accuracy |
| CNN from the scratch | 0.0529 | 0.9821 | 0.0240 | 0.9886 | 0.0060 | 0.9981 |
| VGG16 | 0.0284 | 1.0000 | 0.0102 | 1.0000 | 0.0159 | 0.9875 |
| InceptionV3 | 0.7774 | 0.9375 | 3.6680e-08 | 1.0000 | 0.0436 | 0.9727 |
| MobileNetV2 | 0.0279 | 0.9932 | 0.0323 | 0.9943 | **0.0029** | **1.0000** |
| Xception | 2.5851e-05 | 1.0000 | 0.0201 | 1.0000 | 0.0188 | 0.9980 |

Table 3. Models computational time performance (Plant village Datasets).

| Models | Train time [sec] | Test time [sec] |
| --- | --- | --- |
| CNN from the scratch | 354.9859 | 0.6309 |
| VGG16 | 1374.4456 | 7.1726 |
| InceptionV3 | 314.2792 | 1.6804 |
| MobileNetV2 | **135.0196** | **0.78016** |
| Xception | 594.6289 | 3.13671 |

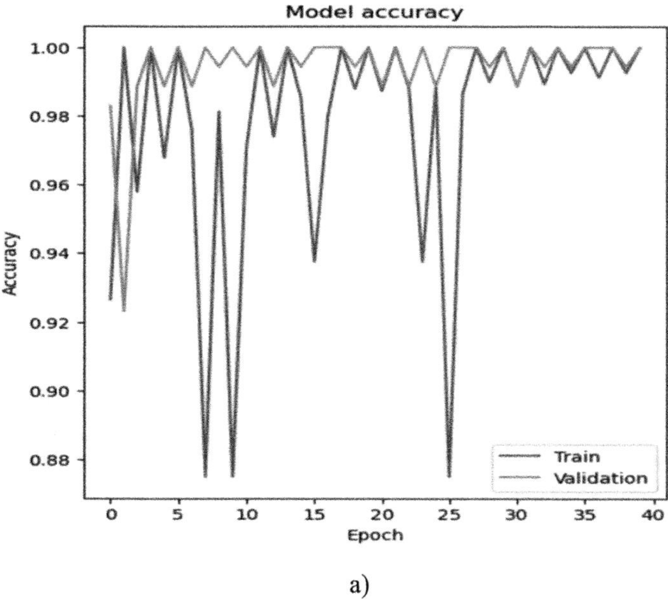


a)

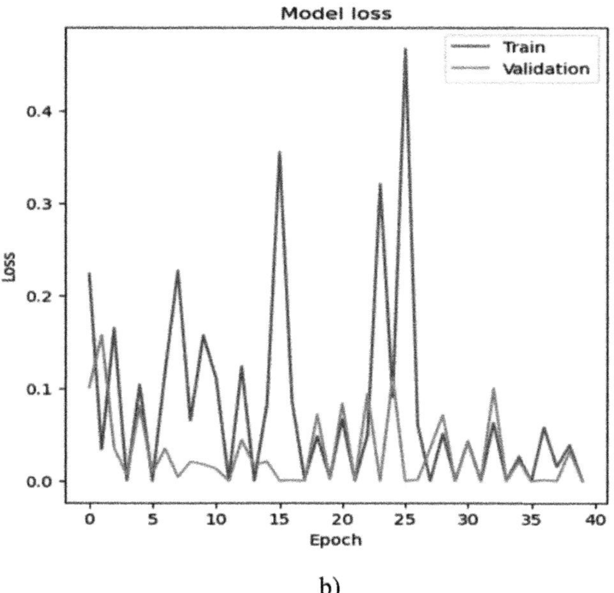


b)

**Fig. 2.** The Plot of Accuracy and Loss using MobileNetV2 (Plant village Datasets).

**Table 4.** Accuracies and Loss obtained with in the plant pathology dataset (training/validation/test set).

|  | Train Data Set | | Val Data Set | | Test Data Set | |
|---|---|---|---|---|---|---|
|  | Loss | Accuracy | Loss | Accuracy | Loss | Accuracy |
| CNN from the scratch | 0.0602 | 0.9721 | 0.1171 | 0.9653 | **0.1114** | **0.9716** |
| VGG16 | 0.3584 | 0.8678 | 1.2672 | 0.7153 | 0.8240 | 0.7474 |
| InceptionV3 | 0.1825 | 0.9375 | 0.3149 | 0.8333 | 0.7891 | 0.9049 |
| MobileNetV2 | 0.1368 | 0.9708 | 1.3201 | 0.9236 | 0.4255 | 0.8949 |
| Xception | 0.9812 | 0.8125 | 0.2723 | 0.8333 | 0.5657 | 0.8939 |

**Table 5.** Models computational time performance (Plant pathology Datasets).

| Models | Train time [sec] | Test time [sec] |
|---|---|---|
| CNN from the scratch | **299.1190** | **1.1129** |
| VGG16 | 1263.29106 | 7.23000 |
| InceptionV3 | 725.3115 | 4.52200 |
| MobileNetV2 | 600.9792 | 3.92101 |
| Xception | 893.7637 | 5.38900 |

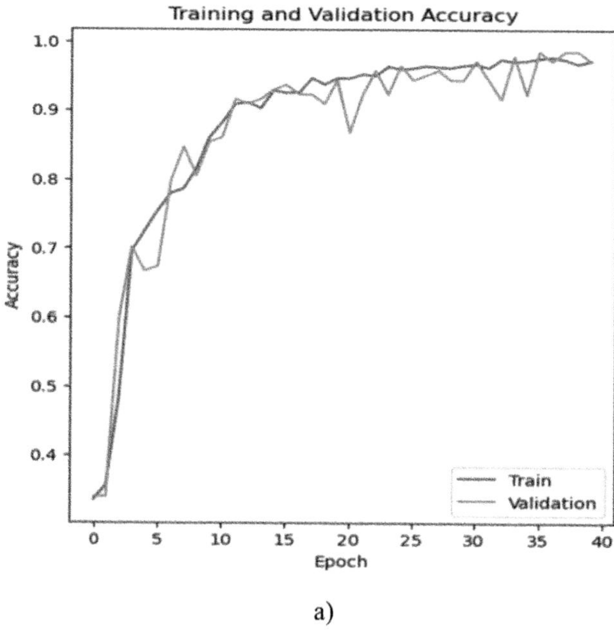


a)

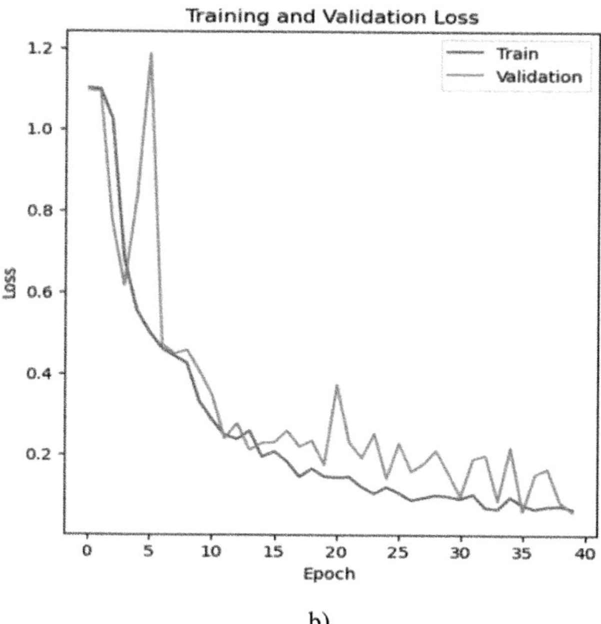


b)

**Fig. 3.** The Plot of Accuracy and Loss using CNN from the scratch (Plant pathology Datasets).

## 4  Conclusion

In this paper, we investigated the effectiveness of various CNN architectures, including both transfer learning models (VGG16, InceptionV3, MobileNetV2, and Xception) and a custom CNN model developed from scratch to perform apple leaf disease classification from both uncluttered images and images with complex background. The findings indicate that the MobileNetV2 model excels in both performance metrics and computational efficiency when applied to the Plant Village dataset, making it a strong choice for this type of data. In contrast, when the models are tested on the Plant Pathology dataset, the custom CNN model developed from scratch demonstrates a more balanced performance, suggesting that it is better suited to the challenges of realistic field images.

## References


1. Fenu, G., Malloci, F.M.: Forecasting plant and crop disease: an explorative study on current algorithms. Big Data Cogn. Comput **5**, 2 (2021)
2. Padol, P.B., Yadav, A.: A SVM classifier based grape leaf disease detection. In: Proceedings of the 2016 Conference on Advances in Signal Processing (CASP), Pune, India, pp. 175–179, (2016)
3. Islam, M., Dinh, M., Wahid, A., Bhowmik, K.P.: Detection of potato diseases using image segmentation and multiclass support vector machine. pp. 1–4 (2017)
4. Al-Hilary, H., Bani-Ahmad, S., Reyalai, M., Braik, M., Rahamneh, A.L.: Fast and accurate detection and classification of plant diseases. Int. J. Comput. Appl. (2011)
5. Al Bashidh, D., Braik, M., Ahmad, S.B.: Detection and classification of leaf diseases using k-means-based segmentation and neural-networks-based classification. Inf. Technol. J. (2011)
6. Xu, P., Wu, G., Guo, Y., Chen, X., Yanh, H., Zhang, R.: Automatic wheat leaf rust detection and grading diagnosis via embedded image processing system. Procedia Comput. Sci. **107**, 836–841 (2017)
7. Lopez-Morales, V., Lopez-Ortega, O., Ramos-Fernandez, J., Munoz, L.B.: JAPIEST: an integral intelligent system for the diagnosis and control of tomatoes disease and pests in hydroponis greenhouses. Expert Syst. Appl. **35**, 1506–1512 (2008)
8. Chen, J., Chen, J., Zhang, D., Sun, Y., Nanehkaran, Y.A.: Using deep transfer learning for image-based plant disease identification. Comput. Electron. Agric. **173**, 105393 (2020)
9. Sladojevic, S., Arsenovic, M., Anderla, A., Culibrk, D., Stefanovic, D.: Deep neural networks-based recognition of plant diseases by leaf image classification. Comput. Intell. Neurosci. (2016)
10. Liu, B., Zhang, Y., He, D., Li, Y.: Identification of apple leaf diseases based on deep convolutional neural networks. Symmetry **10**, 11 (2018)
11. Yan, Q., Yang, B., Wang, W., Wang, B., Chen, P., Zhang, J.: Apple leaf diseases recognition based on an improved convolutional neural network. Sensors **20**, 3535 (2020)
12. Zhang, K., Wu, Q., Liu, A., Meng, X.: Can deep learning identify tomato leaf disease? Adv. Multimed. (2018)
13. Amara, J., Bouaziz, B., Algergawy, A.: A deep learning-based approach for banana leaf diseases classification. In: Proceedings of the BTW (Workshops), Stuttgart, Germany, pp. 79–88 (2017)
14. Ramcharan, A., Baranowski, K., McCloskey, P., Ahmed, B., Legg, J., Hughes, D.P.: Deep learning for image-based cassava disease detection. Front. Plant Sci. **8**, 1852 (2017)

15. Fujita, E., Kawasaki, Y., Uga, H., Kagiwada, S., Iyatomi, H.: Basic investigation on a robust and practical plant diagnostic system. In: Proceedings of the 2016 15th IEEE International Conference on Machine Learning and Applications (ICMLA), Anaheim, CA, USA, pp. 989–992, (2016)

16. Durmus, H., Günes, E.O., Kırcı, M.: Disease detection on the leaves of the tomato plants by using deep learning. In: Proceedings of the 2017 6th International Conference on Agro Geoinformatics, Fairfax, VA, USA, pp. 1–5 (2017)

17. Rangarajan, A.K., Purushothaman, R., Ramesh, A.: Tomato crop disease classification using pre-trained deep learning algorithm. Procedia Comput. Sci. **133**, 1040–1047 (2018)

18. Fenu, G., Malloci, F.M.: Evaluating impacts between laboratory and field-collected datasets for plant disease classification. Agronomy **12**(10), 2359 (2022)

19. Thapa, R., Zhang, K., Snavely, N., Belongie, S., Khan, A.: The plant pathology challenge 2020 data set to classify foliar disease of apples. ArXir preprint, arXir: 2004. 11958 (2020)

20. Fenu, G., Malloci, F.M.: Using multi-output learning to diagnose plant disease and stress severity. Complexity **2021**, 18 (2021)

21. Chollet, F.X.: Deep learning with depth wise separable convolutions. In: Proceedings of the IEEE conference on computer vision and pattern recognition. pp. 1251–1258 (2017)

22. Deng, J., Dong, W., Socher, R., Li, L., Li, K., Fei-Fei, L.: Imagenet: a large-scale hierarchical image database. In: Proceedings of the IEEE Conference on Computer Vision and Pattern Recognition, pp. 248–255, Miami, FL, USA (2009)

# Modeling and Analysis of the Landing Gear System with the Generalized Contracts


A. Abdelkader Khouass[1,2(✉)], J. Christian Attiogbé[1], Mohamed Messabihi[2],
and Abdelkrim Benamar[2]

[1] University of Nantes, LS2N CNRS UMR 6004, Nantes, France
abderrahmaneabdelkader.khouass@univ-tlemcen.dz,
christian.attiogbe@univ-nantes.fr
[2] University Abou Bekr Belkaid, LRIT, Tlemcen, Algeria
{mohamedelhabib.messabihi,a_benamar}@univ-tlemcen.dz



**Abstract.** Complex systems exist in different fields such as aeronautics, autonomous vehicles, nuclear power plants, etc. These systems do not have a precise perimeter, they are open and made of various specific components built with different languages and environments. The modeling, assembly and analysis of such open and complex heterogeneous systems are challenges in software engineering. The Minarets method proposes solutions to these challenges. This paper describes how the Minarets method decreases the difficulty of modeling and analysis of the case study of the landing gear system. Applying the method consists in: equipping individual components with *generalized contracts* that integrate various *facets* related to different concerns, composing these components according to their facets and verifying the resulting system with respect to the involved facets as well. We have therefore taken into account the different facets in the case study, like data for the communication between the components, time, safety and liveness. This study helps us to validate the Minarets method, and can also serve as a working guide for other developers.

**Keywords:** Heterogeneous systems · Modeling · Verification · Contracts · Landing gear system


## 1 Introduction

Complex systems exist in various fields such as (air traffic control, nuclear power plants, railway, autonomous vehicles, etc.); these systems are assembled by a huge number of hardware/software components. Their complexity forces one to have a wide variety of heterogeneous components, due to the requirements.

The components may cover different *facets* and they may be composed in different manners. Therefore, if the integration of components into a global system is not mastered, it may generate considerable time losses and overcharges due to inconsistency of requirements, incompatibility of meaning and properties, late detection of composition errors, etc. For these reasons, the modeling, the composition and formal analysis of such heterogeneous systems are challenging. The use of efficient methods and techniques is required to face these challenges.


© The Author(s), under exclusive license to Springer Nature Switzerland AG 2025
N. Seddari and M. Redjimi (Eds.): ICMSCT 2024, CCIS 2606, pp. 332–347, 2025.
https://doi.org/10.1007/978-3-032-01922-6_28

We aim at studying and alleviating the difficulties of practical modeling and integration of heterogeneous components.

This paper deals with a simplified version of the landing gear system (LGS) presented in [6]. The LGS is a complex system composed from various components that communicates and it covers different *facets* (data, functionality, time, safety, liveness). Moreover, some tools/languages are not made to model and analyse some facets like time and liveness. The modeling and analysis of such system are challenging. We need a method and appropriate tool to assist the users to model, compose and analyse the heterogeneous systems. In the previous work [8] we proposed a method (named "ModelINg And veRifying heterogeneous sysTems with contractS" (Minarets)), which consists in modeling and verifying a system with the concept of *generalized contracts*. The contract is generalized in the sense that it allows one to manage the interaction with the components through *given facets*: the properties of the environment, the properties of the concerned components, the communication constraints, non-functional properties (quality of service for example), etc.

In this paper, we apply our Minarets method on the LGS to face the challenges. In this case study, we describe the main components, we *normalize, compose* and *verify* them using the *generalized contract* principles; this contributes to decrease the difficulty of the modeling and verification of the LGS. After that, we compare our approach of modeling and verification of the LGS with the related approaches. The challenge here is not to do the complete case study as required in [6].

The rest of the article is structured as follows. Section 3 introduces the materials and an outline of the Minarets method. In Sect. 4 we model and analyse the LGS with our Minarets method. Section 2 provides an overview of related work, and finally, Sect. 5 gives conclusions and future work.

## 2   Related Work

The landing gear system was widely presented by the software engineering community; some of these works are mainly based on Event-B such as [2, 11, 12]. If we focus on the analysis of the various properties, we remark that the safety, time and liveness properties were formalized and proved, but from our point of view the analysis of time and liveness properties is heavy using Event-B; it is not possible to directly verify the properties of these facets, but it is necessary to improve Event-B method and enriching its tools to could analyse these facets.

The LGS was also presented in the work [4] using multi-machine hybrid Event-B, an hybrid modeling that combines software, hydraulic and mechanical parts; but without the verification steps.

The work [3] presented LGS with different method; they used the Abstract State Machine (ASM) and ASMETA framework for modeling and analysis. They take into consideration only the weaker versions of properties without taking into consideration the real time aspects.

Frédéric Boniol in the work [6] made a benchmark for techniques and tools of some approaches that dealt with LGS, described more the case study in [6] and made different experimentations in [7] using different tools like UPPAAL, Lustre, SMV, ...etc. (the case study is completely experimented in each tool).

In addition, unlike the most approaches that present the LGS, in our work we deal with heterogeneous components; our system is a combination of components modelled with different languages and in different environments, the component *interface* is modelled with the SPIN tool with the ProMeLa language, and the components *Gear*, *Door*, *Actuator* are modelled with UPPAAL. Moreover, we specify the various properties with the wide expressive language PSL to could specify formally all the needed facets.

## 3    Generalized Contracts and Outline of the Minarets Method

We present in this section the materials, the outline of our Minarets method and some notations to deal with our challenges which are: the composition, the composability (stability of component structures across integration) and the compositionality (the inferring of global system properties from the local properties) [13].

### 3.1    Materials

In the Minarets method [8], we extended the traditional Assume-Guarantee (A-G) contract with the purpose of mastering the modeling and verification of heterogeneous systems.

**Definition 1 (Generalized contract).**    *A generalized contract is a multi-faceted Assume-Guarantee contract. It is an extension of contract, structured on the one hand with its assume and guarantee parts, and structured on the other hand according to different clearly identified and agreed-upon facets (data, functionality, time, security, quality, etc.) in its assume or guarantee.*

The generalized contract are layered during modeling steps to facilitate properties analysis. Every facet may have a priority which is a natural number ($n \in \mathbb{N}$). Therefore, an analysis of a facet may be done prior to another facet. A layer is denoted by a facet $F$ and its priority $n$; thus, we will have a couple $(Facet, n)$ for each layer; we will write simply $Facet[n]$.

An example of a generalized contract (already layered) is as follows:

    (Assume = {(**DATA**[1], Predicate),
            (**TIME**[2], Predicate), ...},
    Guarantee ={(**TIME**[2], Predicate),
            (**SECURITY**[3], Predicate), ...})

**Definition 2 (Normalized component).**    *A normalised component is a component which is equipped with a generalized contract, acting as the components interface with other components.*

In this work we chose the PSL language [1] to specify properties and contracts. PSL is a formal language for specifying properties and behaviour of systems. It is an extension of the Linear Temporal Logic (LTL) and the Computation Tree Logic (CTL).

PSL could be used as input for formal verification, formal analysis, simulation and hybrid verification tools. We use the ALDEC Active-HDL[1] tool that supports PSL.

For experimentation purpose, we use ProMeLa and SPIN [14], the model checker UPPAAL [5] to model and verify components. SPIN is an automated model checker which supports parallel system verification of processes described with its input PROtocol MEta LAnguage (ProMeLa).

## 3.2 Outline of the Proposed Method: Minarets

The working hypothesis is that a heterogeneous system should be an assembly of *normalized components* (see Def. 2). Each component has a generalized contract and a behaviour expressed with *labelled transition system* (LTS); after that, for the sake of the verification efficiency, we may add a priority to each facet.

Several issues are considered for heterogeneous systems:

*i)* Elementary components are from various languages and cover different facets.

*ii)* Global properties are heterogeneous; they should be clearly expressed, integrated and analysed.

*iii)* Composition of elementary components should preserve their local requirements and should also be weakened or strengthened with respect to global level properties.

*iv)* Global properties require heterogeneous formal analysis tools; this generates complexity. We choose to separate the concerns, so as to target various tools and try to ensure the global consistency.

The Minarets method integrates solutions to these issues.

The working flow of the Minarets method is depicted in Fig. 1 with 13 steps (S1, S2, $\cdots$ S13). For the sake of brevity, we present the different steps along the case study; more details can be found in our works [8,9].

## 3.3 Notations

We denote by $C_i=(GC_i, B_i)$ a normalized component; where $GC_i$ is the generalized contract of $C_i$ and $B_i$ is the behaviour of $C_i$. The behaviour $B_i$ is represented with a LTS. Let consider $\mathscr{C}$ as a set of components, $\mathscr{P}$ a set of properties, $\mathscr{F}$ a set of facets.

A generalized contract GC is denoted by $GC=(A, G)$ where $A$ is the assumption and $G$ is the guarantee. The assumption $A$ or the guarantee $G$ is a set of properties $P_k \in \mathscr{P}$; where each property has a facet $F_i$; $F_i \in \mathscr{F}$ and each facet $F_i$ has a priority $n_j$; then, $A=\{(F_i[n_j], P_k)\}$, $G=\{(F_i[n_j], P_k)\}$.

**Simplifications:** The landing system [6] is in charge of operating the landing gear and the associated doors. The landing system is made up of 3 landing sets: front, right and left. Each landing set contains a door, a landing gear and associated hydraulic cylinders. In [6], there are 3 sensors; each sensor sends a value, the three values must be the same, otherwise, if one value of the three ones is different it will be eliminated, only the two remaining values are considered. In this case study we consider only the front door and the front landing gear without taking into consideration the hydraulic cylinders. We also consider only one sensor to prevent the state of each mechanical parts instead of 3 sensors as mentioned in the original version [6].

---

[1] https://www.aldec.com/en/products/fpga_simulation/active-hdl.

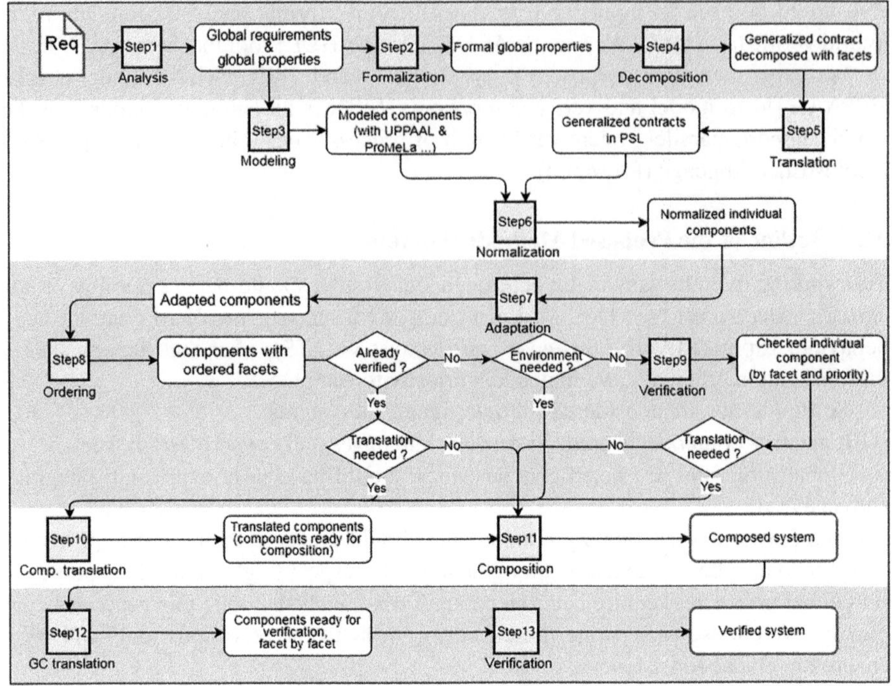



**Fig. 1.** The Minarets process chart

# 4   Modeling and Verifying the Landing Gear System with **Minarets**

In the following, we present the steps of modeling and verification of the LGS.

**Step 1** *(modeling).* In this step we should express the global requirements and global properties in an informal way. We will not repeat all the requirements and the properties of the LGS. We present only a limited number of properties and we always refer to our research report [10] that contains a complete description of this case study.

**Global Requirements:** The landing gear system is composed from four components (*gear*, *door*, *interface* and *actuator*); these components communicate to manage the outgoing and the retraction of the gears. When the actuator is moved from up to down, the door opens and eventually the landing gear is extended, after that, the door will be closed. If the actuator is moved from down to up, the door opens and eventually the landing gear is retracted; after that, the door will be closed. Furthermore, If the gear or the door is manoeuvring, the interface orange light will be turned on; If the gear is fully extended, the interface green light will be turned on. More details can be found in our report [10]. A simplified schema of a landing system is shown in Fig. 2

**Global properties (selected properties):**
**P1:** In landing mode, eventually the gear must be fully extended and locked in down position.

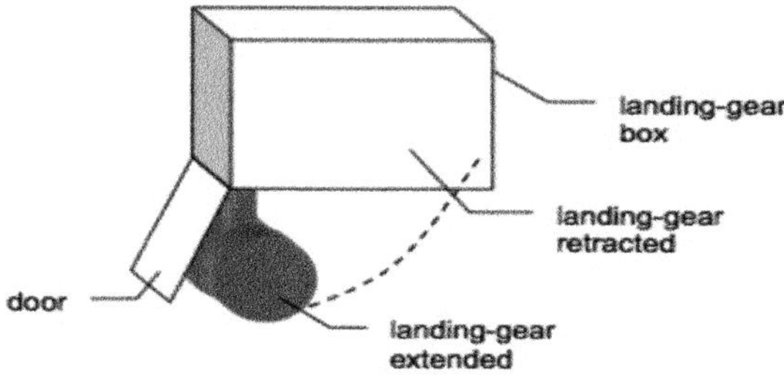


**Fig. 2.** Simplified view of the landing system [6]

**P2:** In cruise mode the gear must be retracted and the door is locked in high position (the cruise mode means that there is no ongoing landing or retraction operation) .

**P3:** If the gear is manoeuvring from high to down, then the door must be completely open.

**P5:** If the landing gear actuator has been pushed DOWN, and stays down, then eventually the door latching will be unlocked in 4 s.

**P6:** If the landing gear actuator is pushed down and stays DOWN, then eventually the door will be completely opened in 16 s.

**P7:** The gear never manoeuvring from down to high and the door is not open.

**P9:** If the landing gear actuator is DOWN and the door is completely open, then eventually the gear will be unlocked in high position in 8 s.

**P22:** If there is a failure in the door or in the gear, then the pilot interface shows that the red light is on.

The components communicate, each component may send and receive informations. Then, the properties $(P_i*)$ are needed for communication between components. For instance, the door must be completely open to could extend the gear.

**P1*:** The door is completely open.

/* The door must be manoeuvring so that the interface orange light turn on*/

**P7*:** The door is manoeuvring from high to down. **P11*:** There is a failure in the door.

/* The gear must be locked in high or down position to could open the door */

**P15*:** The gear is locked in high position. **P17*:** The gear is locked in down position.

**P31*:** The actuator position is up.

/* The actuator position must be down to could execute the outgoing sequence */

**P32*:** The actuator position is down.

...

**Step 2 (*modeling*).** Formalisation of the required global properties with any formal specification language. In our case we choose PSL language. The point of this step is to express formally the global properties.

```
/*Formalisation of properties with PSL */
/* ---Assumptions--- */
/* ---The door must be completely open--- */
P1*: door_open == true;
/* ---The door must be in manoeuvre (operation)--- */
P7*: door_m_highdown==true;
/* ---There is a failure in the door--- */
P11*: door_failure==true;
/* ---The components that receive informations assume that
the needed variables must be boolean--- */
P341 : boolean door_locked; P342 : boolean door_open;
...

/* ---Guarantees--- */
/* In landing mode, eventually the gear must be fully
extended and locked in down position.*/
P1 : landing==true ->eventually! full_gear_extension==true
and gear_locked_down==true;
P2 : AG landing==false and retraction==false ->
full_gear_retraction==true and door_locked==true;
P3 : AG gear.man_highdown ->door_open==true;
......
```

**Step 3** *(modeling)*. The goal here is to show that we could select/reuse pre-modelled components even if they are in different languages and in different environments. From our analysis of the case study, we distinguish the following components:

- The door, it is open or closed depending on the actuator position and the gear position as well.

- The gear that models the landing gear, it may be extended in the landing mode or retracted after the take-off.

- The actuator that is used by the pilot to launch the outgoing sequence if it is moved to down position, and to launch the retraction sequence if it is moved to up position.

- The interface that displays the corresponding light on the pilot screen, depending on the state of the door and the gear (retracted= no light, manoeuvring= orange light, extended= green light, failure= red light).

We use ProMeLa to model the component *interface*, ProMeLa doesn't support time, that's why we chose the component *interface* that doesn't contain time facet; we use UPPAAL to model the components *door*, *gear*, *actuator*[2]. The gear component is depicted in the Fig. 3.

---

[2] The detailed models of the components are given in the appendix (Figures 3,7, 8).

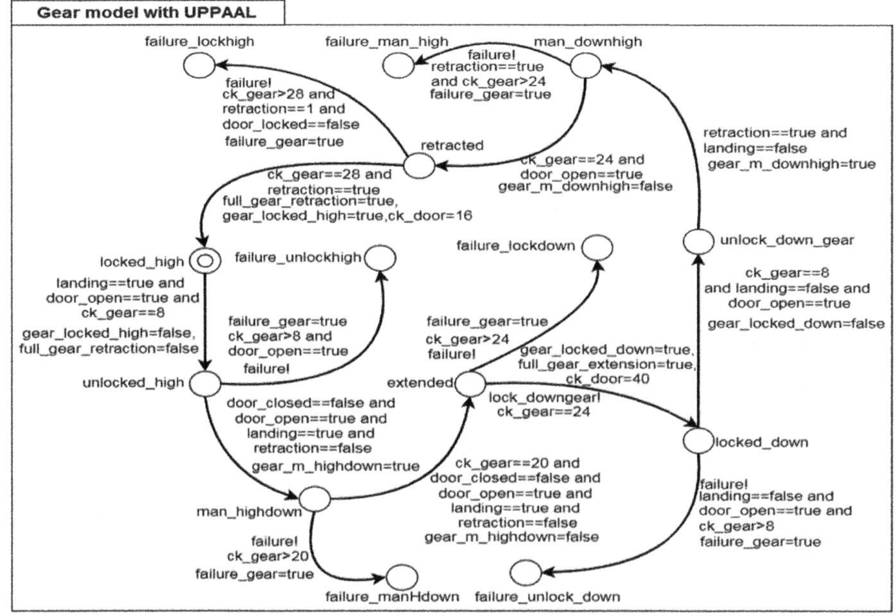


**Fig. 3.** The gear model with UPPAAL

**Step 4** (*modeling $M_4$*). We have to decompose the global properties with respect to the chosen facets; we consider the following ones: data, safety, liveness, functionality, time). Then, we obtain the *generalized contract* decomposed with the facets.

```
/*Data facet*/
/* ---Assumptions--- */
/* ---The components that receive informations assume that
the needed variables must be boolean--- */
P341:boolean door_locked; P342:boolean door_open;
P343:boolean door_closed; P361:boolean full_gear_extension;
...

/* ---Guarantees--- */
/*The components that provide informations guarantee that the
variables are boolean, boolean array, clock, etc.*/
/*boolean variables*/
P341: boolean door_locked; P364: boolean gear_locked_down;
/*channel variables : The channel type used in UPPAAL
model-checker is a very specific type, that is why we
represent it like a boolean array.*/
P371 : boolean gear_extend[];
/*Clock variables*/
```

```
P39 : clock ck_gear;
...
```

```
/*Safety facet*/
/* --- Assumptions --- */
/* ---The door component assumes that the gear must be
locked in high position so that it can be opened (to work
correctly)--- */
P17*: gear_locked_down==true;
/* ---The interface component assumes that the gear is
manoeuvring so that it can turn on the appropriate light
color --- */
P19*: gear_m_highdown==true; P21*: gear_m_downhigh==true;
/* ---The door assumes that the gear is fully extended or
retracted to could be opened (to work correctly)--- */
P25*: full_gear_extension==true;
...
/* --- Guarantees --- */
/*In cruise mode the gear must be retracted and the door
is locked (the cruise mode means that there is no ongoing
landing or retraction operation)*/
P2 : AG landing==false and retraction==false ->
full_gear_retraction==true and door_locked==true;
P3 : AG gear.man_highdown ->door_open==true;
P7 : AG gear.man_highdown ->!door_open==false;
...
```

**Step 5 (*modeling $M_5$*).** Unlike in Step 2, here we should express the structured and formalized properties with the PSL language, because it supports the LTL and CTL, then, it is important to get the appropriate properties that are very similar to UPPAAL syntax; this initiates the Step 12 of the properties translation from PSL to UPPAAL.

**Step 6 (*modeling $M_6$*).** Normalizing the individual components obtained from Step 3. We describe the assumptions and guarantees of each individual component. Figure 4 shows the normalized individual components. Here the attribution of each property to the appropriate component is made manually, in an empirical manner.

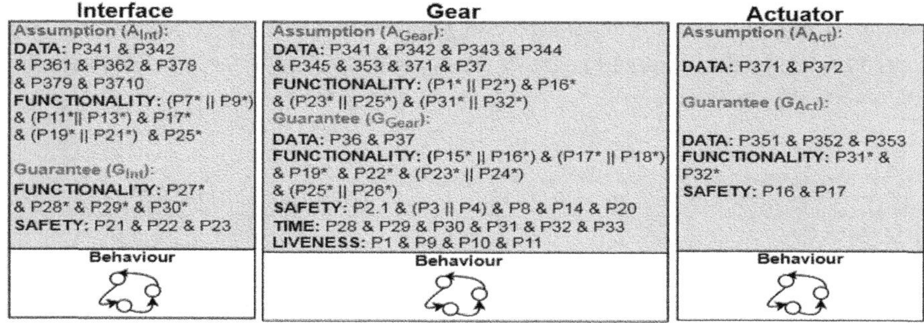


**Fig. 4.** Normalized components

**Step 7** *(modeling $M_7$)*. According to the result of *Step 4* (the required contracts), we may add a facet of the global property to a component, or ignore some of its facets, if necessary. This step is important when we select/reuse pre-modelled components; in this step we could adjust them according to our needs. In the current case study we didn't add or ignore a facet, because we modelled all the components from scratch and according to our needs.

**Step 8** *(modeling $M_8$)*. We should prepare the verification step by layering the properties; then, we attribute the following priorities to each facet (data=1, functionality=2, safety=3, time=4, liveness=5); where 1 is the highest priority. We obtain ordered layers with respect to facets and properties as well. The verification by layer allows one to verify the contracts by priority; from a very important layer (primary) to a less important layer (secondary). If the behaviour of our system does not satisfy a primary layer of contract, then, it is not necessary to continue the verification with the other layers. This step is important to save time and to be more efficient during verification.

**Step 9** *(verification $V_1$)*. In this step we have to check the appropriate functioning of each normalized individual component, As we use ProMeLa we have an adequate tool (SPIN) but, the only component *Interface* modelled with ProMeLa cannot be verified without composition with its environment; moreover, in our study we have the facet time (we use clocks) but ProMeLa language doesn't support time; consequently, a translation from ProMeLa to UPPAAL is needed. Note that the method works in both cases, with the component translation if it is necessary, and without translation, if each component stays in its environment.

**Step 10** *(modeling $M_9$)*. The involved components should be translated into the input language of the verification tool. Here, we translate only the ProMeLa component *Interface* to the targeted tool UPPAAL using the algorithm presented in our research report [9], we obtain a component *Interface* ready for composition.[3]

**Step 11** *(modeling $M_{10}$)*. We compose the generalized contracts of the components (*Interface, Actuator*), facet by facet. The Fig. 5 shows the composition of the generalized contracts of the components *Interface* and *Actuator* to build the *Cockpit* com-

---

[3] The translated component is in the appendix. See Fig. 10.

ponent. A component resulting from a composition should be in the normalized form; therefore, we need to compute its generalized contract. We use the logical conjunction to compose the facets.

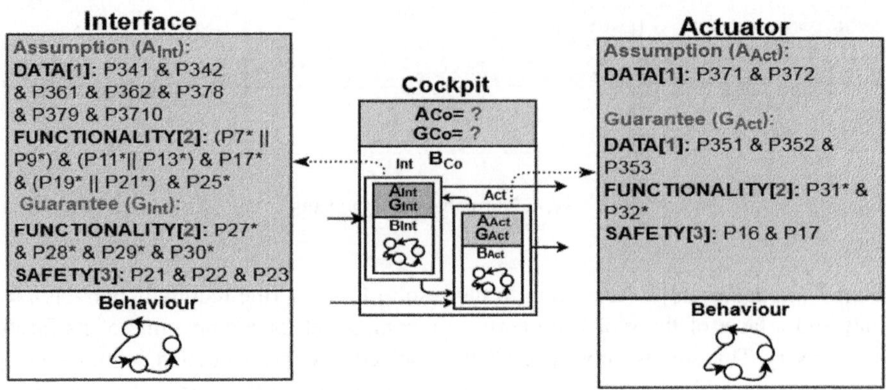


**Fig. 5.** The composition of the generalized contracts of the components Interface and Actuator

As depicted in the Fig. 5, the generalized contracts $GC$ of the component *Interface*, $GC_{Int} = (A_{Int}, G_{Int})$ is:

$A_{Int} =$
$\{(\textbf{DATA}[1] : P341, P342, P361, P362, P378, P379, P3710),$
$(\textbf{FUNCTIONALITY}[2] : (P7* \vee P9*) \wedge (P11* \vee P13*) \wedge P17* \wedge (P19* \vee P21*)$
$$\wedge P25*)\}$$

$G_{Int} =$
$\{(\textbf{FUNCTIONALITY}[2] : P27* \wedge P28* \wedge P29* \wedge P30*),$
$(\textbf{SAFETY}[3] : P21 \wedge P22 \wedge P23)\}$

The generalized contracts $GC$ of the component *Actuator*, $GC_{Act} = (A_{Act}, G_{Act})$ is:

$A_{Act} =$
$\{(\textbf{DATA}[1] : P371, P372)\}$

$G_{Act} =$
$\{(\textbf{DATA}[1] : P351, P352, P353),$
$(\textbf{FUNCTIONALITY}[2] : P31* \wedge P32*), (\textbf{SAFETY}[3] : P16 \wedge P17)\}$

The components *Interface* and *Actuator* run simultaneously and communicate between each other through shared channels to perform their tasks; the components are dependent.

We use the logical conjunction between the properties, facet by facet; then, the composition of the generalized contract $GC_{Int} \oplus GC_{Act}$ is:

$A_{Cockpit} =$
$\{(\textbf{DATA}[1] : (P341, P342, P361, P362, P378, P379, P3710), (P371, P372)),$
$\quad (\textbf{FUNCTIONALITY}[2] : (P7* \vee P9*) \wedge (P11 * \vee P13*) \wedge P17 * \wedge (P19 * \vee P21*)$
$$\wedge P25*)\}$$

$G_{Cockpit} =$
$\{(\textbf{DATA}[1] : P351, P352, P353),$
$\quad (\textbf{FUNCTIONALITY}[2] : (P27 * \wedge P28 * \wedge P29 * \wedge P30*) \wedge (P31 * \wedge P32*)),$
$\quad (\textbf{SAFETY}[3] : (P21 \wedge P22 \wedge P23) \wedge (P16 \wedge P17))\}$

**Step 12** *(verification $V_2$)*. For the sake of verification; we translate the generalized contracts (only the desired properties to be verified) from *PSL* to the UPPAAL language. The following properties are an extract of the translated properties (P7, P22)[4].

```
A[] gear.man_highdown imply !door_open==false

A[] failure_door==true or failure_gear==true imply interface.red
```

**Step 13** *(verification $V_3$)*. After the composition of the system in *Step*11, we verify the properties of the same layer together, facet by facet; for instance, considering the safety facet, we have verified the properties (P2, P3, P7, P8, P13, P14, P16, P17, P18, P19, P20); also, the primary properties before the secondary properties; for instance, the verification of the safety facet that has the priority 3, before the verification of the time facet that has the priority 4. At the end of this step, we obtain the verified components. Figure 6 shows the verification status of some translated properties (P19, P20, P21, P22, P23, P24, P25, P26).

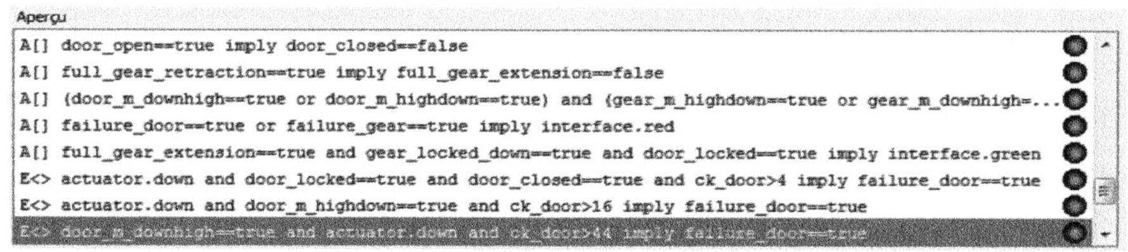


**Fig. 6.** Status of properties verification with UPPAAL

Folowing the Minarets method, we presented successively, each step and its result, starting from the LGS requirements until the modeling and the verification of its components and their composition.

---

[4] were "A [ ] Prop" denotes the "always property".

## 5    Conclusion

In this paper, we have shown how we applied our Minarets method on the landing gear system; we described step by step the modeling and analysis procedure using the generalized contracts. We specified the contract with PSL; PSL gives one the possibility to specify formally all the needed facets, even the complex one, like time and liveness facets. We modelled the components *Gear*, *Door* and *Actuator* with UPPAAL and the component *Interface* with ProMeLa, this heterogeneity shows that our method is efficient even if we reuse pre-modelled components (components in different languages and environments). We normalized these components and layered their generalized contracts to could compose them by facets. Finally, we gave a priority to each facet and verified the generalized contracts of each component by order, from the primary facet to the secondary facet. We tackled the LGS by facets instead of the whole system at once. The Minarets method helps one to increase the efficiency and decrease the difficulty of the modeling and verification of heterogeneous systems.

Future works address not only the scalability but also the study of various policies for the composition of contracts; it is important to analyse the components composition as we done through the used tools but, to improve the method we are defining in a formal way the semantics of sequential composition, parallel composition and additive composition of our normalized components. This will furthermore strengthen the foundations of the proposed method, and enable the contract management tool construction; for this purpose, we will develop a tool to guide the users in modeling and analysis of heterogeneous systems, it helps in some steps of the method; for instance, Step 4 of decomposition, Step 6 of normalisation and Step 11 of composition. For this purpose.

## A    The Components Models with UPPAAL and ProMeLa

```
Interface model with ProMeLa

active proctype pilot_interface() {
no_light:  //-----Manoeuvring
           ((door_locked==false and door_open) or
           (full_gear_extension==false and full_gear_retraction==false)) -> goto orange;
           //---------Failure
           failure? goto red;

orange:    //---------Retracted
           full_retracted? goto no_light;
           //---------- Extended
           full_extended? goto green;
           //---------Failure
           failure? goto red;

green:     //-----Manoeuvring
           ((door_locked==false and door_open==false) or
           (full_gear_extension==false and full_gear_retraction==false)) -> goto orange;
           //---------Failure
           failure? goto red;
           }
```

**Fig. 7.** The pilot interface model with ProMeLa

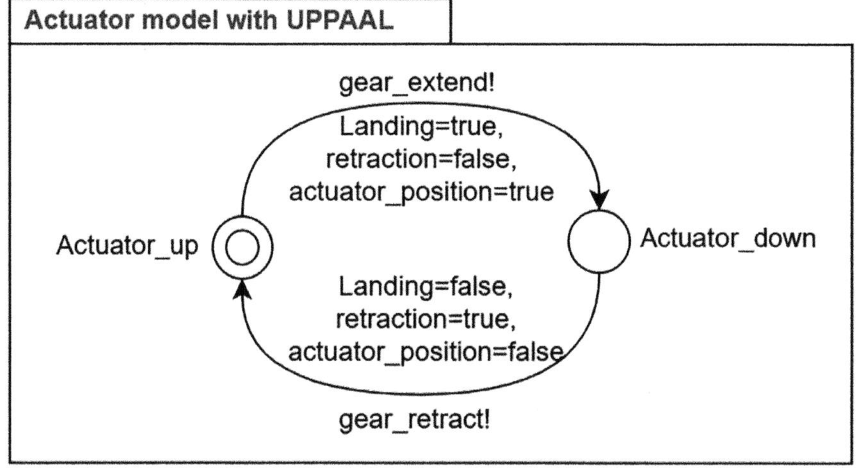


**Fig. 8.** The actuator model with UPPAAL

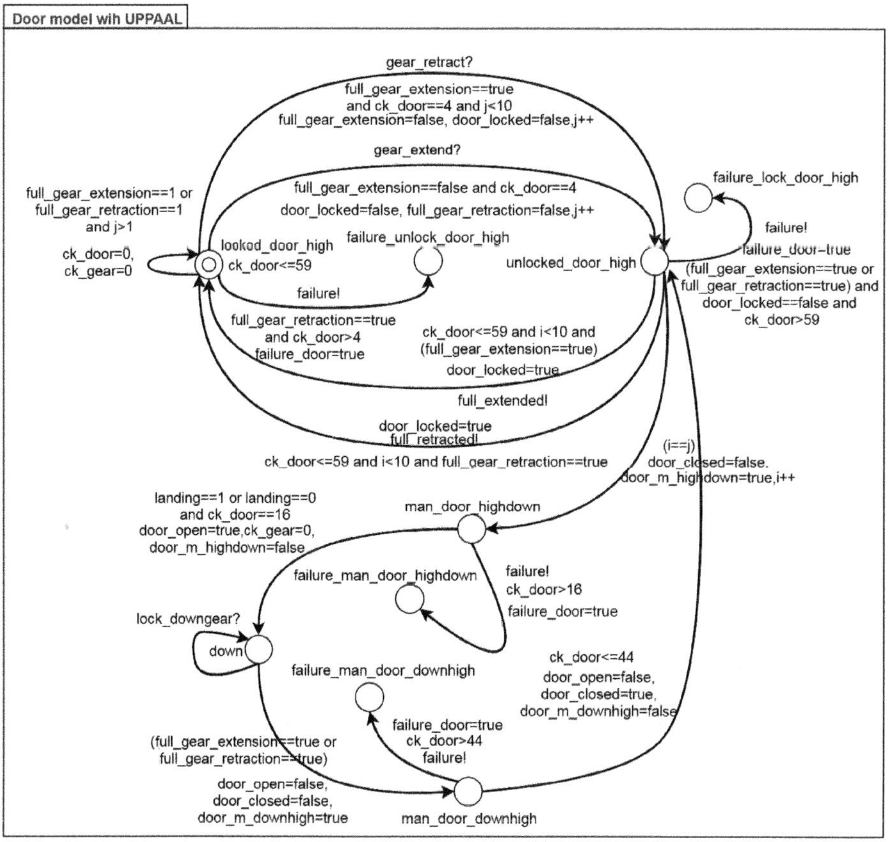

**Fig. 9.** The door model with UPPAAL

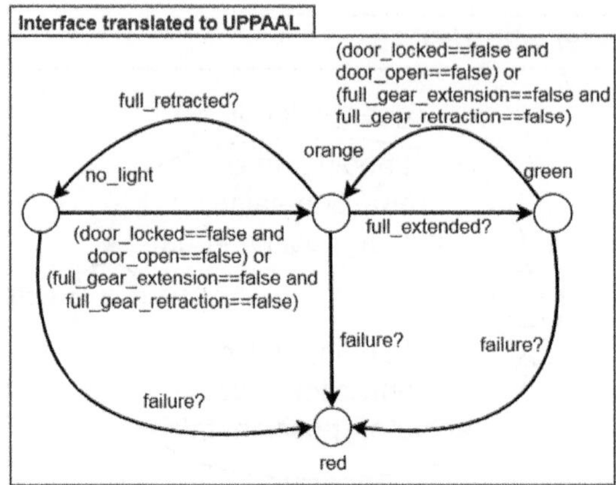


**Fig. 10.** Pilot interface component translated from ProMeLa to UPPAAL

# References


1. IEEE standard for Property Specification Language (PSL). IEEE Std 1850-2010 (Revision of IEEE Std 1850-2005), pp. 1–182 (2010). https://doi.org/10.1109/IEEESTD.2010.5446004
2. André, P., Attiogbé, C., Lanoix, A.: Modelling and analysing the landing gear system: a solution with event-b/rodin (2018)
3. Arcaini, P., Gargantini, A., Riccobene, E.: Modeling and analyzing using ASMS: the landing gear system case study. Commun. Comput. Inf. Sci. **433**, 36–51 (2014). https://doi.org/10.1007/978-3-319-07512-9_3
4. Banach, R.: The landing gear system in multi-machine Hybrid Event-B. Int. J. Softw. Tools Technol. Transfer **19**(2), 205–228 (2015). https://doi.org/10.1007/s10009-015-0409-7
5. Basile, D.: Modelling, verifying and testing the contract automata runtime environment with uppaal. In: Castellani, I., Tiezzi, F. (eds.) The Netherlands, June 17-21, 2024, Proceedings. LNCS, vol. 14676, pp. 93–110. Springer (2024). https://doi.org/10.1007/978-3-031-62697-5_6
6. Boniol, F., Wiels, V., Aït-Ameur, Y., Schewe, K.-D.: The landing gear case study: challenges and experiments. Int. J. Softw. Tools Technol. Transfer **19**(2), 133–140 (2016). https://doi.org/10.1007/s10009-016-0431-4
7. Boniol, F., WIELS, V., Ledinot, E.: Experiences in using model checking to verify real time properties of a landing gear control system. In: Conference ERTS'06. Toulouse, France (2006). https://hal.archives-ouvertes.fr/hal-02270431
8. Khouass, A.A., Attiogbé, J.C., Messabihi, M.: Multi-facets contract for modeling and verifying heterogeneous systems. In: Model and Data Engineering - 10th International Conference, MEDI 2021, Tallinn, Estonia, June 21-23, 2021, Proceedings. LNCS, vol. 12732, pp. 41–49. Springer (2021). https://doi.org/10.1007/978-3-030-78428-7_4
9. Khouass, A., Attiogbé, C., Messabihi, M.: Multi-facets contract for modeling and verifying heterogeneous systems. CoRR **abs/2012.13671** (2020). https://arxiv.org/abs/2012.13671
10. Khouass, A., Attiogbé, C., Messabihi, M.: Modeling and analysis of the landing gear system with the generalized contracts. CoRR **abs/2111.10426** (2021). https://arxiv.org/abs/2111.10426

11. Ladenberger, L., Hansen, D., Wiegard, H., Bendisposto, J., Leuschel, M.: Validation of the ABZ landing gear system using ProB. Int. J. Softw. Tools Technol. Transfer **19**(2), 187–203 (2015). https://doi.org/10.1007/s10009-015-0395-9
12. Mammar, A., Laleau, R.: Modeling a landing gear system in Event-B. Int. J. Softw. Tools Technol. Transfer **19**(2), 167–186 (2015). https://doi.org/10.1007/s10009-015-0391-0
13. Mavridou, A., Sifakis, J., Sztipanovits, J.: Designbip: A design studio for modeling and generating systems with BIP. CoRR **abs/1911.08405** (2019). http://arxiv.org/abs/1911.08405
14. Rashid, M.: Verification of safety of aircraft arrival procedure using SPIN model checker. In: International Conference on Frontiers of Information Technology, FIT 2023, Islamabad, Pakistan, December 11-12, 2023, pp. 262–267. IEEE (2023). https://doi.org/10.1109/FIT60620.2023.00055

# MATSum: Multi-agent Traffic Simulation of Urban Mobility with Consideration of Heterogeneous Vehicles


Oussama Messaoudi[1,2]([✉]) and Akli Abbas[2]

[1] University of Batna2, Batna, Algeria
o.messaoudi@univ-batna2.dz
[2] LIM Laboratory, University of Bouira, Bouira, Algeria
a.abbas@univ-bouira.dz



**Abstract.** As intelligent and connected vehicles advance, evaluating their impact on urban traffic remains challenging due to the lack of real-world data. This paper presents an agent-based simulation framework using JADE and SUMO to analyze heterogeneous vehicle interactions and the effects of deploying advanced driving systems. Our framework enables researchers to explore how conventional and cooperative driving systems influence traffic flow, safety, and emissions, offering a valuable tool for developing intelligent transportation systems. Using our framework, we compare a conventional car-following model with a cooperative model that leverages vehicle-to-vehicle (V2V) communication. The results show that the cooperative model leads to smoother vehicle acceleration and improved driving comfort compared to the conventional model.

**Keywords:** Agent-based modeling and simulation · Microscopic urban traffic simulation · Intelligent vehicles


## 1 Introduction

With the advancement of computing technology and the introduction of more and more intelligent driving assistance systems into our cars, the human driver is slowly being eliminated from the driving loop. In the early days of intelligent driving assistance systems, it started with the conventional cruise control (CC) system, and then the adaptive cruise control (ACC) system was introduced to partially automate the driving task. Then, the cooperative adaptive cruise control (CAAC) system was introduced by incorporating Vehicle-to-Vehicle (V2V) communication into the ACC system [8,12]. Following the 2007 and 2009 DARPA challenges, the focus shifted towards the development of autonomous vehicles (AVs) [3,4], and by integrating V2V communication into these AVs evantually has led to the introduction of cooperative autonomous vehicles (CAVs) [6]. These systems aim to improve traffic flow, safety, and fuel efficiency, and most importantly reduce the negative impacts of internal combustion engine vehicles on our environment by reducing $CO_2$ emissions.

Prior to the deployment of intelligent driving assistance systems, an analysis of their impacts is crucial to ensure that the intended goals are met and maintained. However, since real-world data about such systems is very limited, researchers rely on simulation-based analysis, which is also cost-effective, time-efficient, and risk-free [6,11].

In microscopic urban traffic simulators, each vehicle is individually controlled by two models; the car-following model (CFM) to describe the longitudinal control behavior, and the lane- changing model (LCM) to define the lateral control behavior [5,8]. This enables a more detailed description and control of the vehicle's behavior.

In this paper, we present a versatile framework tailored to the needs of both practitioners and researchers by enabling the creation of realistic agent-based simulations with heterogeneous agents. Our framework models a variety of car-following behaviors to produce a wide range of driving systems, ranging from manual to cooperative automated driving systems. By combining the multi-agent system (MAS) development framework JADE, with the microscopic traffic simulator SUMO, our framework facilitates a broad spectrum of studies. These include the development and analysis of intelligent driving systems. Furthermore, we showcase the implementation of a cooperative automated driving system, with a specific focus on car-following behavior, to enhance road safety in challenging conditions.

## 2   Related Works

Traditional traffic simulation methods struggle to capture the intricate dynamics and the complex behavior and interactions of different entities. The future of transportation demands new simulation tools where heterogeneous agents co-exist to interact, communicate, and collaborate to produce a realistic urban traffic environment.

Multi-agent systems (MAS) have emerged as promising tools in the field of traffic modeling and simulation, offering improved accuracy and efficiency over traditional methods [17]. The autonomous and goal-driven behavior of intelligent agents makes agent models particularly suitable for modeling interactive, complex, and unpredictable behaviors of heterogeneous individuals (e.g. human drivers, different types of automated vehicles, traffic signals, pedestrians, etc.). JADE (Java Agent Development Framework) is a software framework designed to facilitate the development of agent-based applications in Java. It simplifies the creation and management of MAS and provides system services and graphical tools for monitoring, customization, debugging, and deployment [2].

Using the agent-based modeling paradigm, the *Agentpolis* project has proposed a modular framework for executing simulations capable of complex interaction and real-time decision-making [7].

In [15], the authors proposed a two-layered agent-based simulation framework that consists of heterogeneous entities, specifically drivers and traffic light control systems, that can coordinate, collaborate, and compete with each other. The first layer, the tactical layer handles the basic control of vehicles within SUMO, a widely used open-source urban traffic simulator [8], whereas the strategic layer manages the decision-making process in JADE. The proposed simulation framework used TraCI and TraSMAPI to allow SUMO and JADE to communicate with each other and exchange important information.

Similarly, in [1] the authors proposed a multi-agent simulator for traffic light control built with JADE, SUMO, and TraSMAPI. Here, SUMO is used to implement the dynamics of the traffic environment, including vehicles and traffic lights, whereas in JADE each agent controls a traffic light. The TraSMAPI API was used to ensure that agents in JADE have access to the simulation state.

In [10], agent-based simulations have been successfully employed to predict traffic congestion during peak hours and to analyze traffic flow dynamics. Also, in [13] the authors highlighted the need for agent-based modeling and simulation when dealing with autonomous vehicles in the simulation. For this reason, the authors proposed *AGAMAS*, an agent-based traffic simulation framework where JADE agents communicate through TraCI to perceive and act on the simulation environment state SUMO.

To this end, researchers and practitioners require simulation tools capable of reproducing the same conditions as in the real world, where heterogeneous entities capable of interaction, communication, collaboration, and cooperation to accomplish their goals. These tools help in making informed decisions that promote sustainable and efficient transportation systems.

## 3   MATSum: A Multi-agent Traffic Simulation Framework with Consideration of Heterogeneous Vehicles

The aim of this multi-agents based traffic simulation framework is to easily enable researchers and practitioners to conduct realistic microscopic simulations within an environment composed of heterogeneous vehicles with highly complex behaviors. Also, our simulation framework presents the necessary means to control and monitor the simulation state. Analysis studies can leverage our framework to accurately estimate the impacts of deploying intelligent vehicles and car-following models, specifically on traffic flow, safety, fuel efficiency, and $CO_2$ emissions.

### 3.1   MATSum's Architecture

In this paper, we propose a framework for multi-agents simulation of urban mobility with consideration of heterogeneous vehicles. The framework architecture is divided into three layers:

- **Tactical layer**: This layer represents the simulation urban environment SUMO, including the state of all of its components (e.g. roads, traffic lights, vehicles, pedestrians, etc.).
- **Communication layer**: This layer links the strategic layer with the tactical layer to ensure that agents have access to simulation state through *TraCI* and *TraSMAPI*.
- **Strategic layer**: Represented by JADE, it consists of two sub-layers:
  - *Simulation control sub-layer*: Managed by the *SimControlAgent*, this sub-layer is responsible for controlling, perceiving, and acting on the simulation state.
  - *Decision-making sub-layer*: Composed of *DriverAgent*s, each agent is responsible for implementing a car-following model to control the velocity of of a specific vehicle.

### 3.2   The Communication Layer in MATSum

In the strategic layer (JADE), the communication between the simulation control sub-layer and the decision-making sub-layer,specifically between agents, is handled by the Agent Communication Language (ACL). The ACL allows the *SimControlAgent* to forward the vehicle state perceived from the tactical layer SUMO to the *DriverAgent*, and also enable the *DriverAgent* to send its decisions to the *SimControlAgent* to be forwarded to the tactical layer for execution.

On th other hand, for the communication between the strategic layer (*SimControlAgent*) and the tactical layer SUMO, we used TraCI and TraSMAPI, which provide dedicated API calls to JADE agents to interact with SUMO (e.g. *getSpeed(vehicleId)*). However, to facilitate this communication, we made modifications to and recompiled SUMO, TraCI, and TraSMAPI. As a result, with a single API call using *getVehicleState(vehicleId)* the *SimControlAgent* can perceive the full state of any vehicle in the simulation.

This allows researchers and practitioners not only to facilitate the modeling and simulation of complex driving systems, but also to examine their behavior and measure their performance in the simulation on the microscopic level.

### 3.3   SimControlAgent

The *SimControlAgent* plays a central role in the traffic simulation framework, serving as the primary controller for managing and orchestrating the simulation process. Bellow are the key components and behaviors of the *SimControlAgent* (see Fig. 1).

**Graphical User Interface (GUI):** This agent is equipped with a graphical user interface (see Fig. 2) that enables the simulation administrator to:

- Load the simulation scenario, defining road network, vehicles, traffic lights, and any other necessary components.

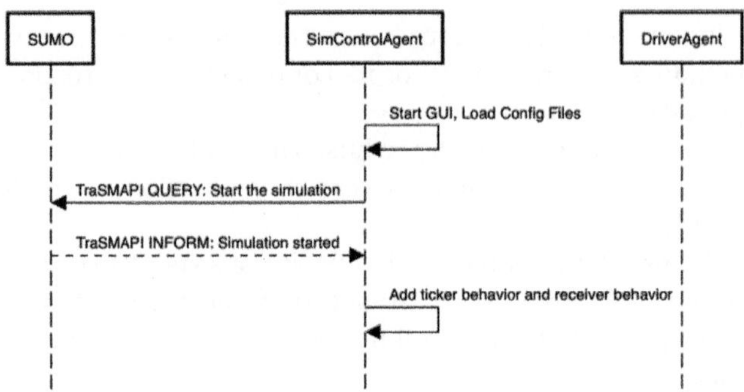

**Fig. 1.** *SimControlAgent* setup sequence diagram.

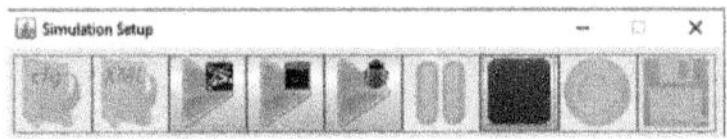

**Fig. 2.** *SimControlAgent*'s graphical user interface.

- Load agent configuration, specifying which driver agent class is implemented to define the *DriverAgent*.
- Launch *SUMO* and start the simulation using the loaded configuration.
- Use the GUI controls to pause, resume, stop, or restart the simulation, and also to save simulation logs.

**Ticker Behavior:** The *SimControlAgent* is equipped with a ticker behavior that executes repeatedly with a predefined interval defined by an argument *delay*. The ticker behavior executes the following key functions in each iteration (See Fig. 3):

- Triggers simulation time increment to drive the simulation one step forward.
- Retrieves the list of active vehicles in the simulation from SUMO.
- Logs, and saves the state of all specified vehicles, lanes, and road sections in agent the configuration file.
- Communicates with and manages *DriverAgents'* life cycle as follows:
  - Creates a *DriverAgent* once the vehicle it will be controlling appears.
  - Terminates *DriverAgents* who's vehicles no longer active in SUMO.
  - Perceives and communicates each vehicle state from SUMO to the *DriverAgent* concerned with its control.

**Receiver Behavior:** The receiver behavior of the *SimControlAgent* is designed to handle communication between the *DriverAgents* and SUMO. It is triggered upon receiving any message signaling a request from a *DriverAgent*, in the form of a vehicle identifier and an acceleration value, to be forwarded to SUMO for execution (see Fig. 4).

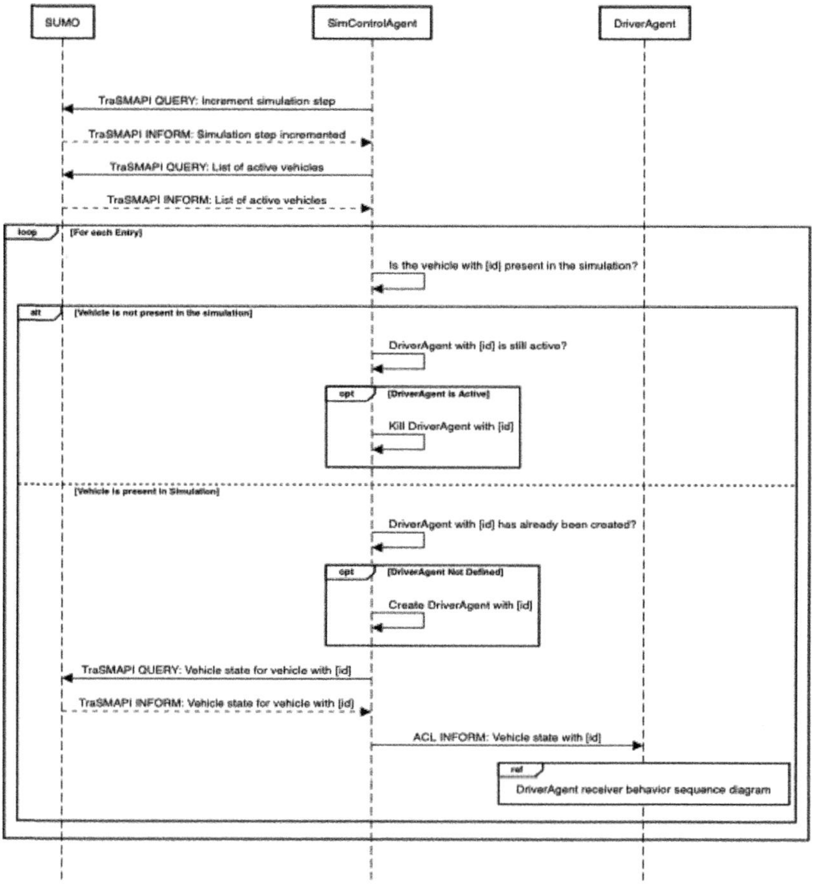

**Fig. 3.** *SimControlAgent* ticker behavior sequence diagram.

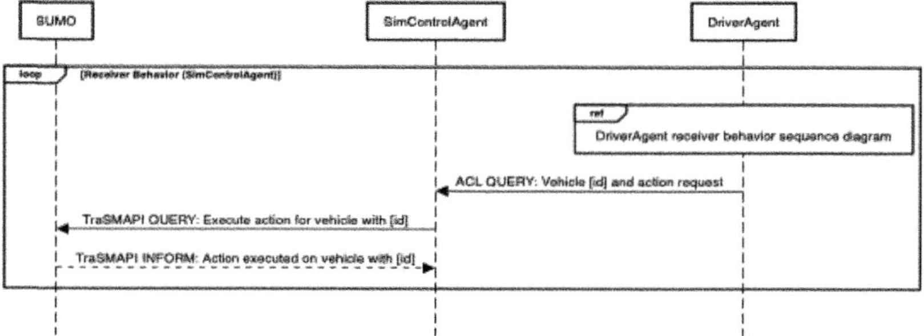


**Fig. 4.** *SimControlAgent* receiver behavior sequence diagram.

### 3.4   Agent Configuration File

The *SimControlAgent* reads the agent configuration file (*agentConfig.XML*) and defines how the simulation should be run. This is achieved through two types of entries in the XML

- **Control entries**: specifically *CFM* entries, each entry defines in the *Model* argument which agent class is used to create the *DriverAgent* whose task is to implement a car-following model to control the velocity of a vehicle specified by the identifier *id*.
- **Simulation state monitoring entries**: specifically *EMS*, *Lane*, and *Edge* entries, provide three different levels of microscopic state monitoring, to log and save the state of the specified vehicle, lane, and road section, respectively.

### 3.5   DriverAgent

A *DriverAgent* is responsible for implementing a specific car-following model to handle the longitudinal control task for the controlled vehicle. In our framework, each *DriverAgent* has a receiver behavior. The latter is triggered after the agent receives a message from the *SimControlAgent* defining the current vehicle state. As a result, this behavior performs the followings (see Fig. 5):

- **Perception**: Receive the message from *SimControlAgent* and use its payload to update the controlled vehicle state.
- **Decision-making**: Implement a car-following model to decide the appropriate amount of acceleration (or deceleration) to apply on the vehicle velocity to maintain the desired goal (e.g. collision-free driving).
- **Action**: Forward its decision to the *SimControlAgent* for execution.

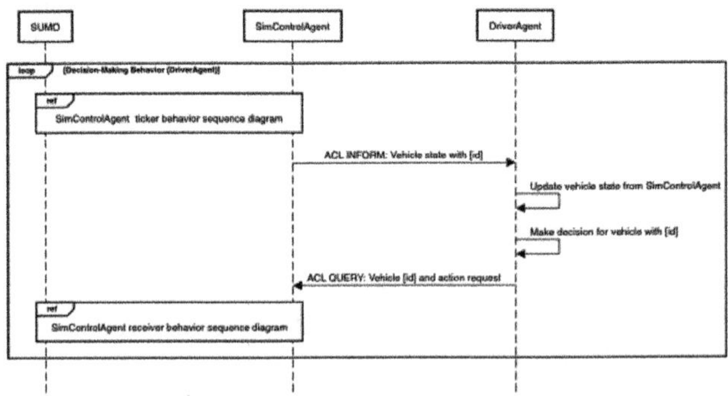


**Fig. 5.** *DriverAgent* receiver behavior sequence diagram.

Finally, by allowing researchers to implement their own models, our framework enables the simulation of various driving dynamics, behaviors, and systems, including human imperfection, driving assistance systems, automated vehicles, and connected automated vehicles. Furthermore, SUMO enables deploying different types of vehicles (e.g. motorcycles, cars, and buses) to run on different types of engines (e.g. internal combustion engines or electric engines). As a result, by combining SUMO with the proposed agent architecture, our simulation framework supports heterogeneous vehicle types enabling researchers to model, simulate, and analyze the behavior of any type of driving system.

## 4    Experimentation and Results

In this section, we present a simulation-based analysis of a cooperative car-following model as a demonstration of how our framework is used to implement, simulate, and analyze the behavioral impact a new longitudinal control behavior, specifically using V2V communication and agent technology.

In the new model, we extend the *DriverAgent*'s perception process using V2V communication, simulated with agent-to-agent communication based on ACL, to perceive the leading vehicle's intent and future state, and as a result, react to any changes to the following vehicle's surroundings at the same time.

In this scenario, we focus on the control of a vehicle approaching an intersection using two different models to highlight the impact of V2V communication. The *DriverAgent A* implements the *krauß* model (see Eq. 1) proposed in [9], whereas the *DriverAgent B* implements the new cooperative car-following model *CoopKrauß* (see Eq. 4).

$$v(t + \Delta t) - \min\{v_{max}, v(l) + u\Delta l, v_{safe}(l)\} \tag{1}$$

$$v_{safe}(t) = v_l(t) + \frac{g(t) - v_l(t)t_r}{\frac{v_l(t)+v(t)}{2b} + t_r} \tag{2}$$

where:
    $v(t + \Delta t)$: following vehicle's velocity at $t + \Delta t$,
    $v(t)$: momentary velocity,
    $v_{safe}(t)$: safe velocity,
    $a$: maximum acceleration $(a > 0)$,
    $b$: maximum deceleration $(-b < 0)$,
    $v_{max}$: vehicle's maximum velocity,
    $\Delta t$: time step,
    $t_\tau$: driver's reaction time,
    $v_l$: velocity of the preceding vehicle,
    $g$: gap to the preceding vehicle.

By solving Eq. 1, the *DriverAgent A* defines the optimal velocity $v(t + \Delta t)$ for time step $t$ that should allow the following vehicle to maintain collision-free driving in the next time step. Also, the *DriverAgent A*'s action is defined by Eq. 3, which is the amount of acceleration or deceleration to be communicated to the *SimControlAgent* to forward to SUMO and act on the following vehicle's velocity.

$$\frac{dv(t)}{dt} = \frac{v(t + \Delta t) - v(t)}{\Delta t} \tag{3}$$

where:

$v(t)$: momentary velocity of the following vehicle,

$v(t + \Delta t)$: optimal velocity of the following vehicle for time step $t + \Delta t$,

$\frac{dv(t)}{dt}$: a deterministic acceleration that allows the model to accurately match $v(t + \Delta t)$ at $t + \Delta t$,

*DriverAgent A* Communicates $s'_A$ to *DriverAgent B*, who in its turn implements the new cooperative model to take into consideration *DriverAgent A*'s current and future state (see Eq. 4).

$$\frac{dv(t)}{dt} = \frac{v'(t + \Delta t) - v(t)}{\Delta t} \tag{4}$$

$$v'(t + \Delta t) = \frac{v(t + \Delta t) + v(t + 2\Delta t)}{2} \tag{5}$$

$$v(t + \Delta t) = \min\{v_{max}, v(t) + a\Delta t, v_{safe}(t)\} \tag{6}$$

$$v_{safe}(t) = v_l(t) + \frac{g(t) - v_l(t)t_r}{\frac{v_l(t)+v(t)}{2b} + t_r} \tag{7}$$

$$v(t + 2\Delta t) = \min\{v_{max}, v(t + \Delta t) + a\Delta t, v_{safe}(t + \Delta t)\} \tag{8}$$

$$v_{safe}(t + \Delta t) = v_l(t + \Delta t) + \frac{g(t + \Delta t) - v_l(t + \Delta t)t_r}{\frac{v_l(t+\Delta t)+v(t+\Delta t)}{2b} + t_r} \tag{9}$$

$$g(t + \Delta t) = g(t) + d_f - d_l \tag{10}$$

$$d_f = \frac{v(t) + v(t + \Delta t)}{2} \times \Delta t \tag{11}$$

$$d_l = \frac{v_f(t) + v_f(t + \Delta t)}{2} \times \Delta t \tag{12}$$

where:

$v'(t + \Delta t)$: following vehicle's optimal velocity with consideration of the leading vehicle's intent.

$v(t)$: momentary velocity of the following vehicle,

$v_l(t)$: momentary velocity of the preceding vehicle,

$v(t + \Delta t)$ and $v(t + 2\Delta t)$: following vehicle's desired velocity at time steps $t + \Delta t$ and $t + 2\Delta t$, respectively.

$v_{safe}(t)$ and $v_{safe}(t + \Delta t)$: following vehicle's safe velocity at time steps $t$ and $t + \Delta t$, respectively.

$g(t)$: gap to the vehicle at time step $t$.

$g(t + \Delta t)$: gap to the preceding vehicle at time step $t + \Delta t$.

$d_f$: traveled distance by the following vehicle between $t$ and $t + \Delta t$.

$d_l$: traveled distance by the leading vehicle between $t$ and $t + \Delta t$.

Based on the simulation results presented in Fig. 6, the *Krauß* model showed significant fluctuations in the action values, reflecting abrupt changes in vehicle velocity and a less comfortable driving style. In contrast, the *CoopKrauß* model demonstrated more stable and controlled actions, leading to smoother changes in vehicle velocity and more comfortable driving style. This difference is attributed to the nature of the cooperative model, which incorporates the preceding vehicle's state and future intent into its decision-making process. This enables more informed and coordinated actions, ultimately enhancing driving comfort and safety.

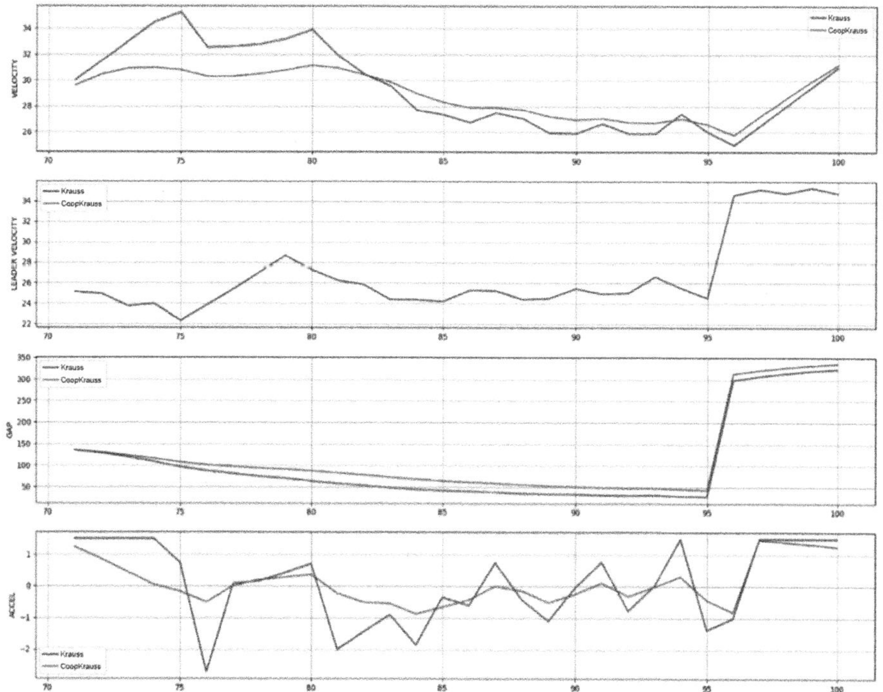

**Fig. 6.** Comparison of Vehicle Dynamics Between *CoopKrauß* and *Krauß* Models: Velocity, Gap, and Acceleration Changes.

## 5   Conclusion

In this paper, we presented a agent-based microscopic simulation framework, focusing on the interactions of heterogeneous vehicle types, including intelligent and connected vehicles. By integrating JADE and SUMO, our framework provides the necessary tools to study the complex dynamics of urban traffic and the impacts of deploying advanced driving systems and models.

The demonstration of the cooperative car-following model highlights the framework's capability to simulate and analyze the complex behavioral changes introduced by V2V communication. The comparative results between the *Krauß* and *CoopKrauß* models reveal significant differences in driving style, with the latter showing smoother and more comfortable driving.

Our framework's flexibility allows researchers to introduce their own models, enabling detailed exploration of various driving dynamics and vehicle types. This serves as an invaluable tool for future studies in traffic flow, safety, fuel efficiency, and emissions reduction. This simulation framework offers a significant contribution to the research and development of intelligent transportation systems, paving the way for more informed decisions in the design and deployment of urban mobility solutions.

## References


1. Azevedo, T., De Araújo, P.J., Rossetti, R.J., Rocha, A.P.C.: Jade, trasmapi and sumo: a tool-chain for simulating traffic light control. arXiv preprint arXiv:1601.08154 (2016)
2. Bellifemine, F., Poggi, A., Rimassa, G.: Developing multi-agent systems with jade. In: Intelligent Agents VII Agent Theories Architectures and Languages: 7th International Workshop, ATAL 2000 Boston, MA, USA, July 7–9, 2000 Proceedings 7, pp. 89–103. Springer (2001)
3. Buehler, M., Iagnemma, K., Singh, S.: The 2005 DARPA grand challenge: the great robot race, vol. 36. Springer Science and Business Media (2007)
4. Buehler, M., Iagnemma, K., Singh, S.: The DARPA urban challenge: autonomous vehicles in city traffic, vol. 56. springer (2009)
5. Fellendorf, M., Vortisch, P.: Microscopic traffic flow simulator VISSIM. In: Fundamentals of traffic simulation, pp. 63–93. Springer (2010)
6. Hult, R., Campos, G.R., Steinmetz, E., Hammarstrand, L., Falcone, P., Wymeersch, H.: Coordination of cooperative autonomous vehicles: toward safer and more efficient road transportation. IEEE Signal Process. Mag. **33**(6), 74–84 (2016)
7. Jakob, M., Moler, Z.: Modular framework for simulation modelling of interaction-rich transport systems. In: 16th International IEEE conference on intelligent transportation systems (ITSC 2013), pp. 2152–2159. IEEE (2013)
8. Krajzewicz, D., Erdmann, J., Behrisch, M., Bieker, L.: Recent development and applications of SUMO–simulation of urban mobility. Int. J. Adv. Syst. Meas. **5**(3&4) (2012)
9. Krauß, S.: Microscopic modeling of traffic flow: investigation of collision free vehicle dynamics. Ph.D. thesis (1998)
10. Luo, Z., et al.: A hybrid method for predicting traffic congestion during peak hours in the subway system of shenzhen. Sensors **20**(1), 150 (2019)

11. Ma, J., Zhou, F., Huang, Z., Melson, C.L., James, R., Zhang, X.: Hardware-in-the-loop testing of connected and automated vehicle applications: a use case for queue-aware signalized intersection approach and departure. Transp. Res. Rec. **2672**(22), 36–46 (2018)

12. Milanés, V., Shladover, S.E., Spring, J., Nowakowski, C., Kawazoe, H., Nakamura, M.: Cooperative adaptive cruise control in real traffic situations. IEEE Trans. Intell. Transp. Syst. **15**(1), 296–305 (2014)

13. Sadeghi Garjan, M., Chaanine, T., Pasquale, C., Paolo Pastore, V., Ferrando, A.: Agamas: a new agent-oriented traffic simulation framework for sumo. In: European Conference on Multi-Agent Systems, pp. 396–405. Springer (2023)

14. Saifuzzaman, M., Zheng, Z.: Incorporating human-factors in car-following models: a review of recent developments and research needs. Transportation research part C: emerging technologies **48**, 379–403 (2014)

15. Soares, G., Kokkinogenis, Z., Macedo, J.L., Rossetti, R.J.: Agent-based traffic simulation using sumo and jade: an integrated platform for artificial transportation systems. In: Simulation of Urban Mobility: First International Conference, SUMO 2013, Berlin, Germany, May 15-17, 2013. Revised Selected Papers 1, pp. 44–61. Springer (2014)

16. Treiber, M., Hennecke, A., Helbing, D.: Congested traffic states in empirical observations and microscopic simulations. Phys. Rev. E **62**(2), 1805 (2000)

17. Wakkumbura, R.T., Hettige, B., Edirisuriya, A.: Real-time traffic controlling system using multi-agent technology. Journal Européen des Systèmes Automatisés **54**(4), 633–640 (2021)

# A Novel Fuzzy Salp Swarm Classifier for Face Identification


Salima Nebti[1]([envelope]) [iD], Kenza Redjimi[2], and Mohamed Redjimi[2]

[1] Department of Communications, Emir Abdelkader University, Constantine 25000, Algeria
s.nabti@univ-emir.dz
[2] Department of Computer Science, 20 Août 1955 University, Skikda 21000, Algeria
{k.redjimi,m.redjimi}@univ-skikda.dz



**Abstract.** The increasing demand for advanced security systems incorporating facial recognition technology has led to the creation of various face identification techniques. However, only a select few have been proven to achieve optimal accuracy under specific situations. This study delved into an analysis of the main classifiers, which encompassed approaches rooted in swarm intelligence, support vector machines (SVMs), and deep learning. Specifically, the research introduces the binary Salp swarm algorithm (BSSA) and the binary grey wolf optimizer (BGWO) as novel pairwise classifiers for face identification. These two approaches, known as DT-BSSA and DT-BGWO, are organized in a 3-node binary decision tree, effectively reducing computing time. Additionally, the research explores the combination of CNN with a statistical classifier based on the salp swarm algorithm (SSA). Furthermore, various techniques for reducing feature space are examined, including PCA, LDA, and KFA. Comparative evaluation with decision tree-based fuzzy SVM classifier and modern methods of recognition using deep learning, such as CNN, shows that swarm-based classifiers displayed the highest efficacy in categorizing modestly sized databases. Nevertheless, the Deep learning solution emerged as the best choice in terms of speed and adaptability for applications requiring real-time processing.

**Keywords:** Binary Salp Swarm Algorithm · Grey wolf optimizer · fuzzy membership function · facial recognition


## 1 Introduction

Being the primary method of biometric identification, face recognition offers numerous advantages. It is widely accepted and easily understood by individuals, and human operators can naturally intervene in machine decisions. In fact, face images are commonly utilized as a secondary verification method by automated fingerprint identification systems [1]. The task of identifying faces is extremely difficult due to several factors, including changes in lighting, different facial expressions and poses, the effects of aging on the face, and occlusions like makeup, glasses, or hair. Developing an automated system that can successfully accomplish this task is a daunting challenge. While there have been numerous systems created that have achieved recognition rates above 90%, achieving a flawless system with a 100% recognition rate still a challenge [2].



N. Seddari and M. Redjimi (Eds.): ICMSCT 2024, CCIS 2606, pp. 360–371, 2025.
https://doi.org/10.1007/978-3-032-01922-6_30

The process of recognizing faces involves three key stages: feature extraction, face detection, and recognition [3]. To accomplish this, various techniques are commonly employed, which can be categorized into three groups [4]. The first group focuses on the entire face's appearance, while the second concentrates on specific facial features or regions. The third group combines both local and global characteristics simultaneously.

Techniques that rely on appearance, such as PCA [8], LDA [5], ICA [6], Kernel PCA, and Kernel LDA [7] …etc., consider the overall attributes of the facial image. These techniques effectively calculate basis vectors to represent the face data. In the subsequent phase, the faces are projected onto these vectors, and the resulting projection coefficients serve as representations of the face images [1].

The second approach, which focuses on local features, has also seen significant research efforts. One prominent technique is Local Binary Pattern (LBP) [8, 9], which is effective in reducing image noise.

The neural approach is based on a fundamental concept of acquiring knowledge about the connections between input faces and their respective categories. After the connections are learned, the weights are stored, allowing for quick testing. Therefore, this approach is the most practical choice due to its prompt responsiveness, [19].

In order to achieve peak performance in a facial recognition system, it is imperative to thoroughly assess the available implementation options, with a particular focus on selecting the appropriate classifier. The choice of classifier plays a critical role in maintaining a high level of performance. To improve recognition accuracy, this study compares various recently introduced classification techniques, analyzing the strengths and weaknesses of each approach to determine the optimal solution. To accomplish this, we extensively evaluate four facial classification methods: a Salp Swarm-based approach, a grey wolf-based classifier, a fuzzy support vector machine (SVM), and a convolutional neuron network (CNN). The motivation behind this research stems from the lack of comprehensive comparisons among facial recognition techniques, including swarm-based systems, support vector machines, decision tree-based approaches, and convolutional neural networks, in terms of both accuracy and speed.

In the realm of literature, there are only a limited number of comparisons that exist. These comparisons typically involve contrasting modern methods with more traditional ones like Eigenface, SVM, and CNN [10]. Another common approach is to utilize CNN as a feature extractor while employing SVM for classification [11, 12]. However, this study takes a comprehensive approach by comparing a DT-BSSA, DT-BGWO, DT-FSVM, and a CNN on databases of moderate size.

The main aim of this work is to expedite the search process in the learning phase by implementing a decision tree-based swarm intelligence classification approach. This approach incorporates the binary salp swarm optimization technique or grey wolf optimization to address the challenge of binary classification between two classes. To achieve this, a three-node decision tree is employed. The decision tree evaluates each of the three face categories in the aforementioned database by using the fuzzy BSSA classifier or the fuzzy BGWO to compare them with the input face. In order to evaluate its efficacy, this approach underwent testing on three widely used databases and was compared to recent and successful classifiers like the decision tree-fuzzy support vector machine classifier and a convolutional neural network CNN.

Furthermore, this study offers readers a clear roadmap to choose the most precise and time-saving approach to facial recognition. It is still difficult to compare the most recent and efficient face recognition systems, even when a plethora of research was done to create novel approaches. The majority of the comparisons that are currently available in the literature compare newer techniques with traditional ones, such as Eigenface, SVM, and CNN [17], or utilizing CNN as a feature extractor and SVM as classifier [11]. This study, however, goes above and beyond by providing an extensive comparison using moderately sized databases across DT-BSSA, DT-BGWO, CNN, and DT-FSVM.

The structure of this document is organized into four sections. It commences with a detailed explanation of the BSSA algorithm and its application in classification. Subsequently, we explore the decision tree-based Salp Swarm classification method and present the accompanying experimental results. Finally, the conclusion provides a cohesive summary of the main findings and offers an analysis of potential future prospects.

## 2   The Binary Salp Swarm Algorithm

The Salp Swarm Algorithm (SSA) is a method of optimization that mimics the movement of salps as they navigate through the water in search of nourishment. These translucent sea creatures gracefully twist and turn as they swim, with one salp leading the way and guiding the others towards the most abundant food source. As the leader salp moves closer to the optimal food supply, the first follower faithfully follows suit and, subsequent followers gradually align themselves with the preceding salps. In our suggested variation of binary SSA, the locations of the salps are converted into binary form using the sigmoid function [13].

In a spiral pattern, semi-translucent salps follow a leader salp to locate the most fitting food source. Through this process, the leader guides the group of followers to locate the best food source [14] (Fig. 1).

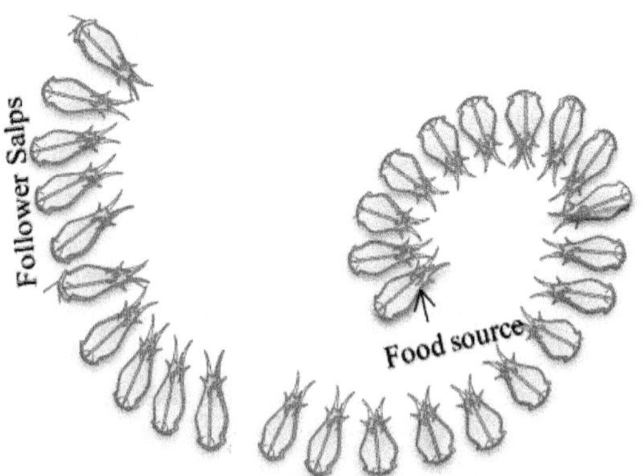


**Fig. 1.** Salps' chain

A two-dimensional matrix $X_{ij}$ ($i = 1...N, j = 1...d$) can be used to specify the salps positions in a d-dimensional issue, where d is the problem's dimensionality and N is the number of salps. The SSA method is used to model the salp colony's displacement process by using the following formulation [14].

$$X_{1j} = \begin{cases} F_j + c_1.r_1\left(Ub_j - Lb_j\right) + c_1.Lb_j, & r_2 \geq 0 \\ F_j - c_1.r_1\left(Ub_j - Lb_j\right) + c_1.Lb_j, & r_2 < 0 \end{cases} \tag{1}$$

where the positions of the food supply and leader in the jth dimension are denoted by $F_j$ and $X_{1j}$, respectively.

$r_1 \& r_2$ are selected randomly from the range of values between 0 and 1 in each iteration.

The algorithm is assessing the salps' fitness, locating the optimal food supply, or position $F_j$, and then using Eq. (1) to direct the leading salp, $X_{1j}$, in that direction [14].

Over the course of the iterations, the variable $c_1$ is steadily reduced and is determined as provided in Eq. (2).

$$c_1 = 2.e^{-\left(\frac{4\,r}{rmax}\right)^2} \tag{2}$$

where r and rmax are the current iteration and the maximum number of iterations, respectively.

The follower salps shift their positions in accordance with Eq. (3)

$$X_{ij} = \frac{1}{2}\left(X_{ij} + X_{(i-1)j}\right) \tag{3}$$

The new Salps' positions in the jth dimension are then restricted by the upper bound $Ub_j$ and the lower bound $Lb_j$.

SSA randomly creates a population of solutions, or salps. Next, an objective function is used to assess the resulting solutions. The Food Source F is the fittest solution in SSA, and it will be pursued by other solutions (follower salps). Every iteration, the positions of the follower salps are updated using Eqs. 1& 3, the $c_1$ variable is updated using Eq. 2, and each dimension in the leader (best salp) is updated using Eq. 1. Every preceding step is repeated until a maximum number of iterations.

In this work, we used a sigmoid function on the newly discovered salps positions to convert them into binary values.

Then if the new position $< 0.5$, the new position $= -1$, else the new position $= 1$.

The proposed BSSA optimizer can be summarized as follows:

---

1. Salps Initialization.
repeat,
2. Salps assessment
3. Select the optimal food source (best salp)
4. update the leader position using eq (1)
5. displace the followers using eq (3)
6. restrict the positions of leaders and followers between
   the range [ $Ub_j$, $Lb_j$]
7. Convert them to binary values
8. till a predetermined maximum number of iterations

---

## 2.1  The BSSA (BGWO) Classifier

In the BSSA classifier, each Salp (grey wolf) is initialized with three training feature vectors (F1, F2, F3) chosen randomly from two candidate classes. The final classification is determined by the best found Salp (the best grey wolf), and this solution classifies each input pattern based on its elements sum: the input pattern's class is the first if the best Salp has all values of "1," and the second if the sum of the best Salp (grey wolf) elements is less than 2.

Each Salp (grey wolf) starts with three randomly selected training feature vectors (F1, F2, and F3) from each of the two classes. The best-found Food source determines the final classification, and classifies the input pattern according to the sum of its elements: the input pattern belongs to class 1 if the best Salp (wolf) has all values of "1," and class 2 if the best Salp's (wolf's) element sum is less than 2.

The BSSA (BGWO) classifier utilizes a fitness function that incorporates the Euclidean distances between the input facial features and the current Salp (wolf) features. The correspondent binary Salp (binary wolf) maintains the target values for the best Salp (best wolf).

In the employed binary SSA approach, Salps are initially assigned feature values. Then, following the methodology outlined in [13], a sigmoid function is applied to convert the solutions into binary form.

The BGWO based approach is quite similar to the BSSA-based one, in this approach the binary version of GWO is used in place of the BSSA while preserving the tree structure and the fuzzy rules. To tackle the problem of face recognition, we implemented the BGWO1 version as suggested by Too et al. [18]. This type of classifiers encompasses other algorithms such as the application of Gabor wavelets to identify the distinguishing features of each facial input and the dimensionality reduction techniques: PCA, LDA, and KFA (Fig. 2).

|                | Selected Features |      |      |
|----------------|-------------------|------|------|
| Salp1          | F1                | F2   | F1   |
| Binary Salp1   | -1                | -1   | 1    |
| ...            | ...               | ...  | ...  |
| SalpN          | F2                | F1   | F1   |
| Binary SalpN   | -1                | 1    | -1   |

**Fig. 2.** An example showing how the positions of Salps relate to their binary positions

To tackle the problem of face recognition, the binary SSA based classifier can be summarized in the following steps:

The BSSA-based classifier solves the facial recognition problem in the following steps:

---

**BSSA classification**

---

1. Configure the BSSA settings.
2. Randomly select faces from the training database belonging to two different classes
3. Give them binary numbers 1 for the first class and —1 for
   the second class.
4. Employ BSSA to optimize the Salps' placements based on
   the Euclidian distance between features as a fitness
   measure.
5. Assign the best binary position to the best-found food
   source.
6. The Sum of the binary best solution elements is computed
   to determine the class:
   In the case that the sum is greater than or equal to 2, the
   input face is assigned to the first class;
   otherwise, it is assigned to the second class.

---

## 2.2  THE DT-BSSA Classifier

Presented in this research is a newly developed classifier that draws inspiration from the DT-FSVM discussed in [15], while still maintaining fuzzy rules in each BSSA classification and substituting the binary SVM classification with the BSSA classifier. Because decision trees need less memory and provide efficient data access, they have become more and more popular in the field of computer science. This study combines the advantages of decision trees with BSSA optimization. First, a predetermined array of possible classes is divided into groups of three, which are subsequently input into the BSSA classifier. The classifier's results are then recursively grouped into additional sets of three. This process continues until each array contains only one class that corresponds to the input face's class.

The DT-BSSA classification can be summarized as follows:

---

**DT-BSSA classification**

1. For each input face $x_i$ do
2. initialize an array of potential classes and another array of the same size to put results.
3. Repeat
4. for k=1:3:Nb-classes do.
5.  Define a sub-group comprising the training data of the three current classes.
6.   For i=1:3
7.    For j=1:3
8.     Use the BSSA classifier to classify $x_i$ into either of the two classes (subgroup (i) or subgroup (j)).
9.     the result is a 3 by 3 matrix $M_{ij}$
10.     Use Equations (4 & 5) to assess $x_i$'s fuzzy membership to the three current classes.
11.     Use Eq (6) to find the most similar class to $x_i$.
12.    End for j
13.   End for i
14. End for k
15. Place the found class in the result-array.
16. replace the class-array by the result-array
17. repeat the previous steps until the result-array solely contains the input face's class.

---

The used fuzzy membership function $m_i(x)$ of $x$ for class $i$ is defined as follows [16]:

$$m_i(x) = \underset{j \neq i, j = 1..n}{min} \left(M_{ij}(x)\right) \tag{4}$$

n is the number of classes

$$\text{with } M_{ij}(x) = \begin{cases} 1, & if M_{ij}(x) \geq 1 \\ M_{ij}(x), & otherwise \end{cases} \tag{5}$$

and $M_{ji}(x) = -M_{ij}(x)$.

As a result, 's class is found by:

$$class(x) = \underset{i = 1..n}{argmax} \, m_i(x) \tag{6}$$

## 3   Combination with Swarm-Based Classifiers

The central idea behind combining classifiers is the ability to decrease errors in classification through the use of complementary classifiers [21]. The research conducted in this study entailed the integration of the output labels generated by a Convolutional Neural Network (CNN) with a classifier employing SSA through a majority vote combination rule.

The SSA based statistical classifier is founded on a principle that involves the random selection of faces, which are represented by their characteristic features, from the learning base. The classifier's objective is to optimize the similarity between the learning faces and the set of input faces. To achieve this, each Salp possesses a collection of facial patterns from the face storage repository. The Salp size is equal to the number of testing faces. The Salp based algorithm prioritizes minimizing the overall sum of Euclidean distances between the training faces and the faces being identified. The most effective solution is the Salp with the lowest fitness value. The best Salp's characteristic vectors are then set as the cluster centers of the learning base for identification purposes. The test classes are determined by minimizing the sum of squared errors (SSE) between the cluster centers and the faces in the training dataset.

## 4 Experiments

A comprehensive study was conducted to evaluate the effectiveness of the suggested techniques by extensively comparing two widely recognized databases, namely ORL and YALE and FERET.

**The Decision Tree Based Approaches**
In the DT-BSSA classifier, the utilization of Gabor wavelet features is implemented. Initially, these features are extracted, but only the most valuable ones are retained through the application of PCA, LDA and KFA subspace techniques.

The initial experiments focused on examining the impact of each subspace generating method on recognition accuracy using the DT-BSSA and DT-FSVM classifiers. The accompanying tables provide a comprehensive explanation of the recognition rates achieved and then compared to FSVM and a CNN based approach.

**Table 1.** DT-BSSA recognition rates

| Classifier | ORL | YALE | FERET |
| --- | --- | --- | --- |
| PCA + DT-BSSA | 95.5 | 84% | 77.75% |
| LDA + DT-BSSA | 98.5% | 100% | 94% |
| KFA + DT-BSSA | 97% | 93.33% | 74% |
| Classifier combination | 99% | 97.33% | 84% |

The recognition rate is consistently improved when employing the majority vote strategy, as evidenced by the findings presented in Table 1. Additionally, the data in Table 1 highlights that the LDA reduction space method is the best method for selecting informative features, resulting in a significant boost in the recognition rate compared to KFA.

By examining of the figures below (Fig. 3), it is apparent that LDA is the superior technique for space reduction, with KFA following close behind, and then PCA. In addition, it has been observed that the greatest recognition rates are achieved when the

**Table 2.** DT-BGWO recognition rates

| Classifier | ORL | YALE | FERET |
| --- | --- | --- | --- |
| PCA + DT-BGWO | 86% | 76% | 60.75% |
| LDA + DT-BGWO | 98.5% | 98.66% | 93.75% |
| KFA + DT-BGWO | 96% | 93.33% | 74% |
| Classifier combination | 98.5% | 97.33% | 81.75% |

**Table 3.** FSVM recognition rates

| Classifier | ORL | YALE | FERET |
| --- | --- | --- | --- |
| PCA + DT-FSVM | 93% | 90.66% | 88.25% |
| LDA + DT-FSVM | 98.5% | 98.66% | 93.5% |
| KFA + DT-FSVM | 96% | 93.33% | 75.25% |
| Classifier combination | 97.50% | 98.67% | 89.50% |

number of classes is less than 15 on ORL database. This is due to the fact that DT-BSSA, DT-BGWO and DT-FSVM are binary classifiers, with the final decision being based on the majority vote of several classes. As a result, distinguishing between great number of classes becomes more difficult, resulting in a lower recognition rate. Furthermore, these three classifiers exhibit similar behavior when changing the number of face classes, particularly when KFA and LDA are utilized as search space reducers.

**The CNN Based Approach**

Following an extensive process of experimentation and modification of parameters, the architecture that found most appropriate was as follows: images of 64 by 64 pixels as input, two convolutional layers: the first with 32 filters of size 5 * 5, followed by a batch Normalization layer, ReLU as the activation function and a 2 * 2 filter for maximum pooling. The second convolutional layer contains 64 filters each with a size of 5 by 5, followed by another batch Normalization layer. The activation function utilized was ReLU, and maximum pooling was achieved using a filter of 2 by 2. Finally, a fully connected layer, followed by a softmax layer, and an output layer were added to complete the classification of the extracted features. The learning rate was 0.004, the minibatch size was 5 for YALE and ORL databases and 4 for FERET, the maximum number of epochs was 10 for YALE and ORL and 25 epochs for FERET. The optimizer Adam (Adaptive moment estimation) was chosen for YALE and ORL while sgdm (Stochastic gradient descent with momentum) was chosen for FERET.

Below are the results obtained.

The results (Tables 1, 2, 3 & 4) indicate that, when utilized for recognizing the YALE ORL and FERET databases, the DT-BSSA classifier outperforms both the DT-FSVM and CNN classifiers in terms of recognition rate. Furthermore, implementing the CNN-based approach presents difficulties due to the multitude of factors that must be taken

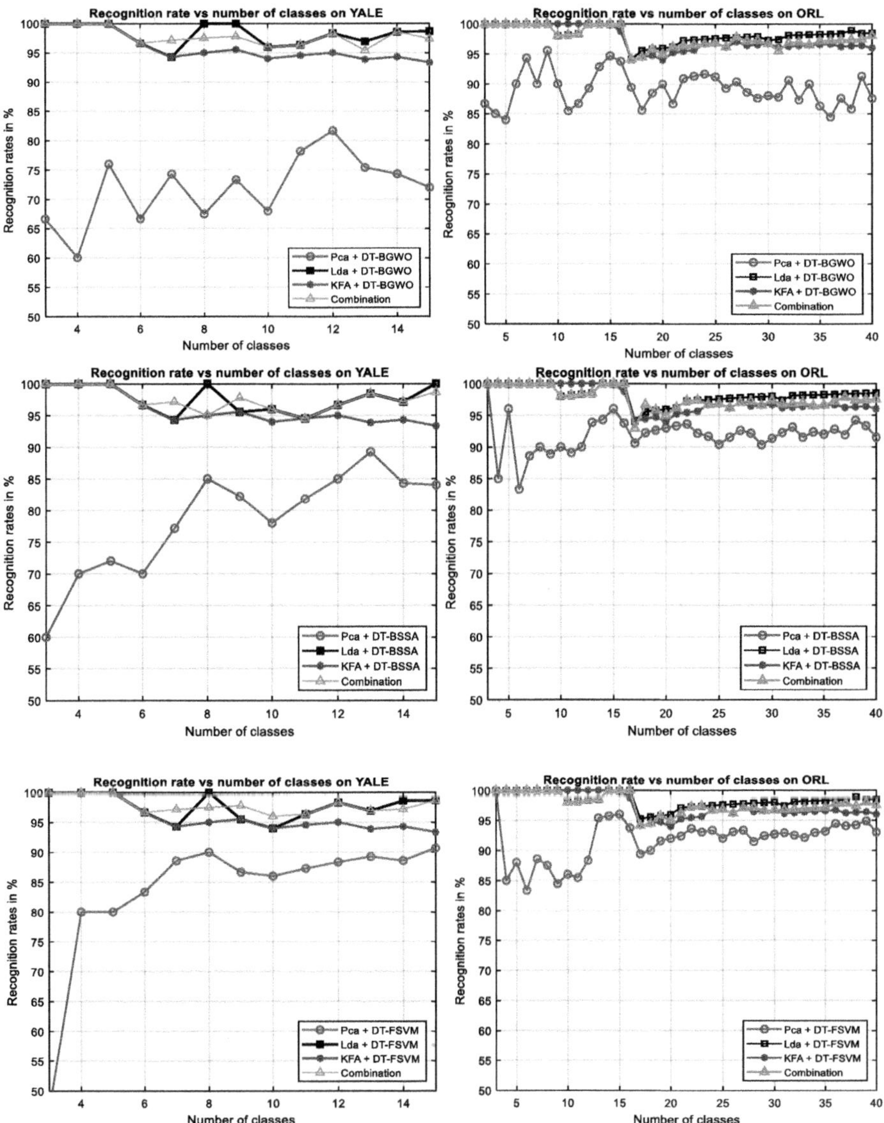

**Fig. 3.** The behavior of each algorithm depending on face categories

into account, such as the size of the patterns to be identified, the number and type of
convolutional and pooling layers, the selection of filters and their dimensions, the stride
of the filters, the choice of activation function, the selection of learning method and its
parameters, the choice of loss function, and the determination of the number of fully
connected or hidden layers, and others.

**Table 4.** CNN recognition rates

| Classifier | ORL | YALE | FERET |
| --- | --- | --- | --- |
| CNN | 91.50% | 90.67% | 65.25% |
| SSA | 92 | 98.67% | 53.50% |
| CNN + SSA | 92.5% | 96% | 51.75% |

## 5  Conclusion

This research focused primarily on the main issue surrounding face recognition. The study delved into an analysis of the key classifiers, which encompassed approaches rooted in swarm intelligence, support vector machines (SVMs), and deep learning. Instead of utilizing swarm optimization techniques for feature selection or parameter optimization, this study introduces new pairwise classifiers called DT-BSSA and DT-BGWO which are based on Binary Salp Swarm Algorithm (BSSA) and binary grey wolf optimization. The face identification problem is solved by DT-BSSA (DT-BGWO) through a sequence of binary classifiers that utilize Binary SSA (BGWO). These classifiers are organized in a 3-node binary decision tree, which reduces computation time when searching across the database.

The proposed research has been evaluated in comparison with a contemporary deep learning classifier (the CNN) and another decision tree-based fuzzy SVM classifier to highlight its importance. Three separate subspace techniques, namely PCA, LDA and KFA, were used to analyze the proposed classifiers.

Through this analysis, it was discovered that the DT-BSSA outperformed the DT-BGWO, the DT-FSVM and CNN in identifying small to medium sized datasets. Additionally, the combination of Gabor magnitude features with LDA space reduction yielded higher accuracy rates compared to using PCA or KFA.

The suggested approaches can be effectively utilized in various biometric authentication scenarios, including the recognition of images in real-time video streams. Additionally, when combined with other swarm-based optimizers that have demonstrated superior, the DT-BSSA classifier may yield even more accurate results. Furthermore, to enhance the accuracy of the DT-BSSA classifier, the inclusion of the Hamming or Cosine distance in the objective function can be considered.

## References


1. Ray, J.: Swarm Optmization Algorithms for Face Recognition. Diss. (2013)
2. Kaur, N., Kaur, R.: A review paper on an enhanced face recognition system using correlation method and abpso. Int. J. Eng. Sci. Res. Technol. (2016)
3. Ding, C., Tao, D.: A comprehensive survey on pose-invariant face recognition. ACM Trans. Intell. Syst. Technol. 7(3), 1–42 (2016)
4. Chihaoui, M., Elkefi, A., Bellil, W., Amar, C.B.: A survey of 2D face recognition techniques. Computers 5(4), 41–68 (2016)
5. Belhumeur, P.N., Hespanha, J.P., Kriegman, D.J.: Eigenfaces vs. fisherfaces: recognition using class specific linear projection. IEEE Trans. Pattern Anal. Mach. Intell. 19(7), 711–720 (1997)

6. Bartlett, M.S., Movellan, J.R., Sejnowski, T.J.: Face recognition by independent component analysis. IEEE Trans. Neural Netw. **13**(6), 1450–1464 (2002)
7. Yang, M.H.: Kernel eigenfaces vs. kernel fisherfaces: face recognition using kernel methods. In: Fifth IEEE International Conference on Automatic Face and Gesture Recognition 2002, Proceedings, pp. 215–220 (2002)
8. Ahonen, T., Hadid, A., Pietikainen, M.: Face description with local binary patterns: application to face recognition. IEEE Trans. Pattern Anal. Mach. Intell. **28**(12), 2037–2041 (2006)
9. Chen, J., Patel, V.M., Liu, L., et al.: Robust local features for remote face recognition. Image Vis. Comput. **64**, 34–46 (2017)
10. Paul, S., Acharya, S.K.: A comparative study on facial recognition algorithms. In: e-Journal-First Pan IIT International Management Conference (2020)
11. Matsugu, M., Mori, K., Suzuki, T.: Face recognition using SVM combined with CNN for face detection. In: Pal, N.R., Kasabov, N., Mudi, R.K., Pal, S., Parui, S.K. (eds.) ICONIP 2004. LNCS, vol. 3316, pp. 356–361. Springer, Heidelberg (2004). https://doi.org/10.1007/978-3-540-30499-9_54
12. Guo, S., Chen, S., Li, Y.: Face recognition based on convolutional neural network and support vector machine. In: 2016 IEEE International conference on Information and Automation (ICIA), pp. 1787–1792) IEEE (2016)
13. Shekhawat, S.S., Sharma, H., Kumar, S., Nayyar, A., Qureshi, B.: BSSA: binary salp swarm algorithm with hybrid data transformation for feature selection. IEEE Access **9**, 14867–14882 (2021)
14. Faris, H., et al.: An efficient binary salp swarm algorithm with crossover scheme for feature selection problems. Knowl.e-Based Syst. **154**, 43–67 (2018)
15. Song, X.N., Zheng, Y.J., Wu, X.J., Yang, X.B., Yang, J.Y.: A complete fuzzy discriminant analysis approach for face recognition. Appl. Soft Comput. **10**(1), 208–214 (2010)
16. Mao, Y., Zhou, X., Pi, D., Sun, Y., Wong, S.T.: Multiclass cancer classification by using fuzzy support vector machine and binary decision tree with gene selection. J. Biomed. Biotechnol. **2005**(2), 160 (2005)
17. Paul, S., Acharya, S.K.: A comparative study on facial recognition algorithms. In e-journal-First Pan IIT International Management Conference (2020)
18. Too, J., Abdullah, AR., Saad, N.M., Mohd Ali, N., Tee, W.: A new competitive binary grey wolf optimizer to solve the feature selection problem in EMG signals classification. Computers (2018)
19. Ajit, A., Acharya, K., Samanta, A.: A review of convolutional neural networks. In: 2020 International Conference on Emerging Trends in Information Technology and Engineering (n-ETITE) (2020)

# Metamodeling of Database Query Languages


Sohaib Hamioud[1,2]([✉]) and Noureddine Seddari[3,4]

[1] LISCO Laboratory, Computer Science Department, Badji-Mokhtar University, Annaba,
Algeria
`sohaib.hamioud@univ-ouargla.dz`
[2] Earth Sciences Department, Kasdi-Merbah University, Ouargla, Algeria
[3] Future Technology Lab, University of Parma, Parco Area Delle Scienze 181/A, Parma, Italy
`n.seddari@univ-skikda.dz`
[4] LICUS Laboratory, Department of Computer Science, Université 20 Août 1955, Skikda,
Algeria



**Abstract.** Database Query Languages are fundamental to perform operations on data and enable users to interact with databases. To ensure efficient and interoperable data manipulation, standard query languages such as SQL provide a common and standardised way to express simple and complex queries for data analysis and manipulation. Unlike SQL, which manipulates relational databases, some query languages, such as MongoDB Query Language (MQL), are better suited for distributed data sources and modern NoSQL systems (Not only Structured Query Language). Specifically, MQL provides powerful capabilities for querying JSON-like structures and handling large volumes of flexible and unstructured data. This paper highlights the importance of metamodeling query languages to enable tool support, extensibility, validation, standardisation, documentation, and interoperability. We present a thorough examination and a systematic assessment of how metamodels facilitate these advantages. For assessment, we implemented examples using our metamodels: SQLM for MySQL and MQLM for MQL. We also employed ATL for model transformations and Xtext for tool generation. Our study demonstrates how metamodels can support query languages, enabling them to achieve greater capabilities and seamless integration across different database systems.

**Keywords:** ATL · Database · Metamodel · MQL · MySQL · Query Language · Xtext


## 1 Introduction

In the realm of data management, query languages play an essential role in facilitating interaction between users and databases. A query language is a computer programming language used to efficiently manipulate and retrieve data from databases. A database is an organised collection of interrelated data stored in a computer system and typically managed by a Database Management System (DBMS). It can contain any type of data



N. Seddari and M. Redjimi (Eds.): ICMSCT 2024, CCIS 2606, pp. 372–384, 2025.
https://doi.org/10.1007/978-3-032-01922-6_31

and is associated with applications ranging from small mobile applications to large-scale enterprise systems. Database management systems can be categorised based on their data models and distribution. The most common model is the Relational DBMS (RDBMS), which stores data in interconnected tables with rows and columns. Although RDBMS enforces ACID (Autonomy, Consistency, Isolation, Durability) properties to ensure data integrity—crucial for transactional and financial systems—it has limitations in handling horizontal scaling and managing both dynamic and unstructured data. To address this limitation, NoSQL [1] was developed to avoid the rigidity of traditional RDBMS schemas, allow unstructured big data storage, and support distributed data.

For each type of DBMS, a query language can be used. For instance, SQL (Structured Query Language) is used with relational databases. It is standardised and allows for the definition, the manipulation, and the querying of data stored in relational schema-based tables. MQL (MongoDB Query Language), on the other hand, is used with the popular NoSQL database MongoDB as a document-oriented query language [1]. MQL is closely related to JavaScript and JSON, as queries are usually written in a syntax similar to JavaScript objects, with data represented in a JSON-like format. There are many commercially available SQL tools with distinct capabilities to help developers, data scientists, and data analysts write and execute SQL queries. Similarly, many free, open-source, and commercial NoSQL databases are available for use. MongoDB, for example, offers both free and paid versions, but it is not fully open source. The choice of database depends on the specific needs of the user.

The need for metamodels in this context is debatable. A metamodel is a higher-level abstraction that defines the structure, rules, and semantics of other models or data schemas. Do we really need an explicit metamodel for the query language? How important is a metamodel of the query language in the process of creating supporting tools? Does a metamodel facilitate language modification along with its supporting tools? To what extent can a metamodel help in promoting interoperability and data migration? To answer these questions, we observe the use of our two metamodels of MySQL and MQL, within an open-source software framework for developing domain-specific languages. The following sections answer the questions using illustrative examples. Section 2 introduces the methodology of inspection and assessment upon which the utility of metamodels is evaluated. Section 2.1 highlights the role of metamodels in developing software tools, using MQLM as a source artifact for tool generation. Section 2.2 sheds light on language extension and evolution through an illustrative extension scenario of the MySQL language. Section 2.4 demonstrates, with an excerpt from our MySQL metamodel, how a model should be structured for validation and documentation, and how it can be integrated into a standardised environment. Section 2.5 introduces an example of migrating SQL queries into MQL queries based on SQLM and MQLM to demonstrate the feasibility of metamodels. Section 3 discusses related works with brief descriptions. Finally, the last section draws conclusions and outlines future work.

## 2 Methodology of Inspection and Assessment

In this study, we examine the role and importance of metamodels in query languages. Our work highlights the benefits of using metamodel-driven design. Our methodology involves a systematic evaluation of key advantages provided by metamodels, framed through the following possible applications: tool support, evolution and extensibility, validation, standardisation, documentation, interoperability and migration. For each application, a concrete example for demonstration is given. Our assessment is based on our two metamodels: SQLM for MySQL and MQLM for MQL.

First, tool support emerges as a significant benefit, as metamodels enable the automatic generation of development tools, such as parsers and editors, which improve developer productivity and reduce errors. In our implementation, we showcase the use of Xtext and MQLM, demonstrating how metamodels facilitate the creation of specialised tools that can be extended and adapted.

Second, evolution and extensibility are critical features of metamodeling, as they allow for the continuous adaptation of languages and systems in response to evolving requirements. Through our implementation, we show how metamodels (e.g., SQLM) can be easily extended to support new query features, proving their flexibility in accommodating changes.

Third, validation, standardisation, and documentation play a crucial role in ensuring the correctness and reliability of query languages. We demonstrate how the use of metamodels (e.g., SQLM) enhances validation mechanisms and promotes standardisation, enabling better documentation practices for both developers and end-users.

Finally, interoperability and migration are key factors in the modern data landscape, where the ability to migrate between different systems and ensure compatibility is paramount. Our implementation highlights the importance of metamodels in facilitating smooth transitions between query languages and systems, showcasing examples where metamodels bridge the gap between SQL and NoSQL systems. Each of these benefits is examined through concrete implementations and examples, which provide practical evidence to support the claims of the metamodel's value in modern software development.

### 2.1  Tool Support

Metamodels enable the development of tools such as syntax checkers, query builders, and code generators. A practical example of the power of metamodels is their use within DSL frameworks, such as EMFText [2] to automatically generate parsers, lexers, and code generators, along with textual and graphical representations via an IDE-compatible interface for the DSL. This includes syntax highlighting, code completion, error checking, debugging, and refactoring. Query languages such as SQL and MQL are DSLs and can be supported by DSL frameworks. Figure 1 illustrates the process of creating tool support for a DSL based on its abstract and concrete syntax.

Figure 2 shows our MQL IDE, which was automatically generated using Eclipse and Xtext [3], based on our MQL Metamodel (MQLM). This IDE offers all the previously mentioned features, such as syntax highlighting and error checking. In this context, a metamodel serves as a blueprint for tool developers. Although Xtext starts with a concrete syntax, the metamodel is still an important artifact, as it is easier to read and use, especially for relatively small languages such as MQL. It would be more natural to use EMFText because it takes metamodels as a starting point to create the tools. However, since we have previously created an Xtext-based concrete syntax for MQL that conforms to our metamodel, we adopted the use of Xtext.

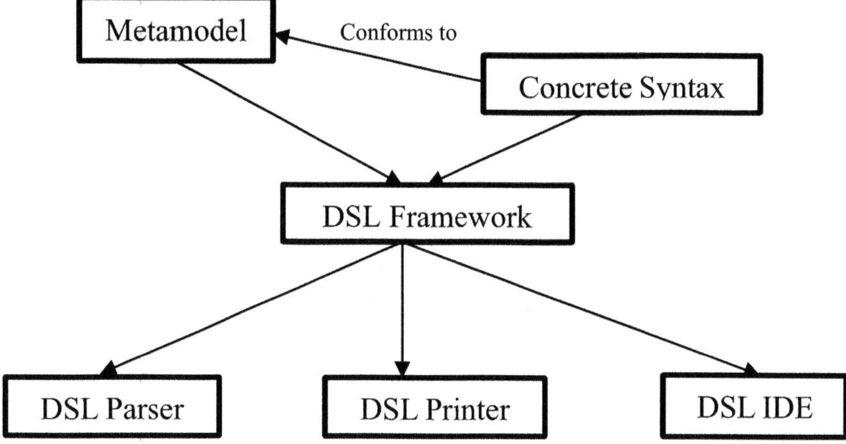

**Fig. 1.** DSL Framework Generation Process.

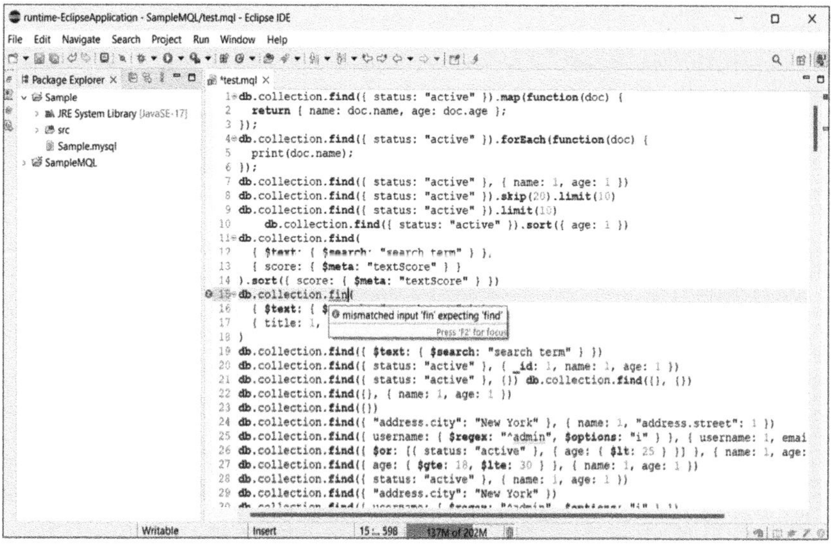

**Fig. 2.** Auto-generated MQL IDE.

## 2.2 Evolution and Extensibility

Metamodels make it easier to modify, update, and evolve query languages over time. New features can be added to the metamodel, and others can be deleted or modified without breaking existing implementations, allowing for smooth transitions and extensions. For example, we consider the scenario of extending SQL to support JSON operators, which were not available before SQL:2016 [4]. JSON data was stored in a text column, and developers had to manipulate the string to extract or modify data. The following SQL code queries JSON data using basic string functions:

```
-- Storing JSON as text and extracting with string functions
SELECT SUBSTRING_INDEX(SUBSTRING_INDEX(json_column, '"name": "', -1), '","', 1) AS name
FROM json_table;
```

Starting with SQL:2016, native support for JSON was introduced, along with its operators:

```
-- Query JSON data using native functions
SELECT json_column->>'$.employee.name' AS employee_name
FROM json_table;
```

A simple modification in this scenario consists of extending the metamodel to include JSON expressions (e.g., Fig. 3), with the corresponding modifications to the concrete syntax:

```
ExpressionAtom
    :
        JsonExpression (COLLATE CollationName)?
;
JsonExpression :
    BitOrExpression ({BinaryOperation.left=current} jsonOperator=JsonOperator right= BitOrExpression)*
;
BitOrExpression :
    BitXorExpression ({BinaryOperation.left=current} bitOrOperator=BIT_OR_OP right= BitXorExpression)*
;
JsonOperator :
        '-' '>' |      '-' '>>'
;

JsonTable:
        JSON_TABLE LR_BRACKET
        STRING COMMA
        STRING
        COLUMNS LR_BRACKET JsonColumnList RR_BRACKET
      RR_BRACKET  (AS? alias=ID)?
;
JsonColumnList:
      jsoncolumns+=JsonColumn (COMMA jsoncolumnd+=JsonColumn)*
;
JsonColumn:
      column_name=(FullColumnName /*|STRING*/) ( FOR ORDINALITY
                        | ( PATH STRING /* =>JsonOnEmpty?*/ jsononerror?=JsonOnError?
                              | EXISTS PATH STRING ) )
    | NESTED PATH? STRING COLUMNS LR_BRACKET JsonColumnList RR_BRACKET
;
JsonOnEmpty:
    (NULL_LITERAL | ERROR | DEFAULT DefaultValue) ON EMPTY
;

JsonOnError: (NULL_LITERAL | ERROR | DEFAULT DefaultValue) ON ERROR
;
TableSource:
        TableSourceItem  joinparts+=JoinPart*
    | JsonTable
;
```

The remainder would involve the automatic regeneration of the tools following our Xtext example.

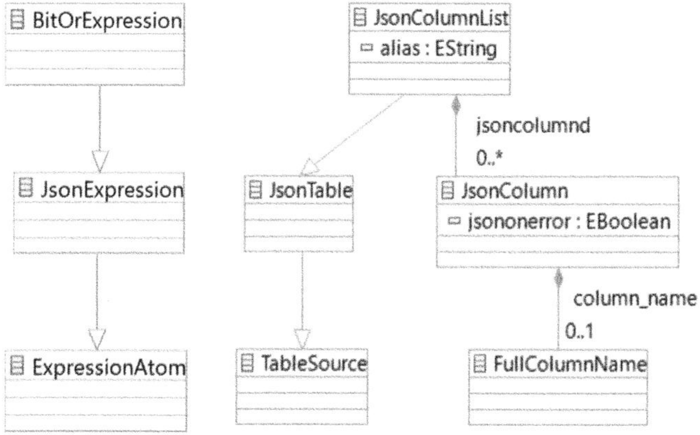

**Fig. 3.** SQLM extension to support JSON expressions.

## 2.3  Validation, Standardisation and Documentation

A metamodel, being the definition of the language structure, allows for the validation
of queries and ensures that they are well-formed before runtime. A query language
can have many implementations. For instance, MySQL [5], T-SQL [6], PL/pgSQL [7],
PL/SQL [8], PostgreSQL [9], and SQLite [10] are variants that generally follow the
basic principles of SQL. A metamodel can serve as a standardised model that ensures
that these implementations are compatible and adhere to the structure and semantics of
the language. Moreover, a metamodel serves as concise and clear documentation of the
query language. It defines the structural elements and their interrelationships within the
language.

Figure 4 represents an excerpt of SQLM. This diagram abstracts away the concrete
constructs of the syntax and focuses on the abstract concepts of the language. For example, an SQL SELECT statement is a DML (Data Manipulation Language) statement,
which is an SQL statement and consists of query specifications. A navigable and graphical representation of the language metamodel helps developers easily understand the
structural features of the language and its capabilities. SQLM can easily be adapted to
any SQL-like language.

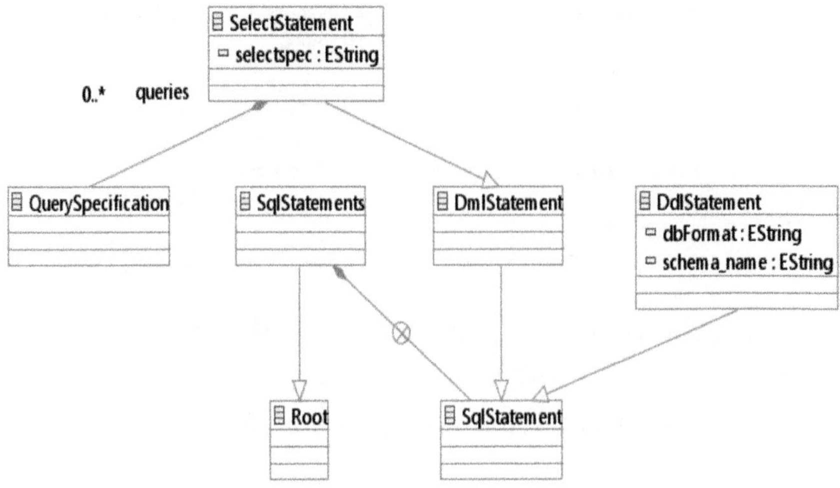

**Fig. 4.** Excerpt of the MySQL Metamodel.

Moreover, a metamodel can easily be integrated with various existing modelling tools, such as EMF (Eclipse Modeling Framework) and used with standardised modelling and transformation languages such as ECORE [11], OCL [12] and QVT [13]. Figure 5 illustrates a representation of an SQL SELECT query based on our MySQL parser and conforming to SQLM. The generated model adheres to SQLM and is displayed in a tree-like format in EMF. It can be read using the metamodel contained in the model, as shown in Fig. 5, and serves as documentation.

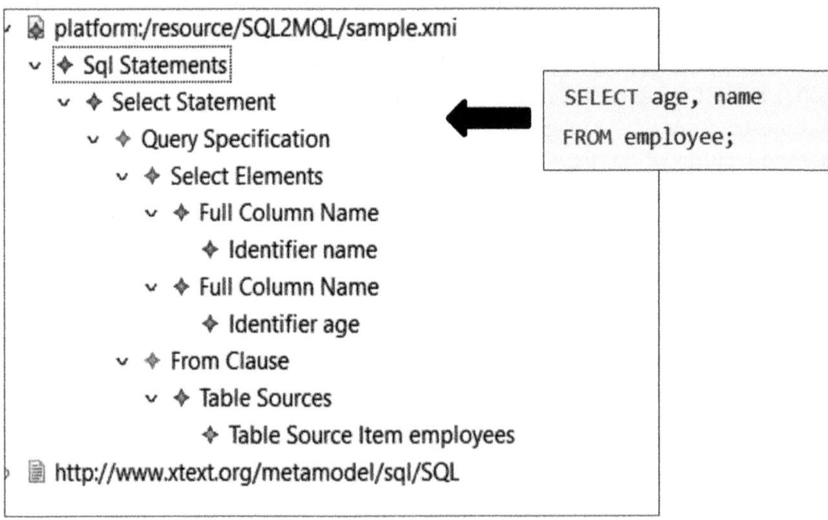


**Fig. 5.** XMI representation of an SQL SELECT query conforming to SQLM.

## 2.4  Interoperability and Migration

A metamodel supports interoperability, as multiple data sources and different query languages can be used in both homogeneous and heterogeneous environment. Metamodels can bridge the gap between distinct systems and overcome this problem by defining a common structure or combining existing metamodels, which facilitates the migration and mapping of queries between different languages. For instance, a migration from SQL to NoSQL can take place for scalability. Using our MySQL and MQL metamodels, the migration can be implemented as a model-based query transformation using ATL [14] or QVT (See Fig. 6).

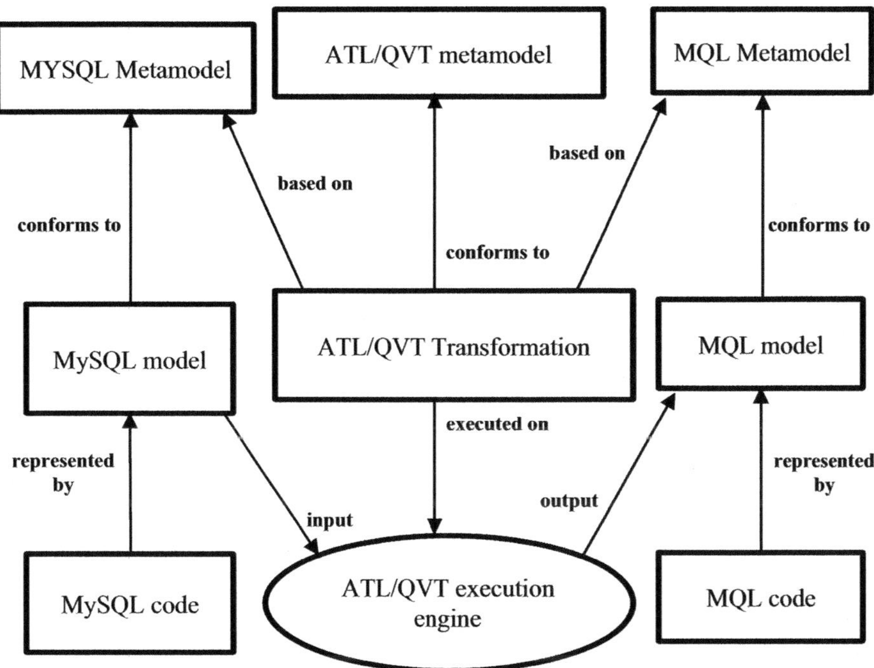


**Fig. 6.** Model-driven MySQL to MQL migration.

MySQL queries will be represented by models conforming to our MySQL Metamodel (SQLM). An ATL or QVT transformation can describe the mapping rules to convert these MySQL models into MQL models that conform to MQLM, as shown in Fig. 6. The following implementation of this example includes a basic ATL transformation rule that enables the mapping of an SQL SELECT query including a FROM clause into an equivalent MQL FIND method query. Both SQLM and MQLM serve as documentation to help understand the transformation.

```
module SQL2MQL;
create OUT : MQLM from IN : SQLM;
helper def : transformToQuery(s : SQLM!SqlStatement) : MQLM!Query =
    if s.oclIsTypeOf(SQLM!SelectStatement) then
        thisModule.SelectQuery2FindQuery(s.oclAsType(SQLM!SelectStatement))
    else
      OclUndefined
    endif;
rule SQLStatementsToModel {
    from
        sqlRoot : SQLM!SqlStatements
    to
        mqlModel : MQLM!Model (
            queries <- sqlRoot.sqlStatements
        )
}
rule BasicSelectQuery2FindQuery {
    from
        selectStatement : SQLM!SelectStatement
    to
        mqlQuery: MQLM!Query (
            collection <- selectStatement.queries->first().clauses->first().fromExpr.tablesources->first().tableName,
            method <- MQLM!FindMethod.newInstance()
        )
    do{      mqlQuery.method.projection <-
        if selectStatement.queries->first().elements.star.oclIsUndefined()then
            thisModule.resolveTemp(selectStatement.queries->first(), 'projection')
        else
          OclUndefined
           endif;
    }
}
rule QuerySpecification2Projection {
    from
        querySpecification : SQLM!QuerySpecification
    to
        projection : MQLM!Projection (

            projectionFields <- querySpecification.elements
        )
}
rule SelectElements2Conditions {
    from
        selectElements : SQLM!SelectElements
    to
        conditionList : MQLM!ConditionList (
        conditions <- selectElements.selectelements->collect(
            selectElement |
                if selectElement.oclIsTypeOf(SQLM!FullColumnName) then
                  thisModule.resolveTemp(selectElement, 'identifierObjectLiteral')
                else
                 thisModule.resolveTemp(selectElement, 'stringObjectLiteral')
                endif
                )
                    )
}
```

```
rule FullColumnName2JSObjectLiteral {
    from
        fullColumnName : SQLM!FullColumnName
    to
        identifierObjectLiteral : MQLM!JSObjectLiteral (
            property <- MQLM!JSPropertyAssignment.newInstance())
            do
            {
                identifierObjectLiteral.property.propertyElement <-
        if fullColumnName.identifier->size() = 1 then
            MQLM!Identifier.newInstance()
        else
            MQLM!StringValue.newInstance()

        endif;
    identifierObjectLiteral.property.propertyElement.value <-
        if fullColumnName.identifier->size() = 1 then
            fullColumnName.identifier->first().value
        else
            fullColumnName.concatenate()
        endif;
                identifierObjectLiteral.property.value <- MQLM!Number.newInstance();
                identifierObjectLiteral.property.value.value <-1;
            }
}

rule StringLiteral2JSObjectLiteral {
        from
            stringLiteral : SQLM!StringLiteral
    to
        stringObjectLiteral : MQLM!JSObjectLiteral (
        property <- MQLM!JSPropertyAssignment.newInstance())

        do
        {
          stringObjectLiteral.property.propertyElement <-MQLM!StringValue.newInstance();
          stringObjectLiteral.property.propertyElement.value <- stringLiteral.value;
        }
}
helper context SQLM!FullColumnName def:concatenate() :String =
let concatenated: String =
    self.identifier->subSequence(2, self.identifier->size())

 ->collect(e | e.value.toString())
        ->iterate(v; acc : String = '' |
            acc.concat(if acc.size() > 0 then '.' else '' endif).concat(v)
        )
    in
        concatenated;
```

## 3  Related Work

The importance of metamodels when dealing with data models and database query languages is evident in many works in the literature. In [15], a unified metamodel for NoSQL and relational databases, called U-Schema, was presented. The authors defined the mappings between their metamodel and the data models defined for each database paradigm, including columnar, document, key-value, graph, and relational schemas. U-Schema can be used to create a generic language for managing NoSQL and relational stores.

An extension of SQL called SQL++ was presented in [16]. It is a unifying semi-structured query language designed to standardise querying across different database systems, including NoSQL, SQL-on-Hadoop, and NoSQL databases like MongoDB. The extension aims to give SQL the capability to handle semi-structured data, addressing the interoperability challenges between diverse database query languages. Its metamodel abstracts away the complexities of data models, enabling users to write queries in a consistent manner across different systems.

Cosentino in [17] used a SQL and PL/SQL metamodel to extract business rules for relational databases. The metamodel represents the grammar to parse SQL and PL/SQL source code for business rules extraction.

Metamodels are used in language and database transformations and migrations. A mapping between SPARQL and NoSQL was introduced in [18]. The transformation system is based on a SPARQL metamodel and several NoSQL metamodels, including Cassandra QL and MongoDB QL.

The authors in [19] outline a system called KDA, which acts as an intermediary between users and SQL-like databases. They used a simplified SQL query metamodel to generate SQL queries after mapping ontologies' constructs into relational databases for knowledge extraction.

## 4   Conclusion and Future Work

The need for an explicit metamodel of the query language is essential, as it forms the foundation for query validation, standardisation, documentation, extending the language, mapping it to other query languages, and developing tool support. Without a metamodel, model-driven approaches, such as transforming queries between different paradigms, become infeasible. In our work, we demonstrated the effectiveness of the MySQL and MQL metamodels in performing such mappings. These metamodels not only facilitated the transformation but also enabled the generation of supporting tools through a DSL workbench, which further allowed for the easy modification and extension of the languages.

To validate the metamodels' effectiveness, we used MQLM and SQLM as the foundational metamodels for implementing examples that demonstrated the practical utility of metamodels. These implementations proved the usefulness of the metamodels by showing how they supported not only language extensions and transformations but also provided real-world examples of how the metamodels facilitated tool support, improved documentation, and enabled consistent validation of queries across different paradigms. Through these implementations, the metamodels proved to be versatile, supporting various tasks from query validation to migration between MySQL and MongoDB.

In the future, we plan to enhance our approach by implementing a full model-based transformation of MySQL queries to MQL using advanced transformation languages such as ATL (Atlas Transformation Language) or QVT (Query / View / Transformation). We aim to employ the EMFText framework to manage the transformation process while using SQLM and MQLM as the source and target metamodels, respectively. This would offer a robust and automated method for translating relational database queries (SQL) into NoSQL document-based queries (MQL).

Moreover, recognizing the diversity of NoSQL databases, we aim to expand our metamodel-based approach to encompass other NoSQL paradigms, including key-value stores, column-family stores, and graph databases. By creating comprehensive metamodels for the query languages of these database types, we intend to enable seamless interoperability across the NoSQL spectrum. These future efforts will build on our previous work, ultimately providing a unified metamodel-driven approach for query manipulation and transformations across both SQL and NoSQL databases.

Thus, the introduction and development of metamodels not only enables query language flexibility and extensibility but also forms the backbone for achieving cross-database interoperability. This approach paves the way for a standardised model-driven transformation process that can accommodate the increasingly complex and heterogeneous data landscape across modern database systems.

# References


1. Husain, M.S., Khan, M.Z., Siddiqui, T.: Big Data Concepts, Technologies, and Applications. CRC Press, Boca Raton (2023)
2. Heidenreich, F., Johannes, J., Karol, S., Seifert, M., Wende, C.: Model-based language engineering with EMFText. In: Lämmel, R., Saraiva, J., Visser, J. (eds.) Generative and Transformational Techniques in Software Engineering IV. GTTSE 2011. LNCS, vol. 7680, pp. 322–345. Springer, Berlin, Heidelberg (2011). https://doi.org/10.1007/978-3-642-359 92-7_9
3. Bettini, L.: Implementing Domain-Specific Languages with Xtext and Xtend. Packt Publishing, Birmingham (2016)
4. Michels, J., et al.: The new and improved SQL: 2016 standard. ACM SIGMOD Rec. **47**(2), 51–60 (2018)
5. Christudas, B.: MySQL. In: Christudas, B. (ed.) Practical Microservices Architectural Patterns, pp. 877–884. Apress, Berkeley (2019)
6. Zhang, P.: Practical Guide for Oracle SQL, T-SQL, and MySQL. CRC Press, Boca Raton (2017)
7. Shaik, B., Chemuduru, D.K.: PL/pgSQL essential extensions In: Procedural Programming with PostgreSQL PL/pgSQL: Design Complex Database-Centric Applications with PL/pgSQL, pp. 293–309. Apress, Berkeley (2023)
8. Pribyl, B., Feuerstein, S.: Learning Oracle PL/SQL. O'Reilly Media Inc., Sebastopol (2002)
9. Worsley, J., Drake, J.D.: Practical PostgreSQL. O'Reilly Media Inc., Sebastopol (2002)
10. Owens, M., Allen, G.: SQLite. Apress, New York (2010)
11. Budinsky, F.: Eclipse Modeling Framework: A Developer's Guide. Addison-Wesley Professional, Boston (2004)
12. Cabot, J., Gogolla, M.: Object constraint language (OCL): a definitive guide. In: Bernardo, M., Cortellessa, V., Pierantonio, A. (eds.) Formal Methods for Model-Driven Engineering. SFM 2012. LNCS, vol. 7320, pp. 58–90. Springer, Berlin, Heidelberg (2012). https://doi.org/10.1007/978-3-642-30982-3_3
13. Kurtev, I.: State of the Art of QVT: a model transformation language standard. In: Schürr, A., Nagl, M., Zündorf, A. (eds.) Applications of Graph Transformations with Industrial Relevance. AGTIVE 2007. LNCS, vol. 5088, pp. 377–393. Springer, Berlin, Heidelberg (2008). https://doi.org/10.1007/978-3-540-89020-1_26
14. Ouault, F., Allilaire, F., Bézivin, J., Kurtev, I., Valduriez, P.: ATL: a QVT-like transformation language. In: Companion to the 21st ACM SIGPLAN Symposium on Object-Oriented Programming Systems, Languages, and Applications, pp. 719–720 (2006)
15. Candel, C.J.F., Ruiz, D.S., García-Molina, J.J.: A unified metamodel for NoSQL and relational databases. Inf. Syst. **104**, 101898 (2022)
16. Ong, K.W., Papakonstantinou, Y., Vernoux, R.: The SQL++ Query Language: Configurable, Unifying, and Semi-Structured. arXiv preprint arXiv:1405.3631 (2014)

17. Cosentino, V.: A Model-Based Approach for Extracting Business Rules Out of Legacy Information Systems. Doctoral dissertation, École des Mines de Nantes (2013)
18. Banane, M., Erraissi, A., El Khalyly, B., Belangour, A., Azzouazi, M.: ScalSPARQL: a new scalable system for the mapping of SPARQL queries to NoSQL languages based on MDE approach. In: 2021 6th International Conference on Renewable Energy: Generation and Applications (ICREGA), pp. 101–105. IEEE (2021)
19. Lahoud, I., Monticolo, D., Hilaire, V.: Mapping the semantic web to SQL query to extract knowledge. J. E-Technol. **3**(2), 69 (2012)

# Visual Explanation of Deep Learning Models for Wildfire Detection: A Grad-CAM Approach

Hamza Touati[(✉)] [iD], Said Labed [iD], and Hadjir Zemmouri [iD]

MISC Laboratory, University of Constantine 2 - Abdelhamid Mehri, Ali Mendjeli
Constantine, 25000 Constantine, Algeria
{hamza.touati,said.labed,zemmouri.hadjir}@univ-constantine2.dz
http://www.misc-lab.org

**Abstract.** Wildfires pose a significant threat to ecosystems and human life, making early detection critical. Recent advances in deep learning, particularly Convolutional Neural Networks (CNNs), have shown promise in automating wildfire detection using camera and drone imagery. However, these models often operate as "black boxes," providing little insight into their decision-making processes, which raises concerns about their reliability in real-world applications. This paper tackles this issue by evaluating three CNN models—ResNet-50, Inception-V3, and Xception—on the DeepFire dataset, not only in terms of accuracy but also in interpretability. We apply Grad-CAM, an Explainable AI (XAI) technique, to visualize where each model focuses when predicting fire presence. Our results show that while Xception achieves the highest accuracy, Inception-V3 demonstrates more consistent attention to critical fire-related areas. By integrating Grad-CAM, we enhance model transparency, offering a more reliable and interpretable approach to wildfire detection.

**Keywords:** Convolutional neural networks · wildfire detection · explainable AI

## 1 Introduction

Wildfires pose a significant and growing threat to ecosystems, human lives, and infrastructure worldwide. The frequency and intensity of wildfires have increased due to climate change, resulting in severe environmental and economic damage. Early and accurate detection of wildfires is crucial for minimizing their impact, allowing for rapid intervention by fire management teams [1].

In recent years, convolutional neural networks have emerged as powerful tools for automating wildfire detection using image data. CNNs excel at recognizing patterns and features in images, making them well-suited for identifying visual cues associated with wildfires, such as flames, smoke, and thermal anomalies [2].


N. Seddari and M. Redjimi (Eds.): ICMSCT 2024, CCIS 2606, pp. 385–397, 2025.
https://doi.org/10.1007/978-3-032-01922-6_32

Despite their success, CNN-based models often function as black boxes, providing little insight into how they arrive at specific predictions. This lack of interpretability is particularly problematic in safety-critical applications like wildfire detection, where understanding the reasoning behind a model's decision is essential for building trust and ensuring reliable deployment in real-world scenarios [3].

The field of Explainable Artificial Intelligence seeks to address this issue by making machine learning models more transparent and interpretable. XAI techniques such as Grad-CAM (Gradient-weighted Class Activation Mapping) and LIME (Local Interpretable Model-agnostic Explanations) offer visual and local explanations for CNN predictions, helping users understand which regions of an image are most important to the model's decision-making process [4,5]. By providing these insights, XAI enables fire management teams to trust the outputs of detection models, especially in ambiguous situations involving smoke, fog, or partial fire visibility [6].

Despite the increasing interest in XAI for computer vision tasks, few studies have explored its application to wildfire detection, especially on camera networks which is one of the extensively investigated methods in the wildfire detection community.

In this paper, we begin by providing an overview of the latest technologies for wildfire detection, followed by an investigation of the most commonly used algorithms and datasets in camera networks. We then examine recent works that apply explainable AI techniques to wildfire detection. Afterward, we utilize Grad-CAM on one of the most widely used datasets, alongside established algorithms, to visualize the inner workings of the models.

## 2    Related Work

Recent advancements in AI and monitoring tools have improved fire detection through four main approaches: sensor nodes, unmanned aerial vehicles, camera networks, and satellite surveillance [7,8]. Each has advantages and limitations, with camera networks being preferred for continuous, high-resolution monitoring [9].

Several studies have focused on using camera networks and developing custom deep learning architectures for wildfire detection. Khan and Khan [10] introduced FFireNet, a model based on the MobileNetV2 architecture, classifying images as forest fire or non-fire. Akyol [11] emphasized optimizing classification accuracy by extracting deep features from ResNet50, which were integrated into a DNN-3 classifier and then evaluated on the DeepFire dataset. Wang et al. [12] proposed Reduce-VGGNet, a modified VGG architecture, along with an optimized CNN for detecting wildfire regions, utilizing the FLAME dataset that includes manually annotated fire regions. Idroes et al. [13] developed TeutongNet, a modified ResNet50-V2 model tailored for high-precision forest fire detection, trained on a manually collected dataset, with additional validation using ROC curve analysis.

In addition to these custom architectures, hybrid and combined techniques have been explored. Wei et al. [14] proposed an intelligent wildfire detection system that combines MobileNetV3 with YOLOV5S, integrating multi-scale feature extraction and bi-directional feature fusion networks for improved detection. Zhang and Zhang [15] used two CNN models for real-time wildfire detection, optimizing separate models for daytime and nighttime scenarios, which resulted in 98% precision, 8% higher than traditional single-model methods. Habbib and Khidhir [16] focused on optimizing network input sizes and leveraging transfer learning for real-time fire identification, employing EfficientNet and ResNet-50 models trained on a dataset of 4,250 images. Finally, Li et al. [17] tackled the challenge of smoke detection by introducing SMWE-GFPNNet, an edge-focused network that incorporates the Swin Multidimensional Window Extractor and Guillotine Feature Pyramid Network, tested on the Forest Fire Smoke Complex Background Detection Dataset.

While the above mentioned works have significantly improved the accuracy of wildfire detection systems, the inner workings of deep learning algorithms often remain obscure. In recent years, some researchers have turned to explainable AI to address this issue in the context of wildfire detection. For instance, Apostolopoulos et al. [18] used Inception and techniques like Grad-CAM++ and LIME on a dataset of 53,585 images, showing that models effectively focused on flames and smoke, thus enhancing trust in predictions. Fernandes et al. [19] developed a smoke plume detection system using surveillance camera images, combining a multidimensional output method with Grad-CAM, achieving an AUROC of 0.949. Meanwhile, Khan and Park [20] introduced FireXplainNet, a CNN model that integrates LIME for decision insights, outperforming existing models but needing further real-world validation.

Other work leveraging explainable AI for wildfire detection, identified through the Scopus and Google Scholar databases, primarily focus on satellite imagery and aim to provide semantic explanations for the model's behavior. For example, Ahmad et al. [21] developed FireXnet for real-time detection in resource-limited areas, enhancing feature attribution and transparency. Liu et al. [22] used SHAP to analyze wildfire mapping via remote sensing, revealing the impact of terrain, vegetation, and climate on fire risks. Similarly, Abdollahi et al. [23] employed SHAP to identify environmental factors, such as humidity and wind speed, that influence wildfire susceptibility.

## 3   Materials and Methods

In the field of wildfire detection, previous research has extensively leveraged CNNs, with models such as VGG and ResNet being particularly prominent. These models are favored due to their proven accuracy in a variety of computer vision tasks, including wildfire detection. While CNN accuracy has improved, few studies explore how they interpret images and update gradients, especially for wildfire detection.

In this study, we evaluate three commonly used CNN architectures for wildfire detection using a widely recognized dataset within the wildfire detection community. Beyond assessing their performance, we employ an explainability technique to visualize the specific regions of the images that capture the model's attention, providing insights into how these networks make predictions.

The methodology employed in this study is summarized in Fig. 1.

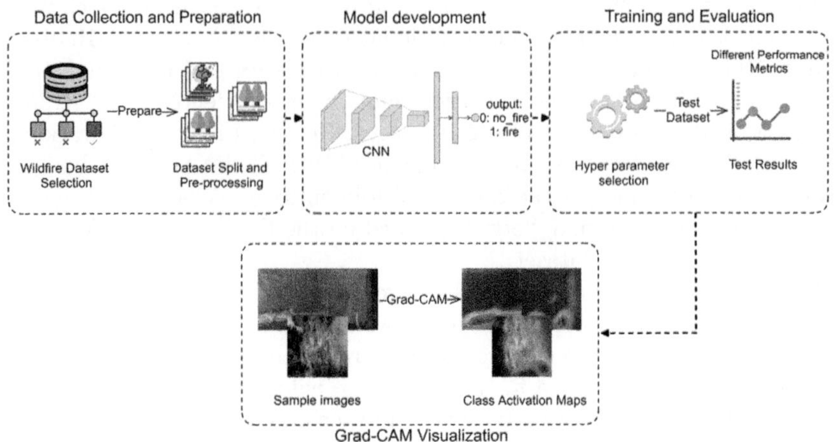


**Fig. 1.** Overview of the methodology.

## 3.1   Dataset

The absence of a standard benchmark dataset in wildfire detection from image data significantly hinders researchers in evaluating their methodologies [24]. Variations in characteristics such as image resolution, scene diversity, and annotation quality can greatly affect algorithm performance and complicate cross-study comparisons [25]. Despite this issues, several widely used datasets have emerged as vital resources in this field. Examples of these are the *Mivia Lab Fire Detection* Dataset [26], which includes videos of real-world fire scenes, varying in resolution and content, mostly obtained through internet; the *Portuguese Firefighters Portal* Database [27], consisting of rich firefighting data, both statistical and environmental; Fire Luminosity Airborne-based Machine learning Evaluation (*FLAME*) dataset [28], that includes fire classification and segmentation with over 47,000 labeled aerial imagery; the *Corsican Fire Database* [24], which contains images both in visible and near-infrared spectrums in different conditions; *DeepFire* dataset [25], with 1900 images in the visible spectrum; finally, *Fire Dataset* [29], created during the NASA Space Apps Challenge in 2018, with images categorized into fire and no-fire classes.

In this study, we used the *DeepFire* dataset, introduced by Khan et al. [25]. This dataset is relatively new but has gained increasing traction in recent years due to its comprehensive and diverse collection of wildfire-related imagery.

The DeepFire dataset contains two primary classes: fire and no_fire, each with 950 images. The fire class features various wildfire scenarios with diverse conditions, including different times of day, fire sizes, and smoke presence, making it suitable for robust model training. The no_fire class consists of forest images without fires but includes elements like clouds, sunsets, and reddish trees, which could be confused with fire. A selection of images from the DeepFire dataset is displayed in Fig. 2.

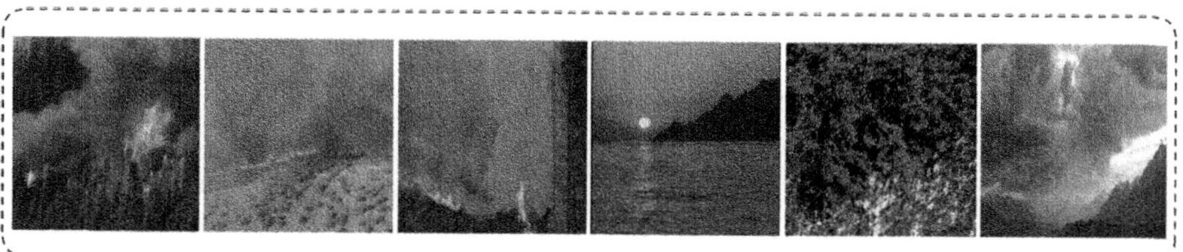

**Fig. 2.** Sample images from the DeepFire dataset [25].

The dataset required no preprocessing as all images were uniformly sized at $250 \times 250$ pixels and consistently formatted. A quality check found no damaged files, ensuring the dataset's high quality. The data was then randomly split into training (64%), validation (16%), and test (20%) sets, following a common approach used in related research using the same dataset.

### 3.2   Models

To assess the performance of different CNNs on wildfire detection, we selected three state-of-the-art CNN architectures that have demonstrated strong performance across various computer vision tasks. Specifically, we used:

- **ResNet-50**: A deep residual network that uses skip connections to combat the vanishing gradient problem, allowing for training deeper networks without performance degradation [30].
- **Inception-V3**: Known for its ability to capture multi-scale features using different filter sizes, Inception-V3 has been widely adopted for complex image classification tasks [31].
- **Xception**: An extension of Inception architecture with depthwise separable convolutions, which has proven highly effective in both accuracy and computational efficiency [32].

All three models—ResNet50, InceptionV3, and Xception—were used without their top layers, and the same custom classification head was added to each. The custom layer begins by flattening the output of the base model, followed by a Dense layer with 128 neurons and Rectified Linear Unit (ReLU) activation, which is sufficient for our binary classification task given the small dataset size. A Dropout layer with a rate of 0.4 is included to prevent overfitting. Finally, a single neuron Dense layer with sigmoid activation is added for binary classification output.

For all models, the base layers are frozen, meaning their weights are not trainable, and pre-trained ImageNet weights are applied to leverage existing feature representations. This approach helps prevent overfitting, as these models have a large number of parameters, which could otherwise lead to overfitting on our small dataset.

### 3.3  Training and Evaluation

Each CNN model was trained on the DeepFire dataset, which was randomly split into training (64%), validation (16%), and test (20%) sets. No data augmentation techniques were applied to the dataset. The models were trained using the Adam optimizer with a learning rate of 0.0001, and binary cross-entropy was used as the loss function. All models were trained for 20 epochs.

For evaluation, the following performance metrics were used:

- **Accuracy**: The proportion of correctly predicted instances out of the total instances.
- **Precision**: The ratio of true positive predictions to the total predicted positives.
- **Recall**: The ratio of true positives to the actual number of positives in the dataset.
- **F1-Score**: The harmonic mean of precision and recall, providing a balanced assessment of the model's performance.
- **True Negative Rate (TNR)**: The proportion of actual negatives that are correctly identified.
- **False Positive Rate (FPR)**: The proportion of actual negatives incorrectly classified as positives.
- **False Negative Rate (FNR)**: The proportion of actual positives incorrectly classified as negatives.

After training, the ResNet-50 model achieved an accuracy of 85.54% on the training set and 88.16% on the validation set, indicating no overfitting. However, both Inception-V3 and Xception exhibited slight overfitting. Inception-V3 scored 100% on the training set and 98.68% on the validation set, while Xception achieved 99.78% on the training set and 98.68% on the validation set. This overfitting could be attributed to the larger capacity of these models, which, given the relatively small size of the DeepFire dataset, allows them to memorize the training data rather than generalizing well to unseen data. The absence of data augmentation or other regularization techniques likely contributed to this overfitting as well.

The remaining results from the test set are summarized in Table 1.

**Table 1.** Comparison of classification results on the DeepFire dataset

| Model | Acc. | Prec. | F1 | Recall | TNR | FPR | FNR |
|---|---|---|---|---|---|---|---|
| ResNet50 | 86.84% | 79.47% | 85.79% | 93.20% | 82.11% | 17.88% | 06.79% |
| Inception-V3 | 97.89% | 97.36% | 97.88% | 98.40% | 97.39% | 02.60% | 01.59% |
| Xception | 99.47% | 100% | 99.47% | 98.95% | 100% | 0.00% | 01.04% |

## 3.4 Explainability with Grad-CAM

While traditional performance metrics like accuracy, precision, and recall are crucial, they do not provide insights into the internal decision-making process of models. To address this, we applied Gradient-weighted Class Activation Mapping [4] to visualize the regions of the image that contributed most to the models' predictions.

Grad-CAM generates a heatmap that highlights important regions in the input image by leveraging the activation maps from the final convolutional layers of the models—ResNet-50 (conv5_block3_out), Inception-V3 (mixed10), and Xception (block14_sepconv2_act). Instead of operating on the final Fully Connected (FC) layers or the sigmoid output, Grad-CAM works by calculating the gradients of the target class with respect to the activation maps. The process includes the following steps:

- Activation Maps: The final convolutional layers output activation maps, which capture spatial features of the image.
- Compute Gradients: Gradients of the predicted class are computed with respect to these activation maps to understand their influence on the prediction.
- Global Average Pooling: Gradients are globally averaged to assign weights to the activation maps.
- Weighted Combination: The feature maps are weighted by these gradients and summed to produce a coarse heatmap.
- ReLU: The ReLU function is applied to highlight only the positive contributions, generating the final Grad-CAM heatmap.

This heatmap is then overlaid on the original image, revealing the areas that the model focused on during its decision-making.

Figure 3 illustrates the Grad-CAM process.

## 3.5 Grad-CAM Application on DeepFire Images

We utilized Grad-CAM to visualize the attention focus of each model—ResNet-50, Inception-V3, and Xception—on a subset of images from the DeepFire dataset. For the fire class, we selected three scenarios: a close-range fire, a medium-range fire, and a distant fire. In the no_fire class, we used one image to assess how each model interprets non-fire scenarios. Grad-CAM heatmaps were

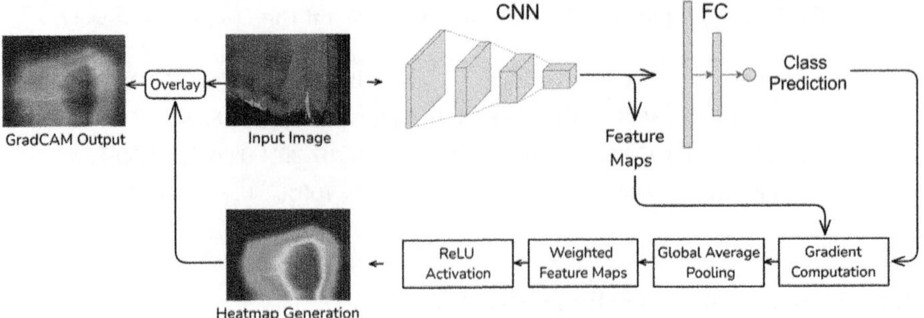



**Fig. 3.** Grad-CAM process.

generated for each prediction, providing insights into the models' interpretability and attention focus.

Table 2 summarizes the Grad-CAM results for the three models across the three fire images and the no_fire image (last image).

## 4   Discussion

### 4.1   Discussion of Model Results

The results in Table 1 show that Xception outperforms both ResNet50 and Inception-V3 on the DeepFire dataset, achieving the highest accuracy of 99.47% and perfect precision (100%). Xception did not produce any false positives (FPR: 0.00%) and maintained a high recall of 98.95%, indicating its strong ability to detect fires while minimizing false alarms. Inception-V3 also performed well with an accuracy of 97.89%, high precision (97.36%), and recall (98.40%), demonstrating a balanced performance in detecting fire events and avoiding false classifications. On the other hand, ResNet50 showed a decent but lower accuracy (86.84%) and had a relatively higher false positive rate (17.88%), suggesting that it struggles with misclassifying non-fire images as fires, leading to more false alarms.

Both Inception-V3 and Xception benefited from their advanced architectures, which allow for more effective feature extraction compared to ResNet50. Xception's perfect precision makes it the most reliable model for wildfire detection in scenarios where false positives must be minimized. However, its slightly lower recall and the near-perfect training scores for both Inception-V3 and Xception suggest slight overfitting, likely due to the lack of regularization techniques such as data augmentation. Nonetheless, Xception stands out as the most robust model overall, offering the best balance of precision and recall, making it the most suitable candidate for real-world wildfire detection where both accuracy and minimizing false alarms are critical.

**Table 2.** Saliency visualizations across different models

| Original image | ResNet50 | Inception-V3 | Xception |
| --- | --- | --- | --- |

## 4.2  Discussion of Explainability and Model Focus

Through the Grad-CAM visualizations, we observed that the three models—ResNet-50, Inception-V3, and Xception—focused on different regions of the images when making predictions, particularly around the fire or its immediate surroundings. However, in the no_fire class, ResNet-50 failed to make the correct prediction. The model misinterpreted elements such as rocks, clouds, and trees in the distance as potential fire indicators, focusing its attention on a far-off mountain where no fire was present. This suggests that ResNet-50 may be overly sensitive to certain textures or color patterns, such as the white areas (possibly snow or other bright regions), leading to false positives.

The models exhibited varied attention patterns:

ResNet-50 often concentrated on the upper regions of the fire, particularly the smoke, indicating that the model relied heavily on the presence of smoke as a key feature for detection. However, this focus on smoke contributed to its failure in the no_fire class, as the model misinterpreted non-fire-related features like clouds and bright regions as fire-related elements, leading to a high False

Positive Rate (FPR: 17.88%). Inception-V3, with its ability to handle multi-scale features, displayed more robustness in differentiating between smoke and similar textures, such as clouds. Although it occasionally struggled with ambiguous cases, its attention generally covered larger regions, including both fire and smoke. This model achieved a False Positive Rate of 2.60% and a True Positive Rate of 98.40%, reflecting its strong balance in detecting fires while avoiding false alarms. Xception demonstrated the most focused attention but, interestingly, in some cases, it highlighted regions not directly associated with the fire or smoke. In the medium- and far-distance fire images, the model concentrated on surrounding areas that were not immediately related to the flames or smoke, suggesting that it may occasionally rely on context rather than focusing solely on the fire itself. Despite this, Xception produced the best precision (100%) and a perfect True Negative Rate (TNR: 100%), indicating that it was extremely effective at avoiding false positives. However, its focus on irrelevant regions raises questions about whether this precision comes at the cost of missing subtle fire details. These findings reveal that while traditional metrics such as accuracy, precision, recall, and even True Positive Rate and False Positive Rate provide valuable insights into the models' statistical performance, they are not always sufficient to fully assess a model's real-world suitability for wildfire detection. For instance, although Xception achieved the highest accuracy (99.47%), it occasionally focused on irrelevant areas, whereas Inception-V3—with a slightly lower accuracy (97.89%)—appeared to identify critical regions more effectively in certain scenarios. This highlights the importance of using explainability tools like Grad-CAM, which provide crucial insights into how models operate and whether their predictions are based on relevant features.

Thus, for critical applications such as wildfire detection, where lives and resources are at stake, interpretability should be considered alongside standard performance metrics. It ensures that models not only perform well statistically but also focus on the correct image regions, thereby improving the trust and reliability of these systems in real-world deployments.

## 5   Conclusion

In this study, we evaluated the performance of three state-of-the-art CNN models ResNet-50, Inception-V3, and Xception—on the DeepFire dataset for wildfire detection, comparing their classification accuracy and interpretability using Grad-CAM visualizations. While Xception demonstrated the highest accuracy (99.47%) and perfect precision (100%), Inception-V3 also performed exceptionally well with a balanced accuracy (97.89%), precision, and recall. ResNet-50, though effective in detecting fires, exhibited a higher false positive rate, making it less reliable for real-world deployment.

The Grad-CAM analysis revealed important insights into the attention mechanisms of each model. ResNet-50 often focused on non-relevant regions in the no_fire images, contributing to its higher false positive rate. Inception-V3 displayed more consistent attention on fire-related regions, while Xception, despite its superior performance metrics, sometimes focused on areas not directly related to fire or smoke. These findings emphasize the importance of incorporating explainability tools like Grad-CAM in evaluating models for critical applications such as wildfire detection, where both performance and interpretability are essential.

Ultimately, this study underscores the need to go beyond traditional metrics like accuracy and precision when assessing model performance. Explainability is crucial in ensuring that models not only make correct predictions but also focus on the right features in complex, real-world scenarios. For future work, integrating additional regularization techniques and exploring more advanced explainability methods could further enhance the robustness and transparency of wildfire detection systems. Leveraging multiple XAI tools together will provide a richer understanding of model behaviors and contribute to more reliable and interpretable AI solutions.

# References


1. Carta, F., Zidda, C., Putzu, M., Loru, D., Anedda, M., Giusto, D.: Advancements in forest fire prevention: a comprehensive survey. Sensors **23**(14), 6635 (2023). https://doi.org/10.3390/s23146635
2. Tsalera, E., Papadakis, A., Voyiatzis, I., Samarakou, M.: Cnn-based, contextualized, real-time fire detection in computational resource-constrained environments. Energy Reports **9**, 247–257 (2023). https://doi.org/10.1016/j.egyr.2023.05.260
3. Zhang, Q., Zhu, S.: Visual interpretability for deep learning: a survey. Front. Inf. Technol. Electron. Eng. **19**(1), 27–39 (2018). https://doi.org/10.1631/FITEE.1700808
4. Selvaraju, R.R., Cogswell, M., Das, A., Vedantam, R., Parikh, D., Batra, D.: Grad-cam: visual explanations from deep networks via gradient-based localization. Int. J. Comput. Vis. **128**, 336–359 (2020). https://doi.org/10.1007/s11263-019-01228-7
5. Ribeiro, M.T., Singh, S., Guestrin, C.: Why should i trust you? Explaining the predictions of any classifier. In: Proceedings of the 22nd ACM SIGKDD International Conference on Knowledge Discovery and Data Mining, pp. 1135–1144 (2016). https://doi.org/10.48550/arXiv.1602.04938
6. Rubab, S.F., Ghaffar, A.A., Choi, G.S.: Firedetxplainer: decoding wildfire detection with transparency and explainable ai insights. IEEE Access (2024). https://doi.org/10.1109/ACCESS.2024.3383653
7. Saleh, A., Zulkifley, M.A., Harun, H.H., Gaudreault, F., Davison, I., Spraggon, M.: Forest fire surveillance systems: a review of deep learning methods. Heliyon (2024). https://doi.org/10.1016/j.heliyon.2023.e23127
8. Mohapatra, A., Trinh, T.: Early wildfire detection technologies in practice—a review. Sustainability **14**(19), 12270 (2022). https://doi.org/10.3390/su141912270
9. Labed, S., Touati, H., Herida, A., Kerbab, S., Sairi, A.: An AI-based image recognition system for early detection of forest and field fires. Eur. J. Forest Eng. **9**(2), 48–56 (2023). https://doi.org/10.33904/ejfe.1322396
10. Khan, S., Khan, A.: FFireNet: deep learning based forest fire classification and detection in smart cities. Symmetry **14**(10), 2155 (2022). https://doi.org/10.3390/sym14102155

11. Akyol, K.: A comprehensive comparison study of traditional classifiers and deep neural networks for forest fire detection. Clust. Comput. 1–15 (2023). https://doi.org/10.1007/s10586-023-04003-z
12. Wang, L., Zhang, H., Zhang, Y., Hu, K., An, K.: A deep learning-based experiment on forest wildfire detection in machine vision course. IEEE Access **11**, 32671–32681 (2023). https://doi.org/10.1109/ACCESS.2023.3262701
13. Idroes, G.M., et al.: Teutongnet: a fine-tuned deep learning model for improved forest fire detection. Leuser J. Environ. Stud. **1**(1), 1–8 (2023). https://doi.org/10.60084/ljes.v1i1.42
14. Wei, C., Xu, J., Li, Q., Jiang, S.: An intelligent wildfire detection approach through cameras based on deep learning. Sustainability (Switzerland) **14**(23) (2022). https://doi.org/10.3390/su142315690
15. Zhang, A., Zhang, A.: Real-time wildfire detection and alerting with a novel machine learning approach: a new systematic approach of using convolutional neural network (CNN) to achieve higher accuracy in automation. Int. J. Adv. Comput. Sci. Appl. **13**(8), 1–6 (2022). https://doi.org/10.14569/IJACSA.2022.0130801
16. Habbib, A., Khidhir, A.: Fire recognition using EfficientNet and ResNet 50. In: International Conference on Engineering, Science and Advanced Technology (ICESAT), pp. 18–22 (2023). https://doi.org/10.1109/ICESAT58213.2023.10347300
17. Li, R., et al.: SMWE-GFPNNet: a high-precision and robust method for forest fire smoke detection. Knowl.-Based Syst. **289**, 111528 (2024). https://doi.org/10.1016/j.knosys.2024.111528
18. Apostolopoulos, I.D., Athanasoula, I., Tzani, M., Groumpos, P.P.: An explainable deep learning framework for detecting and localising smoke and fire incidents: evaluation of grad-cam++ and lime. Mach. Learn. Knowl. Extr. **4**(4), 1124–1135 (2022). https://doi.org/10.3390/make4040057
19. Fernandes, A.M., Utkin, A.B., Chaves, P.: Automatic early detection of wildfire smoke with visible light cameras using deep learning and visual explanation. IEEE Access **10**, 12814–12828 (2022). https://doi.org/10.1109/ACCESS.2022.3145911
20. Khan, M.A., Park, H.: Firexplainnet: optimizing convolution block architecture for enhanced wildfire detection and interpretability. Electronics **13**(10), 1881 (2024). https://doi.org/10.3390/electronics13101881
21. Ahmad, K., et al.: Firexnet: an explainable ai-based tailored deep learning model for wildfire detection on resource-constrained devices. Fire Ecology (2023). https://doi.org/10.1186/s42408-023-00216-0
22. Liu, J., Wang, Y., Lu, Y., Zhao, P., Wang, S., Sun, Y., Luo, Y.: Application of remote sensing and explainable artificial intelligence (xai) for wildfire occurrence mapping in the mountainous region of Southwest China. Remote Sens. **16**(19), 3602 (2024). https://doi.org/10.3390/rs16193602
23. Abdollahi, A., Pradhan, B.: Explainable artificial intelligence (xai) for interpreting the contributing factors feed into the wildfire susceptibility prediction model. Sci. Total Environ. **879**, 163004 (2023). https://doi.org/10.1016/j.scitotenv.2023.163004
24. Toulouse, T., Rossi, L., Campana, A., Celik, T., Akhloufi, M.A.: Computer vision for wildfire research: an evolving image dataset for processing and analysis. Fire Saf. J. **92**, 188–194 (2017). https://doi.org/10.1016/j.firesaf.2017.06.012
25. Khan, A., Hassan, B., Khan, S., Ahmed, R., Abuassba, A.: DeepFire: a novel dataset and deep transfer learning benchmark for forest fire detection. Mob. Inf. Syst. **2022**, e5358359 (2022). https://doi.org/10.1155/2022/5358359
26. Fire detection dataset – mivia. https://mivia.unisa.it/datasets/video-analysis-datasets/fire-detection-dataset/. Accessed 21 Feb 2024

27. Bombeiros portugueses. https://www.bombeiros.pt/galeria/. Accessed 13 Mar 2024
28. Shamsoshoara, A., Afghah, F., Razi, A., Zheng, L., Fulé, P., Blasch, E.: The flame dataset: aerial imagery pile burn detection using drones (uavs) (2020). https://dx.doi.org/10.21227/qad6-r683
29. Gamaleldin, A., Atef, A., Saker, H., Shaheen, A.: Fire dataset. https://kaggle.com/datasets/phylake1337/fire-dataset. Accessed 01 Mar 2024
30. He, K., Zhang, X., Ren, S., Sun, J.: Deep residual learning for image recognition. In: 2016 IEEE Conference on Computer Vision and Pattern Recognition (CVPR), pp. 770–778 (2016). https://doi.org/10.1109/CVPR.2016.90
31. Szegedy, C., Vanhoucke, V., Ioffe, S., Shlens, J., Wojna, Z.: Rethinking the inception architecture for computer vision. In: 2016 IEEE Conference on Computer Vision and Pattern Recognition (CVPR), pp. 2818–2826 (2016). https://doi.org/10.1109/CVPR.2016.308
32. Chollet, F.: Xception: deep learning with depthwise separable convolutions. In: 2017 IEEE Conference on Computer Vision and Pattern Recognition (CVPR), pp. 1800–1807 (2017). https://doi.org/10.1109/CVPR.2017.195

# Towards a Comprehensive Cloud-Dedicated Data Lifecycle Model

Ilhem Tahiri[(✉)] [iD], Radja Boukharrou[iD], and Ahmed-Chawki Chaouche[iD]

MISC Laboratory, University of Constantine 2 – Abdelhamid Mehri,
Ali Mendjeli Campus, 25000 Constantine, Algeria
{ilhem.tahiri,radja.boukharrou,ahmed.chaouche}@univ-constantine2.dz

**Abstract.** Cloud computing has revolutionized the way that data is stored, shared, and processed. All services and functionalities of cloud computing fundamentally revolve around data and its management. Therefore, it is essential to have a comprehensive model that enables us to understand the data trajectory within the cloud, especially regarding security and confidentiality as data progresses through various stages. By synthesizing existing models, we aim to develop a more unified model adapted to cloud environments. In this work, we propose a comprehensive data lifecycle model that addresses the key characteristics of cloud computing. The results demonstrate that this model not only enhances the understanding of data flow in the cloud but also serves as a foundation for implementing targeted solutions at specific stages of the data lifecycle. This research seeks to enhance overall system efficiency, reduce time and energy consumption, and foster a more secure, privacy-focused cloud computing environment.

**Keywords:** Cloud Computing · Data Lifecycle · Data Security · Data Privacy · Cloud Data Management

## 1 Introduction

In recent years, many studies have explored the data lifecycle, a sequence of interrelated phases from data creation to deletion, with the goal of enhancing understanding across various domains to manage, analyze, and secure data effectively [12]. While many areas have advanced through this research, cloud computing still presents challenges, particularly due to its complexity. Understanding data movement within the cloud—especially regarding data security and privacy—is crucial, as data assumes different states and transformations depending on its context [2,3]. Each data state carries unique risks, underscoring the need to fully comprehend how data transitions within cloud environments. Addressing security concerns requires a clear understanding of when, where, and how data undergoes state changes.

Cloud computing, as defined by different organizations such as National Institute of Standards and Technology (NIST) [9], International Organization


N. Seddari and M. Redjimi (Eds.): ICMSCT 2024, CCIS 2606, pp. 398–412, 2025.
https://doi.org/10.1007/978-3-032-01922-6_33

for Standardization (ISO) [20,21], Amazon Web Services (AWS) [5], Microsoft Azure [6], and Google Cloud Provider (GCP) [13], is a system that provides on-demand remote access to resources like networks, servers, and storage via the internet. These deployments vary by accessibility, ownership, and service models (SaaS, PaaS, IaaS)[1], and are underpinned by key cloud characteristics. Any proposed cloud solution must adhere to these defining traits.

The core problem revolves around the complexity of managing data in cloud computing, particularly ensuring data security and privacy while incorporating essential cloud-specific characteristics. Existing data lifecycle models like [4,7,10,14–17,24] have limitations; they are too domain-specific and they lack the necessary specificity for cloud computing environments. To address these limitations, we introduce a model designed to address real-world challenges, including enhancing data security, ensuring privacy during data transfers, and optimizing resources during storage and usage phases.

In this paper, we propose a comprehensive cloud-dedicated data lifecycle model (CD-DLC) by synthesizing research from the past two decades. We aim to address gaps in understanding cloud data flow by (i) correlating data with cloud characteristics, (ii) proposing a cloud-specific data lifecycle model, and (iii) mapping the model to a practical use case. The CD-DLC model enhances data management, provides effective control mechanisms and forms the basis for developing strong security and privacy practices.

The rest of this paper is organized as follows: Sect. 2 reviews various studies and proposals regarding the data lifecycle. It then explains how data lifecycle phases correlate with cloud computing characteristics, which forms the foundation for our proposed model. Section 3 introduces the Cloud-Dedicated Data Lifecycle (CD-DLC) model, outlining its structure and components designed for cloud environments. Through two use-case scenarios in Sect. 4.1, we demonstrate the effectiveness of our proposed model. Finally, Sect. 5 discusses future research directions and concludes the study.

## 2 Data and Cloud Computing

### 2.1 Data Lifecycle Models

Data lifecycle represents the various stages that data undergoes in the cloud, from its creation to eventual destruction. Several representations have been proposed, where the number of stages can vary widely [11,18,22]. In order to help understand data lifecycle phases, identify the need for conceptual models, evaluate how they impact different domains, and ultimately fill in any gaps in the phases of cloud computing, this review section will take us through a number of data lifecycle models. All the works are presented here in ascending chronological order from 1993 to 2022.

The work [16] investigates the data lifecycle, aiming to clarify activities from creation to utilization and their interrelationships. The study used a conceptual

---

[1] SaaS: Software as a Service, Paas: Platform as a Service, Iaas: Infrastructure as a Service.

modeling methodology that included feedback loops and quality checkpoints. The results showed that data processes are cyclical, with four main cycles being acquisition, usage, and their combination.

The research group Yu and Wen from Beijing University, China, in their contribution [24] explore cloud data security and make a point that existing solutions only addressing certain phases of the data lifecycle are inadequate. They suggest a comprehensive strategy that aligns with the phases of the data lifecycle—creation, storage, use/share, archiving, and destruction, highlighting risks and recommending security precautions such as access control, encryption, and integrity protection. In addition to providing guidance for creating comprehensive data security strategies, the work promotes a comprehensive security solution that addresses the integral data lifecycle.

Demchenko et al. [10] propose the scientific data lifecycle management (SDLM) and the scientific data infrastructure (SDI) models. The study employed the two models to address big data challenges and introduced data lifecycle stages and activities involved in data generation, processing, preservation, and curation. Findings indicated by providing a unified interface for heterogeneous data sources and enabling efficient data access and analysis. Notably, this work aimed to contribute to the development of a common framework and standards for big data in science.

AL-Aswadi and Batarfi [4] discuss the risks to privacy that come with using cloud computing because users rely on cloud providers to manage their personal data. They suggest a framework for assessing a provider's adherence to privacy laws and boosting customer and provider trust. The proposed framework illustrates how cloud computing affects the various stages of the data lifecycle, citing the KPMG$^2$ [1] data lifecycle model.

The work [15] discusses the security threats in cloud computing and details various security techniques to mitigate these risks, focusing extensively on data security, network security, and cloud environment security. In addition, they introduced a data lifecycle model, which includes 5 phases; creation, transfer, execution, storage, and destruction.

Grunzke et al. [14] study the possibility of making complex scientific data easier to manage and analyze and propose a model. The study employed an exploratory investigation of data lifecycle technologies and their adaptations to various infrastructures. This work indicated the importance of having an improved data lifecycle for the scientist's systems to better do their research.

Mei et al. in their contribution [17] explore the complex world of cloud data security, explaining the nuanced nature of data states—in use, in transit, and at rest—and the Cloud Data Lifecycle (DLC) phases: create, store, use, share, archive, and destroy. Through an in-depth examination of these factors, the work highlights the importance of customised security protocols for various data

---

$^2$ KPMG is a global network of professionals who deliver meaningful results through a deep understanding of the issues and operations of the public sector. KPMG firms operate in 143 countries and territories across the globe, offering audit, tax and advisory services.

phases, providing insightful information. This investigation offers a strong basis for assessing and strengthening cloud data security in diverse scenarios.

The growth and development of the data lifecycle model can be divided into five main areas based on the literature review conducted for this work; data management, data security, big data challenges, data privacy, and data quality. While the research has focused on scientific systems, big data, cloud computing, Internet of Things, and smart cities, it seems that none of these works covered a comprehensive DLC model for cloud computing. For example, [10,14] only addressed data management, while [4,7,15,24] summarized the life cycle of data security without taking into consideration other cloud characteristics. On the other hand [14,16,17] did not specify whether their model can be applied on the cloud or not.

## 2.2   Correlating Data with Cloud Computing Characteristics

The enumeration of cloud characteristics can vary, as it is often influenced by the cloud providers' distinct vision of the cloud. However, a core set of fundamental characteristics remains consistent across these differing definitions. Several key works in cloud computing research have identified important cloud characteristics. The NIST [9] framework outlines five essential features, including on-demand self-service, broad network access, resource pooling, rapid elasticity, and measured service, gaining widespread influence with 21,755 citations. R. Buyya et al. [8] offer a more detailed view with 19 characteristics, such as scalability, security, privacy, resource allocation, and failure management, cited 8,802 times. Q. Zhang et al. [25] emphasize multi-tenancy, shared resource pooling, geo-distribution, and dynamic resource provisioning, also receiving 5,615 citations. L.M. Vaquero et al. [23] describe ten characteristics, focusing on user-friendliness, virtualization, scalability, and resource variety, accumulating 4,888 citations. Despite the variations, this research utilizes the categorization framework established by the National Institute of Standards and Technology (NIST) for several reasons.

Firstly, the NIST's characterization of cloud computing has obtained a higher number of citations, indicating its widespread acceptance and recognition within the academic and professional communities. Secondly, NIST is a reputable organization renowned for its authority on standards and technological advancements, lending credibility to its framework. Lastly, the five characteristics identified by NIST encompass, either directly or indirectly, the key attributes outlined in the other three seminal works, thereby providing a comprehensive and inclusive foundation for our research.

Preserving the foundational aspects of cloud computing requires aligning data with its characteristics. This study examines the diverse data states and representations derived from the NIST's cloud characteristics, as outlined in Table 1.

The extracted states can be summarized into 12 phases. Additionally, we introduce the concept of a data deactivation phase, which is the counterpart of data activation, and a data retrieval phase, representing the circumstances of data deletion. These states provide insights into cloud characteristics within the data context and will be further explained in Sect. 3.1.

**Table 1.** Correlation of the data states and representations with NIST's cloud characteristics

| Cloud Characteristics | NIST Description [9] | Data States |
| --- | --- | --- |
| On-demand self-service | A consumer can unilaterally provision computing capabilities, such as server time and network storage, as needed automatically without requiring human interaction with each service's provider. | In data context, this characteristic can be seen as the automated immediate on-demand: 1. *Creating* new data, 2. *Collecting* data as needed, 3. *Using stored data*, 4. *Modifying or updating* existing data to reflect changes or improvements, 5. *Archiving* less frequently used or historical data for future reference and make it *deactivated or activated* for those shared with, 6. *Sharing* data with other customers/services, 7. *Storing* new created/collected data, 8. *Deleting* data, 9. *Destructng* data definitly by ensuring complete removal |
| Broad network access | Capabilities are available over the network and accessed through standard mechanisms that promote use by heterogeneous thin or thick client platforms (e.g., mobile phones, laptops, and PDAs). | In data context, this characteristic can be seen as: 1. The availability of the data, which means *data duplication* in the cloud, 2. The accessibility through various platforms, which can be represented by *data activation and sharing, or transfer* in the cloud |
| Resource pooling | The provider's computing resources are pooled to serve multiple consumers using a multi-tenant model, with different physical and virtual resources dynamically assigned and reassigned according to consumer demand. | In data context, this characteristic can be summarized as: 1. *Sharing* the data between customers or users having the same privileges, 2. *Metadata* helps refer to the data owner |
| Rapid elasticity | Capabilities can be rapidly and elastically provisioned, in some cases automatically, to quickly scale out and rapidly released to quickly scale in. To the consumer, the capabilities available for provisioning often appear to be unlimited and can be purchased in any quantity at any time. | In data context, this characteristic can be seen as the quick scalability for: 1. *Data creation,* 2. *Data collection,* 3. *Data deletion,* 4. *Data destructiion* |
| Measured service | Cloud systems automatically control and optimize resource use by leveraging a metering capability at some level of abstraction appropriate to the type of service (e.g., storage, processing, bandwidth, and active user accounts). Resource usage can be monitored, controlled, and reported, providing transparency for the provider and consumer of the utilized service. | In data context, this characteristic is based on (1) to monitor and analyze the data usage that can be retrieved from (2) and (3). 1. *Metadata* creation, updates timestamps, etc. 2. The *Stored data,* 3. The *Archived data.* |

# 3    Comprehensive Cloud-Dedicated Data Lifecycle Model

In this section, we present our proposed research problem—a new comprehensive model for data lifecycle in cloud computing. Despite the critical importance of data lifecycle considerations in cloud environments, existing models have limitations. They are either too domain-specific, overly general, or lack specificity for cloud computing. To address this gap, we introduce the Cloud Dedicated Data Lifecycle (CD-DLC) comprehensive model, specifically designed to cater to the unique requirements of cloud-based scenarios; shown in Sect. 2.2. Our proposition is informed by a literature review conducted in the Sect. 2.1 and aligns with the distinctive characteristics of cloud computing categorized by NIST [9].

## 3.1    CD-DLC Model

The proposed CD-DLC Model consists of five main phases, each further subdivided into additional sub-phases, resulting in a total of 14 distinct phases. Each phase represents a specific data state, as illustrated in Fig. 1. The data states were derived from Table 1. These phases could be categorized as mandatory, such as creation, collection, metadata creation, duplication, storage, sharing, usage, deletion, and destruction, or optional, like normalization/transformation, update, archive, deactivation/activation, and retrieve. Plus, depending on the use case, the phases could all be executed, or some of them could be skipped. The Normalization/Transformation, Metadata Creation, Duplication, Deactivation/Activation, Update, and Retrieve Phases don't exist on the previously studied data lifecycles.

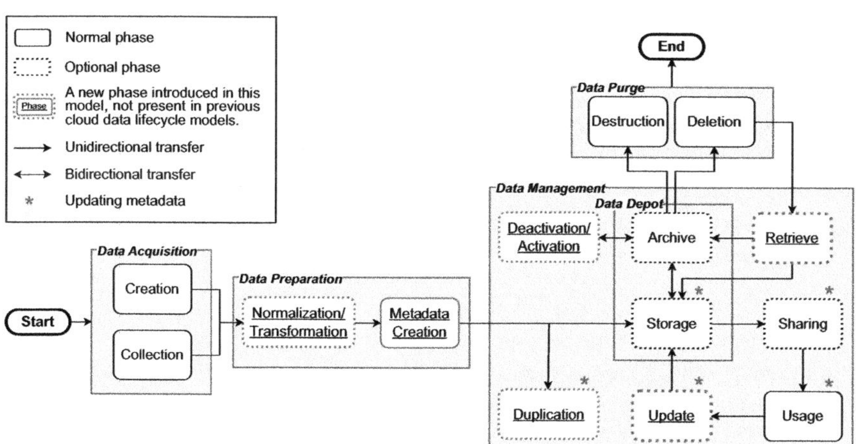


**Fig. 1.** Proposed Cloud Dedicated Data Lifecycle Model (CD-DLC).

**Data Acquisition.** The Data Acquisition phase in cloud computing involves the collection and generation of data from a variety of sources, both external and internal, which is then brought into the cloud infrastructure for processing, storage, and analysis. This phase plays a crucial role in ensuring that all necessary data is available in the cloud environment and is correctly formatted and compatible with cloud-based services.

The collected data can be categorized into three primary types based on the timing and urgency of its acquisition; real-time data, near real-time data, and batch data. The Data Acquisition phase is divided into two key sub-phases:

1. Data Collection: During this sub-phase, data is ingested into the cloud from a wide array of external sources. These external sources may include Third-party systems, Internet of Things (IoT) sensors and devices, and External applications.
2. Data Creation: In this sub-phase, data is generated internally within the cloud environment. This can happen in two primary ways:

    **System-generated data** Cloud platforms or internal systems auto nomously produce data, which can include logs, performance indicators, or usage statistics. This type of data is typically utilized for purposes like managing cloud infrastructure, monitoring performance, and optimizing resources.

    **User-created data** Data generated by users as they interact with cloud-based applications or services. Examples of this data include files, user settings, transactions, or any other information input by end-users or administrators within the cloud environment.

**Data Preparation.** The Data Preparation phase is essential for getting data ready for use or storage within the cloud. Since data is gathered from a variety of sources, it can come in multiple formats, including structured data—Organized data, typically found in relational databases (e.g., SQL)—, unstructured data—Data that doesn't have a predefined structure, such as images, videos, or audio files—, and semi-structured data—Data with some level of organization but not in a fully structured format, like JSON or XML—. This phase is divided into two sub-phases:

1. Data Normalization and Transformation: This sub-phase should take place immediately after data is collected. Data often arrives in a form that doesn't align with the formats required by the cloud's database or processing systems. In such cases, the data needs to be normalized or transformed into the appropriate format. However, if the data already meets the necessary format requirements, this step may be bypassed.
2. Metadata Creation: In this sub-phase, metadata is assigned to the data to ensure it can be properly managed and understood. This includes details like timestamps for when the data was created, collected, or ingested, the source of the data, ownership information, and any relevant descriptions. Metadata

ensures that the data is traceable and well-organized after the normalization process.

**Data Management.** The Data Management phase focuses on handling data after it has been prepared and linked with its metadata, but before it is stored, archived, or made available for further use. This phase is critical for ensuring that data is properly maintained throughout its lifecycle in the cloud environment. This phase is composed of six distinct stages:

1. Data Duplication: In this stage, additional copies of the data, including its associated metadata, are created to ensure availability, fault tolerance, and to provide backups. These duplicated copies are then transferred into storage. Once in the storage environment, the data is replicated across different cloud storage locations to enhance redundancy and prevent data loss.
2. Data Sharing: Data sharing may be optional when the owner has full control over their data. However, when users need to access the data without having the necessary privileges, sharing becomes a critical step. This phase ensures that proper access rights, security controls, and privacy measures are in place. In addition to granting permissions, the ability to revoke access is also a vital part of this stage to protect data integrity and prevent unauthorized use.
3. Data Usage: Based on the permissions granted during the sharing phase, users can now interact with the data. They can visualize, analyze, query, process, or otherwise engage with the data as needed. The level of interaction will depend on whether the user is the data owner or if the data has been shared with them.
4. Data Update: When the data is in an editable state, updates may be made. This stage ensures that any changes made to the data are synchronized across all its duplicates stored in various cloud locations. The goal here is to keep all copies up-to-date and avoid inconsistencies across different storage points.
5. Data Retrieval: Deleted data may not be fully erased and can often be recovered using tools like digital forensics [19]. This phase involves restoring data by leveraging metadata pointers or other recovery techniques.
6. Data Activation/Deactivation: Archived data can be deleted, permanently destroyed, or reactivated for use. To perform earlier management steps, archived data must be made accessible. Activation makes data searchable and retrievable, while deactivation hides it. The data owner is responsible for reactivating shared data as needed.

**Data Depot.** The Data Depot phase is critical for storing data to ensure it can be accessed, updated, shared, or backed up when needed. In cloud environments, data management follows the Information Lifecycle Management (ILM) policy, which organizes data based on importance and usage frequency into three tiers— The hot tier stores frequently accessed data, the warm holds less used data, and the cold is for rarely accessed data in cost-effective, long-term storage.

Once data is no longer needed, it moves to the archive for long-term backup. This transition is governed by the customer's preferences and ILM rules. Data must follow the ILM flow through the tiers before archiving. The Data Depot phase has two sub-phases:

1. Data Storage: Data is stored in the appropriate tier based on its current usage.
2. Data Archive: Data that is no longer frequently accessed is moved to long-term storage, reducing costs while preserving accessibility.

**Data Purge.** In the final phase of the data lifecycle, data that is no longer needed is permanently removed, either by making it inaccessible or completely destroying it. The Data Purge phase has two stages:

1. Data Deletion: Data is made inaccessible by removing references pointers, though it still exists within the system, but is no longer visible or usable.
2. Data Destruction: Ensures permanent and irretrievable removal of data using methods like overwriting, erasing, or destroying the storage medium. This process aligns with security standards like NIST SP 800-53.

## 4    Use Cases

### 4.1    Authentication to a Cloud Service Use-Case

The process of cloud service authentication varies by provider and technology. In this use case, a simple scenario using username and password credentials is mapped to the CD-DLC model. Figure 2 illustrates the scenario's sequence diagram.

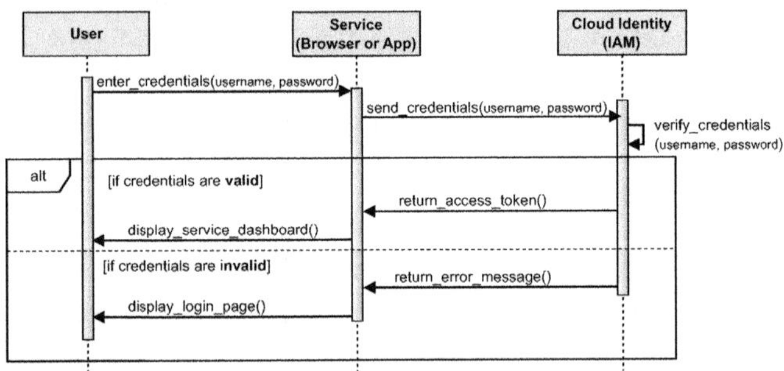


**Fig. 2.** Authentication to a cloud service use-case sequence diagram.

As depicted in Fig. 3, the use case begins with (1) the user (or application) submitting credentials, initiating the collection phase. The cloud service receives the credentials, leading to (2) the metadata creation phase, where the data is (3) duplicated and (4) stored. Next, the cloud service (5) shares the credentials with a cloud identity service for (6) verification, representing the share and use phases. Upon successful authentication, the identity service (7) returns an access token, (8) updating the metadata in the update phase. (9,10,11) The user is then granted access to interact with the service in the use phase. (12,13) If authentication fails, an error is returned, and the user is prompted to retry. (14,15) According to retention policies, unnecessary data is disposed of in the destruction, deletion, or archiving phases.

In Fig. 4 we explore the attack chains in cloud environments, mapped to the CD-DLC model, highlighting critical phases and vulnerabilities. The attack chain comprises three stages: an initial Man-in-the-Middle (MITM) attack, where data may be intercepted or altered, followed by session hijacking, allowing attackers to gain control of active sessions. Subsequently, post-control attacks may occur, including data leakage, ransomware, or malware deployment. By aligning the attack chain with the CD-DLC model, threats can be effectively localized and mitigated by implementing proactive approach based on this model.

### 4.2 Financial Data Management in AWS Use-Case

In this case study, we explore the application of the Cloud-Dedicated Data Lifecycle (CD-DLC) model in a financial institution, leveraging Amazon Web Services (AWS) to securely and efficiently manage sensitive financial data. The institution must process and handle various types of data, including high-volume real-time transactions and large batch-processed reports.

Figure 5 illustrates the mapping of the proposed model with AWS using a combination of advanced services to address key requirements: real-time data ingestion, batch data processing, secure storage, automated archival, and eventual purging of data at the end of its lifecycle. This multi-phase implementation not only optimizes data management but also ensures data security compliance and allows for the implementation of security layer techniques.

- **Data Acquisition** Real-time financial data is ingested using AWS Kinesis Data Streams, enabling continuous streaming of transactions. Amazon S3 is utilized for batch uploads, providing scalable storage for historical records and periodic reports.
- **Data Preparation** AWS Glue performs ETL (Extract, Transform, and Load) operations to convert raw financial data into structured formats, while AWS Lambda automates metadata tagging. Metadata is stored in the AWS Glue Data Catalog for governance and compliance tracking.
- **Data Management** Amazon S3 Cross-Region Replication ensures data redundancy across regions for disaster recovery. Access is controlled through AWS IAM with role-based policies and multi-factor authentication. AWS Redshift facilitates data analysis on large datasets, providing insights for decision-making.
- **Data Depot** S3 Lifecycle Policies automatically transition aged data to Amazon Glacier for cost-efficient long-term storage, ensuring compliance with retention policies.

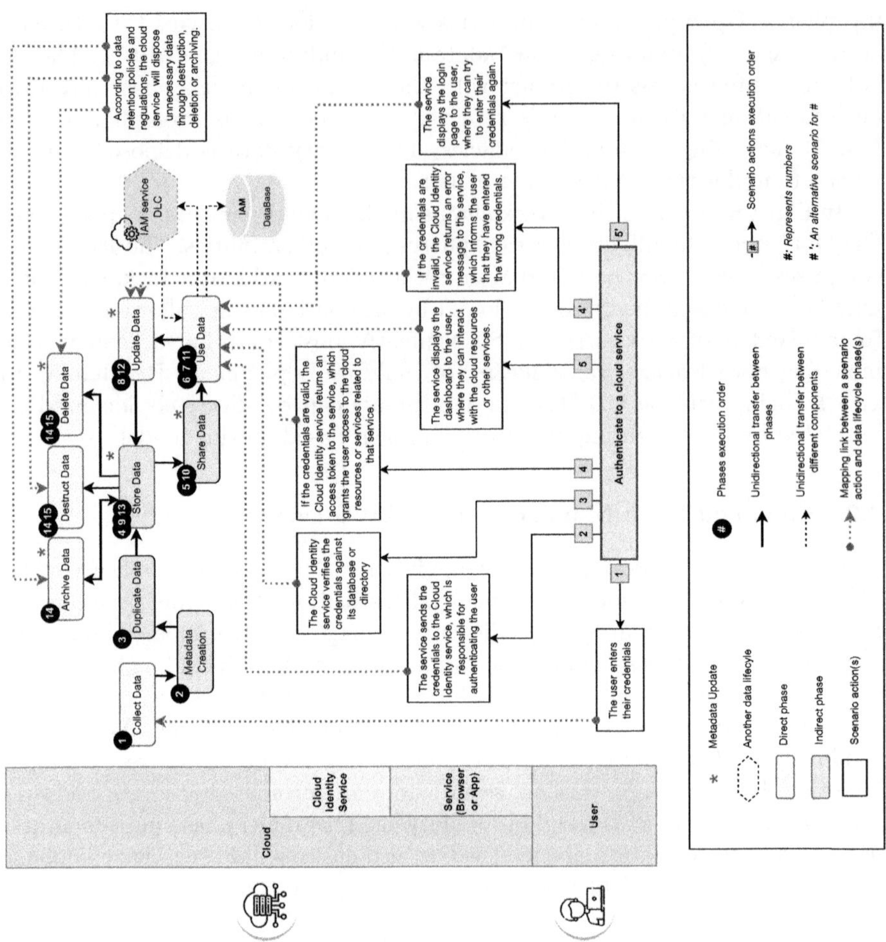

Fig. 3. Authentication to a cloud service use-case mapping with CD-DLC.

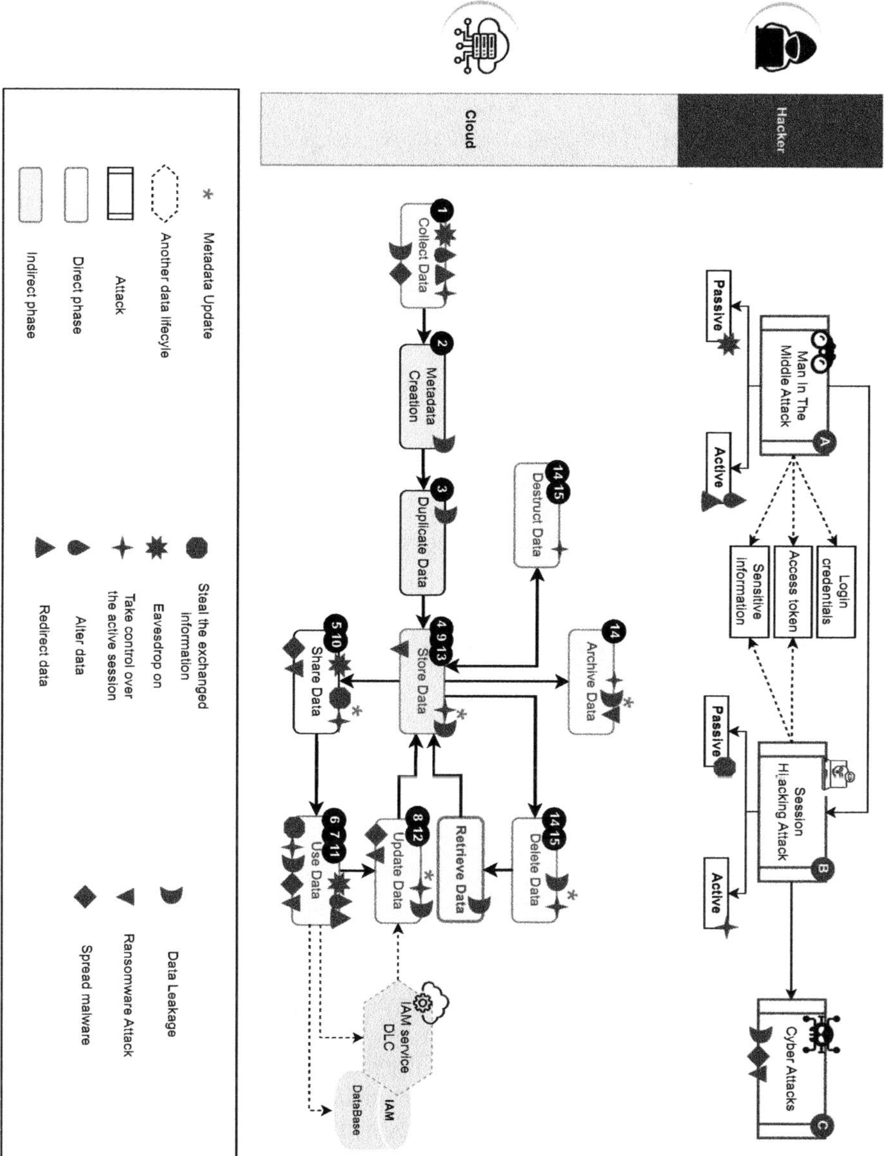


**Fig. 4.** Security threats on authentication to a cloud service use-case mapping with CD-DLC.

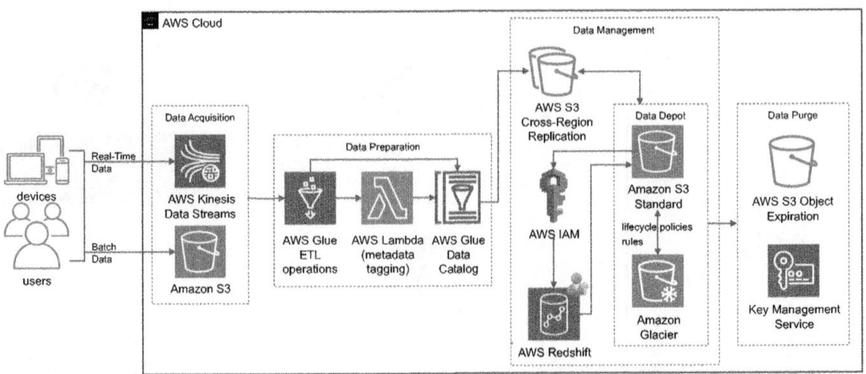


**Fig. 5.** Financial Data Management in AWS use-case mapping with CD-DLC.

– **Data Purge** S3 Object Expiration automates the deletion of expired data,
and AWS KMS ensures secure data destruction by managing and deleting
encryption keys, making data irretrievable.

## 5   Discussion and Conclusion

In this paper, we introduce the Cloud Data Lifecycle model CD-DLC, specifically
designed to improve understanding the data flow within cloud environments.
The CD-DLC incorporates essential cloud-specific features such as data elasticity, on-demand services, and resource pooling—characteristics often absent
in traditional data lifecycle models. These features make the CD-DLC model
more adaptable to the dynamic nature of cloud environments, distinguishing it
from models that primarily address on static data management or isolated security aspects. The existing models only partially cover these crucial cloud-specific
characteristics as discussed in Table 2.

Unlike traditional models that concentrate on specific phases, such as storage
and deletion, the CD-DLC is designed to enable the implementation of security
measures across all phases of the data lifecycle. By providing a foundation for
dynamically addressing security concerns during critical stages, the CD-DLC
model remains highly relevant for managing data in modern, real-time cloud
environments.

Practical use cases in Sect. 4.1 illustrate the model's application, covering
both a general cloud service authentication scenario and a financial AWS use
case. While the CD-DLC does not implement specific security measures, it
serves as a foundation for building future data protection strategies, enabling
a more proactive approach to managing cloud vulnerabilities. Overall, the CD-
DLC advances our understanding of cloud data flow and contributes to more
secure, resilient cloud-based systems.

**Table 2.** Comparison of data lifecycle models based on cloud characteristics

| Work | Comparison Criteria | | | | | |
|---|---|---|---|---|---|---|
| | *Focus on cloud computing* | *On-demand self-service* | *Broad network access* | *Resource pooling* | *Rapid elasticity* | *Measured service* |
| Levitin and Redman model [16] | ? | ✓ | × | × | ? | × |
| CSA model [7] | ✓ | ✓ | ? | × | ✓ | × |
| Beijing University model [24] | ✓ | ✓ | ? | × | ✓ | × |
| Demchenko et al. model [10] | ✓ | ✓ | ✓ | × | ✓ | × |
| AL-Aswadi and Batarfi model [4] | ✓ | ✓ | ✓ | × | ✓ | × |
| Kazim and Ying model [15] | ✓ | ✓ | ✓ | × | ✓ | × |
| Grunzke et al. model [14] | ? | ✓ | × | ✓ | ? | ✓ |
| Mei et al. model [17] | ✓ | ✓ | ✓ | × | ✓ | × |
| Proposed model (CD-DLC) | ✓ | ✓ | ✓ | ✓ | ✓ | ✓ |

✓: exists, ×: doesn't exist, ?: unclear if it exists or not.

Looking ahead, this research paves the way for future exploration in several key areas. One avenue for future work is identifying security vulnerabilities across CD-DLC phases and developing targeted security mechanisms to mitigate these risks. Another promising area is the development of context-aware security measures within the model, supporting more adaptive and proactive data protection strategies.

# References


1. Home - KPMG Global (2023). https://kpmg.com/xx/en/home.html
2. Abdulsalam, Y.S., Hedabou, M.: Security and privacy in cloud computing: technical review. Future Internet **14**(1) (2022)
3. Adee, R., Mouratidis, H.: A dynamic four-step data security model for data in cloud computing based on cryptography and steganography. Sensors **22**(3) (2022)
4. AL-Aswadi, F.N., Batarfi, O.: A framework for enhancing privacy provision in cloud computing. Int. J. Comput. Sci. Inf. Technol. (IJCSIT) (2014)
5. AWS: What is Cloud Computing? - Cloud Computing Services, Benefits, and Types - AWS. aws.amazon.com
6. Azure, M.: What is cloud computing? | Microsoft Azure. azure.microsoft.com
7. Brunette, G., Mogull, R., et al.: Security guidance for critical areas of focus in cloud computing v2. 1. Cloud Security Alliance, pp. 1–76 (2009)
8. Buyya, R., Yeo, C.S., Venugopal, S., Broberg, J., Brandic, I.: Cloud computing and emerging it platforms: vision, hype, and reality for delivering computing as the 5th utility. Futur. Gener. Comput. Syst. **25**(6), 599–616 (2009)
9. Cloud, H.: The NIST definition of cloud computing. Natl. Inst. Sci. Technol. Spec. Publ. **800**(2011), 145 (2011)

10. Demchenko, Y., Zhao, Z., Grosso, P., Wibisono, A., De Laat, C.: Addressing big data challenges for scientific data infrastructure. In: 4th IEEE International Conference on Cloud Computing Technology and Science Proceedings, pp. 614–617. IEEE, Taipei (2012)

11. Elmekki, H., Chiadmi, D., Lamharhar, H.: Open government data: towards a comparison of data lifecycle models. In: Proceedings of the ArabWIC 6th Annual International Conference Research Track, pp. 1–6. ACM, Rabat (2019)

12. Eryurek, E., Gilad, U., Lakshmanan, V., Kibunguchy-Grant, A., Ashdown, J.: Data Governance: The Definitive Guide. O'Reilly Media, Inc. (2021)

13. GCP: What is Cloud Computing? | Google Cloud. cloud.google.com

14. Grunzke, R., et al.: Managing complexity in distributed data life cycles enhancing scientific discovery. In: 2015 IEEE 11th International Conference on e-Science, pp. 371–380. IEEE, Munich (2015)

15. Kazim, M., Ying, S.: A survey on top security threats in cloud computing. Int. J. Adv. Comput. Sci. Appl. **6**(3) (2015)

16. Levitin, A., Redman, T.: A model of the data (life) cycles with application to quality. Inf. Softw. Technol. **35**(4), 217–223 (1993)

17. Mei, R., Yan, H.B., He, Y., Wang, Q., Zhu, S., Wen, W.: Considerations on evaluation of practical cloud data protection. In: Lu, W., Zhang, Y., Wen, W., Yan, H., Li, C. (eds.) CNCERT 2022. CCIS, vol. 1699, pp. 51–69. Springer, Singapore (2022). https://doi.org/10.1007/978-981-19-8285-9_4

18. Michener, W.K., Jones, M.B.: Ecoinformatics: supporting ecology as a data-intensive science. Trends Ecol. Evol. **27**(2), 85–93 (2012)

19. Nikkel, B.: Data deletion challenges and risks of recovery (2019)

20. Recommendation, I.: 3500 (2014) | ISO/IEC 17788: 2014. Information technology–Cloud computing–Overview and Vocabulary

21. Recommendation, I.: 3502 (2014) | ISO/IEC 17789: 2014. Information technology–Cloud computing–Reference architecture

22. Sinaeepourfard, A., Masip-Bruin, X., Garcia, J., Marín-Tordera, E.: A survey on data lifecycle models: discussions toward the 6vs challenges. Technical report (UPC-DAC-RR-2015–18) (2015)

23. Vaquero, L.M., Rodero-Merino, L., Caceres, J., Lindner, M.: A break in the clouds: towards a cloud definition (2008)

24. Yu, X., Wen, Q.: A view about cloud data security from data life cycle. In: 2010 International Conference on Computational Intelligence and Software Engineering, pp. 1–4. IEEE, Wuhan (2010)

25. Zhang, Q., Cheng, L., Boutaba, R.: Cloud computing: state-of-the-art and research challenges. J. Internet Serv. Appl. **1**(1), 7–18 (2010). https://doi.org/10.1007/s13174-010-0007-6

# A Novel Approach to QoS-Aware IoT Service Composition: A Hybrid MGWO-MVCH Model

Macilia Boukhama[1]([✉])[iD], Zoubeyr Farah[1,2][iD], and Lynda Alkama[1][iD]

[1] Laboratoire LITAN, École supérieure en Sciences et Technologies de l'Informatique et du Numérique, RN 75, 06300 Amizour, Bejaia, Algeria
{boukhama,alkama}@estin.dz

[2] Faculty of Exact Sciences, Laboratory of Medical Informatics and Intelligent and Dynamic Environments (LIMED), University of Bejaia, 06000 Bejaia, Algeria
zoubeyr.farah@univ-bejaia.dz
https://estin.dz, https://www.univ-bejaia.dz

**Abstract.** As IoT services continue to evolve, they provide similar functionalities while exhibiting significant variations in Quality of Service (QoS) levels. Despite this increase in services, meeting user satisfaction becomes increasingly complex. This situation underscores the need for the composition of IoT services capable of providing innovative functionalities. A critical challenge in the IoT service composition process lies in selecting the most appropriate atomic services to incorporate into the composition. In this research, we propose a method for modeling the IoT service composition process using the meta-heuristic optimization algorithm MGWO, while considering global QoS constraints. Our model enhances the balance between exploration and exploitation. This balance mitigates the risk of converging on local optima during the optimization process. To effectively address constraint violations associated with user-specified requirements, we introduce a concept termed Modified Violation Constraint Handling (MVCH). This methodology culminates in the development of a novel hybrid model, MGWO-MVCH, specifically designed for IoT service composition with an emphasis on optimizing quality of service.

**Keywords:** IoT Service composition · Metaheuristic optimization algorithm · Modified gray wolf optimizer (MGWO) · Quality of Service (QoS) · Global QoS constraints · Modified Violation constraint-handling (MVCH)

## 1 Introduction

The Internet of Things (IoT) is revolutionizing our daily lives by providing a multitude of personalized and efficient services. Through the interconnectivity of objects, more complex IoT services can be developed, enabling a better response to user needs, which often exceed the capabilities of available IoT services. Consequently, it is frequently necessary to construct more complex services through the aggregation of existing ones to meet user requirements. In this context, IoT services interconnect to create new composite services that deliver functionalities beyond those provided by any individual atomic service alone. This process is referred to as service composition. The growing number of available services has led to the emergence of services that are similar in terms of functional properties but differ in non-functional characteristics, known as Quality of Service (QoS) attributes. Accordingly, a key challenge lies in selecting the most appropriate service from multiple candidates to participate in the composition process. The task of selecting services with optimal QoS attributes constitutes an NP-hard combinatorial optimization problem.

The composition problem has been extensively explored in the literature, with various solutions proposed, encompassing both exact methods and meta-heuristics [2,3,8,9,13,15,20]. Among these approaches, metaheuristic methods —such as evolutionary algorithms (EAs) and swarm intelligence—are particularly notable due to their efficiency in converging and their ability to accelerate problem-solving in large-scale environments. These methods are capable of identifying optimal or near-optimal solutions within a minimal time frame. Initially, meta-heuristic optimization algorithms were developed to address NP-complete problems in the absence of constraints [16]. However, users frequently require that quality of service parameters meet specific threshold values, which are referred to as QoS constraints (e.g., *Localisation* $\approx$ *loc*, *time* $\leq$ 10s, *throughput* $\geq$ 56 kbit/s).

Numerous approaches have been proposed to address the constraint problem. A widely recognized method involves decomposing global constraints into local constraints [1,11], which has been extensively studied. However, selecting services based solely on local constraints may fail to guarantee the satisfaction of overall constraints. Other techniques for managing constraints include penalizing solutions that violate constraints by incorporating a penalty term into the objective function, as demonstrated in studies such as [4,6,19–21]. Nevertheless, achieving an appropriate balance between optimizing the objective function and satisfying the constraints tends to be complex [18], and a significant drawback is that penalty methods often require parameter adjustments [7].

This paper introduces MGWO-MVCH, a novel hybrid model designed for QoS-aware IoT service composition, taking into account global QoS constraints. The model combines the Modified Violation Constraints Handling (MVCH) technique [10] with the Modified Grey Wolf Optimizer (MGWO) [6]. This combination enhances constraints management by providing a distinct alternative to traditional penalty functions. Unlike conventional penalty-based methods, which require complex parameter tuning and may distort the objective function, the MVCH technique evaluates solutions based on their fitness values while independently assessing constraint violations. As a result, MVCH effectively balances the optimization of the objective function with the satisfaction of constraints without requiring parameter tuning, leading to more flexible and efficient constraints handling. Moreover, MGWO contributes to maintaining a dynamic equilibrium between exploration and exploitation, which helps prevent convergence to local optima. To evaluate the performance of the MGWO-MVCH model, a comprehensive simulation study was conducted.

The remaining parts of this paper are as follows: Sect. 2 reviews the related work. Section 3 describes the problem and its modeling. Section 4 introduces the detail of the proposed solution. Section 5 discusses the complexity of MGWO-MVCH. Section 6 shows the experimental results. Finally, Sect. 7 concludes and presents a future scope of the work.

## 2   Related Work

The issue of QoS-aware service composition has been widely explored, with numerous studies proposing various approaches for handling constraints.

In [4], a Genetic Algorithm (GA) was applied for QoS-aware service composition. By employing both static and dynamic penalties, feasible solutions were identified. However, the approach was constrained by the size of the candidate services.

In [21], a hybrid genetic algorithm was developed to enhance solution quality by integrating a local optimizer and a dynamic penalty mechanism. Although this approach improved solution quality, it also increased execution time.

In [6], the Modified Gray Wolf Optimizer (MGWO) was proposed for the QoS-aware service composition problem to balance exploration and exploitation. By incorporating a penalty mechanism, constraint satisfaction was addressed; however, the inclusion of penalty values adversely affected the overall composition quality.

In [20], a hybrid approach, known as HGA, was developed by combining the Genetic Algorithm (GA) with the Fruit Fly Optimization Algorithm (FOA) for the QoS-aware service composition problem. HGA demonstrated improved convergence and optimality while effectively addressing constraint satisfaction through a penalty mechanism.

The work presented in [1] applied mixed integer programming (MIP) to decompose global constraints into local constraints, thereby achieving high levels of optimality. However, this method was computationally expensive; furthermore, the satisfaction of local constraints does not necessarily guarantee the satisfaction of global constraints.

In [11], a Multi-Objective Evolutionary Genetic Algorithm (MOEA/D) was introduced for a QoS fluctuation-aware selection approach of IoT services. This approach decomposed global constraints into local ones, significantly improving optimality and execution speed. Nonetheless, it did not fully ensure the satisfaction of global QoS constraints.

The authors in [19] proposed a hybrid algorithm combining the Genetic Algorithm (GA) and the Artificial Bee Colony (ABC) algorithm, referred to as ABCGA, for QoS-aware service composition. This method employed a penalty mechanism to address constraint violations, resulting in improvements in response time, reliability, and cost. However, balancing utility and constraint satisfaction remained challenging due to the penalty approach.

## 3    Basics of MGWO-MVCH

The theoretical underpinnings of the MGWO-MVCH model are presented in this section.

### 3.1    Service Composition

This section presents theoretical concepts for modeling and expressing the QoS-aware IoT service composition process.

- Abstract service composition ($ASC$) refers to a sequence of interrelated tasks that must be executed in a specific order to achieve a more complex objective. ($ASC$) is represented as $ASC = \mathbf{struct}(T_1, ..., T_i, ..., T_n)$, where n is the total number of tasks required for the composition, $T_i$ denotes the $i^{th}$ task, with $i$ ranging from 1 to $n$, and $\mathbf{struct}$ defines the task composition structure, such as parallel, sequential, loop, or conditional. In our work, we focus on the sequential structure, as other structures can be converted into sequential ones [5].
- An abstract service ($AS$) refers to a collection of concrete services that provide the same functionality but differ in terms of Quality of Service (QoS). $AS_i$ is defined as $AS_i = (s_i^1, ..., s_i^j, ..., s_i^m)$, where $m$ represents the number of concrete services in the $i^{th}$ abstract service, and $s_i^j$ refers to the $j^{th}$ concrete service of the $i^{th}$ abstract service, with $j$ ranging from 1 to $m$. For each task $T_i$ in the Abstract Service Composition ($ASC$), there is a corresponding abstract service, $AS_i$, assigned to its execution.
- A concrete service ($s_i^j$) is a specific instance of an abstract service, defined by its inputs (I), outputs (O), and its Quality of Service (QoS) attributes [12].
- QoS attributes represent the parameters that define the quality and performance of a service, such as cost, availability, time, throughput, security, and others. The QoS attributes of a concrete service are expressed as a vector: $QoS(s_i^j) = (q_{i,1}^j, ..., q_{i,k}^j, ..., q_{i,l}^j)$, where $l$ represents the total number of QoS attributes for service $s_i^j$, and $q_{i,k}^j$ denotes the value of the $k^{th}$ QoS attribute for the $j^{th}$ concrete service, with $k$ ranging from 1 to $l$. These attributes can be categorized into two types: (a) negative attributes, such as cost and time, where higher values indicate lower QoS, and (b) positive attributes, such as throughput and availability, where higher values indicate better QoS.

- The aggregation function $Aggr()$ is used to combine multiple services into a composite service, denoted as $CS = (Q_1, ..., Q_k, ..., Q_l)$. Where, $Q_k = Aggr\left(q^j_{i,k}\right)$, and $i = \overline{1,n}$. The method of aggregation depends on the composition structure and the type of QoS attribute being considered [9].
- Global QoS constraints $(C)$ are upper or lower bounds imposed on the QoS values of a composite service after aggregation. These constraints are represented as $C = (c_1, ..., c_k, ..., c_l)$, where $c_k$ is the constraint for the $k^{th}$ attribute, with the number of constraints matching the number of attributes. For positive attributes, the constraint is $Q_k \geq c_k$, and for negative attributes, it is $Q_k \leq c_k$. A composite service is considered feasible if it satisfies all QoS constraints problem.

  It maps the QoS values of a composite service into a single value called the utility value, it's used to evaluate optimality and performance of the service. The utility function is calculated as follows:

- The utility function $f_{uti}$, as shown in Eq. (1), transforms the composition problem from a multi-objective to a single-objective problem. It consolidates the QoS values of a composite service into a single value, known as the utility value, which is used to assess the service's optimality and performance. The utility function is calculated as follows:

$$f_{uti} = \sum_{k=1}^{l} \text{Norm}(Q_k) \cdot W_k \tag{1}$$

$$\text{Norm}(Q_k) = \begin{cases} \frac{Q_k^{\max} - Q_k}{Q_k^{\max} - Q_k^{\min}} & \text{if the attribute is negative} \\ \frac{Q_k - Q_k^{\min}}{Q_k^{\max} - Q_k^{\min}} & \text{if the attribute is positive} \\ 1 & \text{if } Q_k^{\max} = Q_k^{\min} \end{cases} \tag{2}$$

$$\begin{cases} Q_k^{\max} = \text{Aggr}\left(q_{i,k}^{\max}\right), & i = 1, \dots, n \\ Q_k^{\min} = \text{Aggr}\left(q_{i,k}^{\min}\right), & i = 1, \dots, n \end{cases} \tag{3}$$

where the Eq. (2) represents the normalization value of the $k^{th}$ Qos attribute, of the composite service (CS) that is used to scale QoS values to a standard range between [0,1]. Moreover $W_k$ represents the weight of the $k^{th}$ Qos attribute (users preferences of each QoS attribute), such that $\sum_{k=1}^{l} W_k = 1$.

The parameters $q_{i,k}^{max}$ (respectively, $q_{i,k}^{min}$) is the maximum (respectively, the minimum) value of the $k^{th}$ QoS attribute in all components of the abstract service $AS_i$. $Q_k^{max}$ and $Q_k^{min}$ given in equation (3) denote the aggregation of all $q_{i,k}^{max}$ and $q_{i,k}^{min}$, respectively, in the service **(CS)** [22].

### 3.2   Modified Gray Wolf Optimizer

The Gray Wolf Optimizer (GWO) algorithm, introduced in [17], is inspired by the hunting behavior and social hierarchy of wolves. Wolves typically live in packs of 5–12, with a structured hierarchy. The alpha wolf leads the group, directing activities such as hunting and resting, while beta wolves, second in rank, support the alpha. Delta wolves assist both the alpha and beta wolves, and omega wolves, the lowest-ranking members, are subordinate to the rest of the pack. This hierarchical structure is mirrored in GWO, where the roles of each wolf correspond to different exploration and exploitation strategies. GWO's hunting process is divided into three phases:

- Tracking, chasing, and approaching the prey.
- Pursuing, encircling, and harassing the prey until it stops moving.
- Attack towards the prey.

The mathematical representation of the wolves' social hierarchy begins by assigning the best solution in the population to the alpha wolf, followed by the second and third best solutions assigned to the beta and delta wolves, respectively. The remaining solutions are designated as omega wolves. Once the prey is located, the gray wolves surround it. This process can be mathematically expressed as follows:

$$D = |C \cdot X_p(t) - X(t)| \tag{4}$$

$$X(t + 1) = X_p(t) - A \cdot D \tag{5}$$

where $t$ is the current iteration and $D$ is the distance between the wolf and the prey. $X_p$ and $X$ are the prey position vector and gray wolf's position, respectively. $A$ and $C$ are coefficient vectors, they are calculated as follow:

$$C = 2r_2 \tag{6}$$

$$A = 2a \cdot r_1 - a \tag{7}$$

where $r_2$ and $r_1$ are random numbers in the range [0,1], and $a$ decreases from 2 to 0 over the course of the GWO iterations.

During the hunt, omega wolves follow the same steps as the alpha, beta, and delta wolves. In other words, the search agents in the population, referred to as omega wolves, update their positions based on the positions of the top three search agents—the alpha, beta, and delta wolves. The following formulas represent this process:

$$D_\alpha = |C_1 \cdot X_\alpha - X|; \quad D_\beta = |C_2 \cdot X_\beta - X|; \quad D_\delta = |C_3 \cdot X_\delta - X| \tag{8}$$

$$X_1 = X_\alpha - A_1 \cdot D_\alpha; \quad X_2 = X_\beta - A_2 \cdot D_\beta; \quad X_3 = X_\delta - A_3 \cdot D_\delta \tag{9}$$

$$X(t + 1) = \frac{X_1 + X_2 + X_3}{3} \tag{10}$$

The GWO algorithm primarily facilitates information exchange among the top three solutions within the population. While this approach can be effective during the initial phases of the search, it often results in reduced diversity and limited exploration of the search space, particularly in the later iterations. Consequently, the GWO algorithm faces challenges such as premature convergence and a tendency to become trapped in local optima. To overcome these limitations, a modified version of GWO, termed MGWO, has been proposed by [14].

MGWO enhances the original algorithm by incorporating a genetic crossover operator inspired by the Genetic Algorithm (GA). This operator randomly selects two individuals, denoted as $x$ and $y$, (where $x \neq y$), from the population to generate a new offspring solution, $x_{new}$. If the fitness value of $x_{new}$ exceeds that of the parent $x$, the new solution replaces $x$ in the population. This crossover operator aims to improve the algorithm's performance by fostering a better balance between exploration and exploitation, thereby mitigating risks associated with premature convergence and local optima.

### 3.3   Modified Violation Constraint Handling (MVCH) Technique

Meta-heuristic optimization algorithms were initially developed to address unconstrained NP-complete problems. However, to adapt these algorithms for constrained optimization problems (COP), various constraint management techniques have been introduced. Among these, the penalty method is one of the most widely applied. This approach converts a constrained problem into an unconstrained one by adding a penalty term to the objective function. Despite its popularity, the penalty method can alter the characteristics of the objective function, potentially leading to suboptimal results. Moreover, it does not always guarantee convergence toward the feasible region [16].

Penalty terms in constrained optimization problems modify the objective function to penalize constraint violations, affecting the selection of solutions. Strong penalties can prioritize feasibility over optimality, potentially ranking infeasible solutions lower despite their superior objective values. Conversely, weak penalties may allow infeasible solutions to persist. Over-penalization of small violations may prematurely discard near-feasible solutions, while overly mild penalties might allow infeasible solutions to dominate. Dynamic penalty scaling may facilitate a balance between exploration and convergence; however, inappropriate scaling may lead to premature convergence or impede the discovery of feasible solutions. The penalty term must be carefully calibrated to strike a balance between optimizing the objective and ensuring feasibility. In other words, penalty methods often require parameter tuning, which is challenging to calibrate and can significantly impact the optimization process.

To address these challenges, the authors in [10] introduced the Modified Violation Constraint Handling (MVCH) method, which eliminates the need for penalty terms and parameter adjustments. MVCH maintains a balance between optimizing the objective function and satisfying constraints, ensuring that no compromises are made in the process. In MVCH, solutions are evaluated based on three criteria:

- Fitness value ($f$).
- Degree of Constraints Violated (DCV), calculated using Eq. (11).
- Number of Constraints Violated (NCV), if NCV equal 0, the solution is feasible.

Consider two solutions, $x$ and $y$. If both solutions are feasible, the one with the higher fitness value, denoted as $x$, will be selected. If $x$ is feasible but $y$ is not, $x$ will be preferred as the better solution. When both solutions are infeasible, the decision is made based on the NCV. For instance, if $x$ violates fewer constraints than $y$, $x$ will be chosen. If both $x$ and y violate the same number of constraints, the solution with the lower DCV will be favored. If $x$ and $y$ have equal NCV and DCV, the solution with the higher fitness value will be selected as the final winner.

$$DCV(x) = \begin{cases} \sum_{i=1}^{k} \left| \max(0, g_i(x)) \right| & \text{if } g_i(x) \leq 0 \\ \sum_{i=1}^{k} \left| \min(0, g_i(x)) \right| & \text{otherwise} \end{cases} \tag{11}$$

where $g_i$ represents the constraint, $i \in [1, k]$, and $k$ is the number of constraints.

## 4    MGWO-MVCH Model

MGWO-VCH integrates the strengths of the MGWO algorithm and the MVCH technique to effectively address the QoS-aware service composition problem while considering global QoS constraints. The MGWO-MVCH model is organized into four primary phases: encoding, initialization, evaluation, and iteration.

**Encoding:** To model the service composition problem as an optimization problem, the first step involves defining the encoding scheme. This process entails establishing a connection between the service composition problem and the MGWO algorithm, as follows:

- Set of wolves $pop_f = (X_1, \ldots, X_r, \ldots, X_N)$: Population containing the compositions $pop_{CS} = (CS_1, \ldots, CS_r, \ldots, CS_N)$.
- A wolf $X_r = (x_{r,j}^1, \ldots, x_{r,j}^i, \ldots, x_{r,j}^d)$: Composite service $CS_r = (s_{r,j}^1, \ldots, s_{r,j}^i, \ldots, s_{r,j}^n)$.
- Dimension $d$: The number of abstract services in the composition $CS_r$.
- The position of a wolf: The index of a candidate service in the composition.
- The fitness function $f(X_r)$: The utility value of the composition in terms of QoS $f_{uti}(CS_r)$.
- $\alpha, \beta, \gamma$ wolves: The three best compositions $f(CS_\alpha) > f(CS_\beta) > f(CS_\gamma)$.

**Initialization:** In this stage, all parameters associated with the algorithm are assigned initial values. These parameters include the maximum number of iterations, denoted as $Maxiter$, and the population size, denoted as $N$. Additionally, an initial population, referred to as $pop_{CS}$ consisting of $N$ compositions $(CS_1, .., CS_r, .., CS_N)$ is generated randomly.

**Evaluation:** In this phase, the quality and performance of each composition are evaluated using the utility function, $f_{uti}$, as defined in Eq. (1). To assess the feasibility of these compositions, the Number of Constraints Violated (NCV) is calculated, and the Degree of Constraints Violated (DCV) is computed according to Eq. (11). The MGWO-MVCH algorithm then selects the top three service compositions, denoted as $CS_\alpha$, $CS_\beta$, and $CS_\gamma$, based on the MVCH technique. Rather than solely relying on utility values, MVCH ensures that the selection process considers both the utility of each composition and its ability to satisfy QoS constraints. This method allows the algorithm to effectively balance utility maximization with global QoS constraints satisfaction. Algorithm 1 is used to determine $CS_\alpha$, $CS_\beta$, and $CS_\gamma$, ensuring that the selected service compositions are optimal in terms of utility and feasibility.

---

**Algorithm 1.** Finding $CS_\alpha$, $CS_\beta$, and $CS_\gamma$ Using the MVCH Technique.

---

1: **if** $NCV(CS_i) < NCV(CS_\alpha)$ **then**
2:    $CS_\alpha = f(CS_i)$
3: **else if** $NCV(CS_\alpha) = NCV(CS_i)$ **and** $DCV(CS_i) < DCV(CS_\alpha)$ **then**
4:    $CS_\alpha = f(CS_i)$
5: **else if** $DCV(CS_\alpha) = DCV(CS_i)$ **and** $f(CS_\alpha) < f(CS_i)$ **then**
6:    $CS_\alpha = f(CS_i)$
7: **end if**
8: **if** $f(CS_i) < f(CS_\alpha)$ **then**
9:     **if** $NCV(CS_i) < NCV(CS_\beta)$ **then**
10:        $CS_\beta = f(CS_i)$
11:     **else if** $NCV(CS_i) = NCV(CS_\beta)$ **and** $DCV(CS_i) < DCV(CS_\beta)$ **then**
12:        $CS_\beta = f(CS_i)$
13:     **else if** $DCV(CS_i) = DCV(CS_\beta)$ **and** $f(CS_\beta) \le f(CS_i)$ **then**
14:        $CS_\beta = f(CS_i)$
15:     **end if**
16: **end if**
17: **if** $f(CS_i) < f(CS_\alpha)$ **and** $f(CS_i) < f(CS_\beta)$ **then**
18:     **if** $NCV(CS_i) < NCV(CS_\gamma)$ **then**
19:        $CS_\gamma = f(CS_i)$
20:     **else if** $NCV(CS_i) = NCV(CS_\gamma)$ **and** $DCV(CS_i) < DCV(CS_\gamma)$ **then**
21:        $CS_\gamma = f(CS_i)$
22:     **else if** $DCV(CS_i) = DCV(CS_\gamma)$ **and** $f(CS_\gamma) \le f(CS_i)$ **then**
23:        $CS_\gamma = f(CS_i)$
24:     **end if**
25: **end if**

---

**Iteration Process:** The MGWO-MVCH algorithm operates through both exploration and exploitation phases, achieving a balance between the two via adaptive parameters $\alpha$ and $A$, as described in Eq. (7). When $|A| \geq 1$, approximately half of the iterations are dedicated to exploration, while the remaining iterations focus on exploitation when $|A| < 1$. During these phases, all compositions within the population are updated based on the information from the three best solutions, as described in Eqs. (8)–(10). The primary objective of this update process is to generate new service compositions with enhanced performance. To improve the balance between exploration and exploitation, a two-point crossover method, adapted from Genetic Algorithms (GA), is applied to each composition within the population. This operation produces new compositions, which replace the old ones if they exhibit superior quality and feasibility. Otherwise, the original compositions are retained. This approach ensures population diversity and maintains a robust exploration process throughout the algorithm.

The full steps of the MGWO-MVCH approach are outlined in Algorithm 2.

---

**Algorithm 2.** Pseudo Code of the MGWO-MVCH Algorithm

---

1: **Input:** Initial population $pop_{CS} = (CS_1, CS_2, \ldots, CS_N)$
2: **Output:** Best composite service $CS_\alpha$
3: Initialize the population $pop_{CS} = (CS_1, CS_2, \ldots, CS_N)$ randomly.
4: Initialize parameters $a$, $A$, $Maxiter$, and $C$.
5: Calculate $f(CS_i)$ for each individual $i \in \{1, \ldots, N\}$.
6: Determine $CS_\alpha$, $CS_\beta$, $CS_\gamma$ using Algorithm 1.
7: **while** (t $< Maxiter$) **do**
8:     **for** each $CS_i$ **do**
9:         Update $CS_i$ using Eq. (10);
10:     **end for**
11:     **for** i $= 1$ to size of $pop_{CS}$ **do**
12:         Select randomly $CS_j$ where $CS_j \neq CS_i$;
13:         Produce $CS_i^{new}$ using crossover operator;
14:     **end for**
15:     Update $a$, $A$, and $C$, and $CS_\alpha$, $CS_\beta$, $CS_\gamma$ using Algorithm 1;
16: **end while**
17: **return** $CS_\alpha$;

---

## 5 Complexity of MGWO-MVCH

The complexity of the MGWO-MVCH algorithm can be analyzed by examining the various stages of its execution, which includes initialization, utility evaluation, constraint evaluation, compositions selection, and iterative updates. The overall complexity depends on three primary factors: population size ($N$), constraints number ($k$), and number of iteration ($Maxiter$).

– **Initialization:** The algorithm begins by generating an initial population of $N$ service compositions, which requires $O(N)$.

- **Utility evaluation:** For each composition, the utility function is calculated during each iteration, leading to a complexity of $O(N)$.
- **Constraint evaluation (DCV) and (NCV):** The algorithm evaluates the satisfaction of $k$ QoS constraints for each composition. This step has a complexity of $O(N \times k)$.
- **Selection of Top Solutions:** Identifying the three best compositions ($CS_\alpha$, $CS_\beta$, $CS_\gamma$) involves scanning the population, with a complexity of $O(N)$ per iteration.
- **Crossover and update operations:** During each iteration, crossover and update steps are applied to the population, adding $O(N)$ to the complexity.

Given that these steps are repeated over $Maxiter$ iterations, the overall complexity of MGWO-MVCH is: $O(Maxiter \times N \times k)$.

## 6   Experimental Results and Discussion

To evaluate the performance of our model, we conducted a series of experiments utilizing a dataset comprising five abstract services and five service quality attributes (QoS). The dataset included a variation of concrete services ranging from 2000 to 10000 services for each abstract service. The experimental setup involved a population size of 20 individuals and a maximum of 100 iterations, with each configuration being simulated 10 times to calculate the average utility, average degree of violation of constraints, and the average total number of violated constraints. All experiments were executed on a personal computer equipped with an Intel(R) Core(TM) i7-1065G7 processor (1 30 GHz), 16 GB of RAM, running on a Windows 10 (64-bit) operating system, and utilizing MATLAB R2019a. We compared the performance of the MGWO-MVCH algorithm against MGWO [6] and the standard GWO-P with penalty. This comparison aims to highlight the distinct disadvantages associated with the penalty approach in contrast to the MVCH methodology.

### 6.1   Impact of Concrete Services on Utility Value

Figure 1 presents a performance comparison of MGWO-MVCH with MGWO and GWO-P in terms of utility value as the number of concrete services varies. The results indicate that our approach consistently maintains a significantly higher utility value, ranging from 0.55 to 0.69 even as the number of concrete services increases. In contrast, GWO-P achieves values between 0.50 and 0.60 and MGWO between 0.51 and 0.60. This notable difference underscores the effectiveness of our approach, particularly in large-scale environments, as it achieves a utility value of 0.61 with 10000 concrete services. This performance can be attributed to the MVCH method, which manages constraints satisfaction independently of utility optimization, unlike MGWO and GWO-P, which rely on a penalty mechanism.

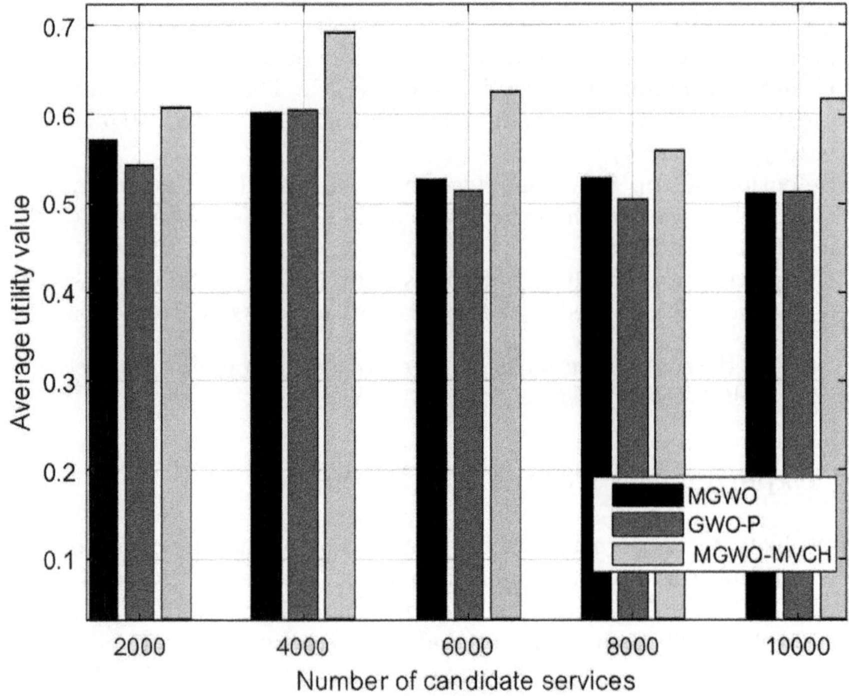

**Fig. 1.** Average utility value on varying number of concrete services.

## 6.2   Impact of Concrete Services on DCV and NCV

Table 1 presents a comprehensive comparison of the Degree of Constraint Violated (DCV) and the Number of Constraints Violated (NCV) for MGWO-MVCH, MGWO, and GWO-P, with the aim of assessing their feasibility. The results demonstrate that MGWO-MVCH consistently achieves the lowest DCV and NCV across different numbers of concrete services. Specifically, the DCV values for MGWO-MVCH are 2.62 for 2000 services, 2.59 for 4000 services, 2.72 for 6000 services, 2.86 for 8000 services, and 2.77 for 10000 services. These low DCV values highlight the superior efficiency of MGWO-MVCH in converging toward feasible solutions, where a solution is considered fully feasible when the DCV equals zero. In contrast, MGWO exhibits DCV values ranging from 2.77 to 3, while GWO-P shows values between 2.81 and 3, indicating that both methods encounter greater difficulty in identifying feasible solutions compared to MGWO-MVCH. The NCV values for MGWO-MVCH are 4.7, 4.8, 4.9, 4.3, and 4.8 for 2000, 4000, 6000, 8000, and 10000 services, respectively. This indicates that across 10 runs, the percentage of feasible solutions found for 2000, 4000, 6000, 8000, and 10000 services are 6%, 4%, 2%, 14%, and 4%, respectively. For MGWO and GWO-P, at most 2% of the solutions found are feasible, and when NCV equals 5, no feasible solutions are found. The efficiency of MGWO-MVCH can be attributed to its ability to effectively balance constraints satisfaction with the optimization of utility values.

**Table 1.** Comparison of DCV and NCV for MGWO-MVCH, MGWO, and GWO-P across different numbers of concrete services.

| Concrete Services | MGWO-MVCH | | MGWO | | GWO-P | |
|---|---|---|---|---|---|---|
| | DCV | NCV | DCV | NCV | DCV | NCV |
| 2000 | 2.62 | 4.7 | 2.77 | 4.9 | 2.77 | 4.9 |
| 4000 | 2.59 | 4.8 | 2.81 | 5 | 2.84 | 5 |
| 6000 | 2.72 | 4.9 | 2.96 | 4.9 | 2.98 | 4.9 |
| 8000 | 2.86 | 4.3 | 2.92 | 4.9 | 2.92 | 4.9 |
| 10000 | 2.77 | 4.8 | 3.01 | 5 | 3 | 5 |

## 7 Conclusion and Future Work

In this paper, we addressed the challenging problem of IoT service composition under QoS constraints. We introduced MGWO-MVCH, a novel hybrid model for QoS-aware IoT service composition, that effectively combines the Modified Grey Wolf Optimizer (MGWO) with a Modified violation constraints handling (MVCH) mechanism. Our model successfully manages the trade-off between utility optimization and constraints satisfaction. By leveraging the MGWO algorithm, known for its ability to balance exploitation and exploration, the model ensures solution diversity and avoids convergence to local optima. Additionally, the application of the MVCH technique, which accurately distinguishes between feasible and infeasible solutions, significantly facilitated the discovery of viable service compositions.

Simulation results demonstrate the effectiveness of our technique in terms of utility, the number of constraints violated, and the degree of constraints violated. Even in large-scale environments, our composition model maintains strong performance. Future work will focus on enhancing MGWO-MVCH by integrating functional constraints and conducting additional experiments with a larger number of abstract services to further evaluate the effectiveness of the MVCH technique.

## References

1. Alrifai, M., Risse, T.: Combining global optimization with local selection for efficient QoS-aware service composition. In: the 18th International Conference on World Wide Web, pp. 881–890 (2009). https://doi.org/10.1145/1526709.1526828

2. Asghari, P., Rahmani, A.M., Javadi, H.H.S.: Service composition approaches in IoT: a systematic review. J. Netw. Comput. Appl. **120**, 61–77 (2018). https://doi.org/10.1016/j.jnca.2018.07.013
3. Boukhama, M., Farah, Z., Lynda, A.: A QoS-aware IoT services composition approach based on the modified violation constraints handling technique and enhanced fireworks algorithm. Cluster Comput. **28** (2025). https://doi.org/10.1007/s10586-024-04859-9
4. Canfora, G., Di Penta, M., Esposito, R., Villani, M.L.: An approach for QoS-aware service composition based on genetic algorithms. In: The 7th Annual Conference on Genetic and Evolutionary Computation, pp. 1069–1075 (2005). https://doi.org/10.1145/1068009.1068189
5. Cardoso, J., Sheth, A., Miller, J., Arnold, J., Kochut, K.: Quality of service for workflows and web service processes. J. Web Semant. **1**(3), 281–308 (2004). https://doi.org/10.1016/j.websem.2004.03.001
6. Chandra, M., Agrawal, A., Kishor, A., Niyogi, R.: Web service selection with global constraints using modified gray wolf optimizer. In: 2016 International Conference on Advances in Computing, Communications and Informatics (ICACCI), pp. 1989–1994 (2016). https://doi.org/10.1109/ICACCI.2016.7732343
7. Chehouri, A., Younes, R., Perron, J., Ilinca, A.: A constraint-handling technique for genetic algorithms using a violation factor. arXiv preprint arXiv:1610.00976 (2016). https://doi.org/10.48550/arXiv.1610.00976
8. Cherifi, A., Khanouche, M.E., Amirat, Y., Farah, Z.: A parallel approach for user-centered QoS-aware services composition in the internet of things. Eng. Appl. Artif. Intell. **123**, 106277 (2023). https://doi.org/10.1016/j.engappai.2023.106277
9. Gavvala, S.K., Jatoth, C., Gangadharan, G., Buyya, R.: QoS-aware cloud service composition using eagle strategy. Futur. Gener. Comput. Syst. **90**, 273–290 (2019). https://doi.org/10.1016/j.future.2018.07.062
10. Jan, M.A., et al.: Hybrid stochastic ranking for constrained optimization. IEEE Access **8**, 227270–227287 (2020). https://doi.org/10.1109/ACCESS.2020.3044439
11. Khadir, K., Guermouche, N., Guittoum, A., Monteil, T.: A genetic algorithm-based approach for fluctuating QoS aware selection of IoT services. IEEE Access **10**, 17946–17965 (2022). https://doi.org/10.1109/ACCESS.2022.3145853
12. Khanouche, M.E., Amirat, Y., Chibani, A., Kerkar, M., Yachir, A.: Energy-centered and QoS-aware services selection for internet of things. IEEE Trans. Autom. Sci. Eng. **13**(3), 1256–1269 (2016). https://doi.org/10.1109/TASE.2016.2539240
13. Khanouche, M.E., Gadouche, H., Farah, Z., Tari, A.: Flexible QoS-aware services composition for service computing environments. Comput. Netw. **166**, 106982 (2020). https://doi.org/10.1016/j.comnet.2019.106982
14. Kishor, A., Singh, P.K.: Empirical study of grey wolf optimizer. In: Pant, M., Deep, K., Bansal, J., Nagar, A., Das, K. (eds.) Proceedings of Fifth International Conference on Soft Computing for Problem Solving. AISC, vol. 436, pp. 1037–1049. Springer, Singapore (2016). https://doi.org/10.1007/978-981-10-0448-3_87
15. Kouicem, A., Khanouche, M.E., Tari, A.: Novel bat algorithm for QoS-aware services composition in large scale internet of things. Clust. Comput. **25**(5), 3683–3697 (2022). https://doi.org/10.1007/s10586-022-03602-6
16. Lagaros, N.D., Kournoutos, M., Kallioras, N.A., Nordas, A.N.: Constraint handling techniques for metaheuristics: a state-of-the-art review and new variants. Optim. Eng. 1–48 (2023). https://doi.org/10.1007/s11081-022-09782-9
17. Mirjalili, S., Mirjalili, S.M., Lewis, A.: Grey wolf optimizer. Adv. Eng. Softw. **69**, 46–61 (2014). https://doi.org/10.1016/j.advengsoft.2013.12.007

18. Runarsson, T.P., Yao, X.: Stochastic ranking for constrained evolutionary optimization. IEEE Trans. Evol. Comput. **4**(3), 284–294 (2000). https://doi.org/10.1109/4235.873238
19. Sefati, S.S., Halunga, S.: A hybrid service selection and composition for cloud computing using the adaptive penalty function in genetic and artificial bee colony algorithm. Sensors **22**(13), 4873 (2022). https://doi.org/10.3390/s22134873
20. Seghir, F., Khababa, A.: A hybrid approach using genetic and fruit fly optimization algorithms for QoS-aware cloud service composition. J. Intell. Manuf. **29**(8), 1773–1792 (2016). https://doi.org/10.1007/s10845-016-1215-0
21. Tang, M., Ai, L.: A hybrid genetic algorithm for the optimal constrained web service selection problem in web service composition. In: IEEE Congress on Evolutionary Computation, pp. 1–8 (2010). https://doi.org/10.1109/CEC.2010.5586164
22. Wang, S., Zhou, A., Yang, M., Sun, L., Hsu, C.H., Yang, F.: Service composition in cyber-physical-social systems. IEEE Trans. Emerg. Top. Comput. **8**(1), 82–91 (2017). https://doi.org/10.1109/TETC.2017.2675479

# Architectural Comparisons and Performance Evaluation of Transformer Models on Sequence Classification: A Case Study on Fake News Detection


Amine Mammasse[1]([✉])(iD), Khaled Bedjou[2](iD), M'hamed Amine Hatem[1](iD), and Faical Azouaou[1](iD)

[1] Laboratoire LITAN, École supérieure en Sciences et Technologies de l'Informatique et du Numérique, RN 75, Amizour 06300, Béjaia, Algérie
{mammasse,hatem,azouaou}@estin.dz
[2] Laboratoire LIMED, Faculté des Sciences Exactes, Université de Bejaia, Bejaia 06000, Algeria
khaled.bedjou@univ-bejaia.dz



**Abstract.** The proliferation of fake news in the digital age poses a significant threat to information integrity and societal stability. This study addresses this critical challenge by presenting a comprehensive comparative analysis of seven diverse language models for fake news detection: LLAMA2, GPT-2, Mistral, RoBERTa, DistilBERT, GEMMA, and MAMBA. Using a carefully curated and balanced dataset, we evaluate these models based on accuracy, precision, recall, F1-score, and macro F1, while also considering their training efficiency and resource requirements. Our results reveal LLAMA2 as the top performer with 92.5% accuracy and 92.7% F1-score, closely followed by Mistral. We observe significant trade-offs between precision and recall across models, with Mistral excelling in precision (96.8%) and RoBERTa leading in recall (97.5%). Smaller models like GEMMA 2B show competitive performance, highlighting the impact of architectural innovations. This study offers valuable insights for selecting models based on various deployment scenarios, balancing performance with computational constraints. It also opens avenues for future research on optimizing AI-driven fake news detection systems, addressing the urgent need for effective tools to combat fake news.

**Keywords:** Fake news detection · Transformer models · Comparative analysis · Language models · Architectural innovations


## 1 Introduction

The rapid advancement of Natural Language Processing (NLP) has led to the new era of sophisticated language models, each with unique architectures and capabilities. Concurrently, the proliferation of fake news on social media and online platforms has created the need for effective fake news detection systems. This intersection of technological progress and societal challenge presents an opportunity to leverage state-of-the-art language models in combating the spread of false information [1]. Recent events underscore the urgency of this task:

1. During the 2023 Maui wildfires in Hawaii, false claims spread on social media about the fires being intentionally set for land grabs, causing panic and mistrust among residents [2].
2. The 2022 Russia-Ukraine conflict saw numerous fake news stories circulating online, including false reports of military actions and casualties, exacerbating tensions and contributing to misinformation about the conflict [3].
3. In 2021, false rumors about the collapse of major banks in Lebanon led to panic withdrawals, further destabilizing the already fragile economy [4].

Our selection of pre-trained language models (PLMs) for this comparative study is consistent with the desire to evaluate a diverse range of architectures and approaches, particularly those that are prominent in the open-source landscape. We primarily relied on the Hugging Face LLMs leaderboards, which provide a valuable ranking of current top performers. Our selection includes both large-scale models like LLAMA2 [5] and Mistral [6], known for their impressive performance and up-to-date availability, as well as smaller, more efficient models like DistilBERT [7], RoBERTa [8] and GEMMA [9] , which demonstrate the impact of architectural innovations on performance relative to model size. We also included GPT-2 [10], a classical model that serves as a valuable benchmark for comparing newer advancements, and MAMBA [11], a model that uses a novel architecture based on selective state spaces, offering an alternative approach to sequence modeling. This diverse set of models allows us to investigate the trade-offs between model size and performance.

Our research aims to evaluate these models on a custom-built dataset of short sequences, designed to simulate different types of content and the amount of content available on social media and news. By assessing their performance, efficiency, and scalability, we aim to provide insight into the strength and limitations of each model in terms of identifying fake news.

This comparative analysis is crucial for several reasons. First, it helps researchers and practitioners in selecting the most suitable model for their specific applications, considering factors such as accuracy, computational requirements, and deployment constraints. Second, it contributes to the broader understanding of how different architectural innovations and training approaches in PLMs translate to performance in real-world classification tasks. Lastly, it lays the groundwork for future research in optimizing language models for the critical task of fake news detection [12].

To provide a clear overview of our study, Fig. 1 outlines the organization and flow of our paper.

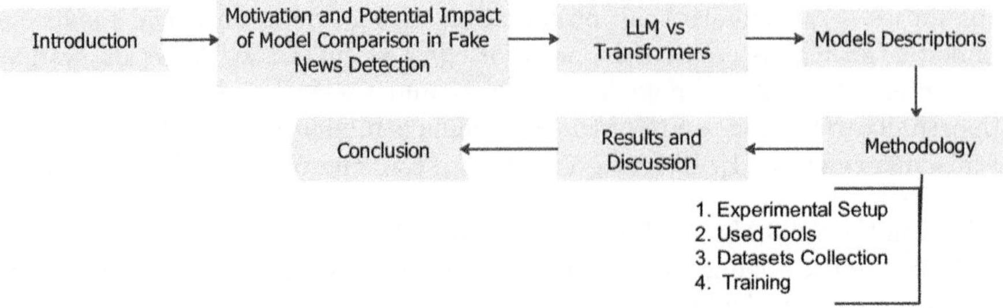

**Fig. 1.** Structural overview of the paper

## 2    Motivation and Potential Impact of Model Comparison in Fake News Detection

We conducted a comparative analysis of various models, including advanced techniques such as LoRA[1], quantization configurations, and customized training approaches. The ultimate goal is to improve the reliability and effectiveness of fake news detection systems, thereby contributing to a more informed public.

The potential impacts of this research include:

- **Enhanced Detection Accuracy:** By identifying the most effective models, this research can lead to more accurate fake news detection systems, reducing the spread of misinformation.
- **Informed Public:** Improved detection systems will help media organizations, fact-checkers[2], and social media platforms filter out fake news more efficiently, ensuring the public receives more reliable information.
- **Technological Advancements:** This research contributes to the broader field of machine learning and natural language processing, potentially spurring further innovations.
- **Computational Efficiency:** We explore the trade-offs between model size and inference speed, which is critical for real-time content moderation scenarios.
- **Generalizability:** By using diverse datasets, we assess how well models generalize across different topics, writing styles, and cultural contexts, which is crucial for maintaining effectiveness as fake news tactics evolve.
- **Resource Optimization:** Comparing models of varying complexity helps identify resource-efficient solutions, making fake news detection more accessible to a wider range of organizations and platforms.

---

[1] Tiny add-on for AI models that helps them learn new things without changing the whole model.

[2] e.g., factcheck.org, snopes.com.

# 3   Models Descriptions

In this study, we evaluate a diverse range of language models, Table 1 presents an overview of the selected language models, their key characteristics, and the rationale behind their inclusion in this comparative study on fake news detection.

**Table 1.** Model Description and Rationale Selection

| Model | Description | Rationale for Selection |
|---|---|---|
| LLAMA2 | Meta's open-source large language model series | Chosen for its state-of-the-art performance in various NLP tasks and high capacity to capture nuanced patterns, which is critical for detecting subtle differences between real and fake news |
| GPT-2 | OpenAI's autoregressive transformer model | Included as a benchmark to compare newer models against a well-known and widely studied baseline in sequence generation and classification tasks |
| Mistral | An efficient language model known for strong performance relative to its size | Selected for its balance between performance and resource efficiency, making it ideal for precision tasks, especially in environments with limited computational resources |
| RoBERTa | A robustly optimized version of BERT with improved training methodology | Chosen to assess the impact of optimized training techniques, like dynamic masking, on model performance for fake news detection |
| DistilBERT | A smaller, faster version of BERT created through knowledge distillation | Selected for its efficiency, offering competitive performance with reduced computational requirements, making it suitable for real-time applications |
| Gemma | Google's open-source, lightweight language model series designed for responsible AI use | Chosen to evaluate the trade-offs between performance and computational efficiency in lightweight models. It serves as an alternative to larger models in resource-constrained scenarios |
| MAMBA | A novel architecture using selective state spaces instead of attention mechanisms for sequence modeling | Selected for its unique architecture, offering potential improvements in long-sequence modeling with efficient resource use, making it an intriguing option for specialized tasks |

The Table 2 showcases the diversity of language models available, with varying sizes, architecture, training data, and key features. The models range from large (LLAMA2, Mistral, GPT-2) to small (RoBERTa, DistilBERT, GEMMA 2B, MAMBA) and while all are Transformer-based, MAMBA uses a unique architecture. Each model is trained on different datasets and possesses unique strengths, impacting their suitability for tasks like fake news detection.

# 4  Methodology

This section details the experimental setup, tools used, datasets collection, training procedures, and evaluation metrics employed in our comparative study of the seven language models for fake news detection.

## 4.1  Experimental Setup

The experiments were conducted using an NVIDIA RTX A5000 with 24 GB of RAM, enabling efficient training and evaluation of the models. The system was configured with CUDA, ensuring compatibility with deep learning frameworks and enabling efficient use of GPU resources.

## 4.2  Used Tools

- Hugging Face Transformers: A library for natural language processing that provides pre-trained models and tools for training and fine-tuning.
- Datasets: A library by Hugging Face for easy access to various datasets and dataset processing tools.
- Evaluate: A Hugging Face library to evaluate machine learning models with standard metrics.
- PEFT: Parameter-Efficient Fine-Tuning library by Hugging Face, which includes methods like LoRA (Low-Rank Adaptation).
- Scikit-learn: A machine learning library for Python, useful for preprocessing and evaluating metrics.
- PyTorch: An open-source deep learning framework used for building and training neural networks.
- WandB (Weights and Biases): A tool for tracking experiments, visualizing results, and managing hyperparameters.
- GPUtil: A Python library for GPU utilization and monitoring.
- BitsAndBytes: A library for efficient training and inference of large-scale models.

## 4.3  Datasets Collection

We compiled our dataset from various publicly available sources, including :

- Social media content (Truth_Seeker)
- Fake news articles and real articles from Reuters (ISOT)
- Fact-checked news from the 2016 US presidential election (BuzzFeed-Webis)
- General labeled fake and real news articles (fakenewsdata1)
- Political content with emphasis on fake vs. real news (TI-CNN)
- Twitter-based information spread during events (TW_info)
- Fine-grained labeled benchmark data (LIAR)

**Table 2.** Comparison of The Chosen Language Models

| Model | LLAMA2 | GPT-2 | Mistral | RoBERTa | DistilBERT | GEMMA | MAMBA |
|---|---|---|---|---|---|---|---|
| Volume (Size) | 13.5GB | 0.5 GB | ~15 GB | ~1.5 GB | 268MB | ~5 GB | 0.5 GB |
| Composition Layers | Transformer-based, 32 layers | 48 layers with 1600 hidden units per layer | composed of 32 transformer layers, with 32 attention heads and 8 key-value heads per layer | 24 layers, 1024 hidden units per layer | 6 layers, 768 hidden units per layer | Transformer-based, 18 layers | stack of specialized Mamba blocks, which combine convolutions, non-linearities, structured state space models |
| Initial Training Data | 2 trillion tokens of data from publicly available online sources | Trained on 40 GB of text data using autoregressive language modeling | N/A | Trained on 160 GB of text, using dynamic masking and larger batch sizes | Trained on the same data as BERT but with knowledge distillation techniques to reduce size | Trained on diverse datasets, focusing on balance between performance and efficiency | Custom training parameters focusing on specific applications |
| Key Characteristics | Large language model, high accuracy, and versatility | High-quality text generation, fluent and coherent text | Efficient and fast, suitable for resource-limited tasks | Robust and optimized BERT variant, high accuracy | Smaller, faster, retains 97% of BERT's capabilities | Balanced performance and efficiency | use of a Structured State Space Model as the core component |
| Architecture | Transformer | Transformer | Optimized Transformer | Transformer | Transformer | Transformer | Custom Transformer |
| Unique Features/Training Methodologies | Incorporates advanced attention mechanisms | Autoregressive transformer with multi-head attention | Emphasis on computational efficiency and speed | Uses dynamic masking and a larger training dataset | Knowledge distillation to reduce size while maintaining performance | Optimized for a balance between size and performance | Tailored for specific applications with custom features |
| Number of Parameters | 7 billion | 1.5 billion | 7.3 billion | 355 million | 66 million | 2 billion | 130 million |
| Computational Requirements | High computational resources for training and inference | Requires moderate computational power for deployment | powerful hardware for both training and inference | High computational resources for training | Lower computational requirements | Moderate computational requirements | lower computational complexity and memory requirements |

We extracted headline and body text from each source, combining them when necessary to avoid null values. The data was unified into a .csv format with two columns: concatenated text and veracity label (0 for true, 1 for fake). Duplicate entries were removed to ensure data quality. The text was tokenized with a maximum sequence length of 512 tokens, aligning with our smallest model's requirements and ensuring standardized input across all models. As illustrated in Fig. 2

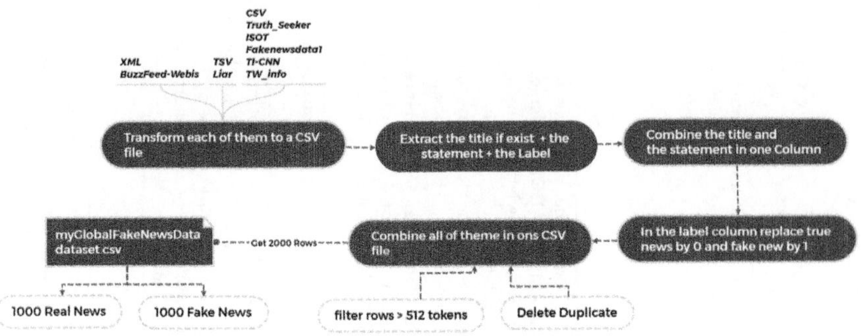


**Fig. 2.** Dataset Collection Workflow

### 4.4 Training

For adapting the pre-trained models to our fake news detection task, we employed Low-Rank Adaptation (LoRA) to efficiently fine-tune specific layers as shown in Table 3. This approach allowed us to adapt large language models with significantly reduced computational requirements. We carefully selected a set of training hyperparameters as described in Table 4. We also employed early stopping to prevent overfitting and to optimize training time.

## 5  Results and Discussion

Each model was evaluated using a custom evaluation function calculating precision, recall, F1-score, accuracy, and macro F1.

Performance was measured using the following metrics:

**Accuracy**: Overall correctness of predictions.

**Precision**: Correct positive predictions over total positive predictions.

**Recall**: Correct positive predictions over actual positives.

**F1-Score**: Harmonic mean of precision and recall.

**Macro F1**: Average F1-score across all classes.

The performance of each model is summarized in Table 5 and Fig. 3.

Our comparative study of various language models for fake news detection yields several important insights, supported by the quantitative results obtained from our experiments:

**Table 3.** Fine-Tuned Layers for Different Language Models

| Model | Fine-Tuned Layers |
|---|---|
| LLAMA2 | Up_proj, o_proj, v_proj, gate_proj, q_proj, down_proj, k_proj |
| GPT-2 | c_attn, c_proj, score |
| Mistral | q_proj, k_proj, v_proj, o_proj, gate_proj, up_proj, down_proj, score |
| RoBERTa | query, key, value, dense, out_proj |
| DistilBERT | q_lin, k_lin, v_lin, out_lin, lin1, lin2, pre_classifier, classifier |
| GEMMA 2B | q_proj, k_proj, v_proj, o_proj, gate_proj, up_proj, down_proj, score |
| MAMBA | / |

**Table 4.** Training Hyperparameters and their Rationale Selection

| Hyperparameter | Value | Reason for Selection |
|---|---|---|
| Learning Rate (lr) | 2e-4 | Selected to ensure a moderate step size for stable learning while avoiding overshooting during optimization. |
| Batch Size (per device) | 16 | A smaller batch size was chosen to balance GPU memory constraints while providing enough data for efficient gradient updates. |
| Number of Epochs (num_epochs) | 10 | Selected to allow enough training for convergence, with early stopping to prevent overfitting. |
| Warmup Ratio | 0.1 | Introduced to stabilize training by slowly increasing the learning rate during the first 10% of training. |
| Weight Decay | 0.001 | Applied to regularize the model and prevent overfitting by penalizing large weight updates. |
| FP16 Precision (fp16) | True | Enabled to reduce memory usage and speed up training by using mixed-precision floating point operations. |
| Save Total Limit | 3 | Limits the number of saved checkpoints to prevent storage overflow. |

- **Model Performance:** The results demonstrate a clear performance hierarchy among the tested models. LLAMA2 emerged as the top performer with an impressive 92.5% accuracy and 92.7% F1-score, closely followed by Mistral (92.2% accuracy, 92.1% F1-score). This suggests that the increased capacity and more advanced training techniques of these larger, more recent models contribute significantly to their ability to discern nuanced patterns in potentially misleading content.
  Interestingly, GEMMA 2B, despite having fewer parameters than LLAMA2 and Mistral, showed strong performance (90.5% accuracy, 90.7% F1-score), indicating that architectural innovations can sometimes compensate for smaller model size.
- **Precision vs. Recall Trade-offs:** Mistral achieved the highest precision (96.8%), while RoBERTa and DistilBERT exhibited the highest recall (97.5% and 97.1% respectively). This highlights an important trade-off in fake news detection: Mistral is less likely to falsely label genuine news as fake, while

**Table 5.** Model Performance Metrics

| Model | Accuracy % | Precision % | Recall % | F1-score % | Macro f1 % |
|---|---|---|---|---|---|
| LLAMA2 | 92.5 | 93.6 | 91.8 | 92.7 | 92.7 |
| GPT-2 | 85.5 | 81.7 | 92.7 | 86.9 | 86.9 |
| Mistral | 92.2 | 96.8 | 87.9 | 92.1 | 92.1 |
| RoBERTa | 83.5 | 76.3 | 97.5 | 85.6 | 85.6 |
| DistilBERT | 80.7 | 73.9 | 97.1 | 83.9 | 83.9 |
| GEMMA 2B | 90.5 | 91.6 | 89.9 | 90.7 | 90.7 |
| MAMBA | 87.2 | 87.2 | 87.2 | 87.2 | 87.2 |

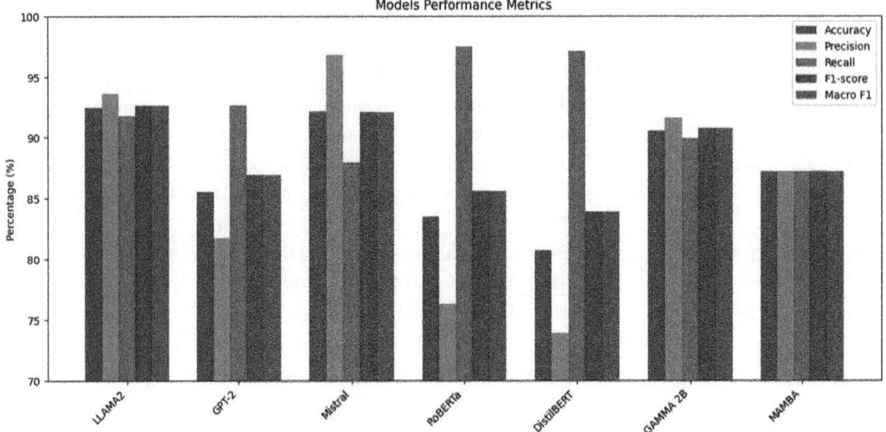

**Fig. 3.** Models Performance Graph

RoBERTa and DistilBERT are more likely to catch a higher proportion of fake news, but at the cost of more false positives.

- **Efficiency vs. Effectiveness:** While larger models like LLAMA2 and Mistral showed superior overall performance, it's noteworthy that some smaller models performed competitively in certain metrics. DistilBERT, despite its lower overall accuracy (80.7%), achieved the second-highest recall (97.1%). This suggests that in scenarios where minimizing false negatives is crucial and computational resources are limited, DistilBERT could be a viable option
- **MAMBA's Performance:** MAMBA's consistent performance across all metrics (87.2%) is intriguing. While not matching the top performers, its balanced performance and unique architecture suggest potential for improvement with further optimization, especially in scenarios requiring processing of longer sequences. GPT-2 and Older Models: GPT-2's performance (85.5% accuracy, 86.9% F1-score) demonstrates that even older, general-purpose language models can be effectively fine-tuned for fake news detection. However,

the gap between GPT-2 and newer models like LLAMA2 and Mistral underscores the rapid progress in the field.

– **Implications for Practical Deployment:** The choice of model for fake news detection should be carefully considered based on the specific use case:

1. For high-stakes applications requiring maximum overall accuracy, LLAMA2 or Mistral would be preferred.
2. In scenarios where avoiding false accusations of fake news is critical, Mistral's high precision makes it a strong candidate.
3. For applications prioritizing the identification of as much fake news as possible, even at the risk of some false positives, RoBERTa or DistilBERT might be more suitable.
4. In resource-constrained environments, the balance of performance and efficiency offered by GEMMA 2B or MAMBA could be optimal.

– **Model Limitations and Error Analyses:** We discuss in Table 6 the limitations of the models and analyze the reasons behind their misclassifications. Despite achieving high accuracy, models like LLAMA2 still have certain weaknesses. Below, we present a table with some examples of incorrect predictions made by LLAMA2, the model with the highest accuracy in our study.

**Table 6.** Performance on Various Statements

| Statement | True Label | Predicted Label | Analysis |
| --- | --- | --- | --- |
| The Trump administration worked to free 5,000 Taliban prisoners. | TRUE | FAKE | Lack of context, potential political bias in training data |
| Foxconn is the largest taxpayer in Racine County. | TRUE | FAKE | Unfamiliar proper nouns, limited local economic data |
| Tommy Thompson created the first school choice program in the nation. | TRUE | FAKE | Incomplete statement, lack of historical context |
| Climate change is a hoax perpetrated by scientists for grant money. | FAKE | TRUE | Convincing but false narrative, common misinformation trope |
| The Earth is flat and NASA is covering up the truth. | FAKE | TRUE | Conspiracy theory, model may struggle with scientific facts |
| Vaccines cause autism in children. | FAKE | TRUE | Persistent medical misinformation, model may be influenced by prevalence of claim |
| 5G networks are responsible for spreading COVID-19. | FAKE | TRUE | Technology-health misinformation, model may struggle with recent events |

## 6   Conclusion and Future Work

Our study of seven language models for fake news detection demonstrates significant progress in AI-driven misinformation identification, while highlighting its inherent complexities. LLAMA2 led in accuracy, but all models exhibited precision-recall trade-offs, emphasizing the importance of application-specific model selection. Notably, smaller models like DistilBERT showed competitive performance in certain areas, indicating that effective detection is possible even with limited computational resources.

These insights are valuable for researchers and practitioners building fake news detection systems. The choice of an appropriate language model should be carefully considered based on the use case. While applications requiring maximum overall accuracy might benefit from models like LLAMA 2 or Mistral, scenarios focusing on detecting as much fake news as possible, even at the risk of some false positives, might favor RoBERTa or DistilBERT.

Building upon these findings, we propose several directions for future research:

1. **Exploration of Longer Sequences:** Given MAMBA's potential advantage in processing longer sequences, future studies should investigate the performance of all models, especially MAMBA, on datasets with varying sequence lengths beyond the 512 token limit used in this study. This could reveal how different architectures scale with input length and potentially uncover MAMBA's full capabilities.
2. **Interpretability and Explainability:** Developing methods to interpret and explain the decisions made by these models is crucial for building trust in AI-powered fake news detection systems. Future research should focus on making model decisions more transparent and understandable to end-users.
3. **Comparison with Newer Models:** Evaluate newer language models (e.g., GPT-4, PaLM 2) for fake news detection. This ongoing comparison would track progress in the field, identifying key architectural and training innovations that enhance performance.

Integrating these AI technologies judiciously within broader fact-checking ecosystems, complementing human expertise, is crucial. This approach enhances our collective capacity to discern reliable information in complex media landscapes.

## References


1. Raiaan, M., et al.: A review on large language models: architectures, applications, taxonomies, open issues and challenges. IEEE Access **PP**, 1–1 (2024). https://doi.org/10.1109/ACCESS.2024.3365742
2. Yan, H.: Missed communications and blocked evacuation routes: new report details problems and heroism from Maui's disastrous wildfires. CNN (2024). Accessed 23 Sep 2024
3. Quinn, B.: Fake UK news sites 'spreading false stories' about western firms in Ukraine. The Guardian (2024). Accessed 23 Sep 2024
4. France 24: lebanon's ex-central bank chief salameh to remain detained amid investigation. France 24 (2024). Accessed 23 Sep 2024
5. Touvron, H., et al.: Llama 2: open foundation and fine-tuned chat models. arXiv preprint arXiv:2307.09288 (2023)

6. Jiang, A.Q., et al.: Mistral 7B. arXiv preprint arXiv:2310.06825 (2023)
7. Sanh, V., Debut, L., Chaumond, J., Wolf, T.: DistilBERT, a distilled version of BERT: smaller, faster, cheaper and lighter. arXiv preprint arXiv:1910.01108 (2019)
8. Liu, Y., et al.: Roberta: a robustly optimized Bert pretraining approach. arXiv preprint arXiv:1907.11692 (2019)
9. Team, G., et al.: Gemma: open models based on gemini research and technology. arXiv preprint arXiv:2403.08295 (2024)
10. Radford, A., Wu, J., Child, R., Luan, D., Amodei, D., Sutskever, I.: Language Models are Unsupervised Multitask Learners (2019)
11. Gu, A., Dao, T.: Mamba: linear-time sequence modeling with selective state spaces. arXiv preprint arXiv:2312.00752 (2023)
12. Mridha, M.F., Keya, A.J., Hamid, M.A., Monowar, M.M., Rahman, M.S.: A comprehensive review on fake news detection with deep learning. IEEE Access 9, 156151–156170 (2021)
13. Dadkhah, S., Zhang, X., Weismann, A.G., Firouzi, A., Ghorbani, A.A.: The largest social media ground-truth dataset for real/fake content: truthseeker. IEEE Trans. Comput. Soc. Syst. (2023)
14. Ahmed, H., Traore, I., Saad, S.: Detecting opinion spams and fake news using text classification. Security and Privacy 1(1), e9 (2018)
15. Ahmed, H., Traore, I., Saad, S.: Detection of Online Fake News Using N-Gram Analysis and Machine Learning Techniques. In: Traore, I., Woungang, I., Awad, A. (eds.) ISDDC 2017. LNCS, vol. 10618, pp. 127–138. Springer, Cham (2017). https://doi.org/10.1007/978-3-319-69155-8_9
16. Potthast, M., Kiesel, J., Reinartz, K., Bevendorff, J., Stein, B.: A stylometric inquiry into hyperpartisan and fake news. arXiv preprint arXiv:1702.05638 (2017)
17. Horne, B., Adali, S.: This just in: fake news packs a lot in title, uses simpler, repetitive content in text body, more similar to satire than real news. In: Proceedings of the International AAAI Conference on Web and Social Media, vol. 11, no. 1, pp. 759–766 (2017)
18. Yang, Y., Zheng, L., Zhang, J., Cui, Q., Li, Z., Yu, P.S.: TI-CNN: convolutional neural networks for fake news detection. arXiv preprint arXiv:1806.00749 (2018)
19. Jang, Y., Park, C.-H., Seo, Y.-S.: Fake news analysis modeling using quote retweet. Electronics 8(12), 1377 (2019)
20. Wang, W.Y.: "Liar, liar pants on fire": a new benchmark dataset for fake news detection. arXiv preprint arXiv:1705.00648 (2017)

# Quantitative Modeling and Evaluation of DNA Multiple Sequence Alignment Tools Through Response Surface Methodology


Mohamed Skander Daas[1]([✉]) [iD], Safia Rouabah[1], Khaled Boulahrouf[1] [iD],
Salim Chikhi[2] [iD], and El-Bay Bourennane[3] [iD]

[1] University of Constantine 1, Constantine 25000, Algeria
daas.skander@umc.edu.dz
[2] University of Constantine 2, Constantine 250000, Algeria
[3] University of Burgundy, Dijon 21000, France



**Abstract.** Multiple sequence alignment plays a vital role in the computational analysis of biological data. Various programs have been developed to analyze sequence similarity. The alignment accuracy of these programs is difficult to analyze and compare, given the noisy effect of several parameters and their interaction on the nonlinear evolution of this performance. This work provides a mathematical modeling analysis and comparison study of many popular sequence alignment programs of DNA sequences. Several tools were also used to generate, simulate sequences and evaluate alignments. Several parameters are considered, namely, number of sequences, sequence lengths, insertion rate and deletion rate. Quadratic mathematical models have been generated and analyzed and can be used for personalized analysis purposes like prediction and optimization of the accuracy of alignment tools. Furthermore, the adequacy of the obtained regression models was tested and analyzed. For most models the coefficient values: $R^2$, adjusted $R^2$, and predicted $R^2$ were found to be good. These models were then employed for performance analysis and comparison of the investigated MSA tools, where thirty 3D surfaces were presented. The obtained models and graphical results provide a clear and easy way to analyze, interpret, optimize and compare MSA tools globally or under personalized configurations.

**Keywords:** Multiple Sequence Alignment · Response surface methodology · mathematical performance modeling · evaluation and comparison of MSA tools


## 1 Introduction

Multiple sequence alignment (MSA) is crucial in many biological analyses, such as motif discovery and phylogenetic reconstruction. Sequence alignment algorithms are powerful tools used to compare DNA or protein sequences, helping researchers search biological databases. They are crucial in uncovering the biological roles of specific genes by identifying similarities across different sequences [5], but achieving accurate alignments is NP-complete, leading to numerous MSA tools being developed over the past three decades [7]. Accuracy evaluations of these tools often rely on reference alignments from benchmarks like BAliBASE. However, these empirical benchmarks are limited, particularly in cases lacking structural data. Simulated datasets, which model evolutionary processes, provide an alternative evaluation method. Simulators like Seq-Gen, Rose, Dawg, and INDELible are commonly used. This study employs AliSim [8], a new tool that simulates biologically realistic alignments using complex evolutionary models.

MSA tools often perform differently across data types–those optimized for amino acids may not work well for DNA sequences [3]. Despite more focus on protein alignments in MSA studies [2], DNA-based alignments remain essential, such as for phylogenetic trees and PCR primer design. No single MSA tool excels universally, making evaluation complex [13]. Performance is influenced by multiple interacting parameters, and heuristic approaches contribute to the inherent noise in results.

To address these challenges, this study applies Response Surface Methodology (RSM) to mathematically model the performance of widely-used MSA tools at the DNA level. These models incorporate parameters such as the number of sequences, insertion and deletion rates, and sequence lengths, offering a comprehensive framework for predicting and optimizing performance. The study employs multiple tools to generate realistic datasets and visualizes the relationships between parameters through three-dimensional plots, providing insights into MSA tool performance.

## 2    Methodology and Experiments

The goal of this study was to model mathematically MSA performance, where mathematical modeling is one of the most powerful ways that makes it easier to analyze and understand complex processes. Thus, the performances of the last versions of the investigated MSA tools indicated in Table 1 are considered for such modeling for effective analysis and comparison.

**Table 1.** Information about the investigated MSA tools.

| Tool | Updated | Reference | Version |
| --- | --- | --- | --- |
| ClustalO | 2017 | [11] | 1.2.2 |
| Muscle | 2021 | [4] | 5.1 |
| Mafft | 2022 | [6] | 7.505 |
| Dialign | 2014 | [1] | 2.2.1 |
| Tcoffee | 2020 | [9] | 13.45.0.4846264 |

First, mathematical modeling of MSA tools is used to generate five mathematical quadratic models using the Response Surface Methodology (RSM) Design. RSM includes the experimental strategy for exploring the process space, empirical modeling to develop an appropriate approximating relationship between a response and process variables, and optimization methods to find the values of the process variables, which produce desirable values of the response. The Box-Behnken design has been used to perform this study by realizing a set of experiments. Default parameters were used for all MSA tools. To achieve each experiment the following process steps are resumed in Fig. 1.

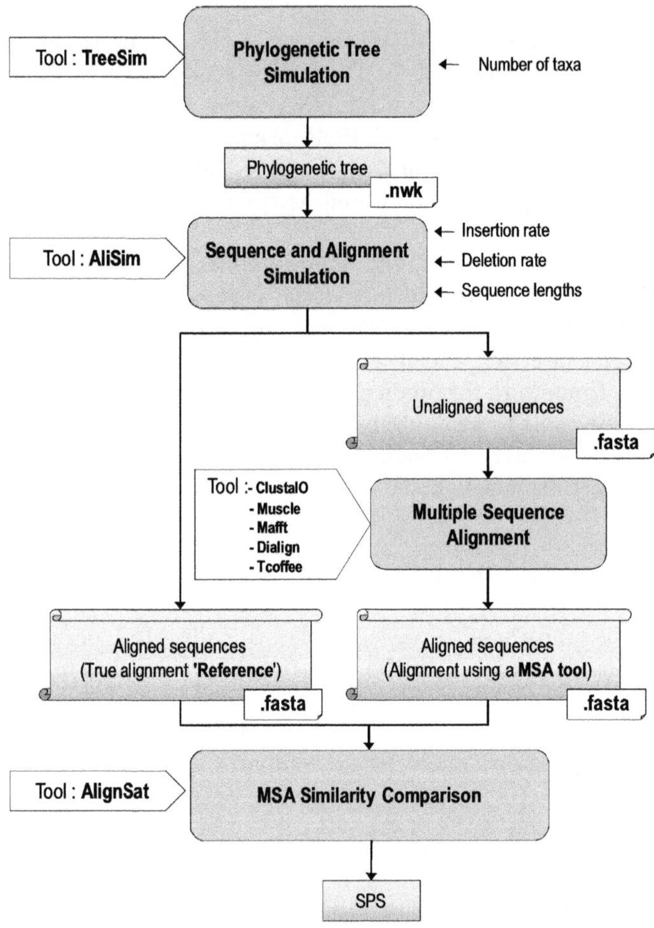


**Fig. 1.** Process of MSA simulation and evaluation.

Second, the obtained models are analyzed where the effects of parameters and its interactions as well as the models' validity and quality are investigated using statistical testing.

Third, many 3D performance plots are provided for a clear and effective analysis, evaluation and comparing MSA tools.

In the following subsections many tools used in this study are presented with examples. The performance calculation is also detailed. Then, the experimental design and the obtained results are described in Subsects. 2.3 and 2.4.

## 2.1   Simulation and Evaluation Tools

**TreeSim** : TreeSim [12] is an R package that combines a set of simulation methods for phylogenetic trees. We used the Simulating birth-death trees on a fixed number of extant taxa (*sim.bd.taxa*) to generate the phylogenetic trees.

**AliSim**: AliSim [8]is a new tool that can effectively simulate biologically realistic alignments using a wide range of complex evolutionary models. It achieves high performance across a large range of simulation conditions. To use AliSim, IQ-TREE tool should be first installed.

**AlignStat**: AlignStat [10] is an R package that can compare two alternative multiple sequence alignments (MSAs) and show how well they align similar residues in the same columns as one another. Similarities and differences are classified into a conserved sequence, conserved gaps, splits, merges, and shifts. Outlining these classes for each column gives information about which columns are matched by the two MSAs and which differ. To compare two alternative multiple sequence alignments of the same sequences we measured the SPS statistic using the R *compare_alignments* function.

The datasets are generated according to the design of experiments as multiple files. First, 3 Newick files are generated containing phylogenetic trees (for 10, 50 and, 90 Taxa) using TreeSim tool. Then, according to the experimental design, 27 files are generated containing the sets of sequences that correspond to each experiment, and 27 other files are generated containing its reference alignments at the step of simulation of alignments using Alisim tool. In addition, a total of 135 Fasta files are generated during the execution step of the alignment tools for each experiment (5*27).

## 2.2   Performance Evaluation Metrics of MSA Tools

To evaluate the performance of a given alignment the **Sum of pairs score (SPS)** is used to determine how well alignment tools align some, if not all, of the sequences. If the $i^{th}$ column of an alignment is $A_{ij}$ ($i = 1...N$) where $N$ is the number of columns, the score $S_i$ is defined for the ith column as follows:

$$S_i = \sum_{j=1,j\neq k}^{N} \sum_{k=1}^{N} p_{ijk} \tag{1}$$

$$p_{ijk} = \begin{cases} 1 & \text{if } A_{ij} \text{ and } A_{ik} \text{ are aligned with each other} \\ & \text{in the reference alignment} \\ 0 & \text{otherwise} \end{cases} \tag{2}$$

The SPS is defined as:

$$SPS = \sum_{i=1}^{M} S_i \Big/ \sum_{i=1}^{Mr} S_{ri} \tag{3}$$

Where $Mr$ and $Sri$ are respectively the number of columns of the reference alignment and its $S_i$ score for the $i^{th}$ column.

## 2.3  Design of Experiments

Experiments were conducted adopting a Box-Behnken Design (BBD) Response Surface Methodology, which is a Response Surface Methodology *RSM*. This type of design was suitable for our objective, which was mainly the modeling of complex processes and their analysis.

Table 2 shows the minimal, midrange, and maximal levels (natural values) used for each parameter, that correspond to $-1$, 0, and $+1$ levels (coded values).

**Table 2.** Experimental study levels of the parameters.

| Parameters → | Number of taxa (N) | Insertion rate (Ins) | Deletion rate (Del) | Sequence length (Len) |
|---|---|---|---|---|
| Level ($-1$) | 10 | 0.001 | 0.001 | 100 |
| Level (0) | 50 | 0.021 | 0.021 | 500 |
| Level ($+1$) | 90 | 0.041 | 0.041 | 900 |

The four parameters considered in this study namely the number of sequences, the size of the sequences, the insertion rate, and the deletion rate were studied. Knowing that the relationship between these parameters and the considered performances is nonlinear (presence of curvature effect), quadratic mathematical modeling is considered using the Box-Behnken design to model the performance of each MSA tool in the function of the four parameters.

The experimental matrix of the Box-Behnken design for the four parameters is provided in Table 3. This design requires 27 runs, where, the central point (0, 0, 0, 0) is repeated three times. Each experiment in Table 3 is carried out following the process mentioned in Fig. 1.

All generated Phylogenetic trees, Unaligned sequences, Reference Alignments and Tools Alignments files are availible in the following link GitHub.

## 2.4  MSA Performance Results

The responses of the SPS measure for each MSA tool of the 27 experiments are provided in Table 4.

The obtained results in this table are brut numerical values that at first glance are difficult to analyze, compare or interpret. This reason leads to the interest in mathematical modelling usage for clear and effective performance analysis.

**Table 3.** Design of experiments matrix.

| Parameter | 1 | 2 | 3 | 4 | 5 | 6 | 7 | 8 | 9 |
|---|---|---|---|---|---|---|---|---|---|
| N | 10 | 90 | 10 | 90 | 50 | 50 | 50 | 50 | 10 |
| Ins | 0.001 | 0.001 | 0.041 | 0.041 | 0.021 | 0.021 | 0.021 | 0.021 | 0.021 |
| Del | 0.021 | 0.021 | 0.021 | 0.021 | 0.001 | 0.041 | 0.001 | 0.041 | 0.021 |
| Len | 500 | 500 | 500 | 500 | 100 | 100 | 900 | 900 | 100 |
| Parameter | 10 | 11 | 12 | 13 | 14 | 15 | 16 | 17 | 18 |
| N | 90 | 10 | 90 | 50 | 50 | 50 | 50 | 10 | 90 |
| Ins | 0.021 | 0.021 | 0.021 | 0.001 | 0.041 | 0.001 | 0.041 | 0.021 | 0.021 |
| Del | 0.021 | 0.021 | 0.021 | 0.001 | 0.001 | 0.041 | 0.041 | 0.001 | 0.001 |
| Len | 100 | 900 | 900 | 500 | 500 | 500 | 500 | 500 | 500 |
| Parameter | 19 | 20 | 21 | 22 | 23 | 24 | *25* | *26* | *27* |
| N | 10 | 90 | 50 | 50 | 50 | 50 | *50* | *50* | *50* |
| Ins | 0.021 | 0.021 | 0.001 | 0.041 | 0.001 | 0.041 | *0.021* | *0.021* | *0.021* |
| Del | 0.041 | 0.041 | 0.021 | 0.021 | 0.021 | 0.021 | *0.021* | *0.021* | *0.021* |
| Len | 500 | 500 | 100 | 100 | 900 | 900 | *500* | *500* | *500* |

# 3  Performance Modelling of MSA Tools

The response models are postulated as quadratic for each response for each MSA performance tool. These models may be expressed in terms of orthogonal parameters as shown in Eq. 4.

$$Y = \beta_0 + \sum_{i=1}^{k} \beta_i x_i + \sum_{i<j} \sum \beta_{ij} x_i x_j + \sum_{i=1}^{k} \beta_{ii} x_i^2 + \epsilon \tag{4}$$

Where $Y$ is the predicted response, $\beta_0$ is the constant coefficient, $\beta_i$ is the $i^{th}$ linear coefficient of input parameter $x_i$, $\beta_{ii}$ is the $i^{th}$ quadratic coefficient of input parameter $x_i$, $\beta_{ij}$ is the interaction coefficients between input parameters $x_i$ and $x_j$, and $\epsilon$ is the error of the model. The estimated coefficients $\hat{\beta}$ are obtained using mathematical regression. In the present work, we considered four variables: N, Ins, Del, and Len. Where the SPS responses are written in the form of the following model:

$$\begin{aligned}
Y(N, Ins, Del, Del) = {} & \beta_0 + \beta_1 N + \beta_2 Ins + \beta_3 Del + \beta_4 Len \\
& + \beta_{12} N Ins + \beta_{13} N Del + \beta_{14} N Len + \beta_{23} Ins C + \beta_{24} Ins Del \\
& + \beta_{34} Del Del + \beta_{11} N^2 + \beta_{22} Ins^2 + \beta_{33} Del^2 + \beta_{44} Len^2 + \epsilon
\end{aligned} \tag{5}$$

FiVE mathematical models were obtained. Each model corresponds to a given MSA tool (Clustal Omega, Muscle, Mafft, Dialign, or T-coffee). These models

**Table 4.** Results of the SPS accuracy of MSA tools.

| Exp | ClustalO | Muscle | Mafft | Dialign | Tcoffee |
|---|---|---|---|---|---|
| 1 | 0,5849853 | 0,6697252 | 0,6787537 | 0,4663395 | 0,6100589 |
| 2 | 0,1149751 | 0,2332363 | 0,2263 | 0,08517699 | 0,1217603 |
| 3 | 0,396425 | 0,528859 | 0,4436602 | 0,4987053 | 0,4962412 |
| 4 | 0,09627714 | 0,1857889 | 0,1274002 | 0,07199865 | 0,09521966 |
| 5 | 0,110148 | 0,0730073 | 0,262686 | 0,1281135 | 0,2757812 |
| 6 | 0,1384689 | 0,2827888 | 0,1856121 | 0,1088788 | 0,1728702 |
| 7 | 0,1546817 | 0,1952783 | 0,1692091 | 0,1036425 | 0,1039737 |
| 8 | 0,1193521 | 0,2063671 | 0,1521642 | 0,09396622 | 0,09811204 |
| 9 | 0,5384436 | 0,5414725 | 0,5855079 | 0,55452 | 0,5582479 |
| 10 | 0,1174466 | 0,4724456 | 0,4443418 | 0,1603076 | 0,2806507 |
| 11 | 0,4020983 | 0,5220682 | 0,5005287 | 0,4451773 | 0,4229699 |
| 12 | 0,1284049 | 0,1802816 | 0,1341424 | 0,08848087 | 0,1189115 |
| 13 | 0,1909182 | 0,2781624 | 0,2071369 | 0,1130365 | 0,2158716 |
| 14 | 0,1285718 | 0,2257347 | 0,2043443 | 0,1144516 | 0,1198151 |
| 15 | 0,119256 | 0,1767599 | 0,1562955 | 0,08288345 | 0,09809946 |
| 16 | 0,1146565 | 0,2013121 | 0,1447153 | 0,0941195 | 0,1094771 |
| 17 | 0,7127931 | 0,7617227 | 0,7115189 | 0,540316 | 0,5961774 |
| 18 | 0,1523116 | 0,2108079 | 0,1905129 | 0,1105431 | 0,1503476 |
| 19 | 0,4926011 | 0,6363019 | 0,4828265 | 0,5113359 | 0,5624491 |
| 20 | 0,1079234 | 0,3119476 | 0,204058 | 0,08985751 | 0,1265332 |
| 21 | 0,1324425 | 0,3652611 | 0,1926993 | 0,1226854 | 0,3011285 |
| 22 | 0,1184954 | 0,2658726 | 0,1527603 | 0,1057534 | 0,1851258 |
| 23 | 0,1799029 | 0,2207463 | 0,1856088 | 0,09941057 | 0,1526402 |
| 24 | 0,1056278 | 0,1574614 | 0,1273364 | 0,09278368 | 0,08640057 |
| 25 | 0,1329215 | 0,228695 | 0,1844252 | 0,1072172 | 0,1291484 |
| 26 | 0,1332076 | 0,2088957 | 0,1918819 | 0,1187224 | 0,1294322 |
| 27 | 0,1248481 | 0,2398784 | 0,1901795 | 0,09270956 | 0,1411012 |

can be used to analyze, optimize, and predict the SPS performance metrics of
the five MSA tools under a given configuration of the four parameters (N, Ins,
Del, and Len). Table 5 shows the coefficients of the models according to Eq. 5.

## 4 Models' Analysis

To better understand the effect of the main factors as well as their interactions
on the SPS measure, F statistical tests are used to test the significance of these
factors and interactions at 95% of the confidence interval. Significant terms are

mentioned in bold among the coefficients of the 5 models. The greater the value (absolute value) of the coefficient $\beta$ the greater impact of its related factor or interaction. As we can see the most significant factor for the SPS measure is the number of sequences (N) represented by the coefficient $\beta_1$ for the five tools.

We have considered three statistical measures $R^2$, adjusted $R^2$ and predicted $R^2$. The coefficients of determination $R^2$, and adjusted $R^2$ show how well a model fits the data.

Table 5. Regression coefficients for SPS model.

| $\beta_i$ | ClustalO | Muscle | Mafft | Dialign | Tcoffee |
|---|---|---|---|---|---|
| $\beta_0$ | *0,130326* | *0,225823* | *0,188829* | *0,106216* | *0,133227* |
| $\beta_1$ | **-0,200834** | **-0,172137** | **-0,173003** | **-0,200836** | **-0,19606** |
| $\beta_2$ | $-0,030202$ | $-0,031572$ | $-0,037215$ | 0,00069 | **-0,03394** |
| $\beta_3$ | $-0,029764$ | 0,005897 | $-0,034978$ | $-0,010755$ | $-0,024535$ |
| $\beta_4$ | $-0,005448$ | $-0,04322$ | $-0,046218$ | **-0,0214** | **-0,0659** |
| $\beta_{11}$ | **0,188039** | **0,214612** | **0,210524** | **0,1963** | **0,19838** |
| $\beta_{22}$ | $-0,00723$ | $-0,003605$ | $-0,024793$ | $-0,013127$ | 0,004914 |
| $\beta_{33}$ | 0,025164 | 0,002804 | 0,003216 | 0,004848 | 0,009622 |
| $\beta_{44}$ | $-0,011857$ | $-0,006839$ | 0,005905 | 0,00642 | 0,025535 |
| $\beta_{12}$ | 0,042466 | 0,023355 | 0,034048 | $-0,011386$ | 0,021819 |
| $\beta_{13}$ | 0,043951 | 0,05664 | 0,060559 | 0,002074 | 0,002478 |
| $\beta_{14}$ | 0,036826 | $-0,06819$ | $-0,056305$ | 0,009379 | $-0,006615$ |
| $\beta_{23}$ | 0,014437 | 0,019245 | $-0,002197$ | 0,002455 | 0,026859 |
| $\beta_{24}$ | $-0,015082$ | 0,009026 | $-0,004583$ | 0,002576 | 0,012441 |
| $\beta_{34}$ | $-0,015913$ | $-0,049673$ | 0,015007 | 0,00239 | 0,024262 |

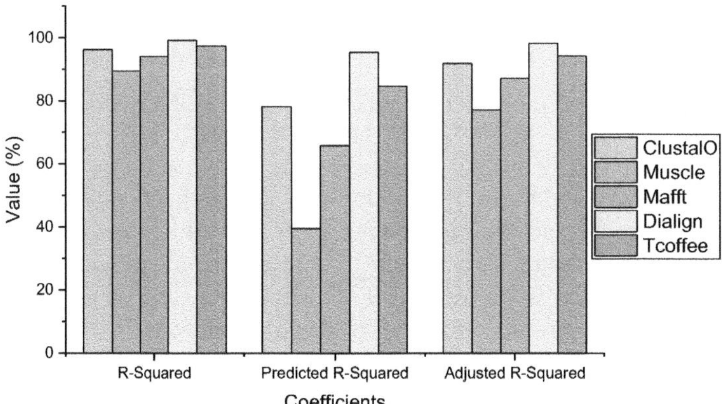


**Fig. 2.** The coefficients of determination for SPS.

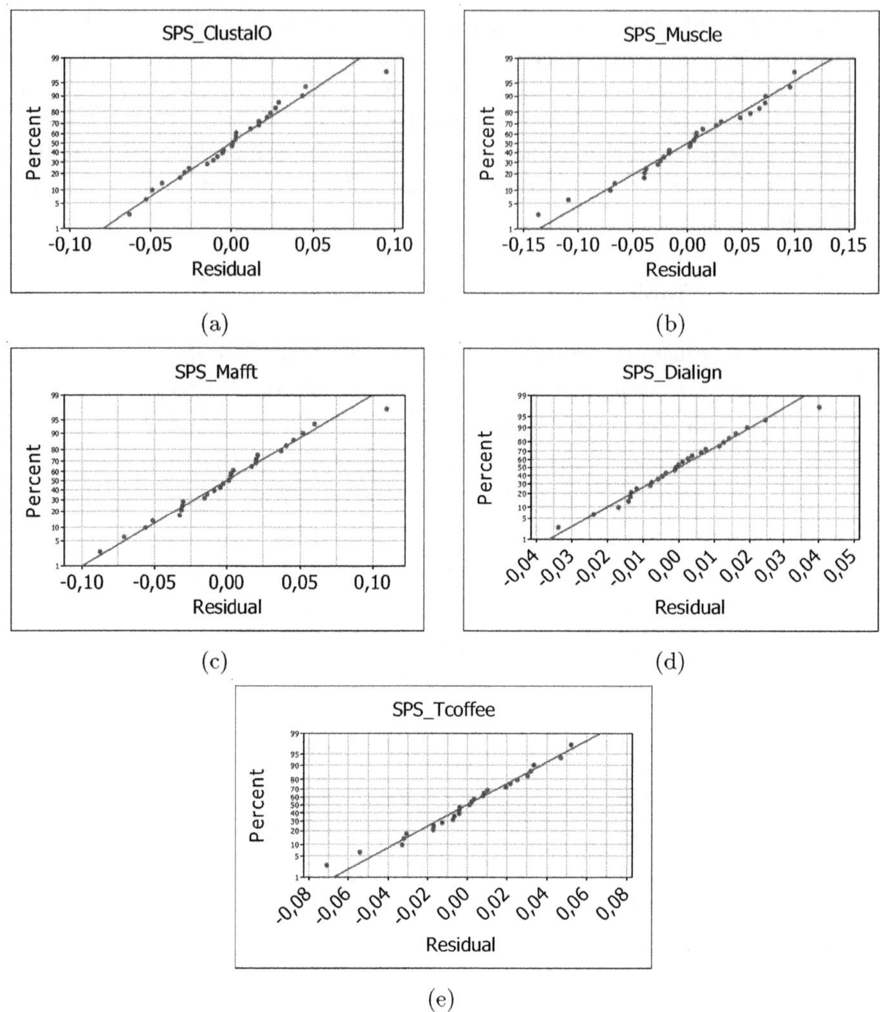

**Fig. 3.** Normal probability plots of residuals for SPS performances.

The coefficients of determination ($R^2$ and adjusted $R^2$) of most tools varies globally between 60 and 99.9% showing high experimental data fitting using these models. The predicted coefficient of determination (predicted $R^2$) of most tools vary globally between 50 and 95% which indicates that the regression models satisfactorily predict response to new observations.

A condition of the linear regression model is that the error terms must be normally distributed. To know if the assumption holds, the normal probability plot of the residuals must be approximately linear.

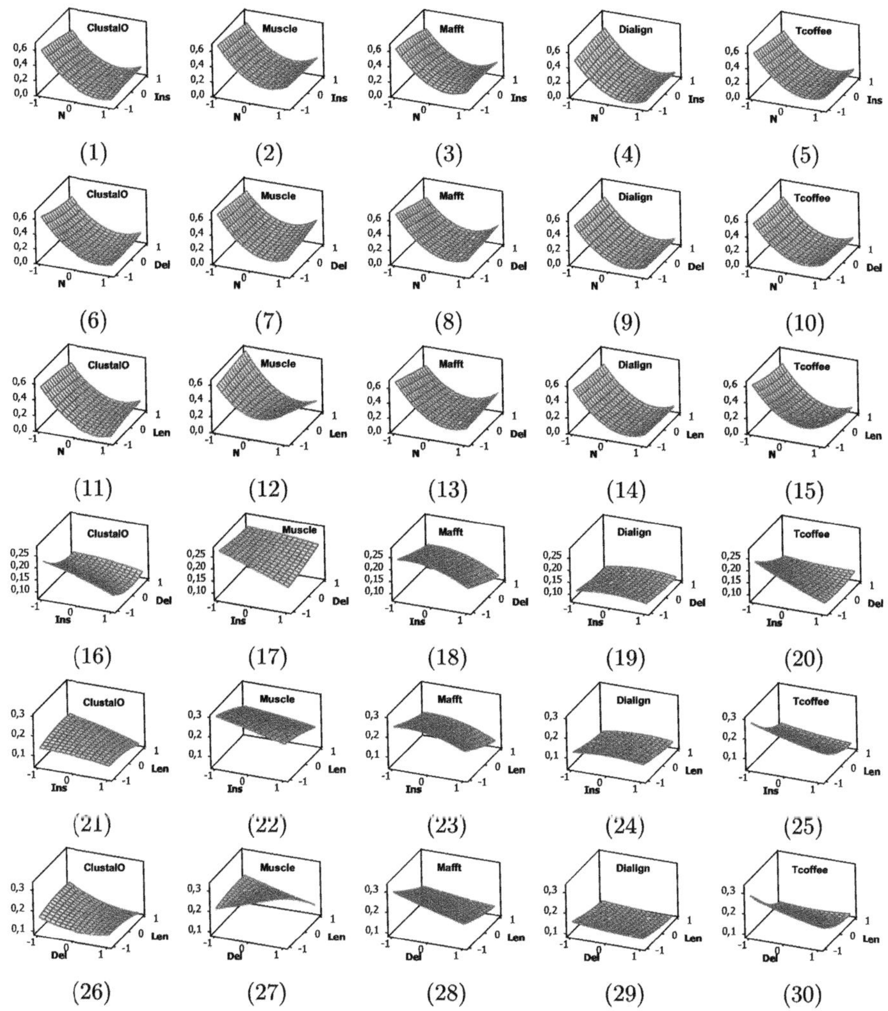

**Fig. 4.** Surface plots for SPS accuracy.

The relationships in Fig. 3 ais approximately linear. We could proceed with the assumption that the error terms are normally distributed for the five investigated models (Fig. 4).

## 5   Analysis and Comparison of MSA Performances

One of the objectives of this study was to provide a clear and easy way to compare, evaluate and analyze multiple alignment programs. Therefore, mathematical modeling of MSA tool performances is performed providing power visual aids of 3D surface plots.

A surface plot contains predictors on the x- and y-axes and a continuous surface on the z-axis that represents the fitted response. Since a surface plot shows only the variation of two parameters at a time, the other parameters are held at a constant level (level 0).

The comparison of the five tools in terms of SPS accuracy performance showed that Muscle outperforms the other tools followed by Mafft. In the third position, ClustalO and Tcoffee showed similar performances, however, for larger sequence lengths, ClustalO exibited better results (4.11 & 4.15, 4.21 & 4.25, 4.26 & 4.30) and it showed better results when the insertion rate is bigger (4.1 & 4.5, 4.16 & 4.20, 4.21 & 4.25), but Tcoffee is better for bigger deletion ratio (4.6 & 4.10, 4.16 & 4.20, 4.26 & 4.30).

Surface plots showed that most parameters have a quadratic (curve) effect on the performances of all the tools but the most significant among these parameters is the number of sequences; these results reflect the values of $\beta_{11}$ coefficient in Table 5. In summary, the best tools in term of precision are Muscle followed by Mafft.

# 6    Conclusion

Five mathematical models predicting the quality performance of many recent versions of popular MSA tools, namely, Clustal Omega, Muscle, Mafft, Dialign, and T-coffee, have been presented. These models include factors such as the number of sequences, insertion rate, deletion rate, and sequence lengths. The adequacy of mathematical models has been tested using analysis of variance (ANOVA) and has proved that most of the generated mathematical models could adequately describe the MSA performances within the factors that are being investigated with 95% confidence interval. The quality of the models is also investigated by the examination of coefficients of determination, $R^2$, adjusted $R^2$, and predicted $R^2$, and showed that most models have an excellent capacity to fit the obtained results and a good prediction capacity. Through which ninety 30 3D response surfaces have been presented providing clear analytical graphics and comparison between the MSA performances of the investigated tools under the influence of various factors. This study gives an up-to-date evaluation and comparison of the most popular MSA tools for DNA datasets and provides a pipeline for performance evaluation and analysis as well as a guiding source to gain further insight into the suitable selection of these tools. It gives another vision of how to evaluate and analyze MSA tools' performance using mathematical modeling.

# References

1. Al Ait, L., Yamak, Z., Morgenstern, B.: Dialign at gobics-multiple sequence alignment using various sources of external information. Nucleic Acids Res. **41**(W1), W3–W7 (2013). https://doi.org/10.1093/nar/gkt283

2. Bininda-Emonds, O.R.: Transalign: using amino acids to facilitate the multiple alignment of protein-coding DNA sequences. BMC Bioinform. **6**(1), 1–6 (2005). https://doi.org/10.1186/1471-2105-6-156
3. Carroll, H., et al.: Dna reference alignment benchmarks based on tertiary structure of encoded proteins. Bioinformatics **23**(19), 2648–2649 (2007). https://doi.org/10.1093/bioinformatics/btm389
4. Edgar, R.C.: Muscle: a multiple sequence alignment method with reduced time and space complexity. BMC Bioinform. **5**(1), 1–19 (2004). https://doi.org/10.1186/1471-2105-5-113
5. Gancheva, V., Stoev, H.: Optimization and performance analysis of cat method for DNA sequence similarity searching and alignment. Genes **15**(3) (2024). https://doi.org/10.3390/genes15030341, https://www.mdpi.com/2073-4425/15/3/341
6. Katoh, K., Misawa, K., Kuma, K.i., Miyata, T.: Mafft: a novel method for rapid multiple sequence alignment based on fast fourier transform. Nucleic Acids Research **30**(14), 3059–3066 (2002). https://doi.org/10.1093/nar/gkf436
7. Kemena, C., Notredame, C.: Upcoming challenges for multiple sequence alignment methods in the high-throughput era. Bioinformatics **25**(19), 2455–2465 (2009). https://doi.org/10.1093/bioinformatics/btp452
8. Ly-Trong, N., Naser-Khdour, S., Lanfear, R., Minh, B.Q.: Alisim: a fast and versatile phylogenetic sequence simulator for the genomic era. Molecular Biology and Evolution **39**(5), msac092 (2022). https://doi.org/10.1093/molbev/msac092
9. Notredame, C., Higgins, D.G., Heringa, J.: T-coffee: a novel method for fast and accurate multiple sequence alignment. J. Mol. Biol. **302**(1), 205–217 (2000). https://doi.org/10.1006/jmbi.2000.4042
10. Shafee, T., Cooke, I.: Alignstat: a web-tool and r package for statistical comparison of alternative multiple sequence alignments. BMC Bioinform. **17**(1), 1–6 (2016). https://doi.org/10.1186/s12859-016-1300-6
11. Sievers, F., Higgins, D.G.: Clustal omega for making accurate alignments of many protein sequences. Protein Sci. **27**(1), 135–145 (2018). https://doi.org/10.1002/pro.3290
12. Stadler, T.: Simulating trees with a fixed number of extant species. Syst. Biol. **60**(5), 676–684 (2011). https://doi.org/10.1093/sysbio/syr029
13. Warnow, T.: Revisiting Evaluation of Multiple Sequence Alignment Methods. In: Katoh, K. (ed.) Multiple Sequence Alignment. MMB, vol. 2231, pp. 299–317. Springer, New York (2021). https://doi.org/10.1007/978-1-0716-1036-7_17

# Simulation-Based Study of WLRFU Policy Performance in Named Data Networking Using Real-World Topology


Samir Nassane[1]([⊠])[iD], Sid Ahmed Mokhtar Mostefaoui[2][iD],
Abdelkader Alem[1][iD], and Bendaoud Mebarek[2][iD]

[1] Laboratoire de Génie Énergétique et Génie Informatique (L2GEGI), University of
Tiaret, Tiaret, Algeria
{samir.nassane,abdelkader83.alem}@univ-tiaret.dz
[2] Laboratoire de Recherche en Intelligence Artificielle et Systèmes (LRIAS),
University of Tiaret, Tiaret, Algeria
{mokhtar.mostefaoui,bendaoud.mebarek}@univ-tiaret.dz



**Abstract.** A novel networking paradigm, known as Named Data Networking (NDN), is proposed as a replacement for the current IP-based internet. With its in-network content caching capabilities, NDN significantly enhances network performance by increasing content availability, lowering latency, and alleviating the load on central producers. This is attained by enabling consumers to retrieve data, based on its name rather than its IP address, from the closest router instead of the farthest producer. In this work, we apply the Window Least Recently Frequently Used (WLRFU) policy as a novel cache replacement strategy in NDN networks. This policy combines recency and frequency based on a sliding access window. Using the ccnSim simulator, we compare the performance of WLRFU with that of the LRFU (Least Recently Frequently Used) and FIFO (First In First Out) policies by employing a real-world network topology and varying several simulation parameters, such as cache size, consumer interest rate, MZipf $\alpha$, number of consumers, and the number of producers. Simulation results reveal that the implemented strategy outperforms others with higher cache hit rate and lower latency. Specifically, WLRFU surpasses LRFU, raising the cache hit rate by 4.70 %–14.87 % and reducing latency by 6.89 %–8.94 %.

**Keywords:** Named Data Networking · In-network caching · Cache replacement strategy · WLRFU · LRFU · FIFO · Cache hit rate


## 1 Introduction

The existing Internet has evolved from a basic IP network to a global platform offering diverse online services. The emergence of these services has triggered new demands among Internet users, encompassing the need for high bandwidth, low latency, and the assurance of data availability and security [1].



N. Seddari and M. Redjimi (Eds.): ICMSCT 2024, CCIS 2606, pp. 452–464, 2025.
https://doi.org/10.1007/978-3-032-01922-6_37

In response to these emerging needs, there is a proposal for a novel network architecture to substitute the current Internet framework; this innovation is referred to as Named Data Networking (NDN) [2].

The current data retrieval model on the Internet, where client requests traverse the Internet backbone to reach the data source, leads to increased delays in data delivery and places additional strain on central servers [3]. These factors negatively influence Internet performance, especially as the number of users continues to rise.

In contrast, the NDN framework allows data to be stored in multiple routers and retrieved from the nearest one, instead of directly from a remote producer. This improves content availability, minimizes latency, reduces network traffic, and notably alleviates the load on central producers [4,5].

Moreover, NDN routers use content names as network addresses rather than relying on IP addresses. Put simply, when seeking data on the NDN network, the consumer just has to mention the content name, as opposed to specifying an IP location address. Following that, the system scans the closest routers for the content name and retrieves the corresponding data [6].

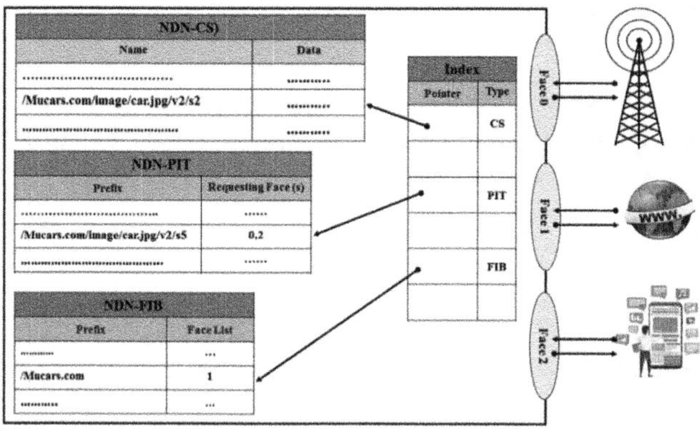

**Fig. 1.** The three components of an NDN router [7]

As shown in Fig. 1, the NDN router comprises three key components: the Pending Interest Table (PIT), which functions as a list for unsatisfied interests; the Forwarding Information Base (FIB), used for routing interest data; and the Content Store (CS), employed to store duplicates of frequently requested data [8]. The steps involved in how an NDN router processes interest and data packets are illustrated in Fig. 2.

In-network caching serves as a fundamental element of the NDN architecture, effectively addressing the challenges of rising traffic by retaining duplicates of named content in routers along the transmission path. This mechanism markedly enhances the overall performance of NDN networks and has consequently attracted significant interest from the networking research community. It comprises two key components: the cache placement strategy and the cache replacement policy [9].

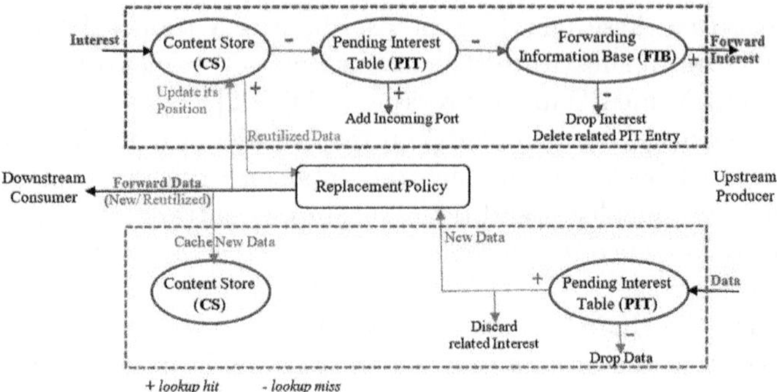

**Fig. 2.** The NDN router's processing for interest and data packets [10]

The objective of the placement strategy is to determine the router for saving content. Examples include LCE (Leave Copy Everywhere), where every router along the path stocks the desired content [11], and LCD (Leave a Copy Down), where only downstream routers duplicate the required data [12]. Regarding the replacement policy, it aims to select content for removal from the cache, allowing space for new entries due to the limited size of the CS [13]. The present paper specifically concentrates on cache replacement policies.

The following sections are organized as follows: Sect. 2 explores related work, Sect. 3 introduces the implemented caching policy, Sect. 4 describes simulation parameters and scenarios, Sect. 5 presents simulation results along with interpretations, and finally, Sect. 6 summarizes the findings and discusses future work.

## 2   Related Work

In this section, we focus on replacement strategies that correspond with the implemented strategy. We can classify these strategies into three categories: those that are based on access frequency, those that prioritize data recency, and those that account for both frequency and recency.

Aubry et al. [13] introduced the Least Frequently Used (LFU) policy for use in NDN networks, based on the premise that content frequently requested is likely to be requested again in the near future. Consequently, this approach involves evicting content with the lowest access frequency from the cache to make room for incoming data.

In [14], Dan and Towsley proposed the Least Recently Used (LRU) policy. Commonly employed in caching, this approach removes contents from the cache based on their least recent usage, making room for new entries.

The Adaptive Replacement Cache (ARC) policy [15] enhances the performance of LRU by maintaining two dynamically-sized LRU lists: T1, which stores data accessed once, and T2, which stores data accessed at least twice. The results indicate that ARC outperforms the LRU policy, achieving a 4% improvement in cache hit rate.

The Window-LFU (WLFU) algorithm was introduced by Karakostas and Serpanos in [16] for web caching. Here, the LFU method is improved by incorporating a sliding access window mechanism. In this approach, access frequencies are stored exclusively for the most recent W accesses, and replacement decisions are made based on this specific window.

Syambas et al. [17] presented the Least Recently Frequently Used (LRFU) policy for application in NDN networks, integrating both LRU and LFU policies through a configurable parameter $\lambda$. When the cache is full, LRFU selects data for eviction based on the minimum CRF (Combined Recency Frequency) value, which is computed by considering both recency and frequency with a weight function associated with $\lambda$. The results demonstrated that LRFU outperformed LRU with a 3.36% higher hit rate and exceeded priority-FIFO by 5.78%.

Bai et al. [18] proposed the Window-LRFU (WLRFU) policy for file systems and databases. This approach combines the LRU (recency) and WLFU (frequency) policies, utilizing the same CRF function as that of the LRFU algorithm. Here, the CRF value is calculated based on a sliding access window. Experimental results demonstrate that the proposed policy surpasses both LFU and WLFU, while performing comparably to LRFU and LRU strategies.

In [19,20], we proposed the WLRFU strategy for NDN networks and conducted a performance comparison with LRFU, LRU, and LFU, using a synthetic Tree topology and focusing on the cache hit ratio. Contrary to the findings reported in [18], our results in the NDN context revealed that the proposed policy outperformed all other strategies in cache hit rate, exceeding LRFU, LRU, and LFU by 1.77%, 1.71%, and 2.62%, respectively.

However, the present study places considerable emphasis on assessing the performance of WLRFU within the context of a real-world network topology. Our contribution involves not only a comparative analysis with the LRFU and FIFO strategies but also highlights the significant impact of other simulation parameters such as cache size, number of consumers, and number of producers, on cache hit ratio and latency.

## 3  WLRFU Policy Implementation

To implement the WLRFU algorithm, we utilize a cache table (CT) to store data with their corresponding attributes, including name, content, $T_{\text{First}}$, $T_{\text{Last}}$, and $CRF_{\text{Last}}$ value. $T_C$ represents the current system time, $T_{\text{First}}(D)$ and $T_{\text{Last}}(D)$ correspond, respectively, to the times of the first access and the last access to a data D within a sliding access window (WIN), and $CRF_{\text{Last}}(D)$ retains the most recent CRF value of data D (see Fig. 3). The weight function is defined as:

$$F(x) = \left(\frac{1}{2}\right)^{\lambda x} \tag{1}$$

where parameter $\lambda \in [0, 1]$. The pseudocode for the WLRFU policy applied in the NDN context is illustrated in Fig. 4.

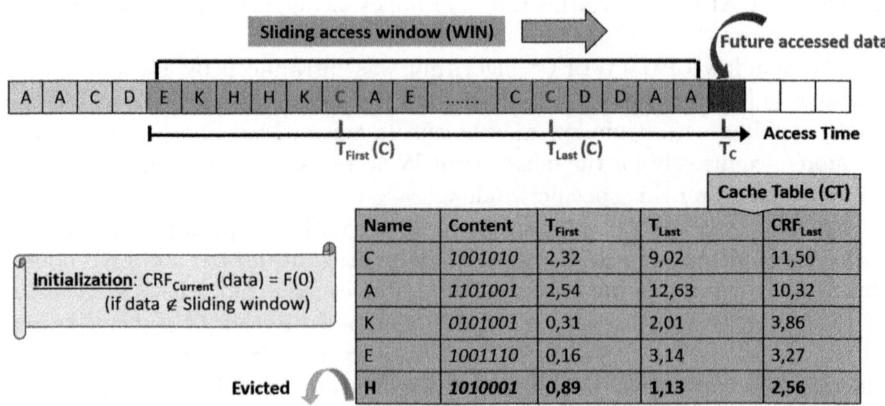


| Name | Content | $T_{First}$ | $T_{Last}$ | $CRF_{Last}$ |
|------|---------|-------------|------------|--------------|
| C | 1001010 | 2,32 | 9,02 | 11,50 |
| A | 1101001 | 2,54 | 12,63 | 10,32 |
| K | 0101001 | 0,31 | 2,01 | 3,86 |
| E | 1001110 | 0,16 | 3,14 | 3,27 |
| H | 1010001 | 0,89 | 1,13 | 2,56 |

**Fig. 3.** Overview of the WLRFU policy architecture

Data **D is accessed** at current system time $T_C$ :

**1) If D ∉ CT:**

- $CRF_{Last}(D) = F(0) = 1$

- $T_{First}(D) = T_{Last}(D) = T_C$

- If CT is full: <u>Evict</u> the data with the <u>minimum</u> $CRF_{Last}$ value.

- <u>Store</u> D with its (name, content, $CRF_{Last}$, $T_{First}$, $T_{Last}$) in CT

- Goto (4)

**2) Else if D ∈ CT and D does not exit WIN :**

- $CRF_{Last}(D) = F(0) + F(T_C - T_{Last}(D)) * CRF_{Last}(D)$

- $T_{Last}(D) = T_C$

- Goto (4)

**3) Else if D ∈ CT and D (duplicate) exits WIN :**

- $CRF_{Last}(D) = F(0) - F(T_C - T_{First}(D)) + F(T_C - T_{Last}(D)) * CRF_{Last}(D)$

- $T_{Last}(D) = T_C$, Update_$T_{First}(D)$

- Goto (5)

**4) If E exits WIN while D is accessed (enters WIN) at time $T_C$ :**

- If E ∈ CT:

    $CRF_{Last}(E) = - F(T_C - T_{First}(E)) + F(T_C - T_{Last}(E)) * CRF_{Last}(E)$

    Update $T_{First}(E)$

**5) $T_C$ = updated system time**

**Fig. 4.** The WLRFU pseudocode in the NDN context

## 4    Simulation Parameters and Scenarios

The NDN is an implementation of the Content-Centric Networking (CCN) architecture [6–8,11] that represents the theoretical framework for data-centric networking. Thus, for our simulations, we use ccnSim [21], a highly scalable simulator that extends the OMNET++ environment [22] for simulating NDN/CCN networks at the data chunk level. In ccnSim, each node comprises three layers that implement NDN router components [21]. Among available simulators, ccnSim is particularly suited for implementing and assessing caching policies in NDN networks [7,21,23]. We evaluate WLRFU in comparison to LRFU and FIFO, after extending ccnSim to support all three policies. The cache hit rate is a key metric for evaluating performance in NDN. A higher cache hit rate signifies that more consumer requests are fulfilled by routers, resulting in reduced latency and decreased load on producers.

The simulations are performed using the Abilene topology [24], a real-world network topology located in the USA that consists of 11 routers and 14 bidirectional links, as shown with link delays in Fig. 5. The producer provides a catalog of $10^6$ different contents for consumers. The consumer sends 20 interest packets per second with randomized inter-interest intervals. The content requested in interest packets is governed by the MZipf [25] probability distribution, which serves as a content popularity model. Table 1 presents the main simulation parameters. We simulate five scenarios that encompass variations in the cache size, consumer interest rate, MZipf $\alpha$, the number of consumers, and the number of producers. These scenarios are detailed in Table 2.

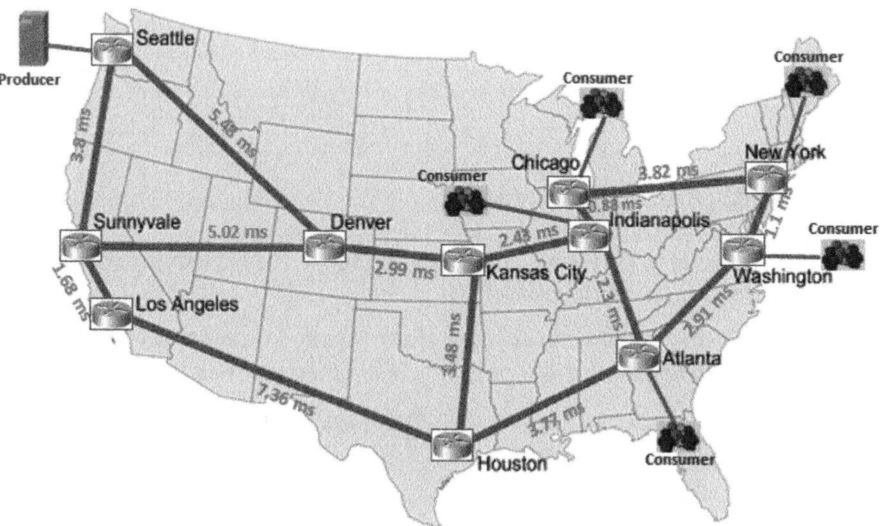


**Fig. 5.** The Abilene topology (11 routers, 1 producer, 5 consumers) [21,24]

**Table 1.** Simulation parameters [5,7,17,23,26]

| Parameter | Values |
|---|---|
| Topology (routers) | Abilene (11) |
| Number of producers (Prod) | 1, 2, 3, 4, 5, 6 |
| Number of consumers (Cons) | 5, 10 |
| Contents catalog size | $10^6$ Diverse contents |
| Content size | 10 KB |
| Cache size (CS) | 0.1% of the catalog size |
| Consumer Interest rate (Int_rate) | 20 int/s (randomized inter-interest intervals) |
| MZipf $\alpha$ | 1 |
| Placement strategy | LCD |
| Forwarding strategy | SPR (Shortest Path Routing) |
| Replacement strategy | WLRFU, LRFU, FIFO |
| WLRFU $\lambda$ value | 0.1 |
| Performance metric | Cache hit ratio, Latency |
| Simulation time | 1000 s |

**Table 2.** Simulation scenarios [5,17,26]

| N° | Scenario | Variation | Values |
|---|---|---|---|
| 1 | Prod = 1, Cons = 5, CS = 1000, $\alpha$ = 1 | Int_rate | 5, 10, 20, 30, 40 |
| 2 | Prod = 1, Cons = 5, CS = 1000, Int_rate = 20 | MZipf $\alpha$ | 0.6, 0.8, 1.0, 1.2 |
| 3 | Prod = 1, Cons = 5, Int_rate = 20, $\alpha$ = 1 | CS | 0.1%–1% |
| 4 | Prod = 1, CS = 1000, Int_rate = 20, $\alpha$ = 1 | Cons | 5, 10 (Max) |
| 5 | Cons = 5, CS = 1000, Int_rate = 20, $\alpha$ = 1 | Prod | 1, 2, 3, 4, 5, 6 (Max) |

## 5   Results and Discussion

This study evaluates performance using two metrics: the average cache hit rate for all caches and the average latency.

*C_Hit_Ratio*: shows the ratio of successfully satisfied interests across all routers to the total number of interests received. It is determined as follows:

$$C\_Hit\_Ratio = \frac{\sum_{k=1}^{N} hits_k}{\sum_{k=1}^{N} total\_interest_k} \times 100 \tag{2}$$

where: N is the total number of routers, $hits_k$ is the number of times a requested data in an interest packet is found in the cache of router k, and $total\_interest_k$ is the total number of interest packets received by router k.

*Latency*: It reflects the average duration for a consumer to access the content after submitting an associated interest, computed by:

$$Latency = \frac{global\ average\ latency}{number\ of\ consumers} \tag{3}$$

### 5.1   Scenario 1: Impact of the Consumer Interest Rate

Figure 6 illustrates the variations in cache hit rate and latency across a range of interest rates for the WLRFU, LRFU, and FIFO policies. WLRFU responds exceptionally well to the increasing interest rate, achieving a higher hit rate and lower latency compared to other policies. For instance, at a rate of 30 int/s, WLRFU outperforms LRFU with an 8.85% higher hit rate and exceeds FIFO by 4.54% (see Fig. 6(a)). Additionally, it reduces latency by 6.89% compared with LRFU and by 3.83% compared with FIFO. (see Fig. 6(b)).

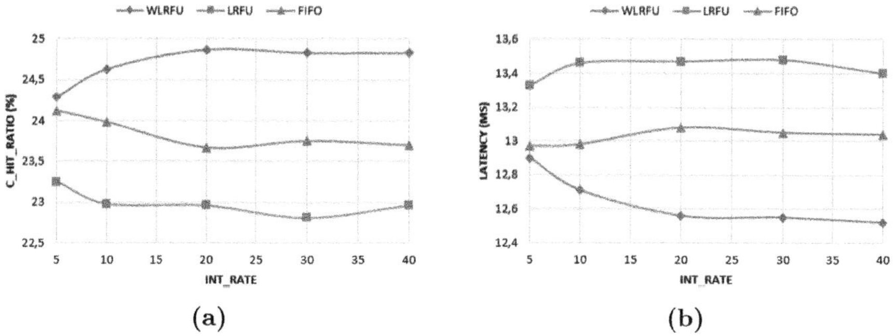


**Fig. 6.** Impact of Interest rates on Cache hit rate and Latency

### 5.2   Scenario 2: Impact of the MZipf $\alpha$

Tables 3 and 4 present the variations in cache hit rate and latency, respectively, across different MZipf $\alpha$ values for the three evaluated policies. The findings demonstrate that WLRFU outperforms the other two policies across all variations of $\alpha$. An increase in the MZipf parameter $\alpha$ enhances the cache hit ratio and reduces latency across all caching policies. Higher $\alpha$ values indicate a smaller set of frequently requested contents, allowing retrieval from the nearest router rather than the original producer.

**Table 3.** Cache hit ratio vs. MZipf $\alpha$

| $\alpha$ | 0.6 | 0.8 | 1.0 | 1.2 |
|---|---|---|---|---|
| **WLRFU** | **2,39** | **8,96** | **24,87** | **45,21** |
| LRFU | 1,75 | 7,8 | 22,96 | 43,18 |
| FIFO | 2,17 | 8,29 | 23,67 | 44,83 |

**Table 4.** Latency vs. MZipf $\alpha$

| $\alpha$ | 0.6 | 0.8 | 1.0 | 1.2 |
|---|---|---|---|---|
| **WLRFU** | **25,37** | **21,21** | **12,56** | **4,65** |
| LRFU | 25,8 | 21,94 | 13,47 | 5,02 |
| FIFO | 25,54 | 21,59 | 13,08 | 4,8 |

## 5.3   Scenario 3: Impact of the Cache Size

Figure 7 depicts the impact of cache size on the performance of WLRFU in comparison to two other strategies. WLRFU consistently outperforms the others across all variations of cache size. As an example, with a cache size of 10,000 contents, WLRFU performs better than LRFU, attaining a hit ratio that is 6.79% higher and a latency that is 8.94% lower.

As the cache size increases, the cache hit ratio improves and latency decreases for all caching policies. A larger cache size allows existing content to remain stored for a longer duration, thereby enhancing the probability of finding the requested content in the cache before reaching the central producer.

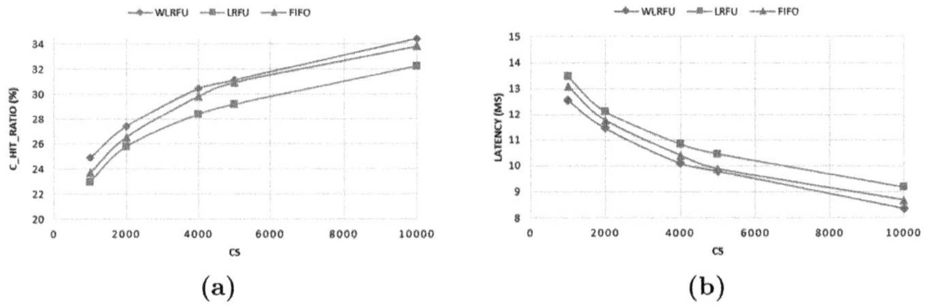

(a)          (b)

**Fig. 7.** Impact of Cache size on Cache hit rate and Latency

## 5.4   Scenario 4: Impact of the Number of Consumers

Figure 8 shows how the number of consumers influences the performance of the three evaluated strategies. WLRFU surpasses the other strategies by attaining a higher hit rate and lower latency.

As the number of consumers augments, the cache hit ratio increases and latency reduces for all caching policies. A greater number of consumers elevates network traffic, thereby increasing the probability of retrieving the requested data from routers closer to the consumers before reaching the farthest producer.

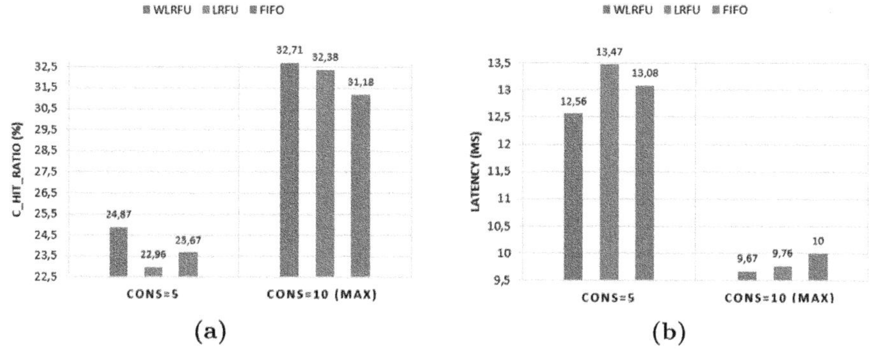


**Fig. 8.** Impact of Consumers' number on Cache hit rate and Latency

## 5.5   Scenario 5: Impact of the Number of Producers

Figure 9 displays how the number of producers impacts the performance of the three evaluated strategies. WLRFU surpasses the other strategies by attaining a higher hit rate and lower latency.

As the number of producers rises, latency decreases across all caching policies. An increased number of producers leads to the formation of multiple shortest paths between consumers and producers, reducing the hop count and thereby shortening content delivery delays.

Regarding the cache hit ratio, the results are unexpected and noteworthy:

- *From 2 to 6 producers*: As the number of producers increases, the cache hit ratio improves for all caching policies. A greater number of producers enhances content distribution across the network through multiple shortest paths, thereby increasing the probability of retrieving the requested content closer to consumers.
- *From 1 to 4 producers*: With just one producer, all three caching policies show higher cache hit ratio than with 2, 3, or 4 producers. This results from the increased network traffic with one producer, which enhances the likelihood of finding requested content from the cache.
- *Between 1, 5, and 6 producers*: The results show that with five or six producers, the cache hit ratio reaches its maximum levels. This occurs because the number of producers equals or exceeds that of consumers, resulting in the formation of a greater number of shortest paths, which in turn enhances content distribution and increases network traffic, thereby improving the probability of accessing the requested data from routers.

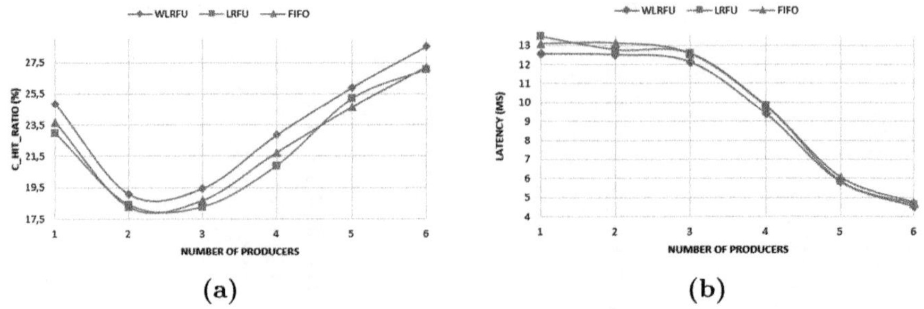

**Fig. 9.** Impact of Producers' number on Cache hit rate and Latency

# 6    Conclusion and Future Work

This work presents WLRFU, a novel caching policy in NDN networks that integrates the WLFU and LRU policies. We used the ccnSim simulator with a real network topology to evaluate the performance of this strategy compared to LRFU and FIFO policies regarding cache hit rate and latency. Five different scenarios were simulated to assess the effects of cache size, consumer interest rate, MZipf $\alpha$, the number of consumers, and the number of producers on both cache hit rate and latency.

In contrast to the findings presented in [18], which indicated that WLRFU performs similarly to LRFU within the context of file systems and databases, our results in the NDN framework demonstrate that the WLRFU strategy outperforms LRFU, reaching a higher cache hit ratio and lower latency across all simulation scenarios. Indeed, WLRFU exceeds LRFU, delivering a 4.70 %–14.87 % increase in cache hit ratio and a 6.89 %–8.94 % decrease in latency. Consequently, the WLRFU strategy proves to be a reliable alternative to the LRFU policy in NDN networks. Additionally, we found that increasing the number of consumers significantly improves network performance. However, an increase in the number of producers does not always lead to optimal performance.

In upcoming studies, we intend to assess this method against other caching approaches using a dynamic content distribution pattern instead of the MZipf model, which is based on static distribution. These evaluations will be performed using other well-known real-world topologies.

# References


1. Roberts, J.: The clean-slate approach to future internet design: a survey of research initiatives. Ann. Telecommun. **64**, 271–276 (2009)
2. Anjum, A., et al.: Towards named data networking technology: emerging applications, use cases, and challenges for secure data communication. Futur. Gener. Comput. Syst. **151**, 12–31 (2024). https://doi.org/10.1016/j.future.2023.09.031
3. Satria, M.N.D., Ilma, F.H., Syambas, N.R.: Performance comparison of named data networking and IP-based networking in Palapa Ring network. In: Proceedings of

the ICWT 2017, 3rd International Conference on Wireless and Telematics, 2017-July, pp. 43–48 (2018). https://doi.org/10.1109/ICWT.2017.8284136

4. Ran, J., Lv, N., Zhang, D., Ma, Y., Xie, Z.: On performance of cache policies in named data networking. In: Proceedings of the 2013 International Conference on Advanced Computer Science and Electronics Information (ICACSEI 2013), pp. 668–671 (2013). https://doi.org/10.2991/icacsei.2013.160

5. Silva, E.T.D., Macedo, J.M.H.D., Costa, A.L.D.: NDN content store and caching policies: performance evaluation. Comput. J. **11**(3), 37 (2022). https://doi.org/10.3390/computers11030037

6. Zhang, L., et al.: Named data networking. ACM SIGCOMM Comput. Commun. Rev. **44**(3), 66–73 (2014). https://doi.org/10.1145/2656877.2656887

7. Alubady, R., Salman, M., Mohamed, A.S.: A review of modern caching strategies in named data network: overview, classification, and research directions. Telecommun. Syst. **84**, 581–626 (2023). https://doi.org/10.1007/s11235-023-01015-3

8. Saxena, D., Raychoudhury, V., Suri, N., Becker, C., Cao, J.: Named data networking: a survey. Comput. Sci. Rev. **19**, 15–55 (2016). https://doi.org/10.1016/j.cosrev.2016.01.001

9. Fan, C., Shannigrahi, S., Papadopoulos, C., Partridge, C.: Discovering in-network caching policies in NDN networks from a measurement perspective. In: ICN 2020 Proceedings of the 7th ACM Conference on Information-Centric Networking, pp. 106–116 (2020). https://doi.org/10.1145/3405656.3418711

10. Nassane, S., Mostefaoui, S.A.M., Mebarek, B., Alem, A.: LPCE-based replacement scheme for enhancing caching performance in named data networking. Int. J. Interact. Mob. Technol. (iJIM) **18**(16), 122 (2024). https://doi.org/10.3991/ijim.v18i16.49185

11. Aboodi, A., Wan, T.C., Sodhy, G.C.: Survey on the incorporation of NDN/CCN in IoT. IEEE Access **7**, 71827–71858 (2019). https://doi.org/10.1109/ACCESS.2019.2919534

12. Laoutarisa, N., Che, H., Stavrakakisa, I.: The LCD interconnection of LRU caches and its analysis. Perform. Eval. **63**(7), 609–634 (2006). https://doi.org/10.1016/j.peva.2005.05.003

13. Aubry, E., Silverston, T., Chrisment, I.: Green growth in NDN: deployment of content stores. In: Proceedings of the 2016 IEEE International Symposium on Local and Metropolitan Area Networks(LANMAN), Rome, Italy, pp. 1–6 (2016). https://doi.org/10.1109/LANMAN.2016.7548850

14. Dan, A., Towsley, D.: An approximate analysis of the LRU and FIFO buffer replacement schemes. In: Proceedings of the ACM SIGMETRICS Conference, vol. 18, no. 1, pp. 143–152 (1990)

15. Singh, P., Kumar, R., Kannaujia, S., Sarma, N.: Adaptive replacement cache policy in named data networking. In: 2021 International Conference on Intelligent Technologies (CONIT), Hubli, India, pp. 1–5 (2021). https://doi.org/10.1109/CONIT51480.2021.9498489

16. Karakostas, G., Serpanos, D. N.: Exploitation of different types of locality for web caches. In: Proceedings ISCC 2002 Seventh International Symposium on Computers and Communications, Italy, pp. 207–212 (2002). https://doi.org/10.1109/ISCC.2002.1021680

17. Syambas, N.R., Situmorang, H., Putra, M.A.P.: Least recently frequently used replacement policy in named data network. In: 2019 IEEE 5th International Conference on Wireless and Telematics (ICWT), Yogyakarta, Indonesia, pp. 1–4 (2019). https://doi.org/10.1109/ICWT47785.2019.8978218

18. Bai, S., Bai, X., Che, X.: Window-LRFU: a cache replacement policy subsumes the LRU and window-LFU policies. Concurr. Comput.: Pract. Exp. (Wiley Online Libr.) **28**(9), 2670–2684 (2016). https://doi.org/10.1002/cpe.3730
19. Nassane, S., Mostefaoui, S.A.M., Mebarek, B., Alem, A.: Window-LRFU scheme in named data networking. In: The First National Conference on Computational Systems and Applications (NCCSA 2024), Khemis Melliana, Algeria (2024)
20. Nassane, S., Mostefaoui, S.A.M., Mebarek, B., Alem, A.: Implementing and evaluating a new caching approach in NDN Netrvorks. In: International Conference on Emerging Intelligent Systems, ICEIS 2024, Aflou, Algeria (2024)
21. Chiocchetti, R., Rossi, D., Rossini, G.: ccnSim: an highly scalable CCN simulator. In: 2013 IEEE International Conference on Communications (ICC), Budapest, Hungary, pp. 2309–2314 (2013)
22. Varga, A. , Hornig, R.: An overview of the OMNeT++ simulation environment. In: Proceedings of the 1st International Conference on Simulation Tools and Techniques for Communications, Networks and Systems & Workshops, SimuTools 2008, Marseille, France (2008)
23. Batool, S., Kaleem, M., Rashid, S., Mushtaq, M.A., Khan, I.: A survey of classification cache replacement techniques in the content-centric networking domain. IJAAS **11**(5), 12–24 (2024). https://doi.org/10.21833/ijaas.2024.05.002
24. Abilene network topology. https://web.archive.org/web/20071218140925/http://abilene.internet2.edu/maps-lists/. Accessed 25 Sept 2024
25. Speranta, C.B., et al.: Some properties of Zipf's law and applications. Axioms **13**(3), 146 (2024). https://doi.org/10.3390/axioms13030146
26. Al-Ahmadi, S.: A new efficient cache replacement strategy for named data networking. Int. J. Comput. Netw. Commun. **13**(5), 19–35 (2021). https://doi.org/10.5121/ijcnc.2021.13502

# Parallel Algorithm for Coloring Large-Scale Graphs Using Pregel API of Graphx

Assia Brighen[1,2(✉)] and Hachem Slimani[1]

[1] LIMED Laboratory, Faculty of Exact Sciences, University of Bejaia,
Bejaia 06000, Algeria
`a.brighen@univ-jijel.dz, hachem.slimani@univ-bejaia.dz`
[2] LAOTI Laboratory, Faculty of Exact Sciences and Computer Science,
University of Jijel, 18000 Jijel, Algeria

**Abstract.** The problem of graph coloring is a well-known problem in graph theory. This problem is to attribute a color to each one of the vertices of a graph in the way that the neighbor vertices have different colors. The graph coloring problem has many practical applications in real-world data analytics, such as scheduling and frequency assignment. Finding optimal solutions for the graph coloring problem is a well-known NP-Hard class problem. However, when it comes to large graph analytics, the speed of graph coloring is very important. In this setting, graph processing systems, such as Graphx and Giraph, are useful and they are considered among the best solutions for getting more processing speed and performance. In this setting, Graphx is massively used for parallel and distributed large graph processing. It acquired popularity due to its superior performance of data analytics on graph-structured data. In this work, we propose a novel Graphx-based algorithm for large graph coloring called the GPA algorithm. We experimented with the GPA algorithm using synthetic random graphs and on a set of real-world graph datasets and it is compared to other concurrent system-based distributed graph coloring algorithms. The results have shown that the GPA algorithm performs much better than other system-based algorithms, in terms of solution quality.

**Keywords:** Graphx · Graph vertex coloring · Distributed computing · Chromatic number · Vertex centric

## 1 Introduction

The vertex graph coloring problem is one of the most studied problems in graph theory for its significant use in several real words applications. It appears straightforward to notice that the map-coloring problem can be modeled as a graph coloring problem [8]. Another example is the problem of assigning frequencies to mobile radios and other users of electromagnetic spectrum [21]. Finding


N. Seddari and M. Redjimi (Eds.): ICMSCT 2024, CCIS 2606, pp. 465–479, 2025.
https://doi.org/10.1007/978-3-032-01922-6_38

an optimal solution for graph coloring problem is well known to be NP-Hard [13]. Due to the latter fact, the sequential approaches are slow and return results in a long time for large graph analytics. One of the best solutions for large graph analytics is the parallel and distributed computing, where the approaches can take advantage of using large computation and memory resources. Due to the efficiency of the parallel and distributed computation, several frameworks were designed for BigData and large graph treatment, such as Hadoop, Spark, Pregel [19], Giraph [1], and Graphx [22].

The first distributed framework designed specifically for large graph processing over clusters of machines is Pregel of Google. The basic Pregel programing model is based on Bulk Synchronous Parallel (BSP) model. Afterward, since Pregel is a Google proprietary system, several Pregel-inspired open source frameworks were emerged. For instance we can cite Giraph [1] and Graphx [22]. These frameworks bring new abstractions for iterative parallel and distributed processing of large graphs on clusters of computer machines. The main provided advantages are their reliability, scalability, and the ease of use. Moreover, they permit to programmers writing user-defined functions that behave on vertices, which the framework can then parallelly apply to any arbitrarily large graph.

The Pregel-inspired frameworks attracted a lot of research in recent years [5, 10,14,23]. However, for the graph coloring problem, few parallel and distributed approaches have been designed [4,6,7,10]. In their previous works, Gandhi and Misra [10] and Brighen et al. [7] presented new greedy algorithms for large graph coloring, based on Giraph framework. These algorithms are efficient but the provided solutions quality are far away from the optimal solution. In order to optimize solution quality of large graph coloring using parallel and distributed large graph frameworks, this paper introduces a novel algorithm, which is capable of coloring vertices of any arbitrary large graph. The algorithm is designed for utilizing the easy parallelization technique provided by the Pregel-API of the framework Graphx.

The rest of the paper is organized as follows. In Sect. 2, we present basic terms and definitions. In Sect. 3, we explain the background where we describe Spark and Graphx. In Sect. 4, we discuss related works. In Sect. 5, we present and describe the proposed algorithm. In Sect. 6, we exhibit our experiment and result discussion. In Sect. 7, we discuss conclusion and future work.

## 2  Basic Terms and Definitions

In this section, we first present different notations used in this work then we recall some definitions on the coloring of graphs.

A graph $G$ is defined as an ordered pair $G = (V, E)$, where $V$ is a finite set of vertices, while $E$ is a finite set of unordered pairs of vertices $(v, u) \in V$ called edges. The cardinality of the set of vertices $V$ is denoted by $n = |V|$. Likewise, the cardinality of the set of edges $E$ is denoted by $m = |E|$. The vertices $u, v \in V$ are called adjacent (or neighbours) if $(u, v) \in E$ and nonadjacent if $(u, v) \notin E$. We assume that $G$ is stored in an adjacency list data structure, where each vertex is assigned a unique *Id* and the vertices are ordered according to their *Ids*. In addition, we denote by $\eta(v)$ the set of vertices that are adjacent to a given vertex $v \in V$ which we call $v$'s neighborhood. Moreover, we denote by *v.deg* and *v.color* respectively the degree of $v$ and the color attributed to $v$.

**Definition 1.** *The degree of a vertex $v$ in a given graph $G = (V, E)$, denoted v.deg, is the number of edges incident to the vertex $v$.*

**Definition 2.** *The vertex-coloring problem of a given graph $G = (V, E)$ consists of coloring all vertices of $G$ using the smallest possible number of colors so that any two incident vertices $u, v \in V$ are assigned different colors.*

**Definition 3.** *The smallest number $k$ for which there exists a $k$-coloring of a graph $G$ is called the chromatic number of $G$ and is denoted by $\chi(G)$.*

## 3   Background

In this section, first we present a summarized description of the framework Spark. Then, we give the main characteristics of Graphx followed by a description of the Graphx's Pregel API, on which our approach is based.

### 3.1   Spark

Spark is an open-source distributed computing framework that follows the master-slave architecture. It is one of the Apache Software Foundation projects. A Spark application consists of one driver and one or more executors. The driver program executes the user-defined function and runs various parallel operations on a cluster. The main abstraction provided by Spark is the Resilient Distributed Dataset (RDD). The RDD is considered as a collection of elements partitioned and can be operated in parallel across the nodes of the cluster. A Spark application contains two kinds of operations, transformations and actions. A transformation applies a function on each element of an RDD and can lead to create one or more additional RDDs. Actions launch the execution of functions associated with one or more transformations to result in meaningful output. For each action in a spark application, a job is performed, which includes several RDD transformations.

### 3.2   Graphx

In order to use Spark framework for large graph data, GraphX [22] has been designed as a distributed graph processing framework on the top of Spark. Graphx has several built in basic operations on graph and provided by several graph algorithms to work with graph analytics, such as (Shortest path, Triangle Counting, Page Rank and Connected Components algorithms). In addition, Graphx has an optimized variant of Pregel API, which is a popular graph processing framework developed by Google in order to solve some of large-scale graph processing problems. Pregel [19] is based on the Bulk Synchronous Parallel (BSP) computation model. It provides the message passing graph-parallel abstraction in which all vertex programs run in a sequence of steps called supersteps.

### 3.3  Pregel API of Graphx

The pregel operator of Graphx is advantageous for graphs parallel computations, that recursively compute properties of vertices. In such computations, the properties of a vertex depend on the properties of their neighbors, which in turn depend on their neighbors, and so on. The computations during an iteration depend on the computation results of the previous iteration. The vertices properties are re-performed in each iteration until a terminate criteria is reached.

Like Pregel, in the Pregel API of graphx (see Fig. 1), a graph computation consists of a sequence of iterations, called supersteps. During a superstep, a user-defined function (called also vertex program) is invoked on each vertex in parallel. A vertex program processes the collected and combined messages sent to a vertex in the previous superstep, updates the properties of a vertex and its outgoing edges, and sends messages to other vertices. The messages sent in a superstep $S$ are processed in the next superstep ($S + 1$). If no messages are transmitted to a vertex during a superstep, the vertex program will not be called on it in the following superstep. All the vertex programs need to perform and finish processing messages of the previous superstep before moving to the next superstep of the algorithm. These supersteps are repeated until there are no messages in transit and all the vertices are inactive or when the algorithm achieves the specified maximum supersteps. In each superstep, when an active vertex completed its task, it becomes inactive. An inactive vertex will be reactivated again when receiving a message sent to it by another vertex.

Unlike Pregel, the vertex program called in each superstep in Graphx has access to the properties of the source and destination vertices as well as the properties of the edge connecting them. In addition, GraphX authorizes messages to be sent only to neighboring vertices.

The pregel API is applied on a graph, takes several arguments and returns the resulting graph. The set of arguments are the following:

◇ the initial message that each vertex will receive in the first superstep;
◇ the maximum number of iterations;
◇ the direction of sending messages;
◇ a user-defined vertex program that will be run in parallel on each vertex in a superstep $S$;
◇ a user-defined send messages function that computes optional messages that will be send to the neighboring vertices;
◇ a user-defined merge messages function that combines or merges messages sent to a vertex in the previous superstep. The framework invokes this function for only the edges of the vertices that received messages in a superstep.

## 4  Related Work

Several works have been done on improving the performance of large graph coloring using parallel and distributed algorithms (for instance see [2,7,10,17]

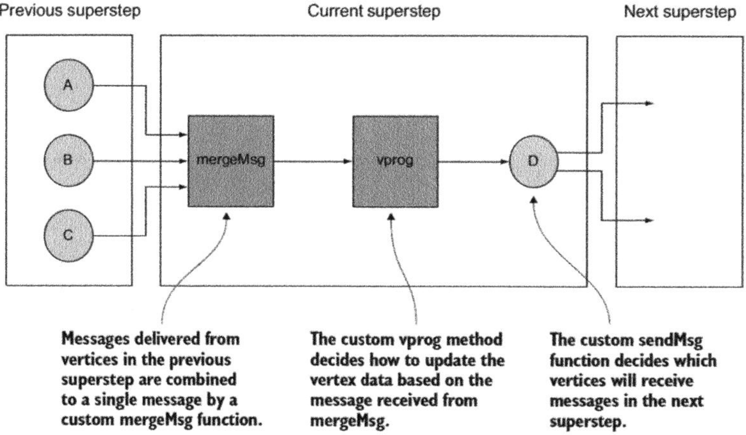


**Fig. 1.** Pregel API of Graphx [18].

and references therein). In fact, a variety of works are based on theoretical models for distributed graph coloring [2,20]. These algorithms are purely theoretical and do not provide any implementations.

Several practical parallel and distributed algorithms were proposed [3,6,7,9, 10,12]. They follow different strategies of vertex coloration, such as speculatively coloring of vertices followed by an other step for potential conflicts resolving [6,17]. Some of these algorithms were implemented using frameworks or existing graph libraries [11]. Some others used many-core architectures such as Xeon Phi and GPU [3,9]. However, the implementation of graph coloring algorithms using many-core architectures is challenging due to the complex data dependencies and irregular access of memory [9].

Another related research field is the parallel and distributed graph computing frameworks such as Pregel [19] and Giraph [1], where the vertex-centric model was used. In this field, few works have been done for graph vertex coloring [6,7,10]. In [10], Gandhi and Misra introduced four new Giraph based algorithms. These algorithms are called: Local Maxima First (LMF), Local Minima-Maxima First (LMMF), Local Largest Degree First (LLDF) and Local Smallest Largest Degree First (LSLDF). Despite the efficiency of these algorithms, they provide an extremely big number of colors that can be immensely further than the optimal solution. In the same context and based on the same distributed framework, Brighen et al. [6,7], designed a distributed large graph coloring algorithm, called DistG. Although the DistG algorithm provides good solutions compared to the Gandhi and Misra algorithms, the returned number of colors is still larger than the optimal solution.

## 5    Graphx Based Pregel-API Algorithm

In this section, we propose a new Graphx based pregel-API graph coloration algorithm (GPA). The pseudo-code of the GPA algorithm is given in Algorithm 1. As we have seen in Subsect. 3.3, the pregel API takes several arguments that were described in the same subsection. The set of arguments of the proposed algorithm are the following:

◇ The initial message:
◇ **Maximum number of iterations:** the maximum number of iteration is illimited and the algorithm converge till all vertices are colored.
◇ **Direction of sending messages:** The direction of sending messages is *out*. This means that a message $m$ is sending from its source vertex to its destination vertex.
◇ **User-defined vertex program:** the proposed vertex program is given by Algorithm 2.
◇ **User-defined send messages function:** the proposed vertex program is given by Algorithm 4.
◇ **User-defined merge messages function:** the proposed vertex program is given by Algorithm 5.

---

**Algorithm 1:** Graphx graph coloring algorithm

---

**Input**: $G$, a Symmetric directed unweighted graph
**Output**: $ColG$, a graph with each node labeled with a color value
1  $NbrItr : Int \leftarrow Int.maxvalue$;
2  $InitMsg : (Boolean, List[Int])$;
3  $InitMsg \leftarrow (false, List(-2))$;
4  $ColG \leftarrow G.pregel(InitMsg, NbrItr, EdgeDirection \leftarrow out, vertexPrg,$
   $sendMsg, mergeMsg)$ ;

---

### 5.1    Vertex Program Function

The pseudo-code of the send message function ($vertexPrg$) is given in Algorithm 2. Each vertex $v$ is characterized by a unique $Id$ and has a value. The value contains two pieces of information: the vertex degree $v.deg$ and the vertex color $v.color$. In the first superstep, every vertex $v \in V$ executes $vertexPrg$. In this case, each vertex $v$ takes as value its degree ($v.deg$) and $-1$ as initial color ($-1$ is to indicate that the vertex is not colored yet).

In the other supersteps, the $sendMsg$ can be executed by only an active vertex $v$, after merging all received messages that sent in the previous superstep. If the merging result of all received $m.state$ is $true$ and the vertex $v$ is not colored, this means that $v$ has the priority to be colored in this superstep. In this case, $v$ looks for and gets the min color that is not already used by its neighbors. The research for the min color is performed by the $getMinColor$ function (see Algorithm 3).

---

**Algorithm 2:** Function vertexPrg

```
1 Function vertexPrg(v: vertex, m: message)
2 { if m = InitMsg then
3 |   v.setValue(v.deg, −1)
4 else
5 |   if (m.state and (v.color = −1) then
6 |   |   v.setValue(v.deg, getMinColor(m.listofcolor))
7 }
```

---

**Algorithm 3:** Function getMinColor

```
1  Function getMinColor(ll:listofcolor): color
2  {
3  i : Int ← 0; max : Int ← −1;
4  while (i < ll.size) do
5  |   if ((max < ll(i) and ll(i) ≠ (−1))) then
6  |   |   max ← ll(i)
7  |   i ← i + 1
8  trouve : boolean ← false; i ← 0;
9  while (i ≤ max and not trouve) do
10 |   if (not ll.contains(i)) then
11 |   |   trouve ← true
12 |   i ← i + 1
13 if (trouve ) then
14 |   return i − 1
15 else
16 |   return max + 1
17 }
```

## 5.2  Send Message Function

The pseudo-code of the send message function ($sendMsg$) is given in Algorithm 4. Before describing the send message function, first we need to define the data structure of each message.

The sent messages at each iteration have the same format. The sent information through an edge $e = (v, u)$ in a message $m$ is defined as follow:

⋄ $m.state$: is a boolean information that indicates if the destination vertex $u$ (that will receive the message $m$) is more priority to be colored in the next superstep than the source vertex $v$.

⋄ $m.color$: is the color value of the source vertex $v$ (that sent the message $m$).

If the vertex $v$ is already colored ($v.color \neq -1$), $m.state$ takes $true$ otherwise the vertex $v$ checks the information of destination vertex $u$. If the destination vertex $u$ is not colored and $u.deg > v.deg$, then $m.state$ takes $true$ to indicate that $u$ is more priority than $v$. However, there is only one case where the $m.state$ takes $false$, which is when $v$ is not colored and it is more priority to be colored than $u$.

---

**Algorithm 4:** Function sendMsg

```
1 Function sendMsg(e: Edge (v:vertex,u: vertex))
2 { if (v.color≠ −1) and (u.color≠ −1) then
3   |   Do not send any messages
4 else
5   |   if ((v.deg < u.deg) and (v.color = −1) ) or ((v.deg=u.deg) and (v.Id <
      u.Id) and (v.color = −1)) then
6   |   |   send_Message(true, v.color) to u
7   |   else
8   |   |   if v.color ≠ −1 then
9   |   |   |   send_Message(true, v.color) to u
10  |   |   else
11  |   |   |   send_Message(false, v.color) to u
12 }
```

## 5.3   Merge Message Function

The pseudo-code of the merge message function ($mergeMsg$) is given in Algorithm 4. The merge function is responsible for merging all messages received by a vertex $v$. Those messages are sent to $v$ at the end of the previous superstep by its neighbors. The $mergeMsg$ returns two values. The first value is the result of applying the $and$ operator on the all received $m.state$. The second value is a list which contains all received colors. The combined and returned values of the $mergeMsg$ function will be used by the $vertexPrg$ function as a one received massage $m$. This latter contains two values: $m.state$ and $m.listofcolor$.

---

**Algorithm 5:** Function mergeMsg

```
1 Function mergeMsg(a: message, b: message)
2 {
3 return (a.state and b.state, a.color ∪ b.color)
4 }
```

# 6    Experimentation

In this section, we measure the performance of the proposed algorithm using synthetic random graphs as well as the SNAP library [16], that is the most widely used benchmark of large scale graphs.

## 6.1    Datasets

We have used several datasets to evaluate our algorithm, including (1) synthetic random graphs generated according to the Lancichinetti et al. [15] model, and (2) publicly available graph datasets from the SNAP library [16].

**Table 1.** Statistics of LSG networks

| $n$ | $m$ | $\Delta$ | Avg-d |
|---|---|---|---|
| 1000 | 4630 | 10 | 5 |
| 6000 | 27916 | | |
| 12000 | 56120 | | |
| 18000 | 83920 | | |
| 24000 | 112196 | | |
| 30000 | 140084 | | |

**Table 2.** Statistics of LDG networks

| $n$ | $m$ | $\Delta$ | Avg-d |
|---|---|---|---|
| 1000 | 15318 | 20 | 15 |
| 6000 | 91042 | | |
| 12000 | 182184 | | |
| 18000 | 273014 | | |
| 24000 | 360414 | | |
| 30000 | 455168 | | |

1. **Computer-generated networks:** we performed experiments on two kinds of undirected and unweighted graphs, according to the max degree of vertices ($\Delta$) and their average degrees (Avg-d). We have generated Large Sparse Graphs denoted LSG with $\Delta = 10$ and Avg-d $= 5$, and Large Dense Graphs denoted LDG with $\Delta = 20$ and Avg-d $= 15$, by using the random graph generator provided by Lancichinetti et al. [15]. For each case, we have generated six graphs by augmenting vertices numbers by 6000, varied from 1000 vertices to 30,000 vertices. These graph datasets are listed, respectively, in Tables 1 and 2.

2. **Real graph datasets:** Table 3 summarizes basic statistics of the used real graph datasets. For each dataset, the number of vertices ($n$), the number of edges ($m$), the maximum degree of vertices ($\Delta$), and the average degree of vertices (Avg-d) are reported. The average degree of vertices is calculated by $m/n$. The directed graphs are transformed into symmetric graphs by adding additional edges, whereby if an edge $(u, v) \in E$, we add edge $(v, u)$ if $(v, u) \notin E$.

**Table 3.** Statistics of the used real graph datasets to perform the experiments

| Dataset | $n$ | $m$ | $\Delta$ | Avg-d |
|---|---|---|---|---|
| Facebook | 4039 | 176468 | 1045 | 43.69 |
| Wiki-vote | 7115 | 201524 | 1065 | 28.32 |
| Cit-Hep | 27770 | 704609 | 2468 | 25.37 |
| Email-Enron | 36692 | 367662 | 1383 | 10.02 |
| Amazon | 262111 | 1799584 | 420 | 6.86 |
| web-Stanford | 281903 | 3985272 | 38625 | 14.13 |
| web-Google | 875713 | 8644102 | 6332 | 9.87 |
| roadNet-PA | 1088092 | 3083796 | 9 | 2.83 |
| com-Youtube | 1134890 | 5975248 | 28754 | 5.26 |
| roadNet-TX | 1379917 | 3843320 | 12 | 2.78 |
| roadNet-CA | 1965206 | 5533214 | 12 | 2.81 |

## 6.2    Experimental Results and Discussion

In the first phase, we have verified the scalability of our algorithm by gradually increasing the number of input vertices and examining the corresponding increase in execution time. The results are shown in Figs. 2 and 3. From these results, we notice that the execution time of the GPA algorithm grows linearly as the size of the input (number of vertices) increases.

In the second phase and in order to analyze the performance of the proposed algorithm, we have compared the results of the GPA algorithm with those obtained by the Gandhi and Misra's [10] algorithms (LMF, LMMF, LLDF and LSLDF algorithms) and the DistG algorithm [6,7] on each dataset listed in Table 3. The codes of the existing algorithms are written in the programming language Java using Giraph framework. The code of the GPA algorithm is written in the programming language Scala using Graphx framework. The performances of the algorithms were compared on the basis of their solution quality (number of the returned colors by each algorithm).

Table 4 presents the obtained experimental results by each algorithm. In addition, to show the solution quality of the proposed algorithm, Fig. 4 visualizes the

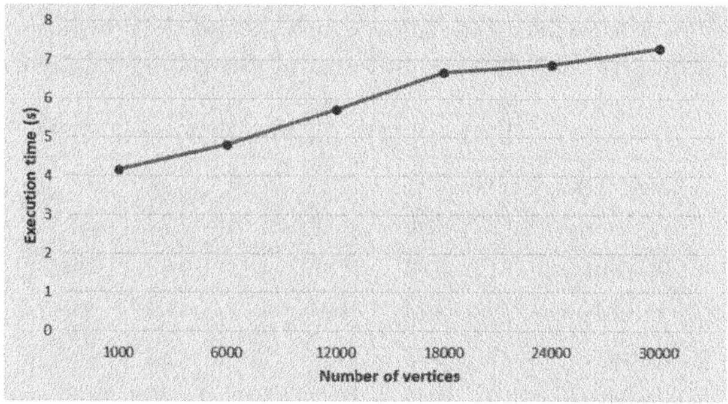

**Fig. 2.** Execution time versus number of vertices using LSG networks

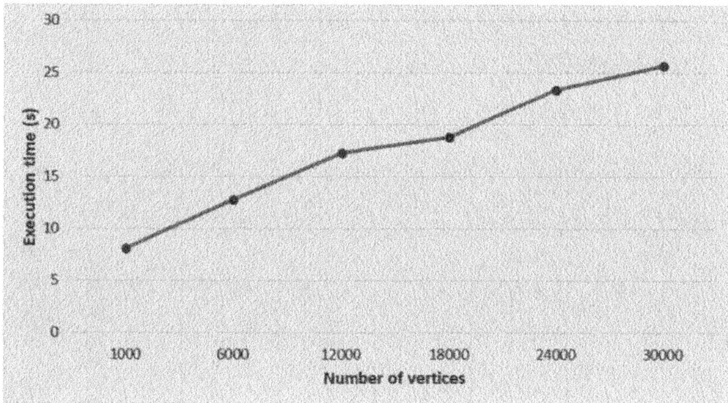

**Fig. 3.** Execution time versus number of vertices using LDG networks

comparison of the proposed GPA algorithm against the existing Giraph based graph coloring algorithms on the number of used colors for each graph dataset of Table 3.

The obtained results show that, for all datasets, GPA algorithm out perform LMF, LMMF, LLDF, LSLDF, and DistG. This is because LMF, LMMF, LLDF, and LSLDF give a number of colors that extremely depends on the necessary supersteps number for coloring a graph. In fact, in each superstep a new color is generated and there is no attempt to select an appropriate color from the unused color. Furthermore, in contrary to LMF, LMMF, LLDF, and LSLDF, the GPA algorithm does not generate a new color unless if it is necessary.

In addition, the DistG algorithm speculatively color all vertices in the second superstep with the absence of synchronization. This may lead to involve more colors being needed. Moreover, the conflicts correction step of DistG can also lead to involve more colors without attempt to minimize the number of colors. Unlike DistG, GPA algorithm colors all vertices step by step with a separate synchronization phase between each step. In the synchronization phase, each active vertex sends a message which contains its degree and its color to its neighbors. Afterward, the vertex that has the priority to be colored can choose the smallest

**Table 4.** Experimental results of the GPA algorithm and the previous systematic graph coloring algorithms on graph datasets in terms of the quality solution

| Dataset | LMF | LMMF | LLDF | LSLDF | DistG | **GPA** |
|---|---|---|---|---|---|---|
| Facebook | 347 | 347 | 254 | 263 | 123 | **76** |
| Wiki-vote | 407 | 405 | 172 | 173 | 74 | **28** |
| Cit-Hep | 312 | 313 | 177 | 183 | 65 | **29** |
| Email-Enron | 393 | 393 | 160 | 157 | 69 | **29** |
| Amazon | 177 | 177 | 32 | 29 | 12 | **7** |
| web-Stanford | 194 | 195 | 149 | 145 | 104 | **64** |
| web-Google | 82 | 83 | 84 | 87 | 64 | **45** |
| roadNet-PA | 472 | 473 | 101 | 97 | 6 | **5** |
| com-Youtube | 704 | 705 | 267 | 261 | 128 | **33** |
| roadNet-TX | 344 | 345 | 126 | 123 | 6 | **5** |
| roadNet-CA | 364 | 365 | 121 | 128 | 6 | **5** |

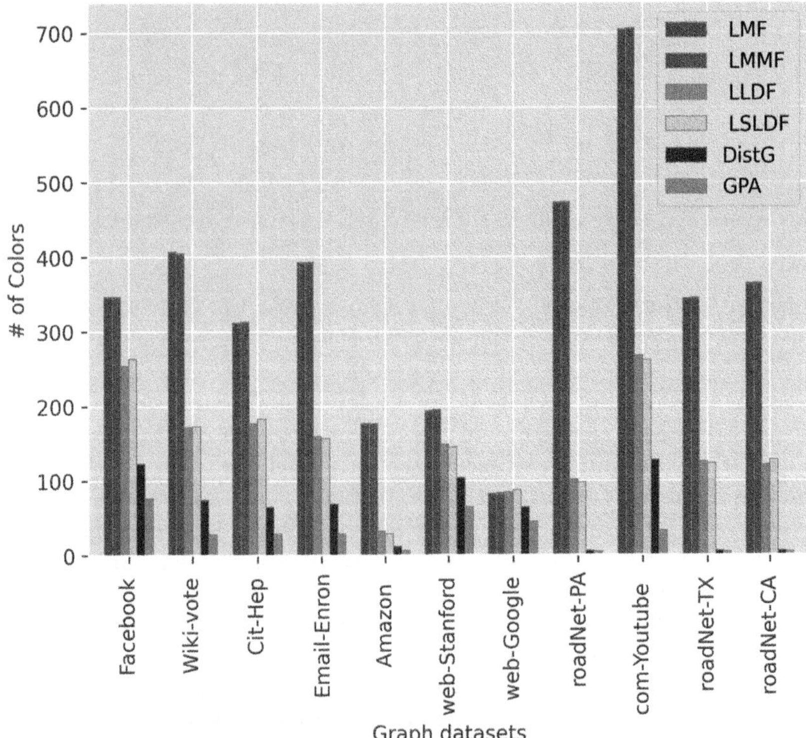


**Fig. 4.** Comparison of GPA algorithm against the existing distributed system-based graph coloring algorithms on the number of colors used for each dataset

integer that has not been selected as a color by any of its neighbors. Therefore, this strategy can easily minimize the number of used colors and returns a solution that is better than speculatively coloring strategy.

## 7    Conclusion

Due to the fact that graphs are effective for modeling different data structures, they are largely used in these huge datasets. However, when the input graph size becomes huge, the traditional graph algorithms present limitations and become inefficient. Thus, using or designing algorithms that can be easily parallelized over cluster of machines is a necessity.

This paper has introduced and described a novel algorithm, denoted GPA, for large graph vertex coloring. The algorithm has been designed for the Graphx framework, which is a Pregel-inspired framework developed specifically for large graph processing. We have experimented the presented algorithm using synthetic random graphs as well as a set of real world graph datasets and it has been compared to existing system-based large graph coloring algorithms. The obtained results have shown that the GPA algorithm out performs much better existing system-based algorithms, in term of the provided solution quality.

As future work, it would be interesting to strengthen the obtained results and conclusions by taking into consideration other metrics, especially CPU time metric and scalability of the GPA algorithm. Furthermore, it is worthwhile to continue research on large graph coloring by enhancing the quality solution of the GPA algorithm and developing other large graph coloring algorithms using other methods, such as parallel meta-heuristic approaches.

## References


1. Avery, C.: Giraph: large-scale graph processing infrastructure on Hadoop. In: Proceeding of the 2011 Hadoop Summit, Santa Clara (2011)
2. Barenboim, L., Elkin, M.: Deterministic distributed vertex coloring in polylogarithmic time. J. ACM **58**(5), 1–25 (2011). https://doi.org/10.1145/2027216.2027221
3. Bogle, I., Slota, G., Boman, E., Devine, K., Rajamanickam, S.: Parallel graph coloring algorithms for distributed GPU environments. Parallel Comput. **110**, 102896 (2022). https://doi.org/10.1016/j.parco.2022.102896
4. Brighen, A., Slimani, H.: Giraph-based distributed algorithms for coloring large-scale graphs. Int. J. Parallel Program. **53**(1) (2025). https://doi.org/10.1007/s10766-024-00781-0
5. Brighen, A., Slimani, H., Rezgui, A., Kheddouci, H.: Listing all maximal cliques in large graphs on vertex-centric model. J. Supercomput. **75**, 4918–4946 (2019). https://doi.org/10.1007/s11227-019-02770-4
6. Brighen, A., Slimani, H., Rezgui, A., Kheddouci, H.: A distributed large graph coloring algorithm on giraph. In: 2020 5th International Conference on Cloud Computing and Artificial Intelligence: Technologies and Applications (CloudTech), Marrakesh, Morocco, pp. 1–7 (2020). https://doi.org/10.1109/CloudTech49835.2020.9365872

7. Brighen, A., Slimani, H., Rezgui, A., Kheddouci, H.: A new distributed graph coloring algorithm for large graphs. Clust. Comput. **27**, 875–891 (2024). https://doi.org/10.1007/s10586-023-03988-x

8. Chen, H., Zhou, P.: An ant algorithm for solving the four-coloring map problem. In: The Ninth International Conference on Natural Computation (ICNC), pp. 491–495 (2013). https://doi.org/10.1109/ICNC.2013.6818026

9. Deveci, M., Boman, E., Devine, K., Rajamanickam, S.: Parallel graph coloring for manycore architectures. In: 2016 IEEE International Parallel and Distributed Processing Symposium (IPDPS), pp. 892–901 (2016). https://doi.org/10.1109/IPDPS.2016.54

10. Gandhi, N., Misra, R.: Performance comparison of parallel graph coloring algorithms on BSP model using Hadoop. In: 2015 International Conference on Computing, Networking and Communications (ICNC), CA, USA, pp. 110–116 (2015). https://doi.org/10.1109/ICCNC.2015.7069325

11. Gebremedhin, A., Nguyen, D., Patwary, M.A., Pothen, A.: ColPack: software for graph coloring and related problems in scientific computing. ACM Trans. Math. Softw. **40**(1), 1–31 (2013). https://doi.org/10.1145/2513109.2513110

12. Hasenplaugh, W., Kaler, T., Schardl, T., Leiserson, C.: Ordering heuristics for parallel graph coloring. In: SPAA 2014: Proceedings of the 26th ACM Symposium on Parallelism in Algorithms and Architectures, pp. 166–177 (2014). https://doi.org/10.1145/2612669.2612697

13. Karp, R.: Reducibility among combinatorial problems. In: Complexity of Computer Computations, pp. 85–103 (1972). https://doi.org/10.1007/978-1-4684-2001-2_9

14. Kulkarni, V., Chaturvedi, A., Naman, P., Simmhan, Y.: To think like a vertex (or not) for distributed training of graph neural networks. In: IEEE/ACM 23rd International Symposium on Cluster, Cloud and Internet Computing Workshops (CCGridW), pp. 351–353 (2023). https://doi.org/10.1109/CCGridW59191.2023.00082

15. Lancichinetti, A., Fortunato, S., Radicchi, F.: Benchmark graphs for testing community detection algorithms. Phys. Rev. E-Stat. Nonlinear Soft Matt. Phys. **78**(4), 046110 (2008). https://doi.org/10.1103/PhysRevE.78.046110

16. Leskovec, J., Krevl, A.: Snap datasets: Stanford large network dataset collection (2014). http://snap.stanford.edu/data. Accessed 01 Mar 2024

17. Luby, M.: A simple parallel algorithm for the maximal independent set problem. In: Proceedings of the Seventeenth Annual ACM Symposium on Theory of Computing, vol. 15, no. 4, pp. 1036–1053 (1986). https://doi.org/10.1145/22145.22146

18. Malak, M., East, R.: Spark GraphX in Action. Manning Publications Co., Shelter Island (2016)

19. Malewicz, G., et al.: Pregel: a system for large-scale graph processing. In: Proceedings of the 2010 ACM SIGMOD International Conference on Management of data, Indiana, USA, pp. 135–146 (2010). https://doi.org/10.1145/1807167.1807184

20. Maus, Y.: Distributed graph coloring made easy. ACM Trans. Parallel Comput. **10**(4), 1–21 (2023). https://doi.org/10.1145/3605896

21. Riihijarvi, J., Petrova, M., Mahonen, P.: Frequency allocation for WLANs using graph colouring techniques. In: Second Annual Conference on Wireless On-demand Network Systems and Services, pp. 216–222 (2005). https://doi.org/10.1109/WONS.2005.19

22. Xin, R., Gonzalez, J., Franklin, M., Stoica, I.: GraphX: a resilient distributed graph system on spark. In: GRADES 2013 First International Workshop on Graph Data Management Experiences and Systems, pp. 1–6 (2013). https://doi.org/10.1145/2484425.2484427
23. Z. Wu, J. Luo, X.: Spark-based label diffusion and label selection community detection algorithm for metagenome sequence clustering. Int. J. Comput. Intell. Syst. **16**(1), 2–12 (2023). https://doi.org/10.1007/s44196-023-00348-w

# Videos Textile Defects Detection Using Deep Learning Algorithms


Halima SI Moussa[1]([✉]) [ID], Mohamed Feyçal Khelfi[2] [ID], Fatima Debbat[3] [ID], and Najia Trache[4] [ID]

[1] University of Oran 1 Ahmed Ben Bella, Computer Science/National Polytechnic School of Oran, 31000 Oran, Algeria
halima.si-mousssa@enp-oran.dz
[2] RIIR Laboratory, University of Oran 1 Ahmed Ben Bella/Higher School of Electrical and Energetic Engineering of Oran, 31000 Oran, Algeria
[3] University of Mascara Mustapha Stambouli, Computer Science, 29000 Mascara, Algeria
[4] RIIR Laboratory, University of Oran, 1 Ahmed Ben Bella, 31000 Oran, Algeria



**Abstract.** Quality control is an area of utmost importance for fabric production companies. By not detecting the defects present in the fabrics, companies are at risk of losing money and reputation with a damaged product. In order to reduce these costs, an automatic textile defect detection embedded system is proposed. To perform the task of defect detection, a Convolutional Neural Network: Mask RCNN and one of the latest YOLO algorithms: YOLOv8 were used in this work. To obtain an embedded system capable of detecting static and moving defects, two datasets of static images and videos were used. In our experiments, we used the best textile defect detection training on static images, from the two pre-trained models in order to detect textile defects in videos. A confidence of 66%–97% was achieved with the two-stage model Mask RCNN and a confidence of 44%–80% was achieved with the one-stage model YOLOv8 in detecting defects within videos. But YOLOv8 was significantly faster than Mask RCNN in terms of speed detection per video.

**Keywords:** Deep Learning · Mask RCNN · Object Detection · Video Textile Defect Detection · YOLOv8


## 1 Introduction

A major issue for fabric quality inspection is in the detection of defaults, it has become an extremely challenging goal for the textile industry to minimize costs in both production and quality inspection. The quality inspection is currently done automatically with intelligent machine vision systems in order to achieve high global quality, uniformity, and consistency of fabrics and to increase productivity. Consequently, Deep learning and computer vision have been largely used in textile fields, and numerous researches are available in on this subject [1–5]. The deep learning approach has recently made an immense contribution to solving a wide number of computer vision challenges. Some


© The Author(s), under exclusive license to Springer Nature Switzerland AG 2025
N. Seddari and M. Redjimi (Eds.): ICMSCT 2024, CCIS 2606, pp. 480–489, 2025.
https://doi.org/10.1007/978-3-032-01922-6_39

methods [5, 7–10] implemented deep learning to identify defects in fabrics. However, while these methods excel in image-based scenarios, adapting them to real-time video analysis poses unique challenges. The dynamic nature of video data demands specialized approaches that balance speed and accuracy, prompting the exploration of novel methodologies [11].

Faster RCNN (Region based Convolutional Neural Network): Pioneering the integration of deep learning into object detection, Faster RCNN introduced region-based approaches, achieving notable accuracy [12]. Faster RCNN is extended by Mask RCNN that focuses on instance segmentation from an image [13]. It is an extension of Faster RCNN and in addition to the class label and bounding box, it also generates the object mask. The accurate detection is essentially required in an instance segmentation task. Therefore, it combines the two important aspects of computer vision task, object detection, which classifies and localizes the objects from an image and semantic segmentation, which classifies and assigns each pixel into a fixed set of categories. YOLO (You Only Look Once): YOLO emerged as a breakthrough with a single-shot detection approach, drastically improving processing speed [14]. While successful in real-time scenarios, YOLO variants still grapple with maintaining high accuracy, particularly in complex video environments. SSD (Single Shot multiBox Detector): SSD addressed the speed-accuracy trade-off by utilizing multiple feature maps for object detection at different scales [15]. Authors in [16] proposed a methodology's innovative approach to real-time video object detection. By combining advanced deep learning architectures with tailored optimization techniques, the model aims to redefine the standards for speed and accuracy in video analytics. The experimental results and analyses demonstrate that the proposed YOLOv8 method represents a significant advancement in real-time object detection, surpassing the performance of existing CNN, YOLOv5, and DenseNet across a diverse set of 100 datasets. This research aims to investigate the effectiveness of textile defect detection in videos, using advanced deep learning models. The main challenge is to detect textile defects in real time without compromising accuracy.

The rest of the article is organized as follows: Section two presents the proposed approach used for detecting manufacturing textile in videos, and the experimental results and discussion are covered in Section three. Finally, the last Section concludes with the scope of the work.

## 2 Approach Proposed

As per our objective, this study is associated with some background ideas and research efforts. Briefly, Mask RCNN model and YOLOv8 model for defects localization and classification and supporting it with image processing has been remarkable ideas to follow. In the context mentioned above, this study followed a process multi step (Fig. 1): from data collection and selection until models training. Each step includes a set of procedures aimed to ensuring optimal learning of the final models. Two sets of data were used, a trained dataset, downloaded from the Roboflow website (https://universe.rob oflow.com/irvin-andersen/jeans-fabric-defect). A second dataset composed of images and videos, collected from TAYAL S.P.A (https://tayal.dz). Deffect location in images (areas of interest) were identified in the annotation process, in order to identify defects.

Meanwhile, the image processing (resize, data augmentation) techniques are important for a good image enhancement, which will be effective for better defect detection at the end. The pretrained object detector models: Mask RCNN and Yolov8 model utilized on our collected dataset were evaluated for both static images and videos.

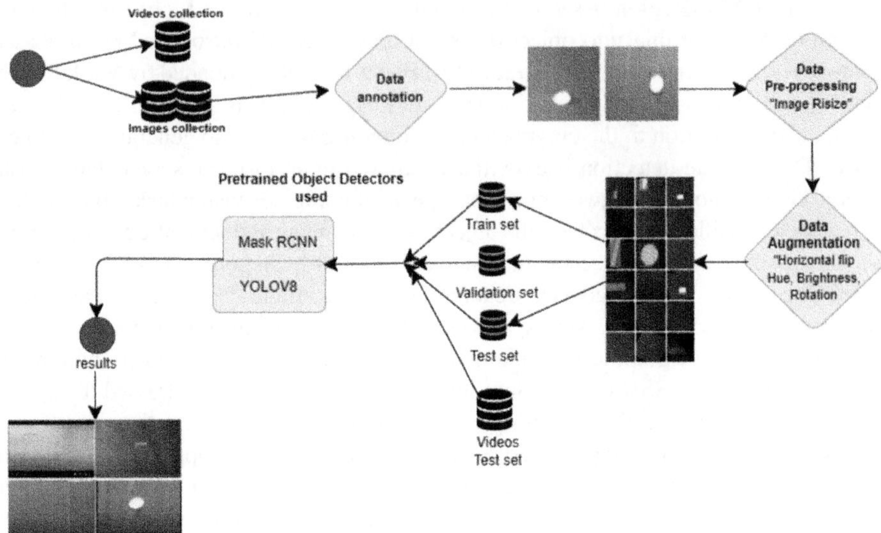


**Fig. 1.** Diagram summarizing the proposed approach.

## 2.1  Dataset

In this study, we used two datasets: the first dataset is a trained dataset, downloaded from the Roboflow website, it contains 865 defects textiles images of two classes of images: images with knot defects and images with broken defects. The second dataset is a collected dataset, from TAYAL factory. We collected 14 videos and 500 images using a phone camera but after a selection operation with the factory's engineers, the 14 videos were validated but only 166 images were validated. This dataset of 166 textile images categorized into four classes (Fig. 2): images with big hole defect (46 images), small hole defect (27 images), knot defect (31 images), and broken defect (57 images). However, these four classes of defect were the only defects shown on the different fabrics for 20 days at TAYAL factory. The factory's engineers note that these four classes of defect are almost the only ones to be detected since the machines were started up.

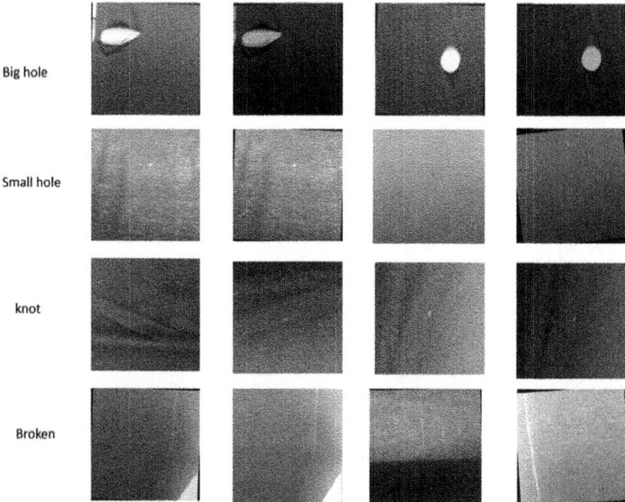

**Fig. 2.** Samples from the collected Dataset.

## 2.2  Data Annotation

Image annotation is the process of marking or labeling images with additional information to describe or identify an object (region of interest) in the image and the assignment of a label corresponding to the object class. The first step was to annotate manually the datasets using roboflow. Roboflow is an end-to-end computer vision platform, this creates text files in « coco, yolov7, PyTorch and yolov8 » format, the text file contains information about the image including the class, x center, y center, width and height of the object. Also, we used LabelImg, a graphical image annotation tool, this creates json files for each image, the «json file» contains information about the image including the image name, the class name and the shape type. The height and the width of the image are recorded under the image data. We should mention labeled, we marked the location of the defect textiles by drawing a bounding box around the defect to be detected.

Following the completion of the annotation process, it was necessary to partition the datasets into distinct subsets for the purposes of training, validation, and testing. With the help of Roboflow, we were able to divide the datasets according to our desired ratios. For the datasets (images), we allocated 70% of the data for training, 20% for validation, and the remaining 10% for testing. This strategic division ensured that our models were trained, evaluated, and tested on separate and representative datasets.

## 2.3  Data Pre-processing

Pre-processing is the method that is applied to the images before the actual processing of the images to enhance the features of the image. The motivation behind image pre-processing is to improve the quality of visual information of each input image. During a data pre-processing phase, several steps can be performing. These steps included adjusting image orientations, enhancing contrasts, resizing. Given that our collected dataset was validated by the engineers of the factory, adjusting image orientations and enhancing contrasts were not of real importance. But we resized the images to a smaller dimension of 640 × 640 pixels for both datasets (Roboflow dataset and collected dataset) before feeding them into the networks.

### 2.4  Data Augmentation

In the literature, data augmentation is a technique that could help fight overfitting, over-fitting can occur when neural network weights excessively memorize the training data instead of generalizing the input to identify patterns within the data. This tendency is particularly prevalent when dealing with smalls datasets. Given that our collected dataset of defect textile images consisted of only 166 samples, it was evident that this quantity was insufficient for achieving robust training results. With such a limited dataset, our neural networks were at a high risk of overfitting. To mitigate the risk of overfitting, we employed data-augmentation techniques.

To train the collected dataset, we implemented four forms of data augmentation including:

- Horizontal flip: Images were horizontally flipped to make the model invariant to subject orientation, allowing it to generalize better.
- Hue adjustment within a range of $-25°$ to $+25°$,
- Brightness adjustment within a range of $0\%$ to $+54\%$
- Rotation within a range of $-15°$ to $+15°$.

This substantial augmentation not only enhanced the diversity of the dataset collected but also significantly increased the number of images at our disposal for training purposes. Subsequently, we proceeded to export our augmented dataset into Coco format, specifically tailored for training the MASK RCNN model. The second version was exported in YOLOv8 format, intended for training the YOLOv8 model, this exportation allowed us to leverage the appropriate formats for training the respective models and further enhance their performance in defect detection tasks. Finally, we also exported the Roboflow dataset into the coco and YOLOv8 formats.

## 3  Experimental and Result

In our experimental setup, we conducted object detection experiments using the two-stage algorithm Mask RCNN and the one stage algorithm YOLOv8 on the collected dataset and Roboflow dataset. The simulations were carried on Windows 10 system, Google Collab run on Python (3.10.0). In the hardware device section for objects detection, the CPU is Intel Core i5 8265U @ 1.90 GHz 1.80 GHz 4cores 8 logical processors. For performance evaluation, we employed several standard metrics such as Precision, Recall, Average Precision, measuring the model's ability to accurately detect and classify objects. Additionally, we compared the two models Mask RCNN and YOLOv8, the comparison involved assessing the trade-off between detection accuracy and speed detection per video in video's defective textiles. Our results demonstrated that the one-stage model YOLOv8 demonstrated a better balance between accuracy and speed detection per video than the Mask RCNN two-stage model.

### 3.1 Experiments with Deep Learning Two-Stage Algorithm and One-Stage Algorithm

**Experiments with Two-Stage Algorithm: Mask RCNN**

The Mask RCNN model used in experiments, utilizes instance segmentation and is specifically built on the 100-layered Feature Pyramid Network (FPN) based Mask RCNN architecture [13]. The Table 1, presents the AveragePprecision (AP) metrics for bounding boxes in the Mask RCNN model. The table provides insights into the model's performance in accurately localizing and classifying objects.

In the Table 1, it can be observed that the APs and APm values are marked as "nan," indicating that the Average Precision for small and medium-sized objects could not be calculated. However, the AP for large objects is reported as 56.662 and 46.293, suggesting that the model performs well in accurately detecting and localizing larger objects.

**Table 1.** Evaluation results for bounding boxes in Mask RCNN

| Metrics | Collected dataset | Roboflow dataset |
| --- | --- | --- |
| AP | 53.084 | 41.470 |
| AP50 | 91.749 | 97.171 |
| AP75 | 48.061 | 27.680 |
| Aps | Nan | Nan |
| APm | Nan | Nan |
| API | 56.652 | 46.293 |

The AP50 value, which represents the Average Precision at an intersection over union (IoU) threshold of 0.5, is higher than all other AP values. This indicates that the model achieves a high level of precision in bounding box detection when considering a moderate IoU threshold.

On the other hand, the AP75 value, representing the average precision at an IoU threshold of 0.75, is lower (48.061 and 27.680). This suggests that the model's performance decreases when stricter criteria are applied to determine the accuracy of bounding box predictions.

In summary, the findings presented in Table 1 indicate that the Mask RCNN model achieves high accuracy in localizing bounding boxes for the collected and Roboflow datasets, especially when dealing with large objects. However, it may face difficulties in accurately detecting smaller and medium-sized objects.

**Table 2.** Average precision per class in Mask RCNN bounding boxes.

| Category | AP "Collected dataset" | AP "Roboflow dataset" |
| --- | --- | --- |
| Big hole | 85.050 | -------- |

(*continued*)

Table 2.  (continued)

| Category | AP "Collected dataset" | AP "Roboflow dataset" |
| --- | --- | --- |
| Knot | 31.199 | 58.581 |
| Broken | 41.040 | 24.360 |
| Small hole | 55.050 | -------- |

Table 2 contains the per-category AP's for the bounding boxes in Mask RCNN. These values provide insights into the model's performance for specific object categories, allowing for a more detailed evaluation of its accuracy and effectiveness in differentiating between different types of defects in the textile datasets:

Collected dataset: The AP for the "Big hole" category is calculated to be 85.050 and Small hole is calculated to be 55.050, indicating a relatively high level of accuracy in detecting and localizing Big hole and small hole within the textile images. On the other hand, the "Broken" category has an AP of 41.040 and "Knot" category has an AP of 31.199, suggesting a lower accuracy in identifying and localizing broken filaments and knot.

Roboflow dataset: The AP for the "knot" category is calculated to be 58.588, indicating a relatively high level of accuracy in detecting and localizing knots within the textile images. On the other hand, the "Broken" category has an AP of 24.360, suggesting a lower accuracy in identifying and localizing broken filaments.

In summary, the findings presented in Table 2 indicate that the Mask RCNN model achieves high accuracy in localizing and detecting "Big hole" category, followed by "Small hole" and knot categories. And a lower accuracy in localizing and detecting "Knot" category.

## Experiments with One-Stage Algorithm: YOLOv8

We trained two versions of YOLOv8: YOLOv8s and YOLOv8x on the datasets. The evaluation results are presented in Table 3.

Table 3.  YOLOv8 evaluation results.

| Datasets | Metrics | YOLOv8s | YOLOv8x |
| --- | --- | --- | --- |
| Collected dataset | mAP50 (B) | 0.69 | 0.64 |
| | mAP50–95 (B) | 0.32 | 0.29 |
| | Precision | 0.79 | 0.71 |
| | Recall | 0.63 | 0.55 |
| Roboflow dataset | mAP50 (B) | 0.78 | 0.73 |
| | mAP50–95(B) | 0.39 | 0.36 |
| | Precision | 0.67 | 0.61 |
| | Recall | 0.73 | 0.74 |

From the table above, we conclude that YOLOv8s outperforms YOLOv8x in terms of mAP50(B), mAP50–95(B), and precision. This indicates that YOLOv8s is able to better predict the location of objects in the images, even when the objects are partially occluded. However, YOLOv8x outperforms YOLOv8s in terms of recall, this indicates that YOLOv8x is better at detecting all objects in the images. Overall, YOLOv8s is a better version for object detection than YOLOv8x.

After conducting a thorough evaluation of the two algorithms: Mask RCNN and YOLOv8, and analyzing the results obtained from our training and testing processes, we can confidently state that Mask R-CNN is an effective two-stage model for detecting textile defects, including small ones, with height accuracy. On the other hand, when it comes to detecting all objects with high accuracy, YOLOv8 proves to be a good single-stage model.

## 3.2 Videos Detection

We trained the Mask R-CNN and YOLOv8 objects detection models on datasets of textile defects images. The two models were able to detect textile defects with high accuracy. We will now present the models performance in detecting defects in videos in real time (see Tables 4 and 5). For this, we tested the two models: Mask RCNN and YOLOv8s on the fourteen collected videos, each video of one minute and twenty-seven seconds.

**Table 4.** Model's confidence results.

|  | YOLOv8s | Mask RCNN |
| --- | --- | --- |
| Big hole | 80% | 97% |
| Small hole | 61% | 92% |
| Broken | 60% | 62% |
| Knot | 44% | 66% |

The Table 4 presents the comparison results of the Mask RCNN and YOLOv8s in term of confidence in detecting defects in video. And the table below presents the comparison between the two models in term of speed detection per video: YOLOv8s is significantly faster than Mask RCNN. However, Mask RCNN is more accurate than YOLOv8s.

**Table 5.** Model's detection speed comparison.

| Models | Speed detection per video |
| --- | --- |
| YOLOv8s | 2 min and 10 s |
| Mask RCNN | 19 min and 12 s |

## 4  Conclusion

The detection of textile defects is becoming increasingly automated thanks to the use of image processing techniques. This growing sector aims to improve the efficiency and precision of Real-Time quality control procedures. Cameras, sensors play a crucial role in the textile industry. Our work is part of the development of an embedded system intended for experts in the field. This system uses the latest Object Detection technology and Image Processing to detect textile manufacturing defects in videos in real time. Our objective was to improve the efficiency of control procedures by providing experts with advanced tools based on Deep learning.

In conclusion, defect textile detection in videos plays a crucial role in ensuring product quality and minimizing the release of defective products into the market. The comparison of the two models: Mask RCNN and YOLOv8 in terms of confidence and speed detection per video, for choosing the better model for this task. Based on the experiments results and analysis, YOLOv8 demonstrates harmonious balance between accuracy and speed detection per video, making it suitable for textile real time quality control applications.

## References


1. Beljadid, A., Tannouche, A., Balouki, A.: Application of deep learning for the detection of default in fabric texture. In: 2020 IEEE 6th International Conference on Optimization and Applications (ICOA), pp. 1–5, (2020). https://doi.org/10.1109/ICOA49421.2020.9094515
2. Deng, D., Wang, R., Wu, H., He, H., Li, Q., Luo, X.: Learning deep similarity models with focus ranking for fabric image retrieval. Image Vision Comput. **70**, 11–20 (2018). https://doi.org/10.1016/j.imavis.2017.12.005
3. Liu, Z., Zhang, C., Li, C., Ding, S., Dong, Y., Huang, Y.: Fabric defect recognition using optimized neural networks. J. Engineered Fibers Fabrics **14** (2019). https://doi.org/10.1177/1558925019897396
4. Zhang, H., Zhang, L., Li, P., Gu, D.: Yarn-dyed fabric defect detection with YOLOV2 based on deep convolution neural networks. In: 2018 IEEE 7th Data Driven Control and Learning Systems Conference (DDCLS), pp. 170–174 (2018). https://doi.org/10.1109/DDCLS.2018.8516094
5. Jing, J., Ma, H., H. Zhang, H.: Automatic fabric defect detection using a deep convolutional neural network. Colorat. Technol. **135**(3), 213–223 (2019). https://doi.org/10.1111/cote.12394
6. Guan, M., Zhong, Z., Rui, Y., Zheng, H., Wu, X.: Defect detection and classification for plain woven fabric based on deep learning. In: 2019 Seventh International Conference on Advanced Cloud and Big Data (CBD), pp. 297–302 (2019). https://doi.org/10.1109/CBD.2019.00060
7. Jing, J., Dong, A., Li, P., Zhang, K.: Yarn-dyed fabric defect classification based on convolutional neural network. Optic. Eng. **56**(09), (2017). https://doi.org/10.1117/1.OE.56.9.093104
8. Banumathi, P., Nasira, D.G.M.: Fabric inspection system using artificial neural networks. Int. J. Comput. Eng. Sci. **2**(5), 20–27 (2012)
9. Guan, J., Lai, R., Xiong, A., Liu, Z., Gu, L.: Fixed pattern noise reduction for infrared images based on cascade residual attention CNN. Neurocomputing **377**, 301–313 (2020). https://doi.org/10.1016/j.neucom.2019.10.054

10. Zhao, Y., Hao, K., He, H., Tang, X., Wei, B.: A visual long-short-term memory based integrated CNN model for fabric defect image classification. Neurocomputing **380**, 259–270 (2020). https://doi.org/10.1016/j.neucom.2019.10.067
11. Bhende, M., Shinde, S.: Deep learning-based realtime discriminate correlation analysis for breast cancer detection. Biomed. Res. Int. **2022**, 1–12 (2022)
12. Kiruthiga, G.: Improved object detection in video surveillance using deep convolutional neural network learning. Int. J. Modern Trends Sci. Technol. **7**(11), 104–108 (2021)
13. He, K., Gkioxari, G., Dollár, P., Girshick, R.: Mask R-CNN. Computer Vision and Pattern Recognition, vol. 1 (2018)
14. Bhatti, M.T., Fiaz, M.J.: Weapon detection in real-time CCTV videos using deep learning. IEEE Access **9**, 34366–34382 (2021)
15. Pavithra, R., Saravanan, V.: Web service deployment for selecting a right steganography scheme for optimizing both the capacity and the detectable distortion. Int. J. Recent Innov. Trends Comput. Commun. **6**(4), 267–277 (2018)
16. Monika, M., Rajinder, U., Tamizhselvi, A., Rumale, A.S.: Real-time object detection in videos using deep learning models. ICTACT J. Image Video Process. **14**(02), 3103–3109 (2023). https://doi.org/10.21917/ijivp.2023.0441

# Recent Advances in Protein Kinase Inhibitors Prediction

Mohamed Abdeldjebbar Acher$^{(\boxtimes)}$ and Hafida Bouziane

Département d'Informatique, Université des Sciences et de la Technologie d'Oran Mohamed Boudiaf USTO-MB, El Mnaouar BP 1505, 31000 Bir El Djir, Oran, Algeria
{mohamedabdeldjebbar.acher,hafida.bouziane}@univ-usto.dz
https://www.univ-usto.dz/

**Abstract.** This review delves into recent advancements in machine learning (ML) techniques within the realm of drug discovery, with a particular emphasis on protein kinase inhibitors (PKIs), a critical class of therapeutic compounds. Kinase inhibitors play a pivotal role in treating diseases such as cancer and viral infections. This review not only depicts the latest ML methodologies designed to enhance the prediction, optimization, and discovery of PKIs but also highlights the integral role of bioinformatics tools and databases in these processes. The synergy between ML models and bioinformatics resources has significantly improved the accuracy and efficiency of PKI identification, enabling more robust predictions across diverse biological datasets. By integrating these tools, researchers can streamline the drug discovery pipeline, leading to the development of more effective therapies. This review aims to provide insights into how this interdisciplinary approach can drive the future of drug discovery and the design of novel anticancer agents, offering promising avenues for future research and development.

**Keywords:** Protein Kinase Inhibitors · Drug Discovery · Bioinformatics · Machine Learning · Chemogenomics · Predictive Modeling

## 1 Introduction

Protein kinases (PKs) belong to a large family of regulatory enzymes responsible for protein phosphorylation in response to specific metabolic signals. This phonomenon results in conformational changes affecting protein function. It is well established that PKs dysregulation, overexpression, or mutation is the hallmark of many human disorders. Given the importance of such proteins and their pivotal roles in cellular signaling pathways, they remain among the most attractive targets for drug development, especially in cancer and other debilitating diseases [1]. Considerable research effort has been devoted to gain a deeper understanding of PKs mechanisms in the cell. This great deal of interest has motivated the development of many novel wet-lab and in silico techniques in


N. Seddari and M. Redjimi (Eds.): ICMSCT 2024, CCIS 2606, pp. 490–501, 2025.
https://doi.org/10.1007/978-3-032-01922-6_40

this context. However, given the high cost of wet-lab experiments, in silico predictions seem to be a rapid and cost-effective alternative.

The convergence of machine learning and bioinformatics in recent years, has catalyzed significant advancements in drug discovery [2]. This synergy has enabled researchers to harness data-driven algorithms and extensive bioinformatics databases to expedite the identification and optimization of potential therapeutic compounds, particularly protein kinase inhibitors (PKIs).

This review provides a comprehensive overview of recent research endeavors that have leveraged ML techniques in tandem with bioinformatics tools for the prediction and characterization of PKIs. By systematically analyzing a series of seminal studies conducted lately, we aim to elucidate the evolving landscape of computational drug discovery in the kinase inhibition domain, emphasizing the role of integrated bioinformatics resources in such particularly active area.

Overall, this review underscores the transformative potential of ML and bioinformatics in accelerating drug discovery efforts. It offers a glimpse into the promising avenues for future research and innovation in the field of computational biology and pharmaceutical sciences. Through interdisciplinary collaboration and the continued refinement of ML algorithms, supported by bioinformatics databases, we aim to foster a deeper understanding of kinase inhibition mechanisms, driving advancements in precision medicine and therapeutic interventions [3].

## 2    Bioinformatics in Drug Discovery and Protein Kinase Inhibitors

Bioinformatics has become an indispensable tool in modern drug discovery, particularly for PKIs. The ability to analyze and interpret vast amounts of biological data has accelerated the identification of therapeutic targets and optimized the drug development process. By leveraging computational tools and specialized databases, researchers can gain deeper insights into the complex interactions between kinases and potential inhibitors, which is crucial for developing effective therapies for diseases like cancer.

### 2.1    Role of Bioinformatics in Drug Discovery

The role of bioinformatics in drug discovery extends beyond data analysis; it is fundamental to understanding the molecular basis of diseases. Through the use of bioinformatics resources, scientists can dissect the pathways and mechanisms involved in disease progression, identifying key proteins such as kinases that can be targeted by drugs. Bioinformatics also supports the design of new inhibitors by predicting their efficacy and potential off-target effects, thereby streamlining the drug development pipeline and enhancing the precision of therapeutic interventions.

**Target Identification and Validation.** Bioinformatics plays a pivotal role in identifying and validating drug targets through the analysis of genomic, proteomic, and transcriptomic data. Techniques such as sequence alignment [4], molecular docking [5], and homology modeling [6] help in predicting potential drug targets. The use of databases like UniProt [7], Gene Ontology (GO) [8], and KEGG [9] allows researchers to annotate and understand the function of proteins and their pathways, aiding in the identification of key therapeutic targets.

**Virtual Screening and Docking.** Virtual screening [10] and molecular docking are essential bioinformatics techniques that predict the binding affinity of compounds to their target proteins. Tools like AutoDock [11] and Glide [12] are widely used in the virtual screening of large compound libraries to identify potential inhibitors. This method reduces the need for expensive and time-consuming wet-lab experiments by prioritizing the most promising candidates for further testing.

**ADMET Prediction.** The absorption, distribution, metabolism, excretion, and toxicity (ADMET) properties of a drug candidate are crucial for its success. Bioinformatics tools such as ADMET Predictor [13] and pkCSM [14] offer computational models to predict these properties early in the drug discovery process, allowing for the selection of compounds with favorable pharmacokinetic profiles.

### 2.2   Bioinformatics in Protein Kinase Inhibitor Discovery

**Kinase Databases and Resources.** The discovery of PKIs has been greatly facilitated by specialized databases and bioinformatics resources. Databases like SARfari [15] and KLIFS [16] provide comprehensive information on kinase structures, inhibitor interactions, and kinase-ligand binding modes. These resources enable researchers to explore the structural diversity of kinases and design inhibitors with improved selectivity and potency.

**Molecular Dynamics and Simulation.** Molecular dynamics (MD) simulations are used to study the dynamic behavior of kinases and their interactions with inhibitors. Tools like GROMACS [17] and AMBER [18] simulate the conformational changes of protein kinases upon inhibitor binding, providing insights into the stability and binding efficacy of potential drugs. These simulations help in understanding the molecular basis of kinase inhibition and guide the design of more effective inhibitors.

**Structure-Based Drug Design (SBDD).** SBDD is a key methodology in the development of PKIs [19], leveraging structural information from X-ray crystallography [20] or cryo-electron microscopy [21]. Bioinformatics tools like Schrödinger's Maestro [22] and Rosetta [23] are used to model and refine the binding interactions between inhibitors and kinases, optimizing drug candidates for enhanced activity and reduced off-target effects.

**Machine Learning and Data Integration.** Recent advancements in machine learning have further enhanced the bioinformatics frameworks for PKI discovery. Algorithms such as Naïve Bayes (NB) [24], Support Vector Machine (SVM) [25], k-Nearest Neighbors (kNN) [26], Random Forest (RF) [27], light Gradient Boosting (lightGBM) [28] and Extreme Gradient Boosting (XGBoost) [29] are used to predict the activity of PKIs based on large datasets of chemical and biological data. Integrating bioinformatics with powerful machine learning models has led to the discovery of novel inhibitors with improved selectivity and potency, as evidenced by studies utilizing databases like ChEMBL[1] [4,30] and BindingDB[2] [31].

As illustrated in Fig. 1, the process of predicting protein kinase inhibitors (PKIs) using machine learning (ML) and bioinformatics tools involves several stages, starting with data collection. Bioinformatics databases such as ChEMBL, PubChem, Binding DB, and CLiFS serve as key sources for chemical and biological datasets that are foundational for PKI prediction. The next stage is feature extraction, where key molecular descriptors, fingerprints, and protein-ligand interaction data are derived from the collected datasets. These extracted features are crucial for training the ML models. Subsequently, feature selection is performed to identify the most relevant variables that can accurately predict PKI activity and selectivity. Following feature selection, the data is split into three subsets: training, validation, and testing sets. The training set is used to develop machine and deep learning models, while the validation set helps in fine-tuning model parameters. Once optimized, the models are evaluated on the testing set to assess their performance. To ensure reliability and robustness, the models are assessed using metrics such as accuracy and Area Under the ROC Curve (AUC) [32]. The top-performing models are then used to predict potential PKIs. This integrated workflow not only streamlines but also accelerates the drug discovery process, guiding further experimental validation and development of PKIs.

## 3    Recent Works in Bioinformatics and Machine Learning for PKI Discovery

Machine learning methods in protein kinase inhibitor discovery have been utilized since the late 1990s and early 2000s. However, the growing fame of ML in recent years, driven by advancements in computational power and the rise of deep learning (DL), has significantly enhanced its application. Here, we describe the major breakthroughs made in PKIs prediction last years:

**JAK2-Based Prediction Method (2019).** Yang et al. [33] developed robust classification models for Janus kinase 2 (JAK2) inhibition, using the XGBoost algorithm [34]. Their approach used a curated dataset from PubChem[3] and

---

[1] https://www.ebi.ac.uk/chembl/ (Accessed: 24 October 2024).
[2] https://www.bindingdb.org/rwd/bind/index.jsp (Accessed: 24 October 2024).
[3] https://pubchem.ncbi.nlm.nih.gov/ (Accessed: 24 October 2024).

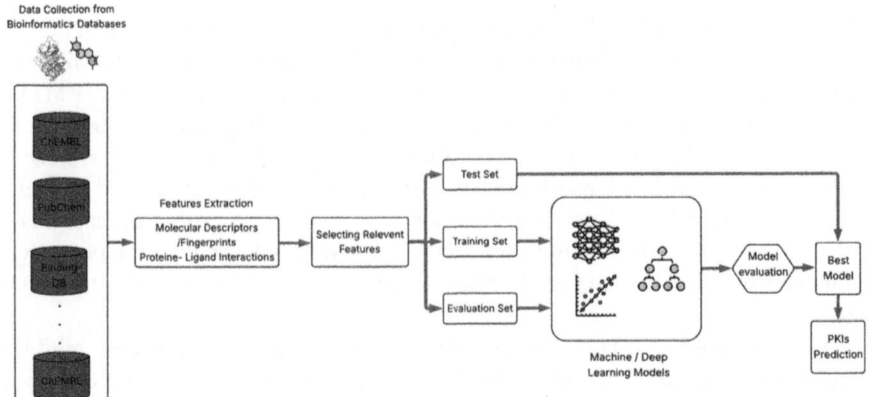


**Fig. 1.** Machine Learning and Bioinformatics Workflow for PKIs Prediction.

Binding DB databases [31], featuring diverse molecular structures and activity data. The integration of these bioinformatics resources allowed for the creation of highly accurate predictive models, with notable AUC values, offering valuable insights into potential therapeutic compounds targeting JAK2.

**Miljković et al. Binding Modes-Based Prediction Method (2020).** The authors in [35] employed machine learning to predict PKI binding modes, focusing on distinguishing various inhibitor types. Using KLIFS[4] database [36], which provides detailed structural information on kinase-ligand interactions. The authors developed models with RF, SVM, and Deep Neural Network (DNN) [37] algorithms. The use of bioinformatics resources like KLIFS enabled the accurate classification of PKIs based on their binding modes, with SVM models achieving high accuracy with balanced accuracy values near 90%.

**MTOR-Based Prediction Method (2021).** Kumari et al. [38] introduced a method for predicting the mammalian target of rapamycin (mTOR) PKIs, essential for cancer treatment. They analyzed the ChEMBL dataset using RF, SVM, Decision Tree (DT), and Artificial Neural Network (ANN) models, achieving up to 94.9% accuracy. The extensive use of bioinformatics databases like ChEMBL allowed for comprehensive analysis and improved model performance, with models focusing on prioritized molecular descriptors, exhibiting superior results.

**Keap1/Nrf2-Based Prediction Method (2021).** Shimizu et al. [39] presented a novel approach to predict protein-protein interaction (PPI) inhibitors of the Kelch-like ECH-associated protein 1 (Keap1) and nuclear factor erythroid 2-related factor 2 (Nrf2), crucial in diseases like cancer. They utilized DLiP[5]

---

[4] https://klifs.net/ (Accessed: 24 October 2024).
[5] https://skb-insilico.com/dlip (Accessed: 24 October 2024).

database with 12,593 molecules, employing RF models trained on datasets from ChEMBL, TIMBAL[6], and 2P2I[7] [40]. The bioinformatics resources provided extensive data on protein-ligand interactions, enhancing the accuracy of the RF models, which identified sixteen hit compounds with >15% inhibition rate at 100 µM.

**JNK1-Based Prediction Method (2022).** Yang et al. [41] integrated machine learning models with traditional computational methods for virtual screening of c-Jun N-terminal kinase 1 (JNK1), which is currently considered as a critical therapeutic target for type-2 diabetes. Using a dataset from ChEMBL, they trained RF, SVM, and ANN models, achieving accuracy over 82%, precision over 79%, and an AUC value of 89%. The incorporation of bioinformatics data from ChEMBL facilitated rigorous data preprocessing and model training, leading to the identification of 22 promising drug-like molecules.

**KUALA Method (2022).** De Simone et al. [42] introduced Kinase drUgs mAchine Learning frAmework (KUALA), a framework for identifying PKIs and facilitating drug repurposing. Using datasets from KLIFS and ChEMBL, they created a dataset of over 80,000 molecules and employed XGBoost and RIDGE regression models, achieving 84% accuracy. The utilization of bioinformatics datasets enabled the extraction of critical molecular descriptors, enhancing predictive accuracy and introducing a multi-target priority score (MTPS) for drug repurposing.

**FLMTS Method (2022).** Zhong et al. [43] developed FLMTS, a computational framework for predicting PKIs by integrating compound-kinase interaction networks. FLMTS uses Random Walk with Restart (RWR) [44]. The method outperformed established models with optimal results on the PKIs dataset. The framework leveraged bioinformatics data to address sparse data challenges and improve prediction accuracy, emphasizing the importance of global network topology.

**VEGFR2-Based Prediction Method (2022).** Salimi et al. [45] focused on Vascular Endothelial Growth Factor Receptor (VEGFR2) PKIs using extensive datasets from OpenCADD[8] project, ChEMBL, and eMolecules[9] database. Employing ML models, virtual screening, and molecular dynamics simulations. They identified two promising molecules with strong VEGFR2 interactions.

---

[6] https://ngdc.cncb.ac.cn/databasecommons/database/id/3188 (Accessed: 24 October 2024).

[7] https://ngdc.cncb.ac.cn/databasecommons/database/id/403 (Accessed: 24 October 2024).

[8] https://volkamerlab.org/projects/opencadd/ (Accessed: 24 October 2024).

[9] https://www.fishersci.com/us/en/brands/e/emolecules-inc.html: 24 October 2024).

The integration of bioinformatics resources allowed for comprehensive analysis, including in silico toxicity and ADMET predictions, offering insights into the therapeutic potential of identified inhibitors.

**TIGIT Method (2022).** Xiong et al. [46] integrated DNA Encoded Library Screening (DEL[10]) and ML for small molecule inhibitors targeting the T cell immunoglobulin and ITIM domain (TIGIT). With a 30-million-member DEL and various ML models, including MLP and lightGBM. They achieved a validation set hit rate of 78%. The use of bioinformatics tools for DEL screening and structural modifications improved compound binding effects, highlighting the potential of ML models in predicting compound activity.

**CDK5-Based Prediction Method (2022).** Di Stefano et al. [47] applied ML to identify inhibitors of Cyclin-dependent kinase 5 (Cdk5), protein kinase well-characterized for its role in the central nervous system. The authors used various models, including RF and MLP. They achieved high precision and identified promising candidates through virtual screening, further validated by biological assays and molecular modeling. Bioinformatics techniques played a crucial role in the identification and analysis of potential inhibitors, enhancing the understanding of pharmacological mechanisms.

**EnsDTI Method (2023).** Yijingxiu Lu et al. [48] introduced EnsDTI, an ensemble model combining ML and DL techniques. The model, leveraging multiple classifiers and four DTI benchmark datasets, namely davis (for binding affinity prediction of kinase inhibitors), kiba (a kinase-inhibitor bioactivity dataset), kinome (originally collected from ChEMBL 25[11]), and human (curated by [49]). It achieved an accuracy of 70.6%. The integration of bioinformatics data from various sources enhanced the model's ability to predict kinase-inhibitor interactions, advancing computational drug discovery efforts.

**PDGFRB-Based Prediction Method (2023).** Ssu-Ting Lien et al. [50] assessed PKIs targeting platelet-derived growth factor receptor-$\beta$ (PDGFRB), using a dataset of over 795,669 compounds and 6.5 million bioactive data points. Their Keras-MLP model achieved high accuracy and AUC values, leading to the identification of novel PDGFRB inhibitors. Bioinformatics tools were integral in curating the extensive dataset and improving the model's predictive performance.

**KIPP Method (2024).** Wu et al. [51] evaluated and compared ML and DL models for kinase profiling using their benchmark dataset referred to as

---

[10] https://novalix.com/chemistry/dna-encoded-libraries/ (Accessed: 24 October 2024).

[11] https://chembl.blogspot.com/2019/03/chembl-25-and-new-web-interface-released.html (Accessed: 24 October 2024).

KinaseNet, built from multiple sources for 354 kinases. A total of 136,290 predictive models were developed based on different data representations and fingerprints, using five mainstream ML algorithms (kNN, NB, SVM, RF, and XGBoost) and seven advanced DL algorithms, including DNN, graph convolutional network (GCN) [52], graph attention network (GAT) [53], message passing neural networks (MPNN) [54], Attentive FP [55], D-MPNN (Chemprop) [56], and FP-GNN [57]. Their study highlighted the potential of multi-task learning in kinase profiling prediction and revealed that descriptor-based ML models generally slightly outperform fingerprint-based ML models in terms of predictive performance. Table 1 summarizes the methods described above.

**Table 1.** Summary of Methods and Bioinformatics Tools/Datasets Used in PKIs Prediction.

| Method/Web-server | Authors | Target Protein | ML Models Used | Bioinformatics Tools/Dataset |
|---|---|---|---|---|
| JAK2-based prediction method [2019] | Yang et al. | JAK2 | XGBoost | PubChem, Binding DB |
| Miljković et al. method [2020] | Miljkovic et al. | Kinase inhibitors | RF, SVM, DNN | KLIFS |
| mTOR-based prediction method [2021] | Kumari et al. | mTOR | RF, SVM, DT, ANN | ChEMBL |
| Keap1/Nrf2-based prediction method [2021] | Shimizu et al. | Keap1/Nrf2 PPI | RF | DLiP, ChEMBL, TIMBAL, 2P2I |
| JNK1-based prediction method [2022] | Yang et al. | JNK1 | RF, SVM, ANN | ChEMBL |
| KUALA[a] method [2022] | De Simone et al. | Kinases | XGBoost, RIDGE regression | KLIFS, ChEMBL |
| FLMTS method [2022] | Zhong et al. | Kinases | Random Walk with Restart (RWR) | Tang, PKIs datasets |
| VEGFR2-based prediction method [2022] | Salimi et al. | VEGFR2 | RF, ANN, SVM | OpenCADD, ChEMBL, eMolecules |
| TIGIT method [2022] | Xiong et al. | TIGIT | MLP, lightGBM | DEL data |
| CDK5-based prediction method [2022] | Di Stefano et al. | CDK5 | RF, SVM, MLP, kNN | ChEMBL 25 |
| EnsDTI[b] Web-server [2023] | Yijingxiu Lu et al. | Kinases | MLP, SVM, RF | Davis, KIBA, Kinome, Human datasets |
| PDGFRB-based prediction method [2023] | Ssu-Ting Lien et al. | PDGFRB | RF, NB, kNN, SVM, MLP | Various public repositories |
| KIPP[c] Web-server [2024] | Wu et al. | Kinases | kNN, NB, RF, SVM, XGBoost and advanced DL architectures | KinaseNet dataset |

[a] https://github.com/molinfrimed/multi-kinases (Accessed: 24 October 2024)
[b] http://biohealth.snu.ac.kr/software/ensdti (Accessed: 24 October 2024)
[c] https://cytoscape.org/ Accessed: 24 October 2024

# 4  Limitations of Current Machine Learning Approaches in PKI Discovery

Despite the advancements in ML for PKI discovery, several limitations persist. The quality and availability of bioinformatics data, such as those from ChEMBL and Binding DB, can be inconsistent, incomplete, or biased, affecting model performance. Moreover, training ML models on large-scale datasets is computationally expensive, requiring high-performance resources, which may limit accessibility. Another challenge is the "black box" nature of many ML models, particularly deep learning, which hinders interpretability—essential in drug discovery for understanding predictions. Finally, generalizability remains a concern, as models trained on specific datasets may not perform well on novel data due to differences in chemical or biological space, limiting their real-world application.

## 5    Improvements and Future Directions

Big data and multi-omics integration have revolutionized bioinformatics in drug discovery by combining genomic, proteomic, and metabolomic data for a comprehensive understanding of disease mechanisms, enabling a more accurate PKIs prediction through tools like Cytoscape[12] and STRING[13]. Artificial intelligence (AI) is also enhancing PKIs discovery, with platforms like AtomNet[14] and DeepChem[15] predicting molecular interactions and optimizing lead compounds, reducing development time and costs. Additionally, cryo-electron microscopy (cryo-EM) has provided high-resolution insights into kinase-inhibitor complexes, enabling the design of inhibitors with improved binding affinity and specificity.

## 6    Conclusion

The synergy between machine learning, bioinformatics and chemogenomics has revolutionized drug discovery, particularly in the realm of PKIs. Through innovative methodologies and rigorous evaluation, researchers have made significant strides in elucidating intricate molecular mechanisms and identifying novel therapeutic candidates. Bioinformatics tools and databases have significantly contributed to these advancements by facilitating the management and analysis of large datasets, essential for the training and validation of ML models. Collaborative efforts that integrate bioinformatics with machine learning are driving progress in precision medicine and therapeutic interventions, ultimately improving patient outcomes and advancing healthcare. We hope that, this review should provide a useful information on PKIs identification field and help in its further progress.

---

[12] https://cytoscape.org/ Accessed: 24 October 2024.

[13] https://string-db.org/ (Accessed: 24 October 2024).

[14] https://www.atomwise.com/how-we-do-it/ (Accessed: 24 October 2024).

[15] https://deepchem.io/ (Accessed: 24 October 2024).

# References


1. Lu, Y., et al.: Ensdti-kinase: web-server for predicting kinase-inhibitor interactions with ensemble computational methods and its applications. bioRxiv (2023)
2. Di Stefano, M., et al.: Machine learning-based virtual screening for the identification of CDK5 inhibitors. Int. J. Mol. Sci. **23**(18) (2022)
3. Lien, S., Lin, T.E., Hsieh, J., Sung, T., Chen, J., Hsu, K.: Establishment of extensive artificial intelligence models for kinase inhibitor prediction: identification of novel PDGFRB inhibitors. Comput. Biol. Med. **156**, 106722 (2023)
4. Needleman, S.B., Wunsch, C.D.: A general method applicable to the search for similarities in the amino acid sequence of two proteins. J. Mol. Biol. **48**(3), 443–453 (1970). https://doi.org/10.1016/0022-2836(70)90057-4
5. Meng, X.Y., Zhang, H.X., Mezei, M., Cui, M.: Molecular docking: a powerful approach for structure-based drug discovery. Curr. Comput. Aided Drug Des. **7**(2), 146–157 (2011). https://doi.org/10.2174/157340911795677602
6. Marti-Renom, M.A., Stuart, A.C., Fiser, A., Sánchez, R., Melo, F., Sali, A.: Comparative protein structure modeling of genes and genomes. Ann. Rev. Biophys. Biomol. Struct. **29**, 291–325 (2000). https://doi.org/10.1146/annurev.biophys.29.1.291
7. Uniprot: A hub for protein information. Nucleic Acids Res. **43**(D1), 204–212 (2015)
8. Ashburner, M., et al.: Gene ontology: tool for the unification of biology. Nat. Genet. **25**(1), 25–29 (2000)
9. Kanehisa, M., et al.: Kegg: integrating viruses and cellular organisms. Nucleic Acids Res. **49**(D1), 545–551 (2021)
10. Macalino, S.J.Y., Gosu, V., Hong, S., Choi, S.: Role of computer-aided drug design in modern drug discovery. Arch. Pharmacal Res. **38**(9), 1686–1701 (2015). https://doi.org/10.1007/s12272-015-0640-5
11. Morris, G.M., et al.: AutoDock4 and AutoDockTools4: automated docking with selective receptor flexibility. J. Comput. Chem. **30**(16), 2785–2791 (2009)
12. Friesner, R.A., et al.: Glide: a new approach for rapid, accurate docking and scoring. 1. method and assessment of docking accuracy. J. Med. Chem. **47**(7), 1739–1749 (2004)
13. Plus, S.: ADMET Predictor
14. Pires, D.E.V., et al.: pkCSM: predicting small-molecule pharmacokinetic and toxicity properties using graph-based signatures. J. Med. Chem. **58**(9), 4066–4072 (2015)
15. (EMBL-EBI), E.B.I.: Kinase SARfari
16. Klifs: A knowledge-based structural database for protein kinases. Nucleic Acids Res. **44**(D1), 365–371 (2016)
17. Abraham, M.J., et al.: GROMACS: high-performance molecular simulations through multi-level parallelism from laptops to supercomputers. SoftwareX **1–2**, 19–25 (2015)
18. Case, D.A., et al.: AMBER 2020. University of California, San Francisco
19. Ferreira, L.G., Santos, R.N., Oliva, G., Andricopulo, A.D.: Molecular docking and structure-based drug design strategies. Molecules **20**, 13384–13421 (2015). https://doi.org/10.3390/molecules200713384
20. Drenth, J.: X-ray crystallography in structural biology: the dark and bright sides. Proc. Nat. Acad. Sci. United States Am. **104**, 11981–11986 (2007). https://doi.org/10.1073/pnas.0705059104

21. Nogales, E., Scheres, S.H.W.: Cryo-em: a unique tool for the visualization of macro-molecular complexity. Mol. Cell 58, 677–689 (2015). https://doi.org/10.1016/j.molcel.2015.02.019

22. Schrödinger, L.: Schrödinger Release 2020-4: Maestro, New York, NY (2020)

23. Rosetta: A software suite for protein modeling and analysis. Biochemistry 43(20), 5995–6008 (2004)

24. Patel, A., Sharma, M.: Naïve Bayes classification algorithms for data classification. Int. J. Adv. Res. Comput. Sci. 11(3), 105–112 (2020)

25. Zhang, L., Wang, L., Zhang, Z.: A comprehensive review of support vector machine classifications. J. Mach. Learn. Res. 22, 1–34 (2021)

26. Li, H., Liu, S.: An overview of k-nearest neighbors algorithm and its variants. J. Artif. Intell. Res. 17, 45–61 (2022)

27. Biau, G., Scornet, E.: A random forests guided tour. TEST 25(2), 197–227 (2016). https://doi.org/10.1007/s11749-016-0481-7

28. Ke, G., et al.: LightGBM: a highly efficient gradient boosting decision tree. In: Proceedings of the 31st International Conference on Neural Information Processing Systems, pp. 3149–3157 (2017)

29. Chen, T., Guestrin, C.: Xgboost: a scalable tree boosting system. In: Proceedings of the 22nd ACM SIGKDD International Conference on Knowledge Discovery and Data Mining, pp. 785–794 (2016)

30. Vanhaelen, Q., Lin, Y.C., Zhavoronkov, A.: The advent of generative chemistry. ACS Med. Chem. Lett. 11(8), 1496–1505 (2020)

31. Aliper, A., Plis, S., Artemov, A., Ulloa, A., Mamoshina, P., Zhavoronkov, A.: Deep learning applications for predicting pharmacological properties of drugs and drug repurposing using transcriptomic data. Mol. Pharm. 13(7), 2524–2530 (2016)

32. Rainio, O., Teuho, J., Klén, R.: Evaluation metrics and statistical tests for machine learning. Sci. Rep. 14, 6086 (2024). https://doi.org/10.1038/s41598-024-56706-x . Accessed 24 Oct 2024

33. Meng, F., Li, Y., Jiang, X.: Recent advances in machine learning-based drug discovery. Front. Pharmacol. 10, 1342 (2019)

34. Wu, J., Chen, Y., Zhao, D., Huang, J., Lin, M., Wang, L.: Large-scale comparison of machine learning methods for profiling prediction of kinase inhibitors. J. Cheminform. 16(1) (2024)

35. Yang, M., et al.: Machine learning models based on molecular fingerprints and an extreme gradient boosting method lead to the discovery of jak2 inhibitors. J. Chem. Inf. Model. 59(12), 5002–5012 (2019)

36. Dzobo, K., Thomford, N.E.: Opportunities provided by advanced microbial and molecular technologies in precision medicine for cancer. In: Microbial and Molecular Technologies in Diagnostic Medicine, p. 171 (2020)

37. McCulloch, W.S., Pitts, W.: A logical calculus of the ideas immanent in nervous activity. Bull. Math. Biophys. 5, 115–133 (1943). https://doi.org/10.1007/BF02478259

38. Pereira, J.C., Caffarena, E.R., Santos, C.N.: Boosting docking-based virtual screening with deep learning. J. Chem. Inf. Model. 56(12), 2495–2506 (2016)

39. Mendez, D., et al.: Chembl: towards direct deposition of bioassay data. Nucleic Acids Res. 47(D1), 930–940 (2019)

40. Palmer, D.S., O'Boyle, N.M., Glen, R.C., Mitchell, J.B.O.: Random forest models to predict aqueous solubility. J. Chem. Inf. Model. 47(1), 150–158 (2007)

41. Miljković, F., Rodríguez-Pérez, R., Bajorath, J.: Machine learning models for accurate prediction of kinase inhibitors with different binding modes. J. Med. Chem. 63(16), 8738–8748 (2020)

42. Vasudevan, S., Patel, K., Warchal, S., McGinnity, D.F., Currie, R.A.: Applications of machine learning in drug discovery and development. Drug Metab. Dispos. **48**(12), 1397–1411 (2020)
43. Kumari, C., Abulaish, M., Subbarao, N.: Exploring molecular descriptors and fingerprints to predict mTOR kinase inhibitors using machine learning techniques. IEEE/ACM Trans. Comput. Biol. Bioinf. **18**(5), 1902–1913 (2021)
44. Jiménez-Luna, J., Grisoni, F., Schneider, G.: Drug discovery with explainable artificial intelligence. Nat. Mach. Intell. **3**(11), 977–987 (2021)
45. Shimizu, Y., Yonezawa, T., Sakamoto, J., Furuya, T., Osawa, M., Ikeda, K.: Identification of novel inhibitors of keap1/nrf2 by a promising method combining protein–protein interaction-oriented library and machine learning. Sci. Rep. **11**(1) (2021)
46. Yang, R., Zhao, G., Yan, B.: Discovery of novel c-jun n-terminal kinase 1 inhibitors from natural products: integrating artificial intelligence with structure-based virtual screening and biological evaluation. Molecules **27**(19), 6249 (2022)
47. De Simone, G., Sardina, D.S., Gulotta, M.R., Perricone, U.: Kuala: a machine learning-driven framework for kinase inhibitors repositioning. Sci. Rep. **12**(1) (2022)
48. Zhong, Y., Shen, C., Wu, H., Xu, T., Luo, L.: Improving the prediction of potential kinase inhibitors with feature learning on multisource knowledge. Interdisc. Sci. Comput. Life Sci. **14** (2022)
49. Liu, H., Sun, J., Guan, J., Zheng, J., Zhou, S.: Improving compound protein interaction prediction by building up highly credible negative samples. Bioinformatics **31**, 221–229 (2015)
50. Salimi, A., Lim, J.H., Jang, J.H., Lee, J.Y.: The use of machine learning modeling, virtual screening, molecular docking, and molecular dynamics simulations to identify potential vegfr2 kinase inhibitors. Sci. Rep. **12**(1), 18825 (2022)
51. Xiong, F., et al.: Discovery of tigit inhibitors based on del and machine learning. Front. Chem. **10** (2022)
52. Kipf, T.N., Welling, M.: Semi-supervised classification with graph convolutional networks. arXiv:1609.02907 (2017)
53. Veličković, P., Cucurull, G., Casanova, A., al.: Graph attention networks. arXiv:1710.10903 (2018)
54. Gilmer, J., Schoenholz, S.S., Riley, P.F., al.: Neural message passing for quantum chemistry. In: Proceedings of the 34th International Conference on Machine Learning, vol. 70, pp. 1263–1272. JMLR.org, Sydney, NSW, Australia (2017)
55. Xiong, Z., Wang, D., Liu, X., al.: Pushing the boundaries of molecular representation for drug discovery with the graph attention mechanism. J. Med. Chem. **63**, 8749–8760 (2020). https://doi.org/10.1021/acs.jmedchem.9b00959
56. Yang, K., Swanson, K., Jin, W., et al.: Analyzing learned molecular representations for property prediction. J. Chem. Inf. Model. **59**, 33703388 (2019). https://doi.org/10.1021/acs.jcim.9b00237
57. Cai, H., Zhang, H., Zhao, D., al.: FP-GNN: a versatile deep learning architecture for enhanced molecular property prediction. Briefings Bioinform. **23**(6), 408 (2022)

# Controller Placement Problem (CPP) Based on POCO Solution for the Algerian Research Network Topology

S. Mamechaoui[(✉)] [iD], I. Benaziza, A. Benhenni, and I. Merzouka

The National Higher School of Telecommunications and Information and Communication Technologies, Ensttic Oran, Es Sénia, Algeria
{sarra.mamechaoui,imad.benaziza,abderrahim.benhenni,
ilyas.merzouka}@ensttic.dz

**Abstract.** The Software Defined Networking (SDN) is a new network paradigm, that is proposed to overcome the weaknesses of traditional networks. SDN revolutionizes network technology by breaking the fundamental idea of traditional networks in which the control plane is completely decoupled from the data plane. The practical aim of SDN is to make networks programmable via a centralized controller called a controller. In this paper, the study focuses on addressing the controller placement problem (CPP), in which placement of multiple controllers at specific locations as well as network segmentation is necessary to ensure network scalability while maintaining performance. The CPP is a critical issue in SDN, as it directly affects the performance and efficiency of SDN Controllers. We propose and implement the POCO (Pareto -optimal) solution on the ARN (Algerian Research Network) topology to tackle this challenging problem.

**Keywords:** Software Defined Networking (SDN) · Controller Placement Problem (CPP) · POCO (Pareto-Optimal COntroller placement)

## 1 Introduction

Traditional networks have faced major challenges in recent years, mainly as a result of the modern applications deployed on them. New technologies and concepts such as IoT (Internet of Things), 5G, VANET (Vehicular ad hoc network), Cloud, virtualization and Data Center, almost all enable devices to connect to each other via the Internet. The Internet as a network is based on traditional equipment and infrastructure such as routers and switches, which have worked well until now. However, the absence of a global view of the entire network, the dynamic changes and modifications to the network infrastructure, the huge, time-consuming and costly administrator task, and the impossibility of programmability are problems that have emerged as the number of devices increases [1].

© The Author(s), under exclusive license to Springer Nature Switzerland AG 2025
N. Seddari and M. Redjimi (Eds.): ICMSCT 2024, CCIS 2606, pp. 502–512, 2025.
https://doi.org/10.1007/978-3-032-01922-6_41

In fact, the emergence of modern applications calls for a new network management policy that the current architecture cannot guarantee. This new policy calls for the improvement of existing parameters such as reliability, availability, throughput, latency reduction, etc., in a way that best facilitates the deployment of the various services mentioned above.

Software Defined Networking (SDN) has emerged as a revolutionary paradigm that offers a solution to the limitations of traditional network architectures. By decoupling the control plane from the data plane, SDN enables flexible and programmable network management, leading to improved network scalability, efficiency, and agility. One of the key aspects in designing an effective SDN network is the placement of controllers.

Controllers in SDN networks act as the decision-makers, providing centralized control and management of network resources. The placement of controllers plays a crucial role in determining network performance, resource utilization, and overall system reliability. The Controller Placement Problem (CPP) refers to the challenge of identifying the optimal placement and number of controllers in an SDN network.

The objective of this paper to delve into the Controller Placement Problem and explore potential solutions to optimize the placement of controllers in SDN networks. By addressing this problem, we aim to enhance network performance, scalability, and fault tolerance. The findings of this study will benefit network operators, researchers, and system designers in effectively deploying SDN networks that meet the evolving demands of modern applications and services.

## 2   Related Work

In order to tackle the issues of scalability and reliability in SDN, an increasing number of researchers have dedicated their efforts to study the Controller Placement Problem (CPP). The CPP focuses on addressing two main questions:

- The determination of the optimal number of controllers to deploy?
- Their respective placement locations?

The first question deals with identifying the appropriate quantity of controllers required in the network to ensure efficient management and control. Having too few controllers may result in overwhelmed resources and performance bottlenecks, while deploying too many controllers could lead to unnecessary overhead and increased deployment costs.

The second question pertains to the strategic placement of controllers within the network infrastructure. The goal is to determine the most suitable locations for controllers that will effectively serve the underlying network elements and minimize communication delays. Proper controller placement contributes to improved network performance, reduced latency, and enhanced reliability.

By delving into the CPP, researchers aim to develop innovative algorithms and methodologies that address these questions and ultimately enhance the scalability and reliability of SDN architectures as shown in Fig. 1.

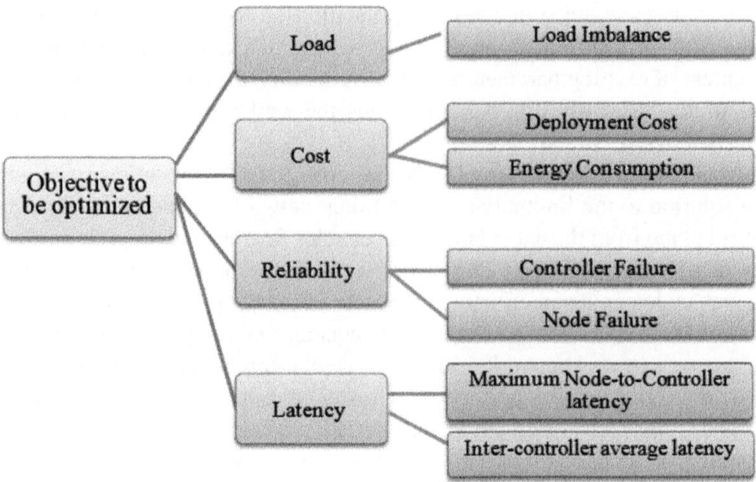



**Fig. 1.** Classification of CPP from the perspective of the optimization objective

Heller et al. [2] conducted a study on optimal solutions for controller placement to minimize the latency in communication between controllers and switches. This included analyzing both average latency and worst-case latency (or maximum latency). In a similar vein, Sallahi et al. [3] considered the costs associated with controllers, such as installation, interconnection, and overall setup expenses. Both studies [2] and [3] employed traversal-based methods to explore various potential solutions and identify the most optimal one. While traversal-based methods offer excellent performance, they tend to be time-consuming, especially when dealing with large-scale networks. In fact, as mentioned in their papers, there are numerous network topologies that couldn't be solved within a 30-h timeframe.

Furthermore, Yao et al. [4] highlighted the importance of ensuring that controllers are not overloaded and introduced the concept of a Capacitated Controller Placement Problem (CCPP). To address this problem, they proposed an advanced capacity K-center algorithm [5] that explores different values of k to identify the minimum k value that meets the capacity requirements. In a subsequent study [6], Yao et al. observed that the controller placement problem is a pre-planning issue influenced by varying flow patterns. They proposed a method for placing controllers at hotspots, where switches carry the most significant flow. Switches with lower flow can then dynamically migrate from overloaded controllers to other controllers.

## 3    Pareto-Based Optimal Controller Placement (POCO)

A comprehensive testbed for solving the CPP in the presence of multiple objectives has been developed by Hock et al. [7, 8]. The mentioned papers openly acknowledge the challenges that arise when trying to formulate a controller placement solution using conflicting metrics. Some of these conflicting challenges are:

### 3.1  Node-to-Controller Latency

The latency between nodes and controllers has a significant impact on the controller placement problem. Higher latency can lead to delays in communication and response times, affecting the overall performance of the network. When considering controller placement, it is crucial to take into account the latency between nodes and controllers to ensure efficient and effective network management. Minimizing node-to-controller latency is often a key objective in controller placement optimization algorithms, as it helps to reduce network overhead, improve responsiveness, and enhance the overall quality of service.

### 3.2  Controller Failures

In order to ensure network resilience in the event of controller failures, it is necessary to have multiple controllers in place. This allows nodes to be reassigned to alternative controllers when their primary controller becomes unavailable. However, it's important to note that the latencies between the reassigned nodes and their new controllers can be significantly higher compared to the latencies with the primary controller. As a result, when evaluating controller placement, it is crucial to consider not only the latencies during normal operation but also the worst-case latencies that may occur during controller failures.

### 3.3  Network Disruption

The failure of network elements, including links or nodes, can have a significant impact on network stability as it modifies the network topology itself. Such outages can potentially result in the isolation of certain parts of the network, where nodes are no longer able to establish connections with any controller. These nodes may remain functional and capable of forwarding traffic, but they lose the ability to request instructions from controllers. In the worst-case scenario, this isolation can lead to disrupted communication and limited network functionality.

### 3.4  Load Imbalance

As the number of nodes controlled by a single controller increases, the workload on that controller also rises. With an increase in node-to-controller requests within the network, there is a greater likelihood of experiencing delays caused by queuing at the controller system. Therefore, in situations where nodes frequently communicate with their respective controllers, it is essential to maintain a well-balanced node-to-controller assignment. This balanced assignment helps to distribute the workload evenly among controllers, minimizing queuing delays and ensuring efficient communication between nodes and controllers.

### 3.5   Inter-Controller Latency

When there are multiple controllers in a network, synchronization becomes crucial to ensure a consistent global state. The latency between individual controllers plays a significant role in determining the effectiveness of synchronization. Therefore, when considering controller placement, it is important to take into account the latency between controllers. The frequency of inter-controller synchronization should also be considered to maintain efficient communication and coordination among controllers, thereby ensuring the consistency of the network's global state.

**The Pareto Front:** When faced with conflicting objectives, it becomes necessary to determine a compromise solution that balances the optimization of various metrics. Even when presented with multiple solutions, it is possible to reach a point where improving one metric inevitably leads to the deterioration of another. This collection of solutions is commonly referred to as the "pareto front." The nature of this front varies depending on the problem at hand, encompassing either discrete values or a continuous vector. Traditionally, most fronts are represented as a trade-off between two or three objectives, enabling visualization through 2D or 3D graphs. However, in reality, there is no limitation on the number of metrics, often referred to as "objective functions," that can be employed to constrain performance.

The previous papers have showcased the wide array of metrics available for evaluating the overall performance of a computer network. To tackle this diversity, the authors introduced the POCO framework, specifically designed to assess network performance across multiple objective functions. This framework takes the form of an open-source GUI program implemented in MATLAB. Figure 2 provides a compilation of common terms and definitions associated with POCO.

The primary purpose of the POCO framework is to assess the optimal positioning of controllers across various network scenarios, focusing specifically on one of the following:

1. Failure free - all nodes of the network operate as normal, including controllers.
2. Up to two node failures - specifically defined as any node not selected to be one of the k controllers.
3. Up to k-1 controller failures - which is designed to test the effects of nodes being forced to fall back on secondary controllers.

Furthermore, the POCO framework incorporates a scenario that enables the identification of the minimum required number of k controllers in a given network. This scenario ensures that all nodes can establish communication with a controller even in the event of any two node failures.

POCO possesses the capability to assess network models across a diverse range of problem instances. The program provides standard problem profiles that can be loaded for a given network topology:

- Calculate best k = 1-to-5 controller placements are determined without considering the risk of controller failure.
- Calculate best k = 3-to-5 controller placements are determined taking into account the risk of up to two node failures. Additionally, there is an option to find the minimum

| Term | Definition |
| --- | --- |
| $G = (\mathcal{V}, \mathcal{E})$ | Graph with $\mathcal{V}$ nodes and $\mathcal{E}$ edges |
| $k$ | Number of controllers |
| $d_{v,w}$ | Distance matrix between all nodes $\{v, w\} \in \mathcal{V}$ |
| $\mathcal{P}$ | Set of placement of controllers in a network |
| $p$ | A single controller placement $\in \mathcal{P}$ |
| $\emptyset$ | Failure free scenario |
| $\mathcal{C}$ | Set of all possible $k - 1$ controller failures |
| $\mathcal{X}$ | Set of all possible 1 to 2 node failures |
| $\mathcal{N}$ | The given node failure scenario for certain metrics |
| $s$ | A given node failure scenario $\in \mathcal{S}$ |
| $e_{v,p}^{s}$ | A matrix with 1 if nodes $v, w$ cannot reach each other in failure scenario $s$, 0 otherwise |
| $\pi$ | A single metric, further specified with superscripts |

**Fig. 2.** A Listing of Terms Used for POCO Framework Equations

'resilient' k value, where resilience is defined as ensuring no node is left without a controller in the event of any two node failures.

- Calculate best k = 1-to-5 controller placements are determined considering the risk of up to k-1 controller failures.

Furthermore, each of these profiles can be customized according to the user's preferences. Users have the option to directly select, using ID numbers, which nodes serve as controllers, which controllers or nodes experience failure.

The original POCO framework functions as an enumerative solution generator. When provided with a network represented by a graph G = (V, E) consisting of N nodes and k controllers, POCO evaluates the performance metrics for all possible combinations of selecting k nodes as controllers. Due to this exhaustive approach, the original POCO framework is ensured to identify the optimal Pareto front for any given network.

## 4 Implementation and Results

In this part, we will implement the POCO framework built on MATLAB on the ARN (Algerian Research Network [9]) topology as shown in Fig. 3 in a failure-free scenario and discuss the optimal controllers placements based on some the previously stated optimization objectives.

**Fig. 3.** Algerian Research Network topology [10]

## 4.1   Node-to-Controller Latency

To quantify the latencies in a network, we take the maximum over all node-to-controller latencies and denote it with $\pi^{\max latency}$ based on a matrix $d_{v,w}$ containing the shortest path distances between all nodes $v$ and $w$ of the set of all nodes $V$, the maximum node-to-controller latency for a placement of controllers $P \in 2^V$ can be defined as:

$$\pi^{\max\,latency}(P) = \max_{vV} \min_{p\in P} d_{v,p} \tag{1}$$

To illustrate the impact of controller placement in minimizing node-to-controller latency, we used the POCO tool to calculate the best and the worst controllers' placements based on the maximum node-to-controller latency with $k = 1$ and $k = 4$ for the ARN topology.

**Single Controller Case $k = 1$**
Figure 4 shows the latencies of all nodes to the controller in two different scenarios (best and worst) with $k = 1$ using a traffic light color scheme (colors change from pure green, yellow, orange to pure red to represent latencies from best to worst).

- In Fig. 4(b), when the controller is placed in node ID 6 (worst placement case), nodes that are far from the controller present very high latencies which leads to delays in communication and response times, affecting the overall performance of the network.
- In Fig. 4(a), the controller is placed in node ID 21 (worst placement case); nodes to controller latencies improve substantially compared to the worst case.

Even though we see an improvement in the node-to-controller latencies in Fig. 4(a), some nodes that are far from the controller (node #6 and #4) still show higher latency

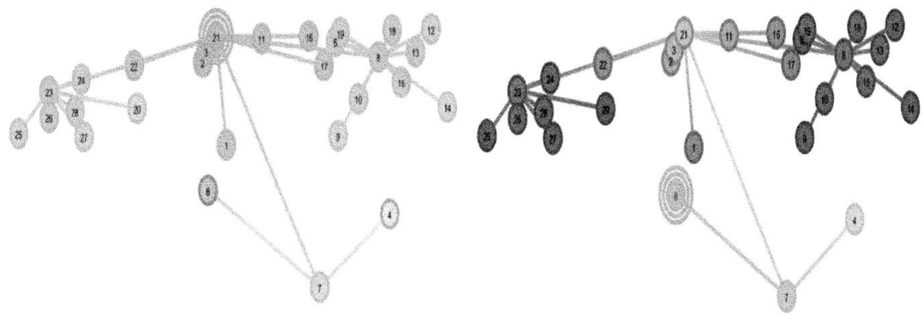

(a) Latency at worst placement                    (b) Latency at best placement

**Fig. 4.** Node-to-Controller latency with k = 1

times, which can slow the network in those areas. For that, having only one controller is not enough, we need to increase the number of controllers in the network.

**Multiple Controllers Case k = 4**

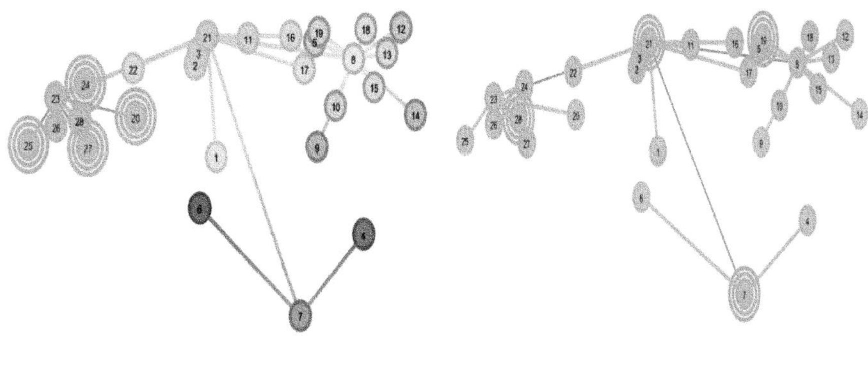

(a) Latency at worst placements                   (b) Latency at best placements

**Fig. 5.** Node-to-Controller latency with k = 4

Figure 5 shows the latencies of all nodes to the controller in two different scenarios (best and worst) with k = 4 using the same traffic lights color scheme to illustrate node-to-controller latency, we notice the following:

- In Fig. 5(a), we have 4 controllers placed in nodes #20, #24, #25 and #27 (worst placement case) which are located in the same area of the network, nodes that are located in far areas from the controllers present high to very high latencies.
- In Fig. 5(b), the controllers are placed in nodes #7, #8, #21, and #23 (best placement case). We can see that latencies improved distinctly when compared to the worst case.

As we analyze Fig. 5, we can conclude that increasing the number of controllers doesn't necessarily guarantee a better node-to-controller latency; these controllers have to be well-distributed across the network to obtain the best latency results and better network performances.

## 4.2  Controllers Load Imbalance

When considering the controllers placement problem, it is not enough to only look at node-to-controller latencies. Depending on some cases, having approximately equivalent workload distributed among all controllers can be advantageous, this ensures that no controller is overloaded while others have a significantly lighter workload.

To calculate controller imbalance, it is assumed that each node is assigned to its closest controller according to the distance matrix $d_{v,w}$. These assignments allow defining assignment matrixes $n_p^s$ containing for each failure scenario $s$ and controller $p$ the number of nodes assigned to this controller. In a failure-free scenario $\pi_\emptyset^{imbalance}$ is defined as follows:

$$\pi_\emptyset^{imbalance} = \max_{(p \in P)} n_p^\emptyset - \min_{(p \in P)} n_p^\emptyset \tag{2}$$

We used POCO framework to calculate the controllers' placements based on two controllers' imbalance scenarios (best and worst imbalance), with k = 4 for the same ARN topology and results are shown in Fig. 6.

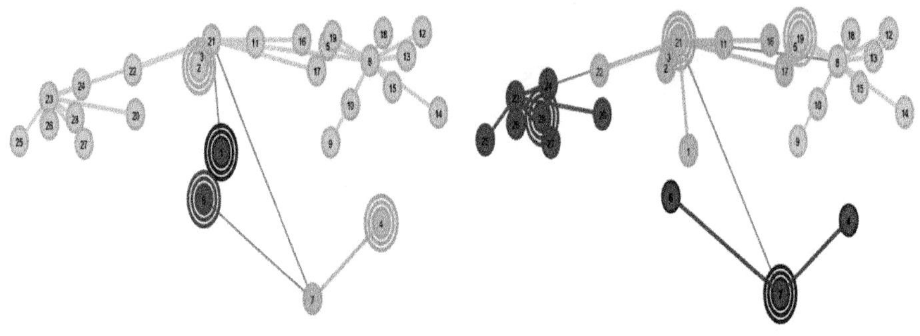

(a) Controller imbalance at worst case          (b) Controller imbalance at best case

**Fig. 6.**  Controller Imbalance with k = 4

Figure 6 highlights the controller imbalance metric, the different color swatches for nodes show which node is connected to which controllers.

- In Fig. 6(a), we can see that the distribution of nodes-to-controllers is highly imbalanced (worst case scenario). Controller #2 is responsible on controlling 23 nodes while controller #4 controls one node, the two other controllers #1 and #6 doesn't have any nodes assigned to them. With this imbalanced nodes-to-controllers distribution, delays on the network will be experienced due to the controller #2 being overloaded by nodes requests.

- In Fig. 6(b), nodes are almost equally distributed over the controllers (best case scenario). Two nodes are assigned to #7, six controllers assigned to #28, seven nodes assigned to #21 and nine nodes assigned to #19.

To decide the optimal controller placement on SDN networks, we have to also ensure that nodes are equally assigned to their nearest controllers.

## 5 Conclusion

In conclusion, this paper focused on the Controller Placement Problem (CPP) in Software-Defined Networking (SDN) networks and introduced the POCO Pareto optimal solution as a means to address this challenging problem. The objective was to optimize the placement of controllers in order to minimize node to controller latency and controller imbalance on the ARN (Algerian Research Network) topology, thereby enhancing network performance and efficiency.

Through an extensive analysis and evaluation, it was observed that the POCO Pareto optimal solution effectively improves the performance metrics of SDN networks. By strategically positioning controllers based on latency considerations and workload balancing, the proposed solution reduces the overall latency between nodes and controllers, leading to improved response times and better network performance.

Furthermore, the evaluation process took into account the concept of controller imbalance, ensuring that the workload is evenly distributed among the controllers. This mitigates the risk of bottlenecks and avoids overburdening individual controllers, thus enhancing the overall stability and scalability of the network.

## References


1. Asian Journal of Research in Computer Science 9(2): 1–18, 2021; Article no. AJRCOS.68725 ISSN: 2581–8260---Comparison of Software Defined Networking with Traditional Networking
2. Heller, B., Sherwood, R., Mckeown, N.: The controller placement problem. ACM Sigcomm Comput. Commun. Rev. **42**(4), 7–12 (2022)
3. Sallahi, A., St-Hilaire, M.: Optimal model for the controller placement problem in software defined networks. IEEE Commun. Lett. **19**(1), 30–33 (2015)
4. Yao, G., Bi, J., Li, Y., Guo, L.: On the capacitated controller placement problem in software defined networks. IEEE Commun. Lett. **18**(8), 1339–1342 (2014)
5. Özsoy, F.A., Pınar, M.: An exact algorithm for the capacitated vertex p-center problem. Comput. Oper. Res. **33**(5), 1420–1436 (2006)
6. Yao, L., Hong, P., Zhang, W., Li, J.: Controller placement and flow based dynamic management problem towards SDN. In: IEEE International Conference on Communication Workshop (2015)
7. Hock, D., Gebert, S., Hartmann, M., Zinner, T., Tran-Gia, P.: POCO- framework for pareto optimal resilient controller placement in SDN-based core networks. In: Network Operations and Management Symposium (NOMS), 2014, pp. 1–2. IEEE (2014)

8. Hock, D., Hartmann, M., Gebert, S., Jarschel, M., Zinner, T., Tran-Gia, P.: Pareto-optimal resilient controller placement in SDN-based core networks. In: Teletra c Congress (ITC), 2013 25th International, pp. 1–9 (2013)
9. Algerian Research Network (arn.dz)
10. http://www.topology-zoo.org/maps/Arn.jpg

# Harnessing CNN and OPTICS Algorithm for Detecting Shilling Attacks in Recommender Systems

Ouahiba Belgacem[1,2]([✉]) [iD], Boudjemaa Boudaa[2,3] [iD], Abderrahmane Kouadria[2] [iD], and Abdelhafid Abouaissa[4] [iD]

[1] Laboratoire de Génie Energétique et Génie Informatique (L2GEGI), University of Tiaret, Tiaret, Algeria
[2] Computer Science Department, University of Tiaret, Tiaret, Algeria
`{ouahiba.belgacem,boudjemaa.boudaa,abderrahmane.kouadria}@univ-tiaret.dz`
[3] LabRI-SBA Lab., Ecole Supérieure en Informatique, Sidi Bel Abbes, Algeria
[4] IRIMAS, University of Haute-Alsace, Mulhouse, France
`abdelhafid.abouaissa@uha.fr`

**Abstract.** In the realm of collaborative filtering recommender systems, detecting shilling attacks, where malicious users artificially influence recommendations, is a critical challenge. Traditional methods often struggle with the complexity and high-dimensionality of feature spaces. To overcome these limitations, we harness the capabilities of CNN-OPTICS, a novel approach that integrates Convolutional Neural Networks (CNNs) with the OPTICS (Ordering Points To Identify the Clustering Structure) algorithm. CNNs are employed to extract deep, nuanced features from user interaction data, revealing subtle patterns and anomalies, while OPTICS clusters these extracted features, identifying meaningful clusters and detecting anomalies indicative of shilling attacks. This synergy between deep feature extraction and flexible clustering allows CNN-OPTICS to outperform traditional methods, providing better resistance to noise and enhanced detection of malicious behavior. Extensive evaluations on the MovieLens and Netflix datasets confirm that CNN-OPTICS delivers superior performance across various attack types, highlighting the potential of integrating deep learning with clustering techniques to secure such systems.

**Keywords:** Shilling Attacks · Convolutional Neural Networks · OPTICS · Recommender Systems · Collaborative filtering

## 1 Introduction

Collaborative filtering recommender systems play a crucial role in enhancing user experiences across digital platforms by predicting and suggesting items based on individual preferences and behaviors. These systems are integral to various applications, including e-commerce, streaming services, and social networks, where they drive user engagement and satisfaction [18]. Despite their

© The Author(s), under exclusive license to Springer Nature Switzerland AG 2025
N. Seddari and M. Redjimi (Eds.): ICMSCT 2024, CCIS 2606, pp. 513–528, 2025.
https://doi.org/10.1007/978-3-032-01922-6_42

advantages, recommender systems are not immune to manipulation, particularly through shilling attacks [19].

Shilling attacks involve malicious users injecting biased or fraudulent ratings to distort recommender system outcomes. These attacks can enhance the ratings of preferred items or diminish those of competitors [16]. They manifest in various forms: random attacks [24] create general noise with arbitrary ratings, average attacks [16] blend manipulated ratings with genuine feedback, bandwagon attacks [13] flood the system with positive ratings to falsely boost popularity, and AOP (Average Over-Promotion) attacks [25] subtly mix average ratings with occasional extremes to influence recommendations inconspicuously. Each attack type presents unique challenges for detection and mitigation.

In this paper, we propose CNN-OPTICS, a novel approach that combines CNNs with the OPTICS algorithm. CNNs are used to extract deep, nuanced features from user interaction data, capturing complex patterns and anomalies that traditional methods might overlook. By leveraging their ability to learn hierarchical feature representations, CNNs can effectively identify subtle variations in user behavior that may indicate the presence of shilling attacks. OPTICS, on the other hand, is a density-based clustering algorithm designed to identify clusters and anomalies in high-dimensional data. Unlike other clustering methods, OPTICS does not require specifying the number of clusters in advance and can handle varying densities within the data. By integrating CNNs with OPTICS, CNN-OPTICS leverages the feature extraction capabilities of CNNs to provide a detailed representation of user data, which OPTICS then uses to effectively identify meaningful clusters and detect anomalies. This combination enhances the robustness and accuracy of recommender systems, enabling them to better manage the noise introduced by different types of shilling attacks and improve overall detection performance.

The remainder of this paper is structured as follows: Sect. 2 provides an overview of collaborative filtering (CF) and examines the impact of shilling attacks on CF systems, then, Sect. 3 reviews related work in the field. In Sect. 4, the methodology is presented, detailing data preparation, feature extraction, feature clustering, and shilling attack detection processes. Section 5 describes the experimental setup, including the datasets, evaluation metrics, clustering analysis, and performance comparison with baseline methods. Finally, Sect. 6 provides a discussion, and Sect. 7 concludes the paper by summarizing the findings and proposing future research directions.

## 2    Preliminaries

Maintaining the effectiveness of collaborative filtering algorithms in the face of shilling attacks is a critical issue in recommender systems. This section delves into the concepts of collaborative filtering and examines the impact and challenges posed by shilling attacks within this framework.

## 2.1   Collaborative Filtering

Collaborative Filtering (CF) is a fundamental technique in recommender systems that predicts user preferences by analyzing patterns in user interactions with items [11]. The core idea behind CF is to leverage the historical behavior of users and items to make personalized recommendations. CF methods generally fall into two categories: user-based and item-based.

In user-based collaborative filtering [9], the system identifies users who have similar historical preferences and recommends items that these users have liked. This approach assumes that if users $u$ and $v$ have a history of similar ratings, they are likely to have similar preferences in the future. The prediction for user $u$ on item $i$ is often based on the ratings of items that user $u$ has liked, weighted by the ratings of other similar users.

Conversely, item-based collaborative filtering [1] focuses on finding similarities between items based on the ratings provided by users. The idea is to recommend items that are similar to those that a user has previously rated highly. For instance, if a user rates item $i$ highly, the system recommends other items that have been rated similarly by users who also liked item $i$.

## 2.2   Shilling Attacks in Collaborative Filtering

Collaborative filtering (CF) is a popular method in recommender systems that predicts user preferences based on historical interactions [14]. However, CF systems are susceptible to shilling attacks, where malicious users introduce deceptive data to manipulate recommendations [20]. They are classified into different types commonly found in the literature [19,26]:

- **Random Attacks:** Malicious users submit ratings with no systematic pattern, aiming to obscure true user preferences and confuse the recommendation system. This randomness dilutes the influence of genuine user data.
- **Average Attacks:** Attackers alter the average ratings of items by inflating or deflating them, which can skew the perceived popularity or quality of items, leading to biased recommendations.
- **Bandwagon Attacks:** By flooding the system with high ratings for certain items, attackers artificially boost the popularity of these items. This manipulation makes them appear more favorable and increases their likelihood of being recommended.
- **Average Over-Popular (AOP) Attacks:** Malicious users generate a high volume of biased ratings to increase the perceived popularity of specific items, making them appear more attractive than they are and distorting the recommendation outputs.

These shilling attack strategies pose significant challenges to the reliability of CF systems, requiring advanced techniques to effectively detect and mitigate their impact.

## 3   Related Work

The field of shilling attack detection in recommender systems has seen significant advancements, employing various methodologies. the study in [27] combined Hidden Markov Models with Hierarchical Clustering to identify suspicious user behaviors, while another introduced the CCPS model, leveraging soft co-clustering for enhanced detection [22]. Classification techniques, paired with k-means clustering, have also been utilized to target push and nuke attacks [6]. More recently, a three-stage detection method for group shilling attacks has been proposed, constructing a weighted user relationship graph and employing clustering to identify dense subgraphs, thereby improving detection accuracy [3]. Additionally, Graph Convolutional Networks (GCNs) have been employed to autonomously extract features for shilling group detection [21], while another study focused on shilling attack detection based on an improved clustering algorithm that uses information entropy to select features and calculate user similarity [12]. Another method in [10] involves the implementation of a two-stage detection strategy that incorporates the partial tagging of user profiles and the training of a detector using Graph Convolutional Networks (GCN).

In addition to these methods, Convolutional Neural Networks (CNNs) have demonstrated versatility in shilling attack detection. For instance, CNNs have been applied for outlier detection in e-commerce recommender systems [5] and unsupervised anomaly detection in collaborative filtering [23]. Other studies have integrated CNNs to detect hurried attacks by analyzing deviations from normal rating patterns [15] and for user classification [7], showcasing the adaptability of CNNs in this domain.

## 4   Methodology

In this study, we address the challenge of detecting shilling attacks in recommender systems by proposing CNN-OPTICS, a novel approach that integrates CNNs with the OPTICS algorithm. Our methodology involves several key steps, including data preparation, feature extraction with CNNs, and clustering with OPTICS. Figure 1 provides an overview of the architecture, depicting the model's structure for classifying users into genuine and fake groups.

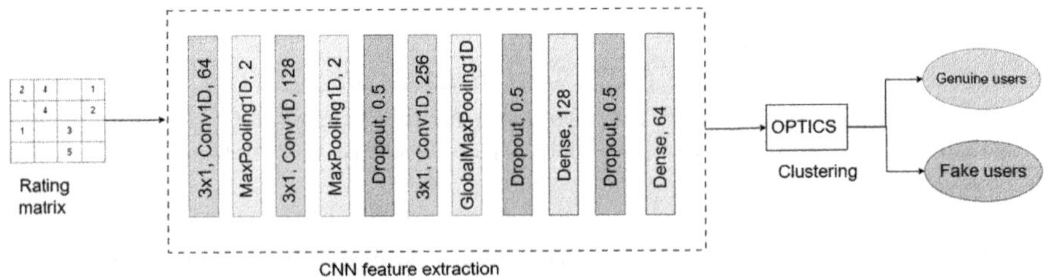


**Fig. 1.** Overview of the CNN-OPTICS architecture

## 4.1  Convolutional Neural Networks for Feature Extraction

Convolutional Neural Networks (CNNs) are adept at automatically learning relevant local and spatial features from the rating matrix due to their layered architecture, significantly reducing the need for manual feature engineering. They also require fewer computational resources and train faster, making them well-suited for the large datasets commonly found in recommendation systems. This architecture is particularly effective for analyzing user-item matrices, facilitating the recognition of complex patterns and anomalies in user interactions. The input for the CNN model is the rating matrix $\mathbf{R}$, represented as follows:

$$\mathbf{R} = \begin{bmatrix} R_{11} & R_{12} & \cdots & R_{1\,m} \\ R_{21} & R_{22} & \cdots & R_{2\,m} \\ \vdots & \vdots & \ddots & \vdots \\ R_{n1} & R_{n2} & \cdots & R_{nm} \end{bmatrix} \tag{1}$$

where $n$ is the number of users, $m$ is the number of items, and $R_{ui}$ denotes the rating given by user $u$ to item $i$.

The CNN architecture processes this rating matrix through several layers to enhance feature extraction. Convolutional layers apply a set of filters $\mathbf{K}$ to the input matrix $\mathbf{R}$. The convolution operation for a filter is expressed as:

$$(\mathbf{R} * \mathbf{K})(i,j) = \sum_{m=1}^{k} \sum_{n=1}^{k} R_{i+m-1,j+n-1} \cdot K_{m,n} \tag{2}$$

where $k$ is the filter size, and $(i,j)$ represents the position in the matrix. This operation detects patterns such as user behavior trends or item preferences.

Following the convolutional layers, pooling layers are employed to reduce dimensionality and computational cost. Max pooling, a common pooling technique, is calculated as:

$$\text{MaxPooling}(i,j) = \max_{m,n} \left( R_{i+m,j+n} \right) \tag{3}$$

where $m$ and $n$ are indices within the pooling window. This reduces the spatial dimensions of the data while preserving essential features.

Dense layers are then used to process the output from the pooling layers. Each neuron in a dense layer computes a weighted sum of its inputs followed by an activation function. ReLU (Rectified Linear Unit) is commonly used as the activation function:

$$\text{ReLU}(x) = \max(0, x) \tag{4}$$

where $x$ is the input to the neuron. ReLU introduces non-linearity into the model, enabling it to learn complex patterns in the data.

To measure the model's performance and incorporate regularization, the mean squared error (MSE) loss function is used:

$$L_{\mathrm{MSE}} = \frac{1}{N}\sum_{i=1}^{N}(R_i - \hat{R}_i)^2 \qquad (5)$$

where $R_i$ and $\hat{R}_i$ represent the true and predicted ratings, respectively.

Regularization is incorporated to avoid overfitting, modifying the loss function to:

$$L_{\mathrm{reg}} = L_{\mathrm{MSE}} + \lambda \sum_j W_j^2 \qquad (6)$$

where $\lambda$ is the regularization parameter and $W_j$ are the weights of the CNN layers. This regularization term helps control the complexity of the model and improve its generalization performance.

## 4.2   Clustering for Anomaly Detection

Clustering is a fundamental technique in unsupervised learning, which involves partitioning a set of data $\mathcal{X} = \{x_1, x_2, \ldots, x_n\}$ into subsets or clusters $C_1, C_2, \ldots, C_k$, where each cluster $C_i$ contains data points that are more similar to each other than to those in other clusters [17].

Clustering methods can be broadly categorized into partitioning, hierarchical, density-based, and model-based approaches, each tailored to different data characteristics, such as distribution, dimensionality, and the presence of noise [8]. The choice of clustering technique is thus guided by these attributes to effectively reveal the underlying structure of the data [4].

For anomaly detection, particularly in detecting shilling attacks in recommender systems, the features $\mathbf{F} = \{f_1, f_2, \ldots, f_m\}$ extracted by a Convolutional Neural Network (CNN) are clustered using the OPTICS algorithm. OPTICS (Ordering Points To Identify the Clustering Structure) [2] is a density-based clustering algorithm designed to discover clusters of varying densities without requiring a predefined number of clusters. It orders data points based on their reachability distance, defined as:

$$\text{reachability\_distance}(p) = \max\left(\text{core\_distance}(p), \min_{q \in \mathcal{N}_\epsilon(p)} \text{dist}(p, q)\right), \qquad (7)$$

where

$$\text{core\_distance}(p) = \min\{r : |\mathcal{N}_r(p)| \geq \text{MinPts}\}, \qquad (8)$$

is the minimum radius $r$ within which the point $p$ has at least $MinPts$ neighbors. $\mathcal{N}_\epsilon(p)$ represents the $\epsilon$-neighborhood around point $p$, and $\text{dist}(p, q)$ is the distance between points $p$ and $q$. The OPTICS algorithm generates a reachability plot by ordering data points according to their reachability distances, which helps in identifying clusters and anomalies. This method is particularly useful for detecting shilling attacks, as it allows distinguishing between genuine users and malicious users by analyzing the density and structure of the data.

### 4.3   Shilling Attack Detection

The strength of OPTICS lies in its ability to accommodate clusters with varying densities, which allows it to distinguish genuine clusters from noise more effectively.

In clustering analysis using OPTICS algorithm, two main groups usually emerge: genuine users and fake users. Genuine users, who represent the majority, tend to cluster into dense, compact groups. In contrast, fake users, often associated with shilling attacks, either appear as isolated points (considered noise) or form smaller, less dense clusters due to their anomalous behavior.

## 5   Experimental Results

The experiments were conducted using the MovieLens 100K and Netflix datasets. Both datasets were subjected to four types of shilling attacks: random, average, bandwagon, and aop. The primary objective of these experiments was to evaluate the performance of the proposed CNN-OPTICS model in detecting and clustering fake users across different attack strategies and datasets. For comparison, baseline models including CNN-KMeans, CNN-Agglomerative Clustering, and CNN-Spectral Clustering were used.

### 5.1   Datasets

To evaluate the effectiveness of our proposed CNN-OPTICS method, we used two datasets: MovieLens 100K[1] and Netflix[2]. The MovieLens 100K dataset consists of 943 users, 1,682 items, and 100,000 ratings, while the Netflix dataset includes 2,000 users, 3,500 items, and 1,000,000 ratings. Table 1 provides a summary of the descriptions of these datasets.

**Table 1.** Dataset Summary

|         | MovieLens 100K | Netflix |
|---------|----------------|---------|
| Users   | 943            | 2,000     |
| Items   | 1,682          | 3,500     |
| Ratings | 100,000        | 1,000,000 |

Data preprocessing involves constructing a user-item rating matrix from the dataset to serve as input for the CNN model. Specifically, the user-item rating matrices for these datasets are denoted as $R_{\text{MovieLens}} \in \mathbb{R}^{m \times n}$ and $R_{\text{Netflix}} \in \mathbb{R}^{m \times n}$, where $m$ represents the number of users and $n$ represents the number of items.

---

[1] https://grouplens.org/datasets/movielens.
[2] https://www.netflixprize.com.

## 5.2   Evaluation Metrics

To assess the performance of our CNN-OPTICS approach for detecting shilling attacks, we use several key evaluation metrics. These metrics provide a comprehensive evaluation of the model's classification accuracy, clustering effectiveness, and overall performance when compared with baseline models. Below, we describe the metrics used, including their definitions and the corresponding formulas.

**Accuracy:** Accuracy measures the proportion of correctly classified instances among all instances.

$$\text{Accuracy} = \frac{TP + TN}{TP + TN + FP + FN} \tag{9}$$

**Precision:** Precision measures the proportion of true positives among all instances classified as positive.

$$\text{Precision} = \frac{TP}{TP + FP} \tag{10}$$

**Silhouette Score:** The Silhouette Score evaluates the quality of clusters by measuring how similar an object is to its own cluster (cohesion) compared to other clusters (separation).

$$\text{Silhouette Score} = \frac{b - a}{\max(a, b)} \tag{11}$$

where:

- $a$ is the average distance between a data point and all other points in the same cluster.
- $b$ is the minimum average distance from the data point to points in a different cluster.

**Normalized Mutual Information (NMI):** NMI is a measure of similarity between two clustering results, normalized to account for chance. It ranges from 0 (no mutual information) to 1 (perfect correlation).

$$\text{NMI} = \frac{2 \times I(U, V)}{H(U) + H(V)} \tag{12}$$

where:

- $I(U, V)$ is the mutual information between clusters $U$ and $V$.
- $H(U)$ and $H(V)$ are the entropies of $U$ and $V$, respectively.

## 5.3  Hyperparameter Tuning

Tuning hyperparameters plays a critical role in optimizing the performance of the CNN-OPTICS model. The main hyperparameters that were tuned during the experiments include:

- **Learning Rate** for CNN training: The learning rate was set to 0.01 after experimenting with values in the range of 0.001 to 0.1.
- **OPTICS $\epsilon$ Parameter:** The $\epsilon$ parameter controls the neighborhood size in OPTICS. A value of $\epsilon = 0.09$ was found to provide the best clustering performance.
- **OPTICS MinPts Parameter:** The MinPts parameter, which defines the minimum number of points to form a cluster, was set to 10 based on grid search.

To further refine these hyperparameters, a grid search approach was applied. The table 2 summarizes the hyperparameter tuning results and the selected values for the final experiments.

**Table 2.** Hyperparameter Tuning Results for CNN-OPTICS Model

| Hyperparameter | Range Tested | Selected Value |
| --- | --- | --- |
| Learning Rate | 0.001–0.1 | 0.01 |
| OPTICS $\epsilon$ | 0.05–0.2 | 0.09 |
| OPTICS MinPts | 5–20 | 10 |

These optimized hyperparameters were crucial in achieving improved model performance and ensuring effective clustering outcomes.

## 5.4  Clustering Analysis

To visually assess the clustering performance of the proposed CNN-OPTICS model under different shilling attack strategies, we plotted the clusters generated for both the MovieLens and Netflix datasets across the four attack types: random, average, bandwagon, and AOP.

Figures 2, 3, 4, 5, 6, 7, 8 and 9 showcase the OPTICS clustering results for each combination of dataset and attack type. The following list details each figure and the specific conditions under which the clustering was performed:

- Figure 2: OPTICS clustering of the MovieLens dataset under Random Attack.
- Figure 3: OPTICS clustering of the MovieLens dataset under Average Attack.
- Figure 4: OPTICS clustering of the MovieLens dataset under Bandwagon Attack.
- Figure 5: OPTICS clustering of the MovieLens dataset under AOP Attack.
- Figure 6: OPTICS clustering of the Netflix dataset under Random Attack.

- Figure 7: OPTICS clustering of the Netflix dataset under Average Attack.
- Figure 8: OPTICS clustering of the Netflix dataset under Bandwagon Attack.
- Figure 9: OPTICS clustering of the Netflix dataset under AOP Attack.

In each plot, the separation between genuine and fake users highlights the clustering capabilities of CNN-OPTICS, with the small yellow clusters representing fake users. This highlights the effectiveness of CNN-OPTICS in achieving clear separations even under complex attack scenarios.

## 5.5   Performance Evaluation with Baselines

To provide a clear understanding of the effectiveness of the CNN-OPTICS model for detecting shilling attacks, Table 3 presents a detailed comparison of performance evaluation metrics across several baseline models, including CNN-

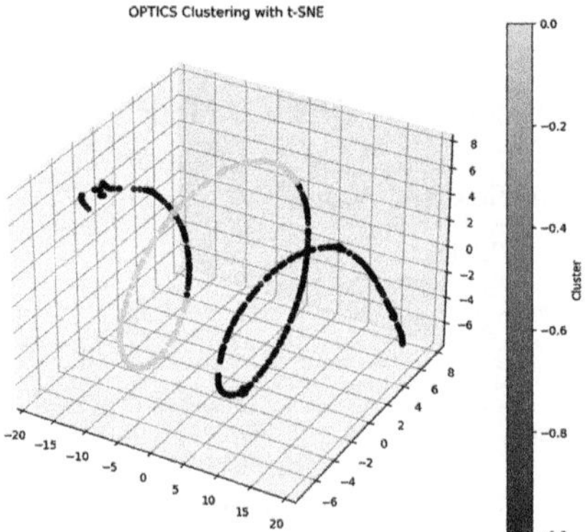

**Fig. 2.** MovieLens: Random Attack

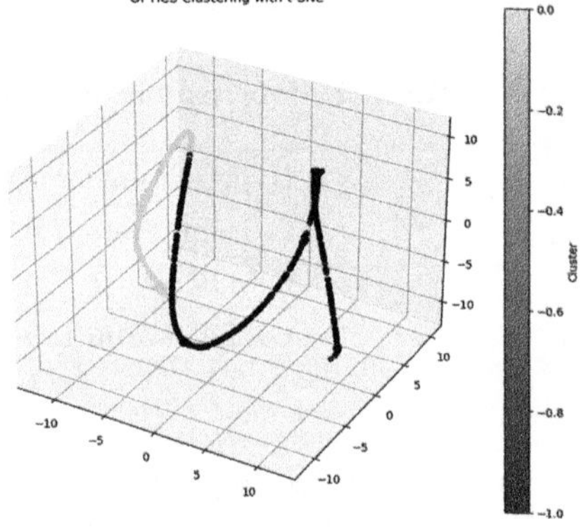

**Fig. 3.** MovieLens: Average Attack

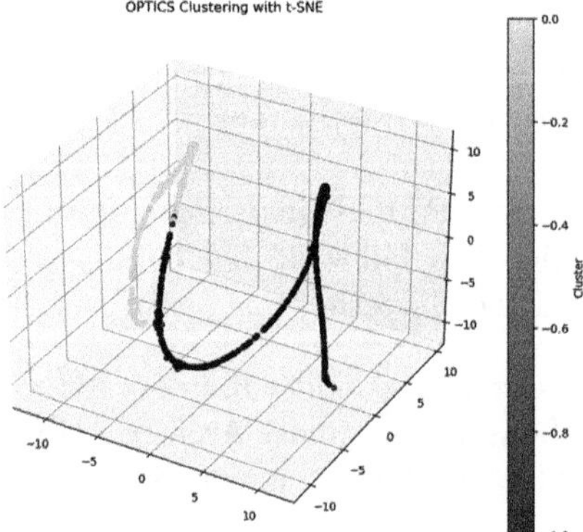

**Fig. 4.** MovieLens: Bandwagon Attack

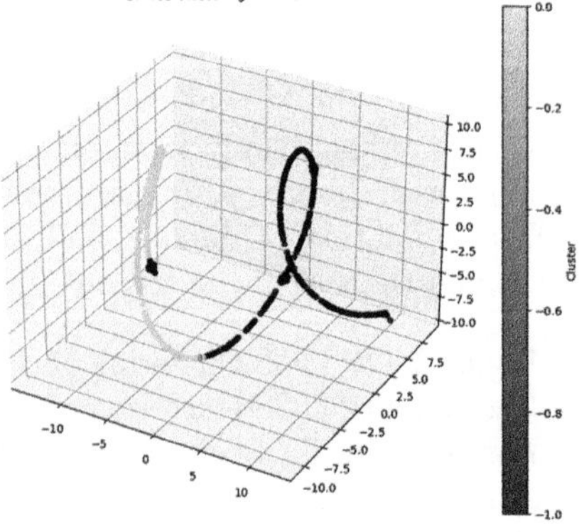

**Fig. 5.** MovieLens: AOP Attack

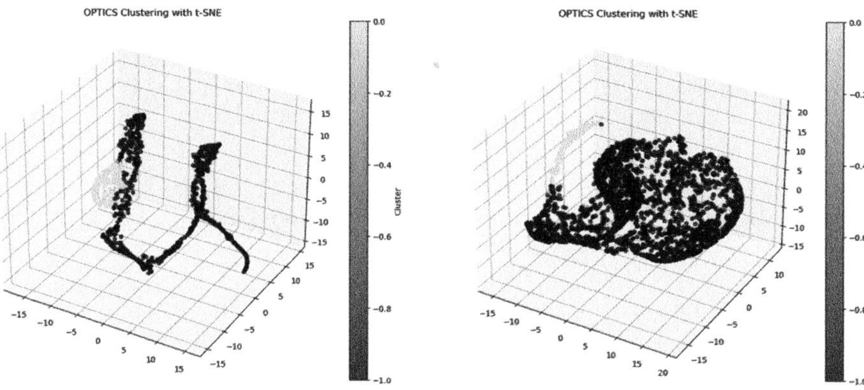

Fig. 6. Netflix: Random Attack    Fig. 7. Netflix: Average Attack

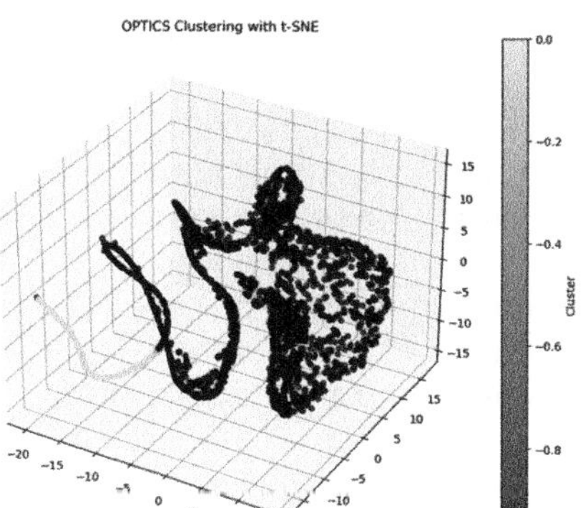
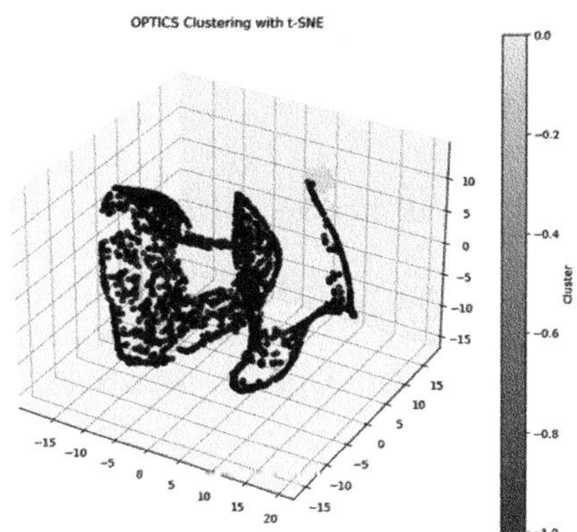

Fig. 8. Netflix: Bandwagon Attack    Fig. 9. Netflix: AOP Attack

KMeans, CNN-Agglomerative, and CNN-Spectral. The table covers key metrics such as Accuracy, Precision, NMI, and Silhouette Score for the MovieLens 100K and Netflix datasets, considering various attack types: Random, Average, Bandwagon, and AOP.

The baseline models were selected to ensure a diverse range of clustering techniques and are described as follows:

- **CNN-KMeans:** Combines CNN feature extraction with KMeans clustering, which assigns users to clusters based on centroids and minimizes within-cluster variance. This method is efficient for spherical clusters but may struggle with complex, non-linear patterns.
- **CNN-Agglomerative:** Integrates CNN with hierarchical agglomerative clustering, which iteratively merges clusters based on similarity until a stopping criterion is met. It offers flexibility by not requiring a predefined number of clusters, but is sensitive to noise and outliers.

- **CNN-Spectral:** Leverages CNN for feature extraction followed by spectral clustering, which uses graph-based methods to cluster users based on eigenvalue decomposition. Spectral clustering excels in detecting non-convex clusters but is computationally demanding and less scalable for large datasets.

**Table 3.** Performance Evaluation Metrics of CNN-OPTICS Compared to Baseline Models

| Dataset | Attack Type | Model | Metrics | | | |
|---|---|---|---|---|---|---|
| | | | Accuracy | Precision | NMI | Silhouette Score |
| MovieLens | Random Attack | CNN-OPTICS | 0.82 | **0.89** | **0.85** | **0.61** |
| | | CNN-KMeans | **0.87** | 0.80 | 0.78 | 0.55 |
| | | CNN-Agglomerative | 0.7 | 0.65 | 0.6 | 0.6 |
| | | CNN-Spectral | 0.8 | 0.79 | 0.5 | 0.45 |
| | Average Attack | CNN-OPTICS | **0.82** | **0.8** | **0.83** | **0.72** |
| | | CNN-KMeans | 0.44 | 0.35 | 0.6 | 0.11 |
| | | CNN-Agglomerative | 0.72 | 0.72 | 0.74 | 0.5 |
| | | CNN-Spectral | 0.51 | 0.55 | 0.42 | 0.33 |
| | Bandwagon Attack | CNN-OPTICS | **0.90** | **0.86** | **0.82** | 0.60 |
| | | CNN-KMeans | 0.68 | 0.71 | 0.75 | 0.55 |
| | | CNN-Agglomerative | 0.8 | 0.5 | 0.29 | 0.51 |
| | | CNN-Spectral | 0.4 | 0.6 | 0.17 | 0.28 |
| | AOP Attack | CNN-OPTICS | 0.87 | **0.9** | **0.81** | 0.56 |
| | | CNN-KMeans | **0.86** | 0.65 | 0.3 | **0.67** |
| | | CNN-Agglomerative | 0.77 | 0.65 | 0.29 | 0.51 |
| | | CNN-Spectral | 0.71 | 0.6 | 0.17 | 0.28 |
| Netflix | Random Attack | CNN-OPTICS | **0.95** | 0.91 | **0.77** | 0.54 |
| | | CNN-KMeans | 0.83 | 0.44 | 0.56 | 0.27 |
| | | CNN-Agglomerative | 0.89 | 0.79 | 0.51 | 0.3 |
| | | CNN-Spectral | 0.91 | **0.92** | 0.39 | **0.55** |
| | Average Attack | CNN-OPTICS | **0.98** | **0.93** | **0.95** | 0.55 |
| | | CNN-KMeans | 0.83 | 0.79 | 0.76 | **0.60** |
| | | CNN-Agglomerative | 0.93 | 0.91 | 0.61 | 0.43 |
| | | CNN-Spectral | 0.92 | 0.91 | 0.59 | 0.46 |
| | Bandwagon Attack | CNN-OPTICS | **0.97** | 0.9 | **0.81** | **0.69** |
| | | CNN-KMeans | 0.52 | 0.49 | 0.45 | 0.2 |
| | | CNN-Agglomerative | 0.94 | **0.91** | 0.72 | 0.3 |
| | | CNN-Spectral | 0.89 | 0.88 | 0.6 | 0.46 |
| | AOP Attack | CNN-OPTICS | **0.97** | **0.95** | **0.9** | 0.43 |
| | | CNN-KMeans | 0.59 | 0.62 | 0.31 | 0.46 |
| | | CNN-Agglomerative | 0.9 | 0.88 | 0.29 | **0.52** |
| | | CNN-Spectral | 0.94 | 0.86 | 0.45 | 0.47 |

These baseline models provide a comprehensive comparison to highlight the strengths of CNN-OPTICS, particularly in handling complex, high-dimensional data while detecting shilling attacks.

The Table 3 illustrates that CNN-OPTICS consistently outperforms the baseline models across multiple performance metrics, particularly in scenarios involving complex shilling attacks. The high NMI and Silhouette Scores achieved by CNN-OPTICS indicate that it not only accurately identifies fake users but also forms well-defined, cohesive clusters. The baseline models, while occasionally showing competitive Accuracy, often struggle with Precision and NMI, resulting in less reliable clustering and higher chances of misclassifying users. Overall, the results confirm that CNN-OPTICS provides a more robust and effective solution for detecting shilling attacks in recommender systems, offering significant improvements compared to the baseline models.

## 5.6   Computational Complexity

The complexity of the CNN used in our model can be expressed as $O(k \times n \times d^2)$, where $k$ is the kernel size of the convolutions, $n$ represents the number of users, and $d$ is the representation dimension (reflecting the number of items or features). This complexity arises from the need to process the user-item matrix, which consists of $n$ users by $m$ items. In contrast, the OPTICS clustering algorithm has a complexity of $O(n^2)$, where $n$ indicates the number of data points (number of users). Therefore, the total computational complexity of the CNN-OPTICS model can be expressed as the sum of these complexities:

$$O(k \times n \times d^2 + n^2)$$

In large datasets, the quadratic term $O(n^2)$ from OPTICS typically dominates the overall complexity, particularly when $m$ and $k$ are considerably smaller than $n$. Thus, the complexity can be approximated as $O(n^2)$, highlighting a potential computational bottleneck from the clustering step. Despite this, OPTICS offers the advantage of identifying clusters with varying densities, making it particularly effective for detecting subtle patterns in data, such as shilling attacks.

## 6   Discussion

Our experiments demonstrate that CNN-OPTICS significantly outperforms the baseline models in detecting shilling attacks across both the MovieLens and Netflix datasets. One of the standout features of CNN-OPTICS is its ability to handle clustering without requiring a predefined number of clusters. This is a major advantage over models like CNN-KMeans, CNN-Agglomerative, and CNN-Spectral, which necessitate specifying the number of clusters and can lead to the misidentification of genuine users into multiple, dispersed clusters.

OPTICS' density-based approach allows CNN-OPTICS to adapt to the varying densities of clusters and effectively identify noise, leading to a clearer and more accurate separation between fake and genuine users. Consequently, CNN-OPTICS not only enhances detection performance but also provides a more coherent and reliable clustering of users, addressing limitations observed in traditional clustering methods.

The sensitivity of OPTICS to its parameters, such as `MinPts` and $\epsilon$, played a crucial role in optimizing clustering performance. Careful tuning of these parameters was essential for maximizing the model's ability to handle different cluster densities and accurately detect shilling attacks. This flexibility underscores the importance of parameter selection in achieving robust anomaly detection and highlights CNN-OPTICS's potential for more effective shilling attack detection in diverse contexts.

## 7   Conclusion

In conclusion, this paper presented CNN-OPTICS, a novel approach that integrates CNNs with the OPTICS clustering algorithm to enhance the detection of shilling attacks in recommender systems. By leveraging CNNs for feature extraction and OPTICS for adaptive clustering, our method effectively identifies hidden patterns and anomalies without the need for predefined cluster numbers or uniform data densities. The experimental results demonstrate that CNN-OPTICS outperforms traditional techniques in terms of robustness and accuracy, providing superior resistance to malicious noise and a more precise detection of fraudulent activities.

This study advances the field of recommender system security and illustrates the benefits of integrating deep learning techniques with clustering algorithms. Moving forward, future research could focus on optimizing the CNN-OPTICS framework and exploring its application to other types of adversarial attacks, potentially broadening its impact across various fields and applications.

## References


1. Ajaegbu, C.: An optimized item-based collaborative filtering algorithm. J. Ambient. Intell. Humaniz. Comput. **12**(12), 10629–10636 (2021)
2. Bhattacharjee, P., Mitra, P.: A survey of density based clustering algorithms. Front. Comput. Sci. **15**, 1–27 (2021)
3. Cai, H., Zhang, F.: An unsupervised approach for detecting group shilling attacks in recommender systems based on topological potential and group behaviour features. Secur. Commun. Netw. **2021**(1), 2907691 (2021)
4. Cappozzo, A., Escudero, L.A.G., Greselin, F., Mayo-Iscar, A.: Parameter choice, stability and validity for robust cluster weighted modeling. Stats 4(3), 602–615 (2021)
5. Chakraborty, S., Karforma, S.: Outlier detection in e-commerce recommender systems using CNN. Comput. Secur. **51**, 169–183 (2015)
6. Davoudi, M., Chatterjee, P.: Detecting push and nuke attacks using classification and clustering. Expert Syst. Appl. **73**, 174–186 (2017)

7. Ebrahimian, M., Kashef, R.: Efficient detection of shilling's attacks in collaborative filtering recommendation systems using deep learning models. In: 2020 IEEE International Conference on Industrial Engineering and Engineering Management (IEEM), pp. 460–464. IEEE (2020)

8. Ezugwu, A.E., et al.: A comprehensive survey of clustering algorithms: state-of-the-art machine learning applications, taxonomy, challenges, and future research prospects. Eng. Appl. Artif. Intell. **110**, 104743 (2022)

9. Fkih, F.: Similarity measures for collaborative filtering-based recommender systems: review and experimental comparison. J. King Saud Univ.-Comput. Inf. Sci. **34**(9), 7645–7669 (2022)

10. Hao, Y., Meng, G., Wang, J., Zong, C.: A detection method for hybrid attacks in recommender systems. Inf. Syst. **114**, 102154 (2023)

11. Koren, Y., Rendle, S., Bell, R.: Advances in collaborative filtering. In: Recommender Systems Handbook, pp. 91–142 (2021)

12. Ma, Y., Su, X., Zheng, W., Liu, L.: Shilling attack detection based on improved clustering algorithm. In: Third International Symposium on Computer Engineering and Intelligent Communications (ISCEIC 2022), vol. 12462, pp. 666–670. SPIE (2023)

13. Moradi, R., Hamidi, H.: A new mechanism for detecting shilling attacks in recommender systems based on social network analysis and gaussian rough neural network with emotional learning. Int. J. Eng. **36**(2), 321–334 (2023)

14. Nadeem, R., Sivakumar, T.: A systematic literature survey on recommendation system (2023)

15. Panagiotakis, S., Papadakis, S., Koutoupis, G.: Enhanced shilling attack detection with CNNs and randomness in item selection (RIS). Data Min. Knowl. Disc. **34**, 215–229 (2020)

16. Rani, S., Kaur, M., Kumar, M., Ravi, V., Ghosh, U., Mohanty, J.R.: Detection of shilling attack in recommender system for YouTube video statistics using machine learning techniques. Soft Comput. **27**(1), 377–389 (2023)

17. Ren, Y., et al.: Deep clustering: a comprehensive survey. IEEE Trans. Neural Netw. Learn. Syst. (2024)

18. Roy, D., Dutta, M.: A systematic review and research perspective on recommender systems. J. Big Data **9**(1), 59 (2022)

19. Si, M., Li, Q.: Shilling attacks against collaborative recommender systems: a review. Artif. Intell. Rev. **53**, 291–319 (2020)

20. Sundar, A.P., Li, F., Zou, X., Gao, T., Russomanno, E.D.: Understanding shilling attacks and their detection traits: a comprehensive survey. IEEE Access **8**, 171703–171715 (2020)

21. Wang, S., Zhang, P., Wang, H., Hongtao, Yu., Zhang, F.: Detecting shilling groups in online recommender systems based on graph convolutional network. Inf. Process. Manage. **59**(5), 103031 (2022)

22. Yang, M., Zhang, L., Wang, H.: CCPS: a soft co-clustering method for detecting shilling attacks. Inf. Sci. **385**, 138–155 (2017)

23. Yang, Z., Wu, J., Liu, X.: Unsupervised anomaly detection using CNNs in collaborative filtering. IEEE Trans. Knowl. Data Eng. **28**(12), 3164–3176 (2016)

24. Zayed, R.A., Ibrahim, L.F., Hefny, H.A., Salman, H.A., AlMohimeed, A.: Experimental and theoretical study for the popular shilling attacks detection methods in collaborative recommender system. IEEE Access **11**, 79358–79369 (2023)

25. Zhang, F., Chan, P.P.K., He, Z.-M., Yeung, D.S.: Unsupervised contaminated user profile identification against shilling attack in recommender system. Intell. Data Anal. (Preprint), 1–16 (2024)

26. Zhang, F., Zhang, Z., Zhang, P., Wang, S.: UD-HMM: an unsupervised method for shilling attack detection based on hidden Markov model and hierarchical clustering. Knowl.-Based Syst. **148**, 146–166 (2018)
27. Zhang, X., Li, Y., Zhang, H.: Detecting shilling attacks using hidden Markov model and hierarchical clustering. Knowl.-Based Syst. **153**, 166–175 (2018)

# Time Series-Based Predictive Monitoring of Mean Blood Pressure in Intensive Care: A Comparative Study of Multiple Linear Regression and LSTM Models


Houcine Aidoun[1(✉)], Fatiha Barigou[1], Zakaria Zine El Abidine Hachemi[1], and Baghdad Atmani[2]

[1] Oran Computer Laboratory (LIO), University of Oran1, Oran, Algeria
aidounefr@yahoo.fr, fatbarigou@gmail.com,
hachemizakariazineelabidine@gmail.com
[2] University of Mostaganem, Mostaganem, Algeria
baghdad.atmani@gmail.com



**Abstract.** Intensive Care Units (ICUs) provide continuous monitoring of severely ill patients who may suffer from numerous health complications that affect morbidity and mortality. For clinicians, interpreting data in real time and making decisions is a challenging task. This study is particularly focused on the prediction and monitoring of Mean Blood Pressure (MBP) as a critical physiological parameter. Our approach included the development of a novel data acquisition system that collects real-world time series data from Dash 4000 monitors connected to patients in intensive care units. The use of real clinical data, cleaned and processed for analysis, distinguishes this work from others, enhancing the practicality of our forecasting models in real-world settings. These data were used to develop two predictive models: one based on Long Short-Term Memory (LSTM) networks and another using Multiple Linear Regression. We applied machine learning and deep learning techniques to predict MBP over various time windows, leveraging the multivariate nature of the collected data. The results demonstrate that the Multiple Linear Regression model outperformed the LSTM model. For example, in the 1-min prediction window, the Multiple Linear Regression model achieved an RMSE of 0.45605 and an $R^2$ score of 0.99053, demonstrating better performance compared to the LSTM model. This system offers promising potential for improving the real-time prediction and monitoring of mean blood pressure in critically ill patients, with practical implications for intensive care settings.

**Keywords:** Intensive Care Unit · real-time vital signs · mean blood pressure · time series · machine learning · deep learning · LSTM · MLR





N. Seddari and M. Redjimi (Eds.): ICMSCT 2024, CCIS 2606, pp. 529–542, 2025.
https://doi.org/10.1007/978-3-032-01922-6_43

## 1    Introduction

The medical intensive care unit (MICU) of Oran hospital (EHU) undertakes continuous monitoring of critically ill patients at risk of severe complications. This is achieved using monitors such as the DASH 4000, which track various physiological parameters. The most monitored physiological data are vital signs, for example: heart rate, respiratory rate, body temperature, blood oxygen saturation, and blood pressure. Those parameters require continuous monitoring to assess the patient's condition. Monitors are constantly generating data every two seconds; as a result, it is difficult for physicians to analyze this massive data in real time to perform timely actions during emergencies. Traditional methods are often limited and time consuming, making it difficult to detect new patterns in vital signs that could lead to earlier identification of conditions such as cardiac problems and sepsis. To address this issue, and building on related work in this context, this paper presents two main contributions. The first contribution involves the development of an acquisition system capable of efficiently collecting and storing large amounts of physiological data from DASH 4000 monitors used in intensive care units. The second contribution focuses on predicting blood pressure problems using the collected data. To achieve this, we will employ machine learning and deep learning techniques, utilizing time series models to analyze the data.

In this study, we utilized time series data, which consists of a sequence of observations indexed by time, typically recorded at successive and often equally spaced intervals [13]. This approach is particularly effective for modeling and forecasting vital sign data, as the temporal nature of these measurements allows for the identification of trends and patterns over time. One of the reasons we chose time series is precisely their ability to forecast future values based on historical data. Generally, time series are divided into two categories, Univariate and Multivariate time series. Univariate data involves single variable observation over time while multivariate time series consist of multiple variables observed simultaneously at each time interval [1]. Numerous algorithms are employed for time series prediction, including Deep Neural Network techniques (GRUs, RNNs, and LSTM) as well as traditional machine learning methods like SVM, Linear Regression (LR), etc. In this paper, we will employ supervised machine and deep learning approaches using Multiple Linear Regression and LSTM models, to analyze real-time data collected from the DASH 4000 monitors. Our analysis will concentrate on forecasting the blood pressure parameter within the multivariate time series data. The rest of the paper is organized as follows. In Sect. 2, we present the literature that has used machine and deep learning models for clinical prediction tasks in the Intensive Care Unit. In Sect. 3, the methodology used in this study is explained. The experimental and the results are given in Sect. 4, and finally, Sect. 5 wraps up the paper.

## 2    Related Work

To assist doctors in making decisions under stress, the work in [1] proposed using deep learning techniques like Long Short-Term Memory (LSTM) and Recurrent Neural Networks (RNN) for predicting multivariate time series with many times-tamped features. The authors also compared the two models and concluded that while predicting multivariate time series is challenging, the LSTM model performs significantly better than the simple RNN.

Recent studies highlight the importance of real time monitoring in ICUs. For instance, a review paper by [12] points out that continuous data collection helps doctors respond more quickly and prevent complications. The review also discusses the growing use of advanced techniques, such as deep learning, in predictive analytics. According to the authors, these techniques are expected to become increasingly common in analyzing health data and identifying potential risks. This review emphasizes the trend towards using advanced methods to improve predictive accuracy and detect anomalies earlier.

In reference [2], the authors propose a heterogeneous architecture combining LSTM and CNN to predict ICU readmissions using the publicly available MIMIC-III dataset. They suggest that their model can aid in resource planning and reduce mortality and length of stay in the ICU. The authors also report that their approach outperforms traditional machine learning methods, such as Support Vector Machines, Random Forests, and Logistic Regression. Their model shows superior AUC, precision, and accuracy.

The work in [10] presents a convolutional neural network based-model to diagnose gallbladder stones by analyzing big medical data from the Internet of Things (IoT). The model aims to identify the chemical composition of gallstones from ultrasound, CT, and MRI images, addressing the challenge of accurate classification and treatment recommendations. The approach leverages IoT data to enhance diagnosis and treatment in smart health systems.

Time series prediction in medical settings has been extensively explored using various algorithms and methodologies. In [5], the authors developed a logistic regression algorithm that uses just 5 min of past physiological data to predict hypotensive events within the next 30 min. They used the MIMIC III database, including six physiological signals: ABP (arterial blood pressure), HR (heart rate), Sys (systolic blood pressure), Dia (diastolic blood pressure), Resp (respiration rate), and SpO2 (peripheral capillary oxygen saturation) for feature extraction. They then derived five additional numerical signals: PP (pulse pressure), MAP (mean arterial pressure), CO, MAP2HR and the mean RR by mathematical formulas. The proposed algorithm was able to detect hypotensive events with 94% accuracy, 85% sensitivity, and 96% specificity.

Ref [4] presented a new model of chronological series of self-regressive events that has the ability to predict the future of multivariate clinical events. More specifically, data on the distant past was calculated using the hidden state space defined by a LSTM-based model. They demonstrated that the new model with the time mechanisms mentioned above improves prediction performance compared to several base lines.

In [3], the authors proposed an efficient real time cardiovascular prediction system where KNN outperformed other algorithms like random forest, decision tree, SVM, and naive Bayes in accuracy, sensitivity, and failure rates. A prototype with sensors was developed to monitor and predict heart disease, achieving 88.52% accuracy. Their model uses KNN to classify patients as "at risk" or "not at risk" of heart disease.

In [11], the authors developed a real time interpretable model using XGBoost to predict cardiac arrest in critically ill patients based on vital signs from the MIMIC-III database. The model predicted 80% of events over 25 min in advance, achieving an AUROC of 0.94, with a sensitivity of 0.86, specificity of 0.85, F1 score of 0.05, and an AUPRC of 0.12.

Ref [8] presented an effective method for detecting fitness states that combines machine learning models including K-Nearest Neighbors, Random Forest algorithm, and other ensemble learning techniques. This approach focuses on leveraging real-time data and machine learning models to forecast cardiac issues. The authors conclude that when employing an ensemble learning strategy on the data, stacking CV classifier with KNN as meta classifier performs better than other classifiers employed in the study, achieving an accuracy rate of 93.39%.

In a study conducted in [6], the optimal machine learning algorithm for forecasting prescription errors in intensive care units was selected through the application of machine learning classification techniques. KNN has the best accuracy (99.13%).

The works reviewed demonstrate significant advancements in using machine learning and deep learning models for real time monitoring and prediction in critical care settings. From cardiovascular risk assessment to ICU readmission predictions, these studies highlight the potential of advanced algorithms to improve patient outcomes. In the next section, we will present our methodology for developing a predictive model tailored to predicting the occurrence of blood pressure problems in ICU patients.

## 3    Methodology

This section outlines the methodology used to develop our proposed system for forecasting mean blood pressure using time series modeling and machine learning, including deep learning techniques. We present a comprehensive framework that covers data collection, storage, and prediction, addressing key challenges to facilitate accurate and timely healthcare decision-making.

### 3.1    System Architecture

The proposed system architecture consists of four key components, each with a specific purpose (see Fig. 1).

- The first component is the data acquisition system, which is responsible for collecting data from various monitoring equipment. This includes real time

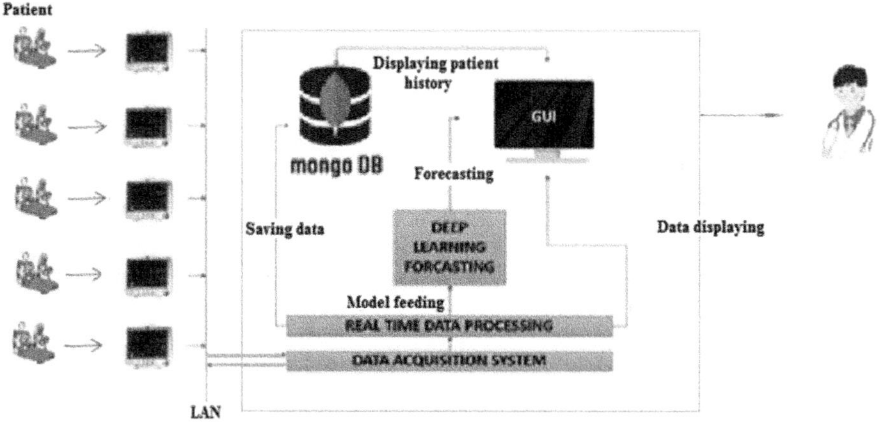


**Fig. 1.** System Architecture

data preprocessing where we convert raw hexadecimal data into meaningful numeric and text values.

- The second component is the storage system, which plays a vital role in efficiently storing the captured data. We use the MongoDB database because of its scalability and flexibility, which allows us to efficiently store and retrieve data for later analysis.
- The third component is the prediction engine, which uses the pre-processed data to predict future values. This predictive capability is particularly valuable because it allows us to predict potential anomalies or critical events. If the predicted value falls outside the expected range, the system immediately triggers an alarm to alert healthcare professionals.
- Finally, the system includes an easy-to-use GUI (Graphical User Interface) that allows real-time visualization of historical health data and alerts. The GUI acts as a central hub for healthcare professionals to monitor patient health, track trends, and respond quickly to any deviations or alerts.

### 3.2  Data Acquisition

The purpose of the data collection component is to gather information from various monitors, specifically the DASH 4000, connected to ICU patients. To facilitate this process, we developed a Python script that manages data acquisition, decodes information into numerical values or signals, and store it in a MongoDB database for further analysis.

As shown in Fig. 2, our system starts by connecting to all ICU monitoring devices. Once communication is established, it acquires vital signs and waveform data from the devices. Data packets are filtered for integrity, with those starting with "172" containing vital signs and others including detailed wave vital signs. We then decode the hexadecimal data into numeric values for analysis. This real

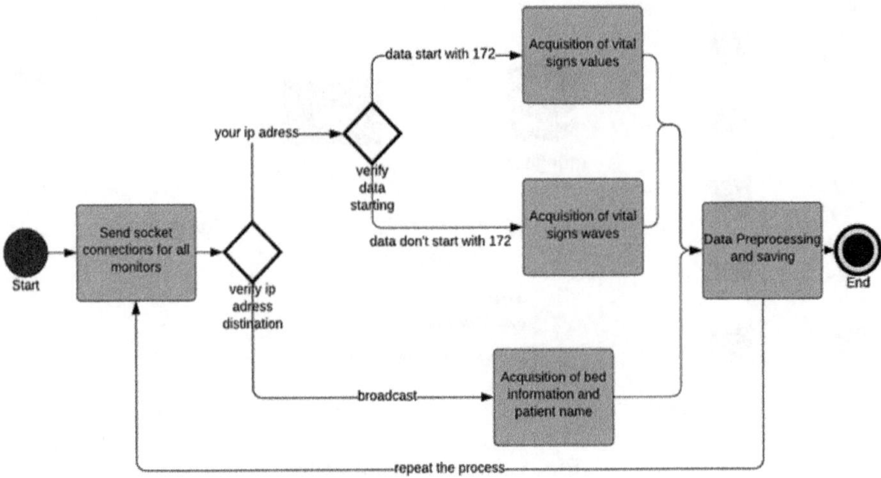


**Fig. 2.** Sequence diagram of the acquisition Component

time process enables continuous monitoring and provides healthcare professionals with accurate, up-to-date patient information for timely decision-making.

### 3.3   Storage

We chose MongoDB as our non-relational database due to its flexibility and scalability. Unlike traditional databases that utilize fixed tables, MongoDB employs
a collections and documents model, enabling adaptable schema design that can
accommodate diverse data structures. This approach facilitates efficient organization and storage of the large volumes of data generated by our system.
Additionally, MongoDB's distributed architecture supports scalability and high
throughput. Our global database schema, detailed in Fig. 3, is designed to effectively manage and store medical data from the ICU.

### 3.4   Data Processing

The data used in this work are vital sign values extracted from the non-relational
database MongoDB, in which, different types of patient related information are
stored. This data is loaded in files in CSV format. Each of these files contains
more than 40,000 observations and 17 features including 16 numerical values of
vital signs and one of Data time type. Table 1 show the vital features exploited
in this work.

The second step, referred to as preprocessing, was cleaning up the data. During this phase, we used the patient identification number to match the collected
attributes from several patient sheets to the patient. After that, a feature range

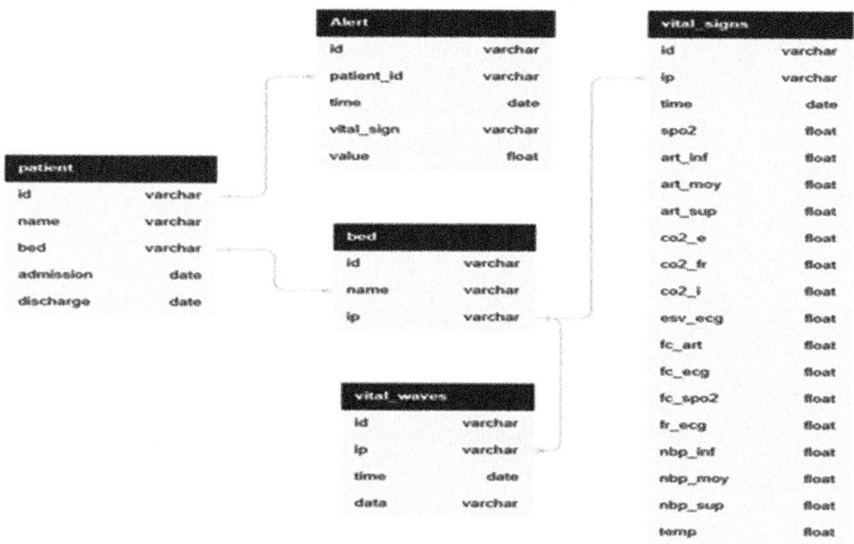

**Fig. 3.** Proposed schema for storing medical data from the ICU of Oran EHU

was defined by eliminating the useless information, such as the time and the number of lines. Ultimately, the null data were replaced with 0. After the pre-processing step, This dataset was separated into 84% training data and 16% test data. As shown in Table 2, six (6) CSV files were used for training and one (1) file for testing.

## 3.5   Development of the Model

We explore continuous monitoring of hospitalized patients' vital signs to predict future values and improve upon the conventional methods used in the ICU at EHU Hospital. Our goal is to develop a model that enables early detection of health deterioration by forecasting vital signs. For that, we used time series. Time series analysis is particularly suitable for modeling and forecasting data that is collected sequentially over time, making it a valuable approach for analyzing vital signs data.

Time series forecasting is particularly suited for predicting future values based on historical data. By using past observations, we can predict vital sign values, enhancing medical monitoring, anomaly detection, and decision-making. In this work, we model the time series prediction problem as a regression task and compare the performance of a deep learning LSTM model (Fig. 4) with traditional Multiple Linear Regression.

In this study, we utilize both Multiple Linear Regression and Long Short-Term Memory (LSTM) networks for time series prediction, as these models are well suited for regression tasks involving numerical outcomes. We constructed the models using real data collected from Dash monitors connected to ICU patients,

**Table 1.** Physiological characteristics used for time series prediction

| Feature | Description |
| --- | --- |
| Spo2 | Oxygen Saturation |
| Fc_spo2 | Heart Rate from SpO2 |
| Fc_ecg | Heart Rate from ECG |
| Esv_ecg | Ectopic Supraventricular Beats |
| Fr_ecg | Respiratory Rate from ECG |
| Co2_e | End-Tidal CO2 |
| Co2_i | Inspired CO2 |
| Co2_fr | Respiratory Rate from CO2 |
| Temp | Temperature |
| Nbp_sup | Systolic Blood Pressure |
| Nbp_inf | Diastolic Blood Pressure |
| Nbp_moy | Mean Blood Pressure (MBP) |
| Art_sup | Systolic Arterial Pressure |
| Art_inf | Diastolic Arterial Pressure |
| Art_moy | Mean Arterial Pressure (MAP) |
| Fc_art | Heart Rate from Arterial Pressure |

**Table 2.** The volume of data for train and test

| Number of recordings for train | Number of recordings for test |
| --- | --- |
| 275058 (84%) | 45822 (16%) |

where each observation is characterized by 16 variables: 15 vital signs and one time variable (see Table 1). The output variable of interest is the mean blood pressure (MBP). We assess the effectiveness of both models using various performance metrics, framing the time series prediction problem as a regression task to accurately predict MBP values over specified prediction windows.

## Multiple Linear Regression (MLR)

This statistical method aims to model the linear relationship between a dependent variable and multiple independent variables by fitting a linear equation to observed data as represented by Eq. (1). In our case, the goal is to predict the value of MBP based on the values of the others vital signs such as oxygen saturation, heart rate, and more.

$$Y = \beta_0 + \beta_1 X_1 + \beta_2 X_2 + \dots \beta_n X_n \tag{1}$$

where :

- Y is the dependent variable (the variable we want to predict, in our case MBP).
- $\beta_0$ is the intercept (the value of Y when all X variables are zero).
- $\beta_1$, $\beta_1$,...,$\beta_n$ are the regression coefficients that represent the impact of each inde-pendent variable on Y.
- $X_1$, $X_2$,..., $X_n$ are the independent variables (in our case the 15 vital parameters we are using as inputs, such as Oxygen Saturation , heart rate, respiratory rate, temperature, etc.).

In order to predict mean blood pressure (MBP), we used the scikit-learn Python library, which offers a practical implementation of linear regression through the Linear Regression class.

**Long Short-Term Memory (LSTM)**

LSTM is a type of recurrent neural network layer that is well suited for modeling sequential data. It can capture temporal dependencies and handle long-term memory. The number of parameters in an LSTM layer depends on the number of hidden units and the input sequence length.

As shown in Fig. 4, the architecture consists of three LSTM layers followed by two dense layers. Each LSTM layer takes an input sequence of length 300 and processes it sequentially. The first LSTM layer has 100 hidden units, the second LSTM layer also has 100 hidden units, and the third LSTM layer has 50 hidden units.

```
Layer (type)                    Output Shape                Param #
=================================================================
lstm_6 (LSTM)                   (None, 300, 100)            40800

lstm_7 (LSTM)                   (None, 300, 100)            80400

lstm_8 (LSTM)                   (None, 50)                  30200

dense_4 (Dense)                 (None, 8)                   408

dense_5 (Dense)                 (None, 1)                   9

=================================================================
Total params: 151,817
Trainable params: 151,817
Non-trainable params: 0
```

**Fig. 4.** LSTM model architecture

## 4 Results and Discussion

In this section, we evaluate the performance of Multiple Linear Regression and the deep learning model LSTM for predicting the patient's Mean Blood Pressure

(MBP). The models were developed and compared using two key metrics: Root Mean Square Error (RMSE) and the Coefficient of Determination ($R^2$ score). These metrics are essential for assessing how closely the prediction results align with the actual data.

- The Root Mean Square Error (RMSE), also known as the Root Mean Squared Deviation, is the square root of the mean of the squared differences between predicted and actual values [9]. In other words, RMSE represents the standard deviation of the prediction errors. It measures how closely the predicted values align with the actual data, indicating how well the line of best fit represents the dataset (see Eq. 2).

$$\text{RMSE} = \sqrt{\frac{1}{N} \sum (y - \widehat{y})^2} \tag{2}$$

- R-Squared ($R^2$) âĂŞ $R^2$, also known as the Coefficient of Determination, indicates the proportion of variance in the dependent variable that can be explained by the independent variables [7,9]. It provides a measure of the goodness of fit, reflecting how well the observed values align with the predicted values in the regression model. The formula of R-squared is provided in Eq. 3, where n denotes the number of predictions, y represents the observed values, $\hat{Y}$ indicates the predicted values, and $\bar{Y}$ is the mean of the observations.

$$\text{Score R2} = 1 - \frac{\sum_1^n \left(y_i - \hat{y}_i\right)^2}{\sum_1^n \left(y_i - \overline{y}\right)^2} \tag{3}$$

The entire dataset is divided into training and testing sets, with 84% allocated for training and 16% for testing. Specifically, 275058 data points are used for training, while 45822 data points are reserved for testing. We employed a time series forecasting strategy during the training phase, considering prediction windows of 1 min, 15 min, and 30 min in advance. Table 3 presents the results of the model evaluation based on RMSE and $R^2$ metrics.

**Table 3.** The performance evaluation of models for predicting MBP parameter

| Models | Prediction windows | | | | | |
| --- | --- | --- | --- | --- | --- | --- |
| | 1 min | | 15 min | | 30 min | |
| | RMSE | R2 score | RMSE | R2 score | RMSE | R2 score |
| Multiple linear regression | 0.45605 | 0.99053 | 0.45940 | 0.99054 | 0.46147 | 0.99055 |
| LSTM | 0.24 | 0.94143 | 0.46 | 0.84623 | 2.28 | 0.88876 |

Figures 5 and 6 shows the actual and predicted data of the MBP feature by applying the Multiple linear regression model and LSTM respectively. In Fig. 5 the model of the Multiple linear regression model gave an RMSE value 0.45605,

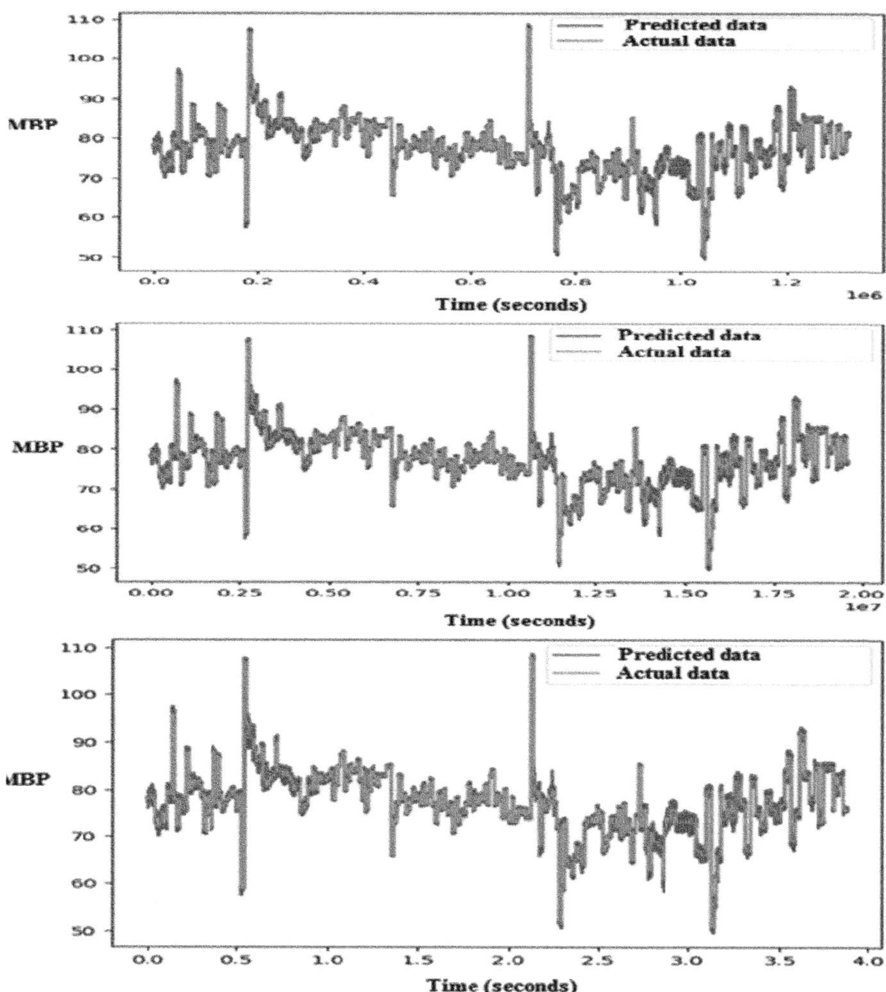



**Fig. 5.** Multivariate time series prediction with Multiple linear regression; Figures (1), (2) and (3) show prediction one minute, 15 min and 30 min in advance respectively.

0.45940, 0.46147 and R2 score value 0.99053, 0.99054 and 0.99055 in the cases of the prediction window of one minute, 15 min and 30 min respectively. Regarding the LSTM model in Fig. 6, we used 20 epochs and a batch size of 40. This model gave an RMSE value 0.24, 0.46, 2.28 and R2 score value 0.94143, 0.84623 and 0.88876 in the cases of the prediction window of one minute, 15 min and 30 min respectively.

Using the data Forecasting method for these two models, we observe that as the prediction window increases, the Root Mean Square Error also increases, while the $R^2$ score improves, reaching its optimal performance at a 30-min prediction window. This advance forecasting time is sufficient for timely

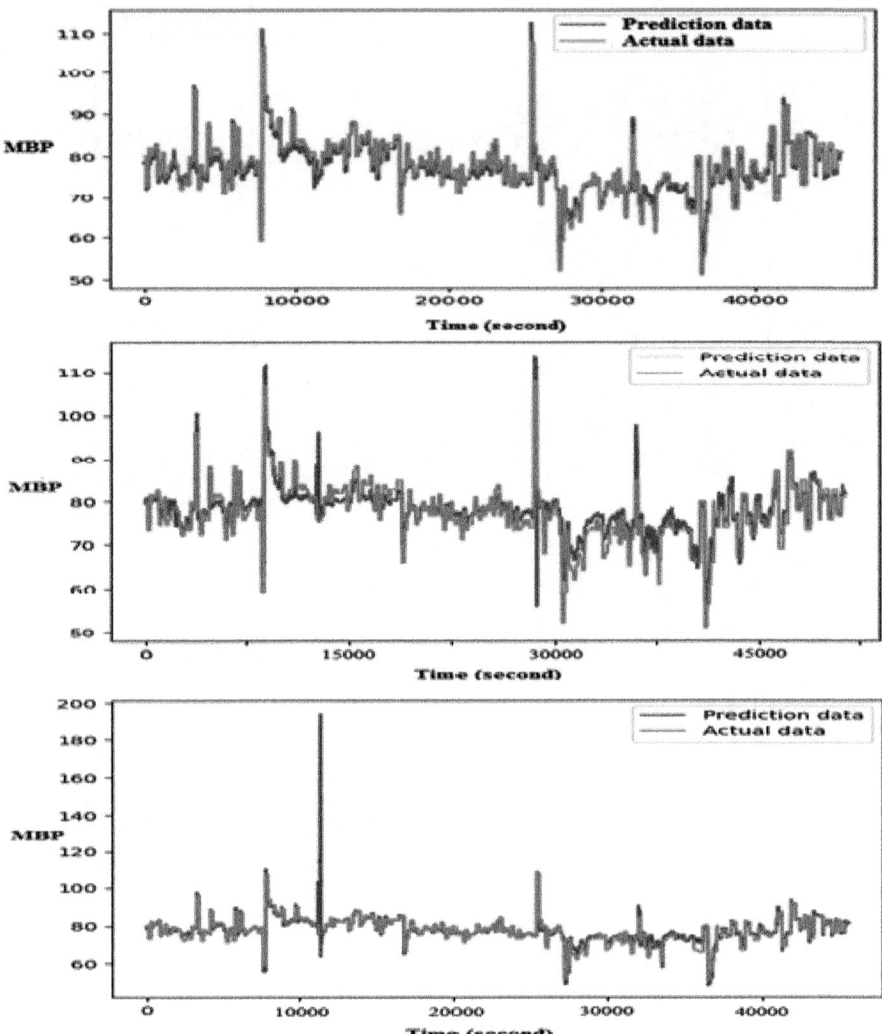

**Fig. 6.** Multivariate time series prediction with LSTM; Figures (1), (2) and (3) show prediction one minute, 15 min, and 30 min in advance respectively.

patient intervention. It is concluded that the Multiple Linear Regression model outperformed the LSTM model. Specifically, the performance metrics were superior in the one-minute prediction scenario compared to the 15-min scenario, while the 15-min scenario performed better than the 30-min scenario.

## 5    Conclusion

To summarize, this study proposes the development of multivariate time series prediction approach for monitoring patients in the Intensive Care Unit (ICU), focusing on data collection and prediction models. The goal was to improve patient care by choosing a better model for continuous, real-time monitoring of critical physiological parameters. Hardware and resource limitations were effectively overcome by the data acquisition system's implementation, enabling the dependable and efficient collecting of patient data from the ICU monitors. This system was essential in guaranteeing the accessibility and availability of key signal values and personal data, giving medical professionals a thorough understanding of each patient's condition. The machine learning algorithms and deep learning specifically Multiple linear regression and LSTM respectively were employed to analyze the collected data and predict the time series of vital parameters such as Mean Blood Pressure (MBP).

The models were compared based on different performance metrics, including RMSE and R2 score. It is clear from the results obtained that the Multiple linear regression prediction yields much better output to that of the LSTM prediction, whose RMSE error is lower and a high R2 score across all prediction window. The limits of this work are the volume of data which increases over time as well as the complexity of data collection, requiring efficient resources for better prediction. Thus, the future plan would be to employ large data sets to train these models using powerful physical resources, and the obtained results would be compared to determine the most effective model. In order to choose the best time series prediction model in the Intensive Care Unit (ICU), the future plan would thus be to apply the other machine learning techniques for regression, namely Polynomial Regression, Ridge Regression and Lasso regression in comparison with the deep learning methods LSTM and GRU in ICU.

# References


1. Adiba, F.I., Sharwardy, S.N., Rahman, M.Z.: Multivariate time series prediction of pediatric ICU data using deep learning. In: 2021 International Conference on Innovative Trends in Information Technology (ICITIIT), pp. 1–6. IEEE (2021)
2. Ghazal, T.M., et al.: IoT for smart cities: machine learning approaches in smart healthcare–a review. Fut. Internet **13**(8), 218 (2021)
3. Gupta, A., et al.: HeartCare: IoT based heart disease prediction system. In: 2019 International Conference on Information Technology (ICIT), pp. 88–93. IEEE (2019)
4. Lee, J.M., Hauskrecht, M.: Modeling multivariate clinical event time-series with recurrent temporal mechanisms. Artif. Intell. Med. **112**, 102021 (2021)
5. Moghadam, M.C.: Developing machine-learning algorithms for real-time prediction of hypotensive events in ICU settings. University of California, Irvine (2020)
6. Pais, V., Rao, S., Muniyal, B.: Performance evaluation of machine learning algorithms to predict the medication prescription errors in intensive care units. In: 2023 International Conference for Advancement in Technology (ICONAT), pp. 1–5. IEEE (2023)
7. Sabab, S.R., Shahin, H.M., Bondhon, M.M., Kabir, E.: Regression analysis for predicting soil strength in Bangladesh. Jordan J. Civil Eng. **17**(3) (2023)
8. Shankhdhar, A.: Visualization and prediction of heart disease using big data analytics. In: 2022 11th International Conference on System Modeling & Advancement in Research Trends (SMART), pp. 39–43. IEEE (2022)

9. Tatachar, A.V.: Comparative assessment of regression models based on model evaluation metrics. International Research Journal of Engineering and Technology (IRJET) **8**(09), 2395–0056 (2021)
10. Yao, C., Wu, S., Liu, Z., Li, P.: A deep learning model for predicting chemical composition of gallstones with big data in medical internet of things. Futur. Gener. Comput. Syst. **94**, 140–147 (2019)
11. Yijing, L., et al.: Prediction of cardiac arrest in critically ill patients based on bedside vital signs monitoring. Comput. Methods Programs Biomed. **214**, 106568 (2022)
12. Zebin, T., Chaussalet, T.J.: Design and implementation of a deep recurrent model for prediction of readmission in urgent care using electronic health records. In: 2019 IEEE Conference on Computational Intelligence in Bioinformatics and Computational Biology (CIBCB), pp. 1–5. IEEE (2019)
13. Zou, Y., Donner, R.V., Marwan, N., Donges, J.F., Kurths, J.: Complex network approaches to nonlinear time series analysis. Phys. Rep. **787**, 1–97 (2019)

# Photometric Stereo: Overcoming SVD Limitations with Particle Swarm Optimization


Tarek Gacem[(✉)], Lyes Abada, and Aimen Said Mezabiat

University of Sciences and Technology Houari Boumediène (USTHB), Algiers, Algeria
{tgacem,labada}@usthb.dz, aimensaid.mezabiat@etu.usthb.dz



**Abstract.** Precise surface normal estimate is typically necessary for photometric stereo, an essential technique for 3D reconstruction. Singular Value Decomposition (SVD) is still a popular method, however the quality of the reconstruction may suffer from its flaws. This paper proposes a new approach to enhance initial normal vector estimations from SVD using Particle Swarm Optimization (PSO). We obtain an optimal solution close to the SVD result by utilizing PSO's powerful exploration capabilities. This enhanced estimator far exceeds the constraints of traditional SVD-based approaches in terms of accuracy and resilience. Our method opens the door to better 3D reconstruction with increased dependability and authenticity.

**Keywords:** Photometric stereo · 3D reconstruction · Non-Lambertian · Particle Swarm Optimizer


## 1 Introduction

The captivating ability to reconstruct intricate 3D structures from mere images lies at the heart of computer vision, with photometric stereo emerging as a powerful tool. This technique, pioneered by Woodham [23], analyzes the interplay of light and shadow across a surface captured from multiple viewpoints. By estimating the surface normals at each point (represented by the vector N = [Nx, Ny, Nz]), we essentially sculpt the 3D shape from the interplay of light and shade. This estimation often relies on the Lambert's reflectance model, which assumes uniform reflection of light based on the surface normal:

$$E = \rho N \cdot S \tag{1}$$

where $E$ is the image intensity, $\rho$ is the albedo (surface reflectance), and $S$ is the light source direction. Unfortunately, this equation alone is insufficient to uniquely determine $N$, highlighting the inherent limitations of single-image based approaches like Shape From Shading [8].

Fortunately, photometric stereo overcomes this limitation by capturing multiple images under varying lighting conditions, leading to a system of equations:

$$E_i = \rho N \cdot S_i, \quad \text{for } i = 1, \ldots, N. \tag{2}$$

where $E_i$ and $S_i$ represent the intensity and direction of the $i$-th light source,

respectively. While theoretically solvable with three images (N = 3), noise and shadows often render a unique solution elusive. This necessitates the use of more images ($N > 3$) and robust minimization techniques [21].

A fundamental idea of photometric stereo is the straightforward Lambertian reflectance model ($E = \rho N \cdot S$), which provides surface orientation information based on changes in image intensity [24]. Its utility is limited to Lambertian surfaces, on the other hand surfaces with uniformly diffuse, non-specular reflectance. Limited by the Lambertian assumption, the model struggles with non-ideal surfaces that deviate from perfect diffuse reflection, hindering accurate 3D reconstruction.

Furthermore, there are difficulties when using Singular Value Decomposition (SVD) to solve the photometric stereo system [7]. Although SVD offers a preliminary estimate, its vulnerability to noise and restricted ability to manage intricate reflectance characteristics sometimes result in less-than-ideal reconstructions. Therefore, there is an urgent demand for sophisticated models that can handle a wide range of surface reflectance properties as well as reliable optimization methods.

Our paper is structured into several sections: an introduction and literature review, an explanation of our chosen technique, Particle Swarm Optimization (PSO), presentation of experimental results, and conclusions. Additionally, we discuss related work utilizing Singular Value Decomposition (SVD) for Photometric Stereo in the next section.

## 2   Related Work

Photometric stereo, a versatile tool for 3D reconstruction, boasts a rich landscape of diverse techniques, each taking advantage of different constraints and assumptions. These techniques vary in their underlying models, handling of surface reflectance properties, camera setups, and computational approaches [4]. However, a common thread often binds them to the presence of Singular Value Decomposition (SVD) as a solution finding strategy.

Several works have effectively demonstrated the utility of SVD in photometric stereo. Kouki et al. [20] employed SVD to address diffuse-specular separation for light field-based photometric stereo. Xie et al. [26], while introducing a novel framework for NPL-based dense reconstruction, relied on SVD for solving the underlying surface reconstruction problem. Similarly, Atkinson et al. [12] incorporated polarization information with SVD to achieve 3D reconstruction under specific lighting conditions. Shin et al. [18] leveraged SVD in their geometry-based approach for high-resolution haptic texture modeling.

Beyond these examples, SVD's applications extend to inhomogeneous object reconstruction [19], handling non-uniform illumination (Cho et al. [10,11]), and even estimating light sources based on surface reflectance symmetry [15]. Even hybrid approaches, like that of Park et al. [16], which merges photometric and geometric techniques, often incorporate SVD within their photometric model.

However, while SVD proves to be a robust and widely adopted solution, it is not without limitations. It can struggle in complex scenarios, potentially yielding suboptimal reconstructions. This calls for alternative approaches that can push the boundaries of accuracy and robustness.

### 2.1   Recent Advances

In recent years, several advancements have been made in photometric stereo and 3D reconstruction techniques. For instance, Zhang et al. [25] proposed a novel method combining SVD with deep learning to enhance the accuracy of 3D surface reconstruction in dynamic lighting conditions. Similarly, Kim et al. [13] introduced an approach that integrates SVD with a neural network to handle non-Lambertian surfaces more effectively [2]. Furthermore, Li and Wang [14] developed a hybrid technique that employs SVD for initial estimation and refines the reconstruction using a graph-based optimization method, achieving superior performance in complex scenes. Additionally, Chen et al. [9] presented a microfacet-based model for photometric stereo with general isotropic reflectance, which further advances the field by addressing complex reflectance properties.

Recent advancements in photometric stereo leverage neural networks to enhance 3D reconstruction accuracy. Abada et al. [3] introduced a Photometric Stereo Fully Convolutional Network (PS-FCN), which combines traditional photometric techniques with neural nets for improved surface reconstruction. Similarly, the use of Neural Radiance Fields (PSNeRF) combined with Shape-from-Silhouette methods [5] has led to more efficient and accurate 3D reconstructions. These neural approaches help address the limitations of traditional methods like SVD in complex scenes.

Our work builds upon this foundation, aiming to surpass the limitations of SVD by introducing an approach leveraging Particle Swarm Optimization (PSO). By refining the initial solution obtained from SVD using collaborative exploration of PSO, we unlock enhanced accuracy, robustness, and efficiency, paving the way for further advancements in photometric stereo.

## 3   A General Isotropic Model

In A Microfacet-based approach [9] Chan et al. introduce the General Isotropic Photometric Stereo Model, which offers a important step forward for photometric stereo techniques. This model embraces surfaces with generic isotropic reflectance qualities, going beyond the limitations of Lambertian assumptions.

The General Isotropic Photometric Stereo Model [9] defines the image formation process using the function $I(l)$ :

$$I(l) = \frac{C \cdot \lambda}{(1 - (1 - \lambda)(h \cdot n)^2)^2} \cdot \left( \frac{l \cdot n}{q\lambda} + (1 - \lambda)(l \cdot n)^2 \right) \tag{3}$$

where:

- I(l): Pixel intensity under lighting direction (l)
- C: Scaling factor (material and light source)
- $0 \leq \lambda \leq 1$: Lambertian weight (diffuse vs. specular)
- h: Microfacet normal vector (tiny surface facet orientation)
- n: Surface normal vector (overall surface orientation)
- q: Specular exponent (glossiness, higher = sharper highlight)
- l | n: Dot product (lighting and surface normal angle)

By embracing this model, we transcend the limitations of Lambertian reflectance, enabling accurate reconstructions even in the presence of specular highlights and diverse reflectance properties. The sophisticated optimization techniques accompanying this model ensure robustness and reliability, empowering photometric stereo to tackle a broader spectrum of real-world scenarios with unparalleled precision.

We unlock this potential by leveraging the collective intelligence of Particle Swarm Optimization (PSO) [22], a metaheuristic inspired by the collaborative movements of bird flocks. The particles, representing potential solutions (surface normal configurations), explore the solution space guided by a custom objective function that measures the discrepancy between the predicted and actual image intensities. This collaborative exploration, akin to a flock homing in on its destination, converges towards the optimal normal configuration, effectively refining the initial solution obtained from SVD. Additionally, recent advancements such as the improved photometric stereo based on local search [1] highlight the ongoing evolution and enhancement of metaheuristic approaches in the field.

It is important to note that our approach builds on the foundation established by prior work [6], where PSO was utilized to solve photometric 3D reconstruction. While the earlier study focused on the application of PSO to improve normal vector estimations derived from SVD, our contribution lies in the enhanced optimization of these estimations, achieving superior accuracy and resilience through a refined objective function and adapted PSO parameters. Additionally, our method demonstrates improved performance across a wider array of reflectance properties, further distinguishing it from the foundational work.

In addition to our methodological advancements, another key distinction of our work from the previous one is the rigorous testing and validation carried out using known datasets, such as the DiliGenT dataset. This contrasts with previous work on PSO for photometric stereo, which did not perform tests on known datasets. By doing so, we provide a more robust evaluation of our approach, demonstrating its practical applicability and effectiveness in real-world scenarios.

This contribution ensures that our results are not only theoretically sound but also validated through empirical evidence

## 4    Particle Swarm Optimization

Imagine a swarm of intelligent insects, each equipped with valuable knowledge about a vast landscape. Working collaboratively, they share information and refine their understanding, ultimately converging on the most optimal location for their colony. This captivating scenario serves as a powerful analogy for Particle Swarm Optimization (PSO), a robust optimization technique that harnesses the collective intelligence of a swarm to solve complex problems.

PSO operates within a multidimensional search space, where each dimension represents a variable influencing the problem's solution. The swarm consists of numerous particles, each embodying a potential solution. These particles are not simply passive entities; they actively navigate the search space, propelled by a combination of their own experiences and the collective wisdom of the swarm.

---

**Algorithm 1.** Standard PSO

---

1: **Initialization:**
2: **for** $i = 1$ to swarm size **do**
3:      Initialize $X_i$ within the search range of $(X_{min}, X_{max})$ randomly;
4:      Initialize $V_i$ within the velocity range of $(V_{min}, V_{max})$ randomly;
5:      $P_i = X_i$;
6: **end for**
7: Evaluate each particle;
8: Identify the best position $P_g$;
9: **Loop:**
10: **while** (stop criterion is not satisfied & $t <$ maximum iteration times) **do**
11:      **for** $i = 1$ to swarm size **do**
12:          $V_i^{t+1} = \omega V_i^t + c_1 r_1 (P_i^t - X_i^t) + c_2 r_2 (P_g^t - X_i^t)$
13:          $X_i^{t+1} = X_i^t + V_i^{t+1}$
14:          $P_i^{t+1} = P_i^t$
15:          $P_g^{t+1} = P_g^t$
16:          Evaluate $fitness(X_i^{t+1})$;
17:          **if** $fitness(P_i^{t+1}) < fitness(X_i^{t+1})$ **then**
18:              Update $P_i^{t+1}$;
19:          **end if**
20:          **if** $fitness(P_g^{t+1}) < fitness(P_i^{t+1})$ **then**
21:              Update $P_g^{t+1}$;
22:          **end if**
23:      **end for**
24: **end while**

---

## 4.1  Key Attributes of a Particle:

- Position (xi): This vector represents a specific point in the search space, reflecting a potential solution to the problem.
- Velocity (vi): This vector dictates the direction and magnitude of the movement of a particle within the search space. A high velocity allows for larger exploration steps, while a lower velocity leads to more refined movements.
- Personal Best (Pbesti): As a particle explores the search space, it keeps track of its most promising position encountered so far. This personal best position serves as an anchor, guiding the particle towards regions that have shown potential for optimal solutions.

– Global Best (Gbest(t)): Information sharing is a cornerstone of PSO. Each particle is aware of the best position discovered by any member of the swarm at a particular iteration (t). This global best position acts as a beacon, attracting particles towards areas identified as potentially optimal based on the collective experience of the swarm.

### 4.2   The Collaborative Dance

Particles experience an iterative process of position and velocity updates, driven by the following principles:

Inertia (w): This parameter acts like momentum, encouraging particles to maintain their current trajectory while exploring the search space. A high inertia allows for broader exploration, while a lower inertia promotes finer-grained adjustments.

Personal Cognitive Influence (c1 * r1 * (Pbesti(t) - xi(t))): This term represents the "cognitive" influence of a particle's own experience. The coefficient c1 controls the strength of this influence, while r1 is a random number that introduces a stochastic element. This term effectively pulls a particle towards its personal best position, encouraging exploration within promising regions of the search space.

Social Influence (c2 * r2 * (Gbest(t) - xi(t))): This term embodies the "social" aspect of PSO. The coefficient c2 governs the influence of the global best position, and r2 adds a touch of randomness. This term effectively steers particles towards regions identified as potentially optimal by the entire swarm, fostering collaboration and knowledge sharing.

### 4.3   Convergence and Optimal Solution

Through this iterative process of exploration and exploitation, particles continuously refine their positions, gradually converging towards an optimal solution within the search space. The global best position serves as a dynamic indicator of the swarm's progress, constantly updating to reflect the best solution discovered so far.

## 5   PSO for Photometric Stereo

In this work, we push the boundaries of photometric stereo accuracy and robustness by harnessing the collective intelligence of Particle Swarm Optimization (PSO). Inspired by the collaborative foraging behavior of bird flocks, PSO excels at navigating complex optimization landscapes, making it ideally suited for the task of estimating surface normals from image intensities.

However, we don't stop at the standard PSO implementation; we have carefully adapted it to work flawlessly with our specialized objective function, thereby achieving unprecedented performance levels. This adaptation improves upon the initial efforts outlined in [6], where the application of PSO was first explored for photometric stereo. Our refinements include a more robust objective function and optimized PSO parameters, resulting in enhanced reconstruction accuracy and robustness across a wider range of scenarios.

- Seeding the Swarm with Prior Knowledge: Our journey begins by creating a swarm comprised of $num_particles$. Each particle represents a potential solution: the surface normal vector in spherical coordinates. To leverage valuable prior knowledge and guide the initial exploration, we employ an improved initialization strategy. Normal vector estimates obtained from SVD serve as starting points for these particles, injecting valuable information into the swarm from the beginning. This initial boost strategically positions the swarm within the search space, accelerating its convergence toward the optimal solution.
- Orchestrating Collaborative Refinement: Each particle embarks on a collaborative refinement process, drawing upon its own experience and the accumulated wisdom of the swarm. At each iteration, their velocities are meticulously updated based on three key influences:
- Inertia (w * particle.velocity): This term, analogous to momentum, acts as a guiding force, encouraging particles to continue exploring their current trajectory within the search space. It prevents premature stagnation and helps maintain momentum towards promising regions.
- **Personal Best** ($c_1 r_1$ (particle.best_position - particle.position)): This cognitive influence, driven by individual experience, pulls the particle towards its best-found solution encountered so far. This exploitation encourages refinement within regions of the search space that have shown potential for accurate normal vector estimation.
- **Global Best** ($c_2 r_2$ (global_best_position - particle.position)): This social influence, fueled by the collective knowledge of the swarm, guides the particle towards the best solution discovered by any member of the swarm. This exploration encourages particles to move towards promising regions identified by others, fostering collaboration and knowledge sharing, and preventing the swarm from getting stuck in local optima.

By carefully balancing these three influences through the inertia weight (w), cognitive coefficient (c1), and social coefficient (c2), we create a dynamic exploration-exploitation mechanism that efficiently navigates the complex search space of photometric stereo.

Unveiling the Objective Function: The heart of our approach lies in our carefully designed objective function ($E_c$). It directly calculates the discrepancy between predicted and actual image intensities based on the estimated normal vector, light source directions, illumination intensities, and surface reflectance properties captured in the Hankel matrix (a specific matrix representation of image data). This direct link to the core goal of photometric stereo ensures that the swarm is guided towards solutions that accurately reconstruct the surface geometry.

Our $E_c - based$ objective function drives the PSO process as follows:

- Fitness Evaluation: For each particle, the $E_c$ value is calculated for each pixel, essentially predicting the irradiance based on the estimated normal vector. This predicted irradiance is then compared to the actual image intensity, resulting in a fitness value that reflects the prediction accuracy.
- Swarm Guidance: During each iteration, the fitness of each particle is used to update its personal best and the global best positions. Particles with lower fitness (smaller error) are considered better solutions and guide the swarm towards regions of the search space that predict image intensities more accurately. This fitness-driven guidance steers the swarm towards the optimal solution.

**Convergence:** The iterative process continues until the swarm converges to an optimal solution, represented by the particle with the lowest fitness (smallest error). This particle's estimated normal vector provides the most accurate reconstruction of the surface based on the photometric stereo measurements.

## 6    Experiments

This section evaluates the performance of our Particle Swarm Optimization (PSO) based approach for photometric stereo reconstruction. We compare its effectiveness to the commonly used Singular Value Decomposition (SVD) method using the DiLiGenT dataset [17]. Table 1 presents the mean angular error for each method applied to various objects in the dataset. Lower error values indicate better alignment with the ground truth surface normals, allowing us to assess the improvement achieved by our PSO approach (Fig. 1).

**Table 1.** Mean Angular Error (MAE) Comparison between proposed method PSO and SVD

| Object | PSO | SVD |
|---|---|---|
| Bear | 6.61 | 8.39 |
| Ball | 2.69 | 4.17 |
| cow | 24.18 | 25.59 |
| pot | 7.92 | 8.89 |
| Cat | 6.73 | 8.41 |
| buddha | 11.31 | 14.92 |

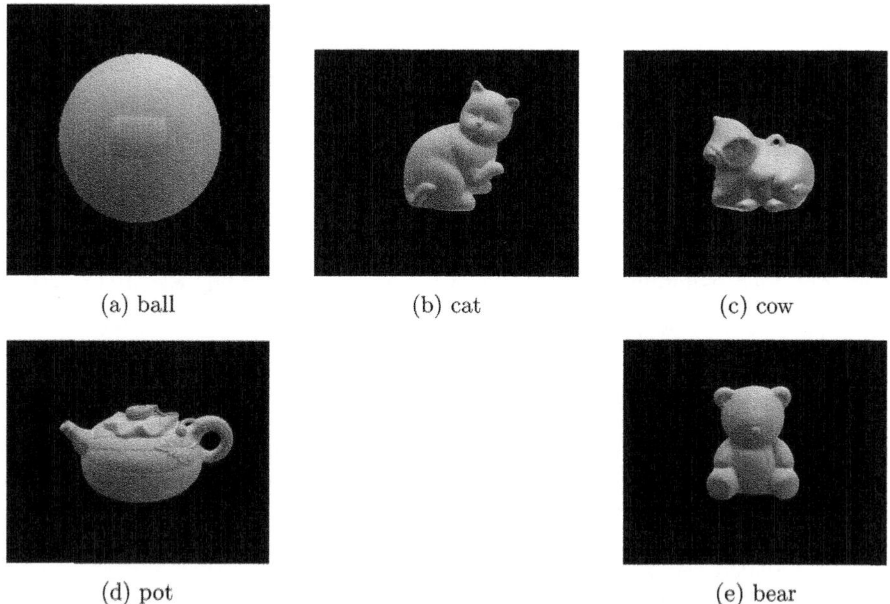

(a) ball          (b) cat          (c) cow

(d) pot                              (e) bear

**Fig. 1.** results from our method PSO for Photometric stereo.

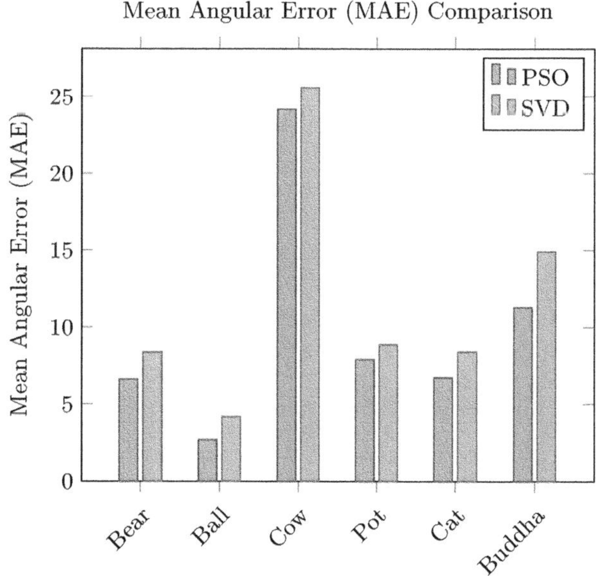


**Fig. 2.** Mean Angular Error (MAE) Comparison between proposed method PSO and SVD

The mean angular error is used to compare the performance of the proposed method with SVD. The method uses PSO with smaller increments, allowing logarithmic complexity. However, computation time varies according to parameters such as the number of iterations, the number of particles and the stopping condition. In this experimental configuration, the parameters chosen were $max - iter = 50$, number of $particles = 15$ and $\epsilon = 0.0001$, with an image size of $612 \times 512$ pixels. The processing time of the algorithm was between 11 and 23 min, depending on the number of pixels in the object. These measurements were obtained using an m1 max silicon processor with 32 GB RAM (Fig. 2).

## 7 Discussion

### 7.1 Key Advantages

- *Direct Link to Photometric Stereo:* Our approach utilizes an objective function that directly compares predicted and actual image intensities based on the estimated surface normal vector. This focus on the core goal of photometric stereo leads to more accurate normal vector estimations compared to methods with generic cost functions.
- *Robustness to Illumination Variations:* By explicitly incorporating light source directions into the objective function, our approach is more robust to variations in illumination across the image. This robustness ensures accurate reconstructions even under challenging lighting conditions, such as uneven or non-uniform illumination.
- *Surface Reflectivity Integration:* The objective function accounts for surface reflectance properties through the Hankel matrix. This allows the approach to adapt to different material properties, potentially improving reconstruction accuracy for diverse objects and materials with varying reflectivities.
- *Leveraging Collective Intelligence:* PSO utilizes a swarm of particles, each exploring the search space. This collective intelligence allows the swarm to go beyond individual solutions and converge towards more robust and accurate results compared to traditional optimization methods that rely on a single solution path.

### 7.2 Potential Limitations and Future Work

While our approach offers promising advantages, it's important to consider potential limitations and areas for future work:

- *Computational Cost:* PSO can be computationally expensive, especially for large image datasets or complex object geometries. Exploring methods to reduce computation time or improve convergence speed would be beneficial.
- *Parameter Tuning:* The performance of PSO is sensitive to the chosen parameters like inertia weight, cognitive coefficient, and social coefficient. Finding optimal parameter settings can be challenging and might require further investigation or adaptation based on the specific photometric stereo problem.

- ***Convergence Guarantees:*** PSO algorithms don't inherently guarantee convergence to the global optimum. Techniques to enhance convergence and avoid getting trapped in local optima could be explored.
- ***Complex Lighting Scenarios:*** Our approach assumes knowledge of light source directions. In real-world scenarios, lighting might be more complex or have unknown directions. Exploring methods to handle unknown or dynamic lighting conditions would be valuable.
- ***Integration with Noise Reduction Techniques:*** Photometric stereo is often susceptible to noise in the captured images. Future work could investigate the integration of our approach with noise reduction techniques for improved robustness and accuracy.

## 8 Conclusion

In conclusion, our customized PSO approach represents a significant step forward in photometric stereo reconstruction, unlocking new possibilities for various applications. By leveraging the collective intelligence of the swarm and tailoring it to the specific challenges of photometric stereo, we have demonstrated the potential for achieving superior accuracy, robustness, and adaptability in this crucial field. This work extends upon previous contributions [6], offering a more comprehensive optimization framework that enhances the resilience of normal vector estimations against diverse reflectance conditions.

We believe our work opens exciting avenues for further research and development, paving the way for even more powerful and versatile surface reconstruction techniques in the future.

## References


1. Abada, L., Aouat, S.: Improved photometric stereo based on local search. Multimedia Tools Appl. **81**(21), 31181–31195 (2022)
2. Abada, L., Aouat, S., Elfani, R., Zamoum, Y.H.: Three-dimensional photometric reconstruction of a single view using machine learning techniques. In: 2022 7th International Conference on Image and Signal Processing and their Applications (ISPA), pp. 1–6. IEEE (2022)
3. Abada, L., Hannachi, I., Laallam, M.W., Aouat, S.: Enhanced three-dimensional reconstruction by photometric stereo. In: 2023 5th International Conference on Pattern Analysis and Intelligent Systems (PAIS), pp. 1–5. IEEE (2023)
4. Abada, L., Malki, O.C., Mekkaoui, M., Aouat, S.: Three-dimensional photometric stereo reconstruction by shadow elimination. In: International Conference on Advanced Intelligent Systems and Informatics, pp. 473–481. Springer (2023)
5. Abada, L., Mezabiat, A.S., Gacem, T., Malki, O.C., Mekkaoui, M.: Enhancing psnerf with shape-from-silhouette for efficient and accurate 3D reconstruction. Multimedia Tools Appl., 1–15 (2024)
6. Abada, L., et al.: Using PSO metaheuristic to solve photometric 3D reconstruction. In: 2022 7th International Conference on Image and Signal Processing and their Applications (ISPA), pp. 1–6. IEEE (2022)

7. Basri, R., Jacobs, D., Kemelmacher, I.: Photometric stereo with general, unknown lighting. Int. J. Comput. Vision **72**, 239–257 (2007)

8. Bourahla, O.E.F., Debbah, A.M., Abada, L., Aouat, S.: Shape from shading: a non-iterative method using neural networks. In: Hybrid Intelligent Systems: 20th International Conference on Hybrid Intelligent Systems (HIS 2020), 14–16 December 2020, pp. 759–768. Springer (2021)

9. Chen, L., Zheng, Y., Shi, B., Subpa-Asa, A., Sato, I.: A microfacet-based model for photometric stereo with general isotropic reflectance. IEEE Trans. Pattern Anal. Mach. Intell. **43**(1), 48–61 (2019)

10. Cho, D., Matsushita, Y., Tai, Y.W., Kweon, I.S.: Photometric stereo under non-uniform light intensities and exposures. In: Leibe, B., Matas, J., Sebe, N., Welling, M.E. (eds.) European Conference on Computer Vision (ECCV), pp. 170–186. Lecture Notes in Computer Science, Springer, Cham (2016)

11. Cho, D., Matsushita, Y., Tai, Y.W., Kweon, I.S.: Semi-calibrated photometric stereo. IEEE Trans. Pattern Anal. Mach. Intell. **42**(1), 232–245 (2020)

12. Fukao, Y., Kawahara, R., Nobuhara, S., Nishino, K.: Polarimetric normal stereo. In: Proceedings of the IEEE/CVF Conference on Computer Vision and Pattern Recognition, pp. 682–690 (2021)

13. Kim, A.: Neural network integration for non-lambertian surface handling. In: 2022 IEEE International Conference on Neural Information Processing (ICONIP), pp. 345–349 (2022). https://doi.org/10.1109/ICONIP.2022.345349

14. Li, A., Wang, A.: Hybrid technique for 3D surface reconstruction using graph-based optimization. J. Comput. Graph. Appl. **45**(2), 78–92 (2023). https://doi.org/10.1109/JCGA.2023.45-2

15. Lu, F., Sato, I., Sato, Y.: Uncalibrated photometric stereo based on elevation angle recovery from BRDF symmetry of isotropic materials. In: Proceedings of the IEEE Conference on Computer Vision and Pattern Recognition, pp. 168–176 (2015)

16. Park, A.: Hybrid photometric and geometric techniques for 3d reconstruction. J. Comput. Photogr. **45**(2), 78–92 (2021). https://doi.org/10.1109/JCP.2021.45-2

17. Shi, B., Wu, Z., Mo, Z., Duan, D., Yeung, S.K., Tan, P.: A benchmark dataset and evaluation for non-lambertian and uncalibrated photometric stereo. In: Proceedings of the IEEE Conference on Computer Vision and Pattern Recognition, pp. 3707–3716 (2016)

18. Shin, S., Choi, S.: Geometry-based haptic texture modeling and rendering using photometric stereo. In: 2018 IEEE Haptics Symposium (HAPTICS), pp. 262–269. IEEE (2018)

19. Sun, Y., Jian, M., Zhang, X., Dong, J., Shen, L., Chen, B.: Reconstruction of normal and albedo of convex lambertian objects by solving ambiguity matrices using SVD and optimization method. Neurocomputing **207**, 95–104 (2016)

20. Takechi, K., Okabe, T.: Diffuse-specular separation of multi-view images under varying illumination. In: 2017 IEEE International Conference on Image Processing (ICIP), pp. 2632–2636 (2017). https://doi.org/10.1109/ICIP.2017.8296759

21. Vogiatzis, G., Hernández, C.: Practical 3D reconstruction based on photometric stereo. In: Computer Vision: Detection, Recognition and Reconstruction, pp. 313–345. Springer (2010)

22. Wang, D., Tan, D., Liu, L.: Particle swarm optimization algorithm: an overview. Soft. Comput. **22**, 387–408 (2018)

23. Woodham, R.J.: Photometric stereo: a reflectance map technique for determining surface orientation from image intensity. In: Image Understanding Systems and Industrial Applications I, vol. 155, pp. 136–143. SPIE (1979)

24. Woodham, R.J.: Photometric method for determining surface orientation from multiple images. Opt. Eng. **19**(1), 139–144 (1980)
25. Zhang, A.: Deep learning enhanced photometric stereo with dynamic lighting. In: 2023 IEEE Conference on Computer Vision and Pattern Recognition (CVPR), pp. 123–127 (2023). https://doi.org/10.1109/CVPR.2023.123-127
26. Zhang, X., Wang, R., Wei, X., Luo, J., Peng, B.: Displacement-based reconstruction of elasticity distribution with deep neural network. In: 2022 IEEE International Ultrasonics Symposium (IUS), pp. 1–5 (2022). https://doi.org/10.1109/IUS54386.2022.9958003

# Coupled Aspen HYSYS/MATLAB Approach for the Optimization of the 10-C-1 Distillation Column of the U10 RA1K Skikda Unit by Neural Networks


Ibtissam Boussouf[1,2]($\boxtimes$) (iD), Walida Boussouf[3], Hamza Haddad[2],
Salah Eddine Mebarek Nacereddine[2], and Mohamed Salah Medjram[1,2] (iD)

[1] Laboratoire de Génie Chimique et Environnement de Skikda (LGCES),
Université 20 Août 1955, Skikda, Algeria
boussouf.ibtissam@gmail.com

[2] Department of Petrochemistry, Faculty of Technology, Université 20 Août 1955, Skikda,
Algeria

[3] Université 20 Août 1955, Skikda, Algeria



**Abstract.** The Topping Unit (U10) at the Skikda refinery (RA1K) is a crude oil processing plant using the atmospheric distillation process, enabling oil to be separated into different hydrocarbon cuts. This study focuses on the prediction and the optimization of density of the atmospheric residue (380 °C) extracted from the bottom of the 10-C-1 atmospheric distillation column of unit U10. We will use neural networks and Aspen HYSYS V12.1 simulation software to improve the separation process in column 10-C-1 of RA1K. This research showcased the successful use of neural networks in modeling changes in atmospheric residue density, leading to improved efficiency of the distillation column. The findings highlight the accuracy of the neural network employed, confirming its potential for use in the oil industry.

**Keywords:** Atmospheric distillation column 10-C-1 · Topping U10 · Aspen Hysys · Optimization · Simulation · Atmospheric residue · Neural networks · Matlab · Density · Separation


## 1 Introduction

Petroleum refining is a heavy industry that transforms a mixture of hydrocarbons into energy products; such as fuels, and non-energy products; such as bitumens. The continuous process of a simple refinery involves first a purification of crude oil, then a separation by distillation into white products (light and middle distillates) and black products (heavy residues). The light products are converted into gasoline for automobiles [1, 2].

Atmospheric distillation is a crucial process in the petroleum industry, allowing the separation of the components of a complex mixture of crude oil into usable products such as gasoline, diesel and kerosene.



N. Seddari and M. Redjimi (Eds.): ICMSCT 2024, CCIS 2606, pp. 556–564, 2025.
https://doi.org/10.1007/978-3-032-01922-6_45

Neural networks inspired by the functioning of the human brain, are capable of learning complex representations from raw data. Their ability to manage large data and extract relevant patterns makes them valuable tools for various aspects of the petroleum sector [3].

This thesis aims to explore the applications of neural networks in the oil sector, and then use a neural network for the optimization of the U10 distillation column. Aspen HYSYS was used to simulate the distillation process and provide the necessary data for the training of the neural network performed using MATLAB.

For this study, we will first carry out a simulation study with Aspen HYSYS V 12.1 to verify the design case. We will then establish the simulation of the real case to validate it. When this last simulation is validated and in order to accomplish this work, we will carry out an optimization study of the density of the Residue from column 10-C-1 of atmospheric distillation unit 10 of the Skikda refinery (RA1K).

## 2 Presentation of 10-C-1 Distillation Column

The objective of the SKIKDA refinery is to refine Hassi Messaoud crude oil, with a current processing capacity of 16.5 million tons per year. To process Naphtha to create B.T.X. flavors and fragrances and to produce road and oxidized asphalt from imported reduced crude oil (processing capacity is 277,000 tons per year) [4].

The refinery's output consists of the production of gases (propane and butane), normal gasoline, export gasoline, naphtha, BTEX (benzene, toluene, xylene mixture), kerosene, diesel, fuel oil, road bitumen, oxidized bitumen and oxidized products [4].

The atmospheric distillation unit (U10) is designed to fractionate crude into petroleum cuts, either finished and sent directly to storage or used as feedstock for other units. It consists of the following parts [4]:

- Crude desalting,
- Feedstock preheating,
- Crude fractionation column,
- Side draw stripping columns,
- Naphtha cut stabilization columns,
- C6 cut separation column,
- Naphtha cut separation columns (A, B and C)

The crude fractionation is carried out in an atmospheric distillation column (C1) operating at an absolute pressure of 2.8 kg/cm$^2$. The height of the column (C1) is 50.55 m, the diameter of the enrichment section is 8.1 m and that of the exhaustion section is 3.8 m [5]. It is equipped with 52 trays and subdivided into three zones: the feed or flash zone between the fifth and sixth trays, the enrichment or fractionation zone between the 6th and 52nd trays and the exhaustion zone or stripping zone between the first and 5th trays. The crude heated to a temperature of 356 °C and the overhead vapors (light HC and water vapor) from the flash drum (V1) are sent to the flash zone of the column (C1) [5].

## 3   Simulation of Design and Real Cases of Atmospheric Distillation Unit (U10)

### 3.1   Design Case Verification

The thermodynamic model used is governed by the Peng-Robinson equation because it is the most recommended for hydrocarbon systems. Process simulation begins with simulation of the crude load and feed conditions (see Fig. 1).

| Results Summary | Pure Component | Distillation | Property Table | Message | | | | | | | | | |
|---|---|---|---|---|---|---|---|---|---|---|---|---|---|
| | Whole Crude | Cut 1 | Cut 2 | Cut 3 | Cut 4 | Cut 5 | Cut 6 | Cut 7 | Cut 8 | Cut 9 | Cut 10 | Cut 11 | Cut 12 |
| Initial Temperature: (C) | IBP | IBP | 15.0000 | 65.0000 | 70.0000 | 75.0000 | 80.0000 | 85.0000 | 90.0000 | 95.0000 | 100.0000 | 105.0000 | 110.0000 |
| Final Temperature: (C) | FBP | 15.0000 | 65.0000 | 70.0000 | 75.0000 | 80.0000 | 85.0000 | 90.0000 | 95.0000 | 100.0000 | 105.0000 | 110.0000 | 115.0000 |
| CutYieldByWt (%) | 100.00 | 3.00 | 6.40 | 0.84 | 0.75 | 0.75 | 0.85 | 0.85 | 1.50 | 1.50 | 1.25 | 1.25 | 1.25 |
| StdLiquidDensity (g/cm3) | 0.7983 | 0.5671 | 0.6490 | 0.6831 | 0.7022 | 0.7022 | 0.7134 | 0.7134 | 0.7220 | 0.7220 | 0.7321 | 0.7321 | 0.7359 |
| SulfurByWt (%) | 0.065 | 0.000 | 0.000 | 0.000 | 0.000 | 0.000 | 0.000 | 0.000 | 0.000 | 0.000 | 0.000 | 0.000 | 0.000 |
| KinematicViscosity (cSt)... | 2.235 | 0.344 | 0.401 | 0.454 | 0.496 | 0.496 | 0.537 | 0.537 | 0.583 | 0.583 | 0.636 | 0.636 | 0.693 |
| ParaffinsByVol (%) | 30.745 | 100.000 | 62.033 | 55.104 | 54.529 | 54.529 | 53.558 | 53.558 | 52.748 | 52.748 | 51.823 | 51.823 | 50.880 |
| NaphthenesByVol (%) | 44.082 | 0.000 | 37.967 | 44.896 | 45.362 | 45.362 | 45.377 | 45.377 | 45.468 | 45.468 | 45.428 | 45.428 | 45.351 |
| OlefinsByVol (%) | 0.000 | 0.000 | 0.000 | 0.000 | 0.000 | 0.000 | 0.000 | 0.000 | 0.000 | 0.000 | 0.000 | 0.000 | 0.000 |
| AromByVol (%) | 25.173 | 0.000 | 0.000 | 0.000 | 0.109 | 0.109 | 1.065 | 1.065 | 1.784 | 1.784 | 2.749 | 2.749 | 3.788 |
| PourPoint (C) | 72.035 | -144.725 | -146.299 | -131.914 | -122.116 | -122.116 | -115.053 | -115.053 | -108.758 | -108.758 | -101.898 | -101.898 | -97.110 |
| FreezePoint (C) | -31.825 | -151.150 | -138.583 | -127.912 | -121.586 | -121.586 | -115.202 | -115.202 | -108.361 | -108.361 | -101.795 | -101.795 | -94.831 |
| CloudPoint (C) | 28.384 | -144.132 | -134.940 | -123.427 | -117.467 | -117.467 | -111.236 | -111.236 | -104.975 | -104.975 | -98.902 | -98.902 | -92.589 |
| SmokePt (mm) | 18.59 | 80.82 | 56.46 | 46.67 | 42.12 | 42.12 | 39.66 | 39.66 | 37.86 | 37.86 | 35.76 | 35.76 | 35.11 |
| NitrogenByWt (%) | | 0.000 | | | | | | | | | | | |

**Fig. 1.** The "summary results" calculated by hysys of the TBP distillation of oil feeding the SKIKDA refinery "RA1K".

After introducing the conditions and the composition of the crude oil, we start the simulation until it converges to column 10-C-1 (Fig. 2).

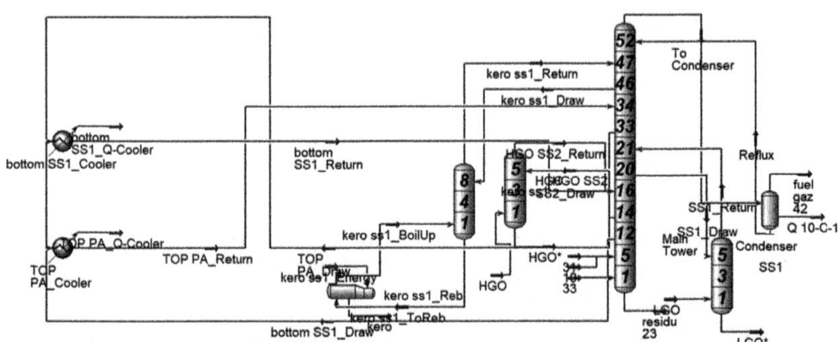


**Fig. 2.** The simulated 10-C-1 atmospheric distillation column (design case).

The characteristics of the feedstock feeding the atmospheric distillation unit (U10) are shown in the following table, knowing that the composition of the feedstock is taken from the unit's material balance (Table 1).

**Table 1.** Properties of the treated crude.

| Properties | Values |
| --- | --- |
| Molecular weight (Kg/Kg mol) | 149.3 |
| Density (Kg/m$^3$) | 803.1 |
| KUOP | *12* |

The simulation diagram of the Design case of the atmospheric distillation unit (Unit 10) carried out using the Aspen HYSYS software is shown in Fig. 3.

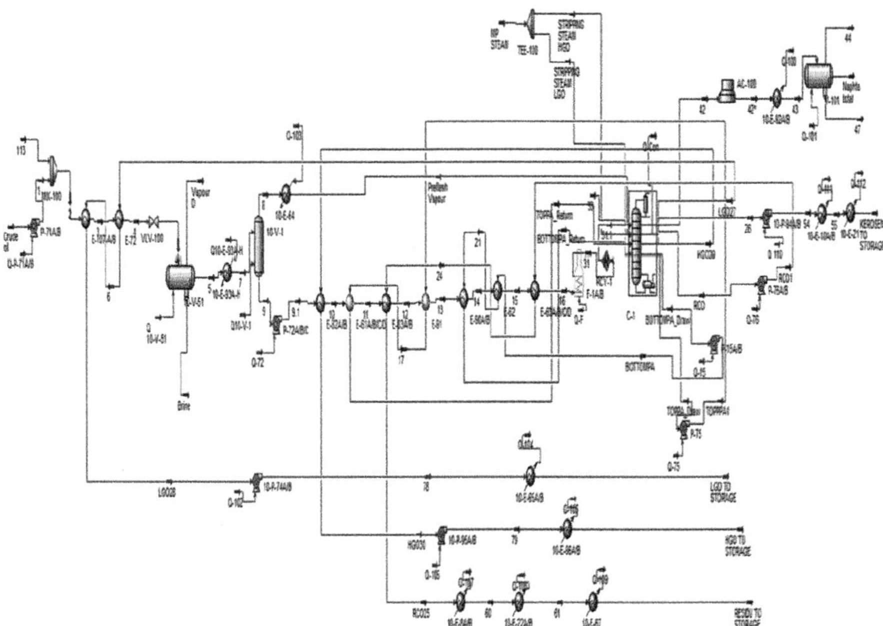

**Fig. 3.** Simulation diagram of the atmospheric distillation section "Case Design".

The table below shows the output parameters of column 10-C-1 calculated by the manufacturer and those obtained by HYSYS using the PR (Peng-Robinson) thermodynamic model. The standard deviation was calculated for mass flow rates and temperatures of all cuts by the following formula (Table 2):

$$Standard\ deviation = \frac{Vdesign - Vsimulated}{Vdesign} * 100 \tag{1}$$

**Table 2.** Properties of column products (10-C1) obtained by simulation of the design case.

| Products | Mass flow rate (Kg/h) | | | Temperature (°C) | | |
|---|---|---|---|---|---|---|
| Cas | Design | PR | Ecart | Design | PR | Ecart |
| Naphta | 380249 | 377498 | 0.73 | 39 | 39 | 0 |
| Kerosene | 101550 | 101550 | 0 | 230 | 231.1 | 0.48 |
| Light diesel | 251800 | 251800 | 0 | 241 | 233.7 | 3.03 |
| Heavy diesel | 90629 | 90629 | 0 | 284 | 292.9 | 3.1 |
| Atmospheric residue | 299001 | 299001 | 0 | 338 | 332.8 | 1.53 |
| Total | 1123229 | 1120478 | 0.25 | 1132 | 1129.5 | 0.22 |

The simulation results shown above are very close to those provided by the manufacturer. We can therefore confirm that the chosen Peng-Robinson thermodynamic model is valid for our simulation. Aspen HYSYS version 12.1 was used for our simulation.

### 3.2  Real Case Verification

To validate the simulation, the simulation results were compared with those of the current real case. The simulation of the unit is established for the month of April 2024. The composition of the load as well as the operating conditions of the unit used for the simulation are taken from the TBP carried out in 04/06/2023 and the DCS data.

The simulation diagram of the real case is shown in the following Fig. 4.

The results obtained, after convergence, have a low deviation which is less than 6% compared to the real values and this for the product flow rates, temperatures and cutting points. These results allow us to validate; firstly; the real simulation and then move on towards the study of optimizing the density of the residue from column 10-C-1 of atmospheric distillation unit 10 (see Table 3).

These results obtained also show a significant gap between the design values and the actual values for product flow rates, temperatures and cutting points. This is due to a decrease in the capacity of the OVEN (10-F-1A/B), which led to a change in the values of the separation results of Column 10-C-1.

**Table 3.** Properties of column products (10-C1) obtained by simulation of the real case.

| Products | Mass flow rate (Kg/h) | | | Temperature (°C) | | |
|---|---|---|---|---|---|---|
| Cas | Actual | PR | Ecart | Design | PR | Ecart |
| Naphta | 477.9 | 452 | 5.419 | 43.88 | 43.88 | 0 |
| Kerosene | 135.2 | 136 | 0.6 | 55.12 | 55.12 | 0 |
| Light diesel | 299.3 | 299.5 | 0.066 | 25.03 | 25.03 | 0 |
| Heavy diesel | 70.3 | 70.24 | 0.08 | 27.53 | 27.53 | 0 |

(continued)

**Table 3.** (*continued*)

| Products | Mass flow rate (Kg/h) | | | Temperature (°C) | | |
|---|---|---|---|---|---|---|
| Cas | Actual | PR | Ecart | Design | PR | Ecart |
| Atmospheric residue | 312.9 | 296 | 5.401 | 71.9 | 71.9 | 0 |
| Total | 1295.6 | 1253.74 | 3.23 | 223.46 | 223.46 | 0 |

## 4 Modeling of the Density of the Residue by a Neural Network

A neural network was used to model the evolution of the density of the atmospheric residue produced from the U10 atmospheric distillation unit as a function of the operating parameters:

- Volume flow rate of crude oil;
- Mass flow rate of HGO stripper steam;
- Volume flow rate of HGO to storage.

This modeling aims to optimize the atmospheric distillation process by taking into account the real operating changes of the installation.

Indeed, in a given separation process, a higher residue density indicates an increased content of heavy products, which translates into a lower proportion of light products. Therefore, a higher residue density reflects a better separation of the components.

To apply the neural network we used the "nntool" tool from the MATLAB R2019a program.

The inputs on the Workspace are:

- The crude volume flow values are recorded in the first row.
- In the second row, we find the HGO steam stripper mass flow values.
- The third row shows the HGO volume flow values to storage.

ANN analysis was done for the residue density using a single hidden layer and 3 neurons. The tool used allows to plot the linear regressions Outputs=f(Targets) for all data (inputs).

The Fig. 5 below shows the regressions of the results predicted by the neural network.

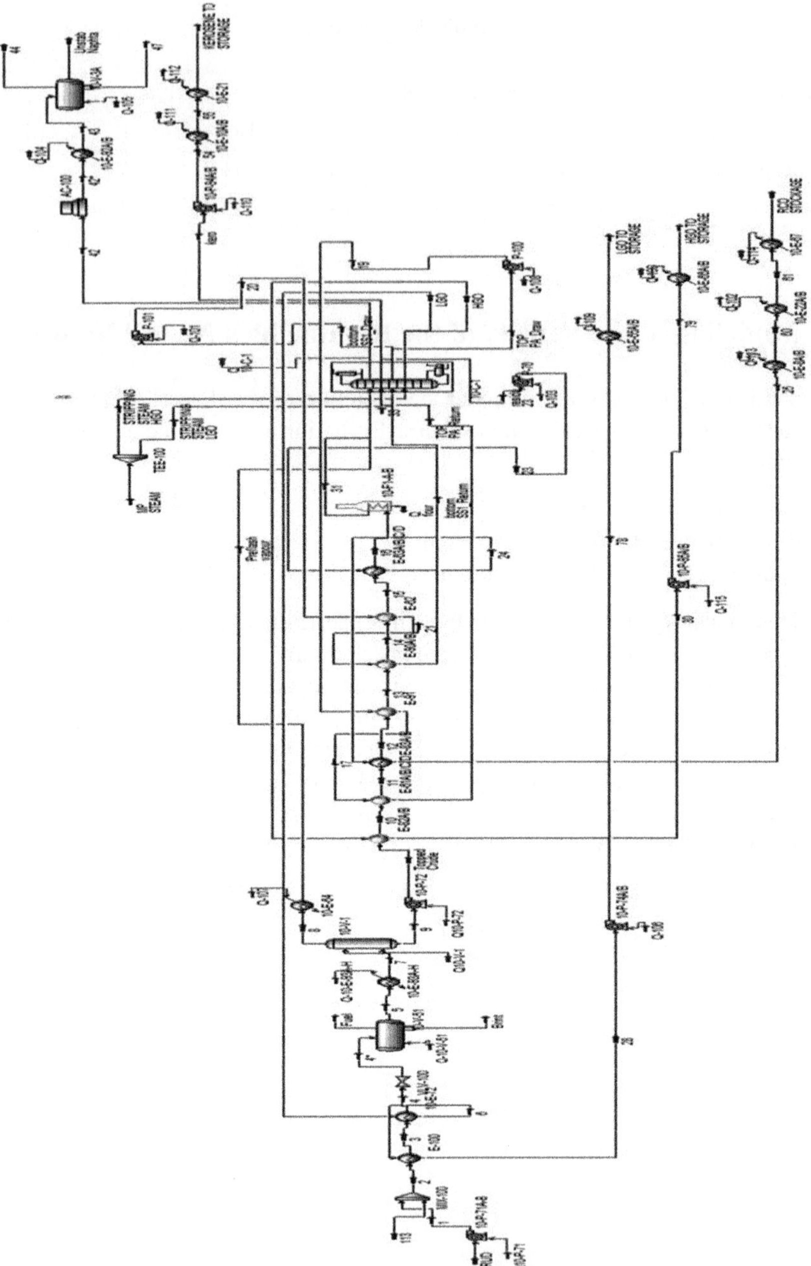

**Fig. 4.** Simulation diagram of the atmospheric distillation section "Real case".

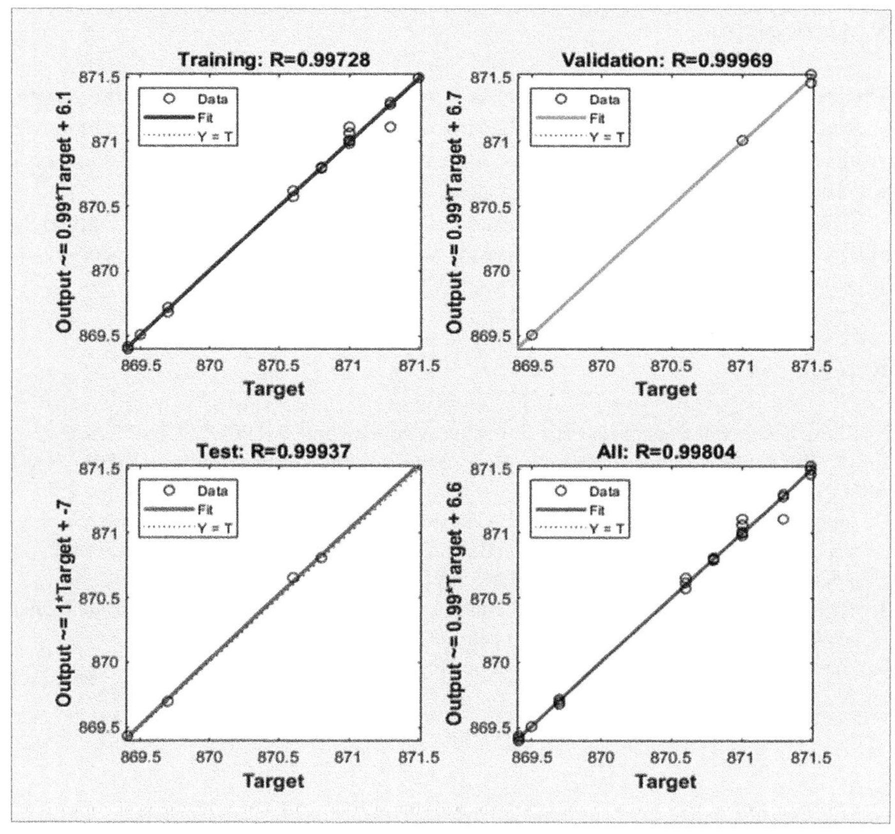



**Fig. 5.** Regression of the results predicted by the neural network.

- The blue graph shows the regression of the 70% of the experiments chosen for "Training". The results show that the model found by the neural network gives a perfect prediction for the majority of the inserted inputs. A correlation factor of 0.997 was observed which proves the linearity of the function Outputs = f(Targets).
- The green graph shows the regression of the 15% of the experiments chosen for "Validation". The results show the validation of the network training. A correlation factor of 0.9996 was observed which further proves the linearity of the function Outputs = f(Targets).
- The red graph shows the regression of the 15% of the experiments chosen for "Testing". The results verify once again the validation of the network training. A correlation factor of 0.999 was observed which proves the linearity of the function Outputs = f(Targets). New simulation results will be tested later in the manuscript to verify the prediction function on other inputs outside those used during training.
- The black graph shows the regression of all the experiments inserted into the neural network. An overall correlation factor of 0.998 was observed.

## 5 Conclusion

This study demonstrated the effectiveness of using neural networks to model the variation of atmospheric residue density and thus possibly optimize the distillation column. The results obtained attest to the reliability of the applied neural network, which confirms its potential for application in the oil industry.

These results highlight the importance of artificial intelligence-based approaches to solve complex problems in the industry and open the way to new opportunities for improving production processes.

## References


1. K. Ed. Rouibet. Raffinage du Pétrole. Institut Algerien Du Petrole IAP, SONATRACH
2. MO-PROO-29. Manuel opératoire Hydrotraitement de Naphta Léger 1 U701- RA1K. SONATRAC.(2014)
3. Ammar, M.Y. : Mise en œuvre de réseaux de neurones pour la modélisation de cinétiques réactionnelles en vue de la transposition Batch/Continu. Thèse de Doctorat de l'institut national polytechnique de – Toulouse – 17 juillet 2007
4. Fares, A., Boudelf, C. : Mémoire fin d'étude, Caractérisation et analyse du gasoil au niveau du laboratoire RA1K. Institut algérien du pétrole, Ecole de Skikda (2017)
5. Manuel opératoire de l'unité de distillation atmosphérique (U10)

# Author Index





N. Seddari and M. Redjimi (Eds.): ICMSCT 2024, CCIS 2606, pp. 565–566, 2025.
https://doi.org/10.1007/978-3-032-01922-6

GPSR Compliance

*The European Union's (EU) General Product Safety Regulation (GPSR) is a set of rules that requires consumer products to be  safe and our obligations to ensure this.*

*If you have any concerns about our products, you can contact us on ProductSafety@springernature.com*

In case Publisher is established outside the EU, the EU authorized representative is:

Springer Nature Customer Service Center GmbH
Europaplatz 3
69115 Heidelberg, Germany

**Batch number: 09166262**

Printed by Printforce, the Netherlands